THE ORIGIN OF OUTFLOWS IN EVOLVED STARS

IAU SYMPOSIUM 366

COVER ILLUSTRATION: HIDDEN IN STARDUST (Katrien Kolenberg)

Astrophysics has many mysteries yet to reveal. Red giants are old stars that eject gaseous material and solid particles through a stellar wind. Some red giants appeared to lose an exceptionally large amount of mass this way. However, new observations reveal that this is not quite the case. The stellar wind is not more intense than normal, but is affected by a partner that was overlooked until now: a second star that circles the red giant. Many old stars do not die alone. The ejected material by the red giant will form the basis for new stars, planetary systems and ultimately possibly life. In this Universe we, Humans, are contemplating our own cosmic origins. We are indeed made of stardust.

This work, created by Prof. Katrien Kolenberg, was made in a series to illustrate the research paper by colleagues Decin et al. (2019), published by Nature Astronomy in February 2019. A painting from this series. "Blown away by binary interaction", was published on the cover of Nature Astronomy in May 2019.

IAU SYMPOSIUM PROCEEDINGS SERIES

INTERNATIONAL ASTRONOMICAL UNION

UNION ASTRONOMIQUE INTERNATIONALE

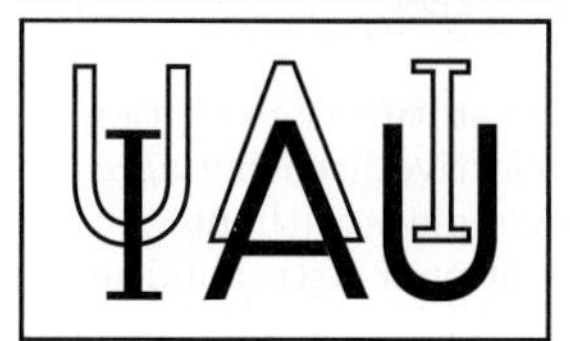

THE ORIGIN OF OUTFLOWS IN EVOLVED STARS

PROCEEDINGS OF THE 366th SYMPOSIUM OF THE INTERNATIONAL ASTRONOMICAL UNION VIRTUAL MEETING, INITIALLY PLANNED FOR LEUVEN, BELGIUM 1–6 NOVEMBER, 2021

Edited by

LEEN DECIN

KU Leuven, Belgium

ALBERT ZIJLSTRA

University of Manchester, UK

and

CLIO GIELEN

KU Leuven, Belgium

CAMBRIDGE UNIVERSITY PRESS
University Printing House, Cambridge CB2 8BS, United Kingdom
1 Liberty Plaza, Floor 20, New York, NY 10006, USA
10 Stamford Road, Oakleigh, Melbourne 3166, Australia

First published 2022

Printed in the UK by Bell & Bain, Glasgow, UK

Typeset in System LaTeX 2ε

A catalogue record for this book is available from the British Library Library of Congress Cataloguing in Publication data

ISBN 9781108490795 hardback
ISSN 1743-9213

Table of Contents

Preface

The prime parameter determining the evolution of a star is its mass. Any modification to the stellar mass over time has large repercussions on its evolutionary path. Both low-mass and massive stars are known to power strong stellar winds at the end of their life. Such winds carry away both mass and momentum from the star's surface, with rates that vary as a function of stellar luminosity, evolutionary phase, and chemical composition (i.e. metallicity). Binary mass transfer is another important mode of mass loss. The mass-loss rate determines the type of the stellar end product and the amount by which these stars contribute to the chemical enrichment of the interstellar medium, hence providing the building blocks of planets and life. A proper understanding of stellar evolution and of the chemical make-up of the building blocks in the Universe near and far can thus not be achieved without a detailed understanding of the wind physics during the late stages of stellar evolution as a function of the cosmic environment and hence metallicity. The goal of this IAUS366 symposium is to propel our understanding of stellar wind physics across stellar mass by bringing together the scientific communities which often focus on either the low-mass or the massive stars. This cross-disciplinary approach will fuel new scientific ideas and insights and will facilitate for new collaborations to grow across communities.

*

* *

Over the past decades, we have witnessed an enormous progress in our understanding of mass loss through stellar winds and binary interaction. Stellar winds govern the evolution during crucial phases of all stars in the universe. The driving force to counteract stellar gravity so that a wind can be launched has been identified for various regions in the Hertzsprung-Russell diagram, including the gas pressure gradient, radiation pressure on metal lines or on dust species, magnetic pressure etc. Although these forces have been established, various factors influencing or even key to the on set of the stellar wind remain elusive. For massive stars, we recall the ground-breaking detection of the first gravitational wave event in September 2015, which was caused by the merging of a binary black hole system with surprisingly high inferred masses, indicating our limited understanding of winds of massive stars. For low mass stars, an esteemed example is the current inability of scientists to predict which little dust seeds will be the first to form when the star is on the Asymptotic Giant Branch (AGB), as a result of which a stellar wind will be launched. As of today, we still do not know the mass-loss rate and the wind velocity that our own Sun will have once it becomes an AGB star. This incomprehension has direct consequences for any evolution model that studies the chances for Earth to survive (or not) the solar AGB evolution.

The outflows of low-mass and massive stars share various similarities, as well as profound differences. Through their winds, evolved low-mass and massive stars are the dominant sources for the chemical enrichment of the interstellar medium. Key species such as carbon, nitrogen, and oxygen are synthesized in the core of low-mass stars with the s-process being responsible for approximately half the atomic nuclei heavier than iron. In massive stars, the temperature and pressure in the core can reach high enough values for fusion of carbon, oxygen, and then even heavier elements like neon, magnesium, and silicon. Low-mass stars will slowly eject their atmosphere via a stellar wind, transiting through the post-AGB phase into a planetary nebula. Massive stars eject mass via a fast stellar outflow and finally via a supernova explosion, the latter potentially providing one of the sources for r-process elements. While chemical elements are further processed into a large variety of molecules and dust species in the stellar outflows of low-mass stars, an analogous chemistry does not occur in the winds of massive stars. The wind driving mechanism between low-mass and massive stars is also highly different with atomic line driving being the cause for the winds of massive stars, dust continuum driving for low-mass stars, and potentially molecular line driving for red supergiant winds. And while the dynamical chemistry seems highly different in low-mass and massive evolved stars, their outflows share various common characteristics including non-local thermodynamic equilibrium conditions, density inhomogeneities, radiative forces and hence acceleration, creation of bow shocks at the interface with the interstellar medium, etc.

Recent observations provide support for 50-90% of all stars being part of a binary (or multiple) system. Binary interaction impacts the evolution of stars and hence can play a decisive role for the stellar end product that is produced. The gravitational waves produced by colliding black holes or neutron stars are one example touching on massive star evolution. Recent high spatial resolution observations also have set aside the long-standing idea that the winds of low-mass stars are spherically symmetric with clear evidence of stellar and planetary companions shaping the wind's morphology.

New observational capabilities were key for a lot of progress in our understanding of stellar winds. Data obtained with MUSE, X-Shooter, Hubble etc. allowed for determining the strong mechanical and radiative feedback from massive stars on their host environments and for scrutinizing the role of density inhomogeneities on the retrieval of wind parameters across metallicity. The study of the winds around low-mass stars has seen an upheaval thanks to new instrumentation allowing for astrochemistry to play a decisive role. Astrochemistry provides us with unique tools to disentangle the phase-transition of atoms to small molecules and larger dust grains in environments unlike any terrestrial laboratories. Data obtained with ALMA, VLB(I), PdBI, SMA, VLT(I) etc. provide us with crucial information to disentangle the prevailing chemical processes occurring in stellar atmospheres, stellar winds, and the surrounding medium. The study of molecular and dust absorption and emission is a key technique in modern astrophysics, particularly through the ability to probe physical environments otherwise hidden from view.

Along with new instrumentation came new developments in HPC facilities allowing for more sophisticated numerical models describing stellar winds. Integrating the highly different timescales involved in various facets of stellar wind physics and chemistry into a unified wind model requires clever use of parallelization and memory sharing. Time-dependent radiation-driven wind models have reached a mature state, although the change from a 1D to a full 3D geometry remains challenging. An analogous challenge holds for incorporating non-equilibrium gas-phase cluster formation in low-mass wind models. As such, models including all aspects related to hydrodynamics, chemistry, and radiation cannot yet be used to directly retrieve parameters from observations, but they serve a key role for proposing more simplified (analytical) wind models and for the a-posteriori interpretation of the observables. New theoretical wind models developed during last few years allowed for some crucial aspects in stellar winds physics to be elucidated and serve as a guide for new instrumentation to be developed.

In view of the exciting new and unexpected results from facilities such as ALMA, VLB(I), PdBI, SMA, VLT(I), MUSE, X-Shooter, GAIA etc. which challenge our understanding of the dominant physical and chemical processes in evolved star's outflows, it was timely to hold an IAU Symposium centring on this theme. Historically, IAU symposia have focused on one of the various classes of evolved stars exhibiting strong stellar winds and their contribution to the galactic evolution. In the past 30 years, there have been 18 Symposia devoted to Massive stars and Supernovae, 7 Symposia on Planetary Nebulae, 3 Symposia on AGB stars, and none specifically oriented toward post-AGB or red supergiant stars. However, no IAU symposium had yet focused on the outflows from evolved stars across stellar mass. As such, common ground on which significant progress can build had remained invisible, and hence not explored. Examples include new diagnostic and numerical methodologies of moving from a 1D to a 3D morphology, dealing with an ensemble of atomic/molecular lines, accurate assessment of radiative forces, solving coupled differential equations, exploiting current HPC facilities, diagnostic tools from detailed observations, statistical model fitting tools, retrieval and forward modelling, optimal use of current and future observing facilities etc. This IAU symposium aimed to bridge the various communities dealing with the outflows of evolved stars with the prime goal being the creation of breeding grounds for new ideas to arise and new collaborations to grow across communities. As such, discussion sessions, education and training were a crucial part of this symposium.

The IAUS366 symposium was endorsed and sponsored by the IAU Division G Steering Committee, with support from the IAU Divisions B, C, and H. The IAU Executive Committee approved the symposium in December 2018. It is a also great pleasure to acknowledge the financial support of our sponsors the ERC and the KU Leuven, and the active support of the members of the SOC and LOC in realizing the numerous details always associated with such a symposium.

This symposium materialized into 7 invited review talks, 6 invited talks, 45 contributed talks, 31 poster papers, and 15 pitch talks, stimulating discussion among some 330 astronomers from 49 countries. May these Proceedings be a scientific reflection of our current knowledge of the outflows of evolved stars and shape a creative research landscape where new ideas arise and enigmas are solved.

Leen Decin chair SOC
Clio Gielen, chair LOC
Leuven, Belgium, 1 November 2021

The Organizing Committee

Scientific

L. Decin (chair, Belgium)
F. Kemper (Taiwan & Germany)
S. Mohamed (South-Africa)
O. de Marco (Australia)
H. Kim (Korea)
M. Boyer (USA)
A. Takigawa (Japan)
K. Kolenberg (Belgium)
A. de Koter (the Netherlands)
T. Millar (Ireland)
A. Zijlstra (UK)
H. Van Winckel (Belgium)
J. Yates (UK)
J. Plane (UK)
N. Soker (Israel)

Local

C. Gielen (chair)
M. Van Meerbeek
M. Dirickx
J. Bolte
S. Maes
D. Gobrecht
L. Decin
K. Kolenberg
T. Danilovich
J. Malfait
F. De Ceuster

Acknowledgements

The symposium is sponsored and supported by the IAU Division G (Stars and Stellar Physics) and supporting divisions Division B (Facilities, Technologies and Data Science), Division C (Education, Outreach and Heritage), and Division H (Interstellar Matter and Local Universe).

The Local Organizing Committee operated under the auspices of the KU Leuven

Funding by the
International Astronomical Union, ERC Consolidator Grant AEROSOL
(P.I. Leen Decin, grant number 646758)
and
KU Leuven C1 Excellence Research Grant: "MAESTRO: Massive Stars Outflows"
(P.I. Leen Decin, grant number C16/17/007)

Address by the Scientific and Local Organizing Committee

Dear colleagues,

It is with great pleasure that I welcome you all to this long-awaited IAU symposium, the IAUS366 that will discuss *The Origin of Outflows in Evolved Stars*. It would have been with even more pleasure if I could have welcomed you all in person, here in Leuven, as originally planned. We tried to escape the impact of the COVID-19 pandemic by moving the symposium from October 2020 to November 2021. But alas, also now we had to decide to change the format of the meeting from an in-person meeting to a virtual conference. As my colleagues of the SOC and LOC can testify, I was the last person accepting that we had to make this inevitable move. It was the only way forward for inclusion at all levels to be guaranteed, since not all countries around the world have the same success-rate in their vaccination strategy. Inclusion and diversity are key elements of this IAUS366 symposium – as you will also see reflected in its rich scientific program. I am proud that I can welcome today some 350 participants from all over the world. Junior and senior scientists, female and male with a diverse ethnic background. It is with pain in the heart that I can't welcome you in person, that I can't show you the nice city of Leuven with its rich history where we planned the symposium to take place, that I can't just take a coffee and chat and discuss with you in person during one of the breaks. It is thanks to the incredibly hard work of the LOC, in particular during these last few weeks, that we could change the format from a live meeting to a virtual one, in a format of which we think is the best to reach you all, to debate science, to foster discussion, to offer excellent training sessions, to let you enjoy the sometimes serendipitous gatherings in Gather Town where you can meet new colleagues.

The idea for this IAU symposium was given birth in my ERC-CoG application aerosol (no. 646758)†. That ERC grant is focussing on the origin of outflows in cool, evolved, low and intermediate-mass stars. It was one of the scientific outcomes of that ERC grant (Decin et al. 2019) which was translated by our colleague and artist Katrien Kolenberg into the very nice painting which we proudly use as cover page for this IAU proceedings, and which was used for the creation of the banner for the IAUS366 website‡. The arts offer a unique way of science communication, of reaching out to the society, of fostering a transdisciplinary discussion beyond our disciplinary silos and so of stimulating creative thinking. I believe that interdisciplinary, and even more transdisciplinary, collaborations are *a conditio sine qua non* for answering the 21st century questions.

This statement is also of relevance even if we stay within our restricted field of astrophysics' research, where we often tend to stay too often, too easily in our comfort zone. Not only do low-and intermediate mass stars lose a substantial fraction of their mass at the end of their life, the same holds true for massive stars. As outlined in the Preface, the outflows of low-mass and massive stars share various similarities and differences. For that reason, we decided to cross and bridge these disciplines to foster cross-fertilization between the research domains of low-mass and massive stars. For that reason, this IAU symposium is also financially supported by a KU Leuven C1 excellence grant, entitled maestro§, which focusses on massive star outflows. The financial support by the ERC and the KU Leuven allows us to offer you this week a symposium on the outflows from evolved stars across stellar mass, a unicum. This IAU symposium aims to bridge the various communities dealing with the outflows of evolved stars with the prime goal being

† https://fys.kuleuven.be/ster/research-projects/aerosol

‡ https://iaus366.be

§ https://fys.kuleuven.be/ster/research-projects/maestro

the creation of breeding grounds for new ideas to arise and new collaborations to grow across communities. As such, discussion sessions and education are a crucial part of this symposium.

We had identified a number of themes to be contained in this 5-days program, each taking roughly 1/2 day. However, the fact that the meeting is now online, and that we wished to accommodate for different time zones made us shuffle the themes across one another; this is also a way of bridging communities. The selection of the daily conference hours was done by weighting with the number of participants from different continents. We apologize if this is not always the most convenient hours for you to attend, but it was the best way for guaranteeing inclusion. Therefore the meeting will be fully recorded to allow you watching some sessions at more appropriate timings.

The same holds true for the skills training sessions which are offered during the symposium. Actually, some excellent sessions were already offered earlier today. I know, one could suggest it would have been more appropriate to welcome you before the start of these first training sessions. However, today 6 pm CET, seemed to us the best timing to reach most of you worldwide, from east to west. Life is never perfect. To continue on that aspect of training, I wish to express my sincere gratitude to all people who have agreed to prepare trainings which tackle a wide range of skills, from numerical modeling to observations, from didactics to career advice. That training is essential to all of us, junior and senior.

There is also time foreseen for social gathering. Yes, I know, it would have been more pleasant if we could have done that with some marvelous Belgian beers and delicious Belgian chocolates and that within the historical city of Leuven which is home to the oldest university of the low countries, inaugurated in 1425 -almost 600 years ago. If you once happen to be in Leuven in the future, I promise to indulge you with these Belgian pleasures. The alternative I can offer this week can be -apologies for that -fully categorized as being non-gastronomic. The LOC has created a Gather Town environment where we can meet ... virtually. Indeed, non-gastronomic but a poll amongst scientists during these COVID-19 times indicates that this virtual environment is often felt to have lower barriers for junior scientists to (virtually) walk to more senior scientists to get into contact. Let's remember the famous Monty Python lyrics 'Always look on the bright side of life'.

And let us now turn again our attention to the rich scientific program that this IAUS366 offers to you. We are honoured by the fact that excellent junior and senior scientists accepted our request for a review or invited talk. We are proud of the numerous submissions for contributed talks, of which we could only select a fraction. That selection was based on an anonymous process. Without knowledge of the identity, seniority, gender of the applicants, the SOC members judged independently the submitted abstracts. The highest ranked abstracts were selected as contributed talks. The outcome of this anonymous selection procedure is also a very nice balance between junior and senior scientists, between male and female scientists across nationalities; something of which I am very proud. In addition, there is a very nice selection of posters which will be displayed via Gather Town. Being an IAU Symposium, the creation of a proceedings book is an integral part of its concept; the result proving the stimulating scientific outcome of the IAUS366 symposium.

Having said this all, there remain two important things to be said. Firstly, I wish to thank my SOC colleagues. I thank them for their scientific advice on all kind of matters starting with shaping the program and submitting the application to the IAU in 2018 to the final selection of speakers. An even greater thanks goes to the LOC. In this challenging COVID-19 times where the rules are continuously changing, these people have been required and have been very flexible to continuously change plans. From in-person to post-poned, to again in-person, and finally virtually. It will not surprise you that my greatest thank goes to Clio Gielen, the chair of the LOC, the person behind the email address iausleuven@kuleuven.be. Only during these last 4 months, we have exchanged above 400 emails and had numerous online and in-person meetings. And this is not counting for the >1000 emails that Clio received from all of you on all kind of organizational matters. Clio, without you, this symposium would not have seen light. The saying is 'Behind every strong man there is a strong woman'. It will not surprise you that I don't like the gender-inequality in this saying and therefore wishes to adapt it to 'Behind every strong person there is another strong person'; and this is you, for the IAUS366 symposium and for our Institute of Astronomy in Leuven.

It is with great pleasure that I finally declare open the IAU symposium no. 366 - *The origin of Outflows in Evolved Stars*. I am honoured and proud to be your chairwoman for this symposium. I wish you all a very constructive and pleasant five working days.

Leen Decin, chair SOC
Clio Gielen, chair SOC
Leuven, Belgium, 1 November 2021

Reference

Decin L., Homan W., Danilovich T., et al. 2019, Nature Astromomy 3, 408

Participants

Mohamed **Abdel Sabour**, NRIAG, Egypt
Iminhaji **Ablimit**, NAOC(National Astronomical Observatories, Chinese Academy of Sciences), China
Anas **Abudhaim**, Al balqa applied university, Jordan
Keshab **Acharya**, TU, Nepal
Swarnadeep **Adhikary**, College student, India
Bhavna **Adwani**, UAB ICE IFAE, Spain
Arief **Ahmad**, Uppsala University, Sweden
Yerlan **Aimuratov**, Fesenkov Astrophysical Institute, Kazakhstan
Muhammad **Akashi**, Technion (IIT) & Kinneret College Israel, Israel
Javier **Alcolea**, Observatorio Astronómico Nacional, Spain
Ali **Aleali**, Payame Noor, Iran
Mats **André**, Swedish Institute of Space Physics, Sweden
Holly **Andrews**, Chalmers Tekniska Högskola, Sweden
Miora **Andriantsaralaza**, Uppsala University, Sweden
Bharti **Arora**, Aryabhatta Research Institute of Observational Sciences (ARIES), India
Jane **Arthur**, IRyA-UNAM, Mexico
Aymard **Badolo**, Universté Joseph KI-Zerbo, Burkina Faso
Naman **Bajaj**, College of Engineering Pune, India
Rahul **Bandyopadhyay**, S. N. Bose National Centre for Basic Sciences, India
Monica **Barnard**, University of Johannesburg, South Africa
James **Barron**, Queen's University, Canada
Sushma **Bashyal**, Tribhuwan University, Nepal, Nepal
Alain **Baudry**, Univ. Bordeaux, LAB, France
Ayushkumar **Bavisha**, Student, Jai Hind College, Mumbai University, India
Emma **Beasor**, NSF's NOIRLab, USA
Brillia **Benny**, Pondicherry University, India
Anirudh **Bharadwaj**, Manipal Institute of Technology, India
Akshara **Binu**, CHRIST(Deemed to be University), Bengaluru, India
John **Black**, Chalmers University of Technology, Sweden
Eric **Blackman**, University of Rochester, United States
Sara **Bladh**, Uppsala university, Sweden
Joris **Blommaert**, Dept. Physics and Astronomy (AARG), Vrije Universiteit Brussel, Belgium
Ronny **Blomme**, Royal Observatory of Belgium, Belgium
Jan **Bolte**, KU Leuven, Belgium
Alceste **Bonanos**, National Observatory of Athens, Greece
Howard **Bond**, Pennsylvania State University, US
Daniela **Boneva**, Space Research and Technology Institute, Bulgarian Academy of Sciences, Bulgaria
Emma **Bossuyt**, UGent, België
Martha **Boyer**, STScI, USA
Sarah **Brands**, Universiteit van Amsterdam, Anton Pannekoek Institute, Netherlands
Valentin **Bujarrabal**, Observatorio Astronomico Nacional (OAN/IGN), Spain
Diego **Calderón**, Charles University, Czech Republic
Emily **Cannon**, KU Leuven, Belgium
Silvina Belén **Cárdenas**, IAFE, Argentina
Antonio **Castellanos**, Instituto de Astronomía, UNAM, Mexico
Thomas **Ceulemans**, KU Leuven, Belgium

Priscila **Chacón**, Departamento de Astronomía, Universidad de Guanajuato, México
Luke **Chamandy**, University of Rochester, United States
Poonam **Chandra**, National Centre for Radio Astrophysics, TIFR, India
Bhawesh **Chandwani**, Student, Jai Hind College, Mumbai University, India
Nguyen **Chau Giang**, Korea University of Science and Technology, Repulic of Korea
Zhuo **Chen**, Tsinghua University, China
Andrea **Chiavassa**, Lagrange, Observatoire de la Côte d'Azur, France
David **Cohen**, Swarthmore College, USA
Akke **Corporaal**, KU Leuven, Belgium
Pierre **Cruzalebes**, OCA-Lagrange, France
Bruno **Curjurić**, Faculty of Science, University of Zagreb, Croatia
Francis **Daly**, Department of Chemistry, University of Leeds, UK
Ashkbiz **Danehkar**, University of Michigan, United States
Taïssa **Danilovich**, KU Leuven, Belgium
Upasana **Das**, National Institute of Science Education and Research, India
Benjamin **Davies**, Liverpool John Moores University, UK
Arjun **Dawn**, Jadavpur University, India
Karyne **de Almeida**, National Observatory, Brazil
Elvire **De Beck**, Chalmers University of Technology, Sweden
Frederik **De Ceuster**, University College London / KU Leuven, Belgium
Alex **de Koter**, Anton Pannekoek Institute of Astronomy, University of Amsterdam, The Netherlands
Orsola **De Marco**, Astronomy Astrophysics and Astrophotonics Research Centre, Macquarie University, Australia
Leen **Decin**, KU Leuven, Belgium Deepali Deepali, University of Hamburg, Germany
Jean-Francois **Desmurs**, IGN-OAN, Spain
José Jairo **Díaz-Luis**, Observatorio Astronómico Nacional (OAN-IGN), España
Reikantseone **Diretse**, University of Cape Town, South Africa
Hoai **Do**, Vietnam National Space Center, Vietnam
Julien **Drevon**, Observatoire de la Côte d'Azur/ Université Côte d'Azur, France
Andrea **Dupree**, Center for Astrophysics — Harvard & Smithsonian, US
Karlis **Dzenis**, University of Bordeaux, France
Ileyk **El Mellah**, IPAG – CNRS, France
Nancy **Elias-Rosa**, INAF – Astronomical Observatory of Padua, Italy
Izumi **Endo**, University of Tokyo, Japan
Dieter **Engels**, Hamburger Sternwarte, Universität Hamburg, Germany
Kjell **Eriksson**, Uppsala university, Sweden
Ana **Escorza**, European Southern Observatory (ESO), Chile
Sandra **Etoka**, Manchester, UK
Paul **Fallon**, University of South Africa, South Africa
Jose Pablo **Fonfria**, IFF-CSIC, Spain
Maximilien **Franco**, University of Hertfordshire, United Kingdom
Juris **Freimanis**, Ventspils International Radio Astronomy Centre of Ventspils University of Applied Sciences, Latvia
Bernd **Freytag**, Uppsala University, Sweden
Iván **Gallardo Cava**, Observatorio Astronómico Nacional (OAN-IGN), Spain
Teresa **García-Díaz**, Instituto de Astronomía UNAM Sede Ensenada, Mexico
Domingo Aníbal **García-Hernández**, Instituto de Astrofisica de Canarias (IAC), Spain
Guillermo **Garcia-Segura**, Instituto de Astronomia – Universidad Nacional Autonoma de Mexico, Mexico
Sabina **Gautam**, Tri-Chandra Multiple Campus, Ghantaghar, Kathmandu, Nepal, Nepal

Aayush **Gautam**, Birendra Multiple Campus, Tribhuvan University, Nepal, Nepal
Krzysztof **Gesicki**, IA UMK, Poland
Clio **Gielen**, Institute of Astronomy, KU Leuven, Belgium
Hila **Glanz**, Technion – Israel Institute of Technology, Israel
David **Gobrecht**, KU Leuven, Belgium Steven Goldman, Space Telescope Science Institute, United States
Miguel **Gómez-Garrido**, Observatorio Astronómico Nacional (OAN-IGN), Spain
Marco A. **Gómez-Muñoz**, IA-UNAM, México
Arturo I. **Gómez-Ruiz**, CONACYT-Instituto Nacional de Astrofísica, Optica y Electrónica, Mexico
Gemma **González-Torà**, European Southern Observatory (ESO), Germany
Hadis **Goodarzi**, School of Astronomy, Institute for Research in Fundamental Sciences (IPM), P.O. Box 19395-5746, Tehran, Iran, Iran
Alex **Gormaz-Matamala**, Universidad Adolfo Ibáñez, Chile
Carl **Gottlieb**, Center for Astrophysics, Harvard & Smithsonian, USA
Mariejo **Goupil**, Observatoire de Paris, France
David **Grant**, University of Oxford, United Kingdom
Mariluz **Graterol Ruiz**, University nucleus "Rafael Rangel" University of Los Andes, Venezuela
Pau **Grébol Tomàs**, INAOE, Mexico
Martin **Groenewegen**, Koninklijke Sterrenwacht van Belgie, Belgium
Maude **Gull**, UC Berkeley, USA
Marcin **Hajduk**, University of Warmia and Mazury in Olsztyn, Poland
Graham **Harper**, Center for Astrophysics and Space Astronomy, CU-Boulder, USA
Xavier **Haubois**, ESO-Chile, Chile
Levin **Hennicker**, Institute of Astronomy, KU Leuven, Belgium
Fabrice **Herpin**, LAB-Bordeaux, France
Erin **Higgins**, Armagh Observatory and Planetarium, UK
Kenneth **Hinkle**, NOIRLab, USA Susanne Höfner, Uppsala University, Sweden
Ward **Homan**, Universite Libre de Bruxelles, Belgium
Bruce **Hrivnak**, Department of Physics & Astronomy, Valparaiso University, USA
Josef **Hron**, Department of Astrophysics, University of Vienna, Austria
Roberta **Humphreys**, University of Minnesota, US
Feven **Markos Hunde**, Ethiopian Space Science and Technology Institute (ESSTI), Ethiopia
Damien **Hutsemékers**, University of Liége, Belgium
Richard **Ignace**, East Tennessee State University, USA
Ashwin **Amalraj J**, Vellore Institute of Technology, India
Prahladsinh **Jadeja**, Marwadi University, India
Kirti **Jadhav**, St. Xavier's College, India
Rukmini **Jagirdar**, Osmania University, India
Alexander (Sandy) **James**, University of Leeds, UK
Atefeh **Javadi**, Institute for Research in Fundamental Sciences, Iran
Manali **Jeste**, Max Planck Institute for Radio Astronomy, Germany
Palmira **Jiménez**, Instituto de Radioastronomía y Astrofísica, UNAM, Mexico
Fran **Jiménez-Esteban**, CAB (INTA-CSIC), Spain
Alain **Jorissen**, Institut d'Astronomie et d'Astrophysqiue, Université libre de Bruxelles, Belgium
Piyush **Joshi**, Tezpur University, India
Moo-Keon **Jung**, Seoul National University, Republic of Korea
Monika I. **Jurkovic**, Astronomical Observatory Belgrade, Serbia

Kay **Justtanont**, Chalmers University of Technology, Sweden
Ankur Jyoti **Kalita**, KU Leuven, Belgium
Anish **Kalsi**, Delhi Technological University, India
Devika **Kamath**, Macquarie University, Australia
Ukesh **Karki**, Tri-Chandra Multiple Campus, Nepal
Amit **Kashi**, Ariel University, Israel
Joel **Kastner**, Rochester Institute of Technology, United States
Cenk **Kayhan**, Erciyes University, Turkey
Franz **Kerschbaum**, Institut für Astrophysik, Universitaet Wien, Austria
Pierre **Kervella**, Observatoire de Paris, France
Theo **Khouri**, Chalmers University of Technology, Sweden
Soon-Wook **Kim**, Korea Astronomy and Space Science Institute, Republic of Korea (South Korea)
Sang Chul **KIM**, Korea Astronomy and Space Science Institute (KASI), South Korea
Hyosun **Kim**, KASI, Korea Jacques Kluska, KU Leuven, Belgium
Sladjana **Knezevic**, Astronomical Observatory Belgrade, Serbia
Kenda **Knowles**, Rhodes University / South African Radio Astronomy Observatory, South Africa
Chiaki **Kobayashi**, University of Hertfordshire, UK
Gloria **Koenigsberger**, UNAM-Instituto de Ciencias Fisicas, Mexico
Katrien **Kolenberg**, Institute of Astronomy – Faculty of Science, KU Leuven, Belgium
Thomas **Konings**, KU Leuven, Belgium Kathleen Kraemer, Boston College, USA
Michaela **Kraus**, Astronomical Institute, Czech Academy of Sciences, Czech Republic
Kateryna **Kravchenko**, MPE, Germany
Brankica **Kubátová**, Astronomical Institute of the Czech Academy of Sciences, Czech Republic
Ranjan **Kumar**, National Institute of Technology, Rourkela – 769008, India, India
Surjakanta **Kundu**, Indian Institute of Science Education and Research, Kolkata, India
Camille **Landri**, Charles University, Czech Republic Ryan Lau, ISAS/JAXA, Japan
Jasmina **Lazendic-Galloway**, TU/e innovation Space, Eindhoven University of Technology, Netherlands
Agnés **Lebre**, University of Montpellier, France
Tiina **Liimets**, Astronomical Institute, Czech Academy of Sciences, Czechia
Ping **Lin**, University of Toronto, Canada
J Alberto **Lopez**, Instituto de Astronomia, UNAM, Mexico
Belén **López Martí**, Centro de Astrobiología (CAB, INTA-CSIC), España Sara Lucatello, INAF, Italy
Adrian **Lucy**, Space Telescope Science Insitute, USA
Walter **Maciel**, University of Sao Paulo, Brazil
Matthias **Maercker**, Chalmers University of Technology, Sweden
Silke **Maes**, KU Leuven, Belgium
Laurent **Mahy**, Royal Observatory of Belgium, Belgium
Jolien **Malfait**, Institute of Astronomy, KU Leuven, Belgium
Grigoris **Maravelias**, IAASARS-NOA, IA-FORTH, Greece
Raffaella **Margutti**, University of California, USA
Paola **Marigo**, Department of Physics and Astronomy, University of Padova, Italy
Louise **Marinho**, LAB, France Lynn Matthews, MIT Haystack Observatory, USA
Iain **McDonald**, Open University and University of Manchester, UK
Marko **Mecina**, Department of Astrophysics, University of Vienna, Austria
Fabian **Menezes**, Lagrange, Observatoire de la Côte d'Azur, France
Jaroslav **Merc**, Astronomical Institute of Charles University, Czech Republic

Dominique **Meyer**, University of Potsdam, Germany
Amir **Michaelis**, Ariel University, Israel
Tom **Millar**, Queen's University Belfast, UK
Ibrahim **Mirza**, University of Tennessee, United States
Aditya **Mishra**, Birla Institute of Technology, Mesra, India
Nicolas **Moens**, Institute of Astronomy, KU Leuven, Belgium
Shazrene **Mohamed**, SAAO/University of Cape Town, South Africa
Nurul Husna **Mohammad Bokhari**, KU Leuven, Belgium
Soumen **Mondal**, S. N. Bose National Centre for Basic Sciences, India
Miguel **Montargés**, LESIA, Observatoire de Paris – PSL, France
Yuvraj **Muralichandran**, University of Potsdam, India
Drishika **Nadella**, National Institute of Technology, Karnataka, India
Chris **Nagele**, University of Tokyo, Japan Tanish Nandre, Fergusson College, India
Ambra **Nanni**, National Centre for Nuclear Research (NCBJ), Warsaw, Poland
Aakash **Narayan**, Osmania university, India
Mahdieh **Navabi**, The institute for research in fundamental science (IPM), Iran
Yael **Naze**, Univ. Liege/FNRS, Belgium
Megan **Newsome**, Las Cumbres Observatory, Unites States
Hervé **Ngremale**, chercheur faculte des sciences, République centrafricaine
Thi Kim Ha **Nguyen**, Université de Nice Côte d'Azur, France
Dieter **Nickeler**, Astronomical Institute, Czech Academy of Sciences, Czech Republic
Walter **Nowotny**, University of Vienna, Department of Astrophysics, Austria Lars Nyman, ESO, Chile
Hiroki **Onozato**, National Astronomical Observatory of Japan, Japan
Roberto **Ortiz**, Universidade de São Paulo, Brazil
Rene **Oudmaijer**, University of Leeds, United Kingdom
Stan **Owocki**, University of Delaware, USA
Estefania **Padilla Gonzalez**, UCSB, United States
Estefania **Padilla Gonzalez**, University of California, Santa Barbara, USA Claudia Paladini, ESO Chile, Chile
Abhishek **Pandya**, Student, India
Toni **Panzera**, University of Denver, United States
Tahere **Parto**, IPM (School of Astronomy), Iran
Beate **Patzer**, TU Berlin, Zentrum für Astronomie und Astrophysik, Germany
Madhu Sudan **Paudel**, Department of Physics, Tri – Chandra Multiple Campus, Tribhuwan University, Kathmandu, Nepal, Nepal
Sivasish **Paul**, COTTON UNIVERSITY, GUWAHATI, ASSAM, India
Susrestha **Paul**, University of Calcutta, India
Günay **Payli**, Akdeniz University, Turkey
Ondřej **Pejcha**, Charles University, Czechia
Nhung **Pham**, Vietnam National Space Center (VNSC/VAST), Vietnam
Asish **Philip Monai**, St. Xavier's College, India
Bannawit **Pimpanuwat**, University of Manchester, United Kingdom
John **Plane**, University of Leeds, UK
Léa **Planquart**, Université Libre de Bruxelles, Belgium
Luka **Poniatowski**, InsInstitute of Astronomy, KU Leuven, Belgium
Nirashan **Pradhan**, University of Padua, Italy
Karlis **Pukitis**, University of Latvia, Latvia
Luis Henry **Quiroga Nunez**, National Radio Astronomy Observatory (AOC), USA
Shiva **Raghav**, College of engineering Guindy, Chennai, India
GB **Raghavkrishna**, Ramakrishna Mission Vivekananda College (Autonomous), India

Ramprasath **Rajkumar**, National Institute of Technology, India
Vlad **Rastau**, University of Vienna, Department of Astrophysics, Austria
David **Raudales**, Universidad Nacional Autónoma de Honduras, Honduras
William **Reach**, USRA/SOFIA, USA William Reach, USRA/SOFIA, USA
Natalie **Rees**, University of Surrey, United Kingdom
Anita **Richards**, University of Manchester, UK
Sanjay **Rijal**, Institute of Science and Technology, Tribhuvan University, Nepal
Sam **Rose**, University of California: Berkeley, United States
Arpita **Roy**, Scuola Normale Superiore di Pisa, Italy, Italy
Gabriel **Rubio**, Universidad de Guadalajara, Mexico Maryam Saberi, University of Oslo, Norway
Gautham **Sabhahit**, Armagh Observatory and Planetarium, UK Laurence Sabin, Instituto de Astronomia, UNAM, Mexico
DSeyedAbdolreza **Sadjadi**, Laboratory for Space Research, Faculty of Science, Department of Physics, The University of Hong Kong, Hong Kong (SAR), China, Hong Kong (SAR)/China
Raghvendra **Sahai**, Jet Propulsion Laboratory, California Institute of Technology, United States
Shambel **Sahlu Akalu**, Ethiopian Space Science and Technology Institute, Ethiopia
Hugues **Sana**, Institute of Astronomy, KU Leuven, Belgium
Carmen **Sanchez Contreras**, Centro de Astrobiología (CAB, CSIC-INTA), Spain
Andreas **Sander**, Zentrum für Astronomie der Universität Heidelberg, Germany
Miguel **Santander-García**, Observatorio Astronómico Nacional (OAN-IGN), Spain
Benjamin **Sargent**, Space Telescope Science Institute, Johns Hopkins University, United States of America
Geetanjali **Sarkar**, Indian Institute of Technology, Kanpur, India
Rafia **Sarwar**, Institute of Space Technology, Islamabad, Pakistan (Last attended), Pakistan
Nobuaki **Sasaki**, Saitama University, Japan Deborah Schmidt, Franklin & Marshall College, USA
Peter **Scicluna**, ESO, Chile
Parth **Sharma**, Monash University, Australia
Shreeya **Shetye**, EPFL, Switzerland
Steve **Shore**, Univ. of Pisa, Italy
Emelie **Siderud**, Uppsala University, Sweden
Lionel **Siess**, Institut d'Astronomie et d'Astrophysique, ULB, Belgium
Arunan **Sinnappoo**, Dr, Sri Lanka
Nathan **Smith**, University of Arizona, USA
Varun **Sohanda**, Fergusson College, Pune, India
Noam **Soker**, Technion, Israel
Maciej **Soltynski**, ASSA, South Africa
Erik **Stacey**, Royal Military College of Canada, Canada
Jon **Sundqvist**, Ku Leuven, Belgium
Ryszard **Szczerba**, Nicolaus Copernicus Astronomical center, Poland
Ali **Taani**, Al Balqa Applied University, Jordan
Sima **Taefi Aghdam**, Institute for Research in Fundamental Sciences (IPM), Iran
Marco **Tailo**, Univesità di Bologna, Italy
Aki **Takigawa**, Department of Earth and Planetary Science, University of Tokyo, Japan
Dai **Tateishi**, Saitama University, Japan
Jesús **Toalá**, IRyA-UNAM, Mexico

Andrea **Torres**, Facultad de Ciencias Astronómicas y Geofísicas, Universidad Nacional de la Plata and Instituto de Astrofísica de La Plata (CONICET-UNLP), Argentina
Alfonso **Trejo-Cruz**, Academia Sinica Institute of Astronomy and Astrophysics, Taiwan
Gia-Bao **Truong-Le**, International University – Vietnam National University, Vietnam
Asif **ud-Doula**, Penn State Scranton, United States
Devendra Raj **Upadhyay**, Department of Physics, Amrit Campus, Tribhuvan University, Kathmandu, Nepal, Nepal
Lucero **Uscanga**, University of Guana juato, Mexico
Stefan **Uttenthaler**, Institute of Applied Physics, TU Wien, Austria
Rama Subramanian **V**, Amrita Vishwa Vidyapeetham, India
Marie **Van de Sande**, University of Leeds, United Kingdom
Griet **Van de Steene**, Royal Observatory of Belgium, Belgium
Sophie **Van Eck**, Université Libre de Bruxelles, Belgium
Hans **Van Winckel**, Institute of Astronomy, KU Leuven, Belgium
Dany **Vanbeveren**, Dept. Physics and Astronomy (AARG), Vrije Universiteit Brussel, Belgium
Juan Luis **Verbena**, Universität zu Köln, Germany Olivier Verhamme, KU Leuven, Belgium
Jorick **Vink**, Armagh Observatory and Planetarium, Northern Ireland
Eda **Vurgun**, UPC-Universitat Politécnica de Catalunya, Spain
Gregg **Wade**, Royal Military College of Canada, Canada
Saakshi **Wadhwa**, Fergusson College, India
Sofia **Wallström**, KU Leuven, Belgium
Jiaming **Wang**, Swarthmore College, United States
Marte Cecilie **Wegger**, University of Oslo, Norway
Patricia **Whitelock**, SAAO & UCT, South Africa
Ethan **Winch**, Armagh Observatory and Planetarium, UK
KaTat **Wong**, IRAM, France
Abhay Pratap **Yadav**, Department of Physics & Astronomy, NIT Rourkela, India
Kadri **Yakut**, University of Ege, Turkey Issei Yamamura, ISAS/JAXA, Japan
Melis **Yardimci**, Max Planck Institut for Astronomy, Germany
Jeremy **Yates**, UKSRC, STFC DIRAC and UCL, United Kingdom
Sun-Chul **Yoon**, Seoul National University, South-Korea
Laimons **Zacs**, University of Latvia, Latvia
Mengfei **Zhang**, Nanjing University, China
Yong **Zhang**, Sun Yat-sen University, China
Albert **Zijlstra**, Department of Physics & Astronomy, Manchester University, UK
Szanna **Zsíros**, Department of Optics and Quantum Eletronics, University of Szeged, Hungary, Hungary

The Origin of Outflows in Evolved Stars
Proceedings IAU Symposium No. 366, 2022
L. Decin, A. Zijlstra & C. Gielen, eds.
doi:10.1017/S1743921322001089

Getting started: How a supersonic stellar wind is initiated from a hydrostatic surface

Stan Owocki

Department of Physics & Astronomy, Bartol Research Institute, University of Delaware, Newark, DE 19716 USA
email: owocki@udel.edu

Abstract. Most of a star's mass is bound in a hydrostatic equilibrium in which pressure balances gravity. But if at some near-surface layer additional outward forces overcome gravity, this can transition to a supersonic, outflowing wind, with the sonic point, where the outward force cancels gravity, marking the division between hydrostatic atmosphere and wind outflow. This talk will review general issues with such transonic initiation of a stellar wind outflow, and how this helps set the wind mass loss rate. The main discussion contrasts the flow initiation in four prominent classes of steady-state winds: (1) the pressure-driven coronal wind of the sun and other cool stars; (2) line-driven winds from OB stars; (3) a two-stage initiation model for the much denser winds from Wolf-Rayet (WR) stars; and (4) the slow "overflow" mass loss from highly evolved giant stars. A follow on discussion briefly reviews eruptive mass loss, with particular focus on the giant eruption of η Carinae.

Keywords. Sun: solar wind; stars: early-type; stars: mass loss; stars: Wolf-Rayet; stars: AGB.

1. Introduction

To set the stage for this symposium's exploration of "The Origin of Outflows from Evolved Stars", I have been asked to review the basic processes underlying the *initiation* of such outflows. Most of a star's mass is bound in a hydrostatic equilibrium for which the outward push of pressure balances the inward pull of gravity. But if at some near-surface layer additional outward forces overcome gravity, this can transition to a supersonic, outflowing wind, with the sonic point, where the outward force cancels gravity, marking the division between hydrostatic atmosphere and wind outflow. This summary reviews general issues with such transonic initiation of a stellar wind outflow, and how this helps set key wind properties like the mass loss rate and wind flow speed.

As summarized in figure 1, much of the discussion, given in §2, focuses on four distinct types of steady-state outflows, namely: (1) the pressure-driven coronal wind of the sun and other cool stars; (2) line-driven winds from OB stars; (3) a two-stage initiation model for the much denser winds from Wolf-Rayet (WR) stars; and (4) the slow "overflow" mass loss from highly evolved giant stars. A follow on discussion in §3 briefly reviews eruptive mass loss (see figure 2), with particular focus on a binary merger model for the giant eruption of η Carinae, perhaps the most famous of the class of eruptive Luminous Blue Variables (eLBV).

2. Initiation of steady winds

The mass $M(r)$ within a local radius r of a spherically symmetric star exerts an inward gravitational acceleration $g = GM(r)/r^2$, with G the gravitation constant; for local mass density ρ, this is balanced by the acceleration associated with a pressure

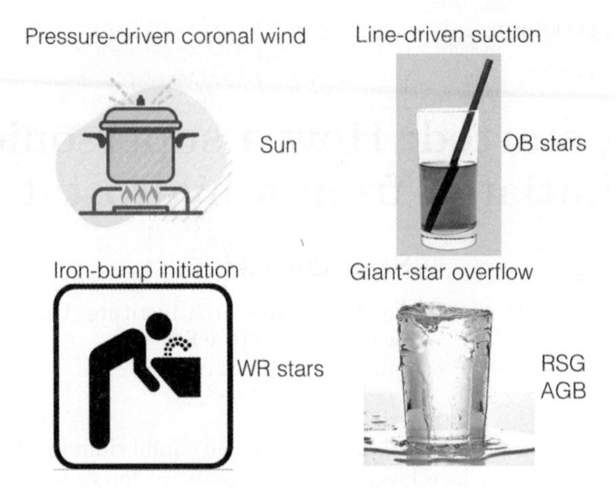

Figure 1. Icons to represent analogies for processes initiating the four different kinds of steady stellar wind outflow.

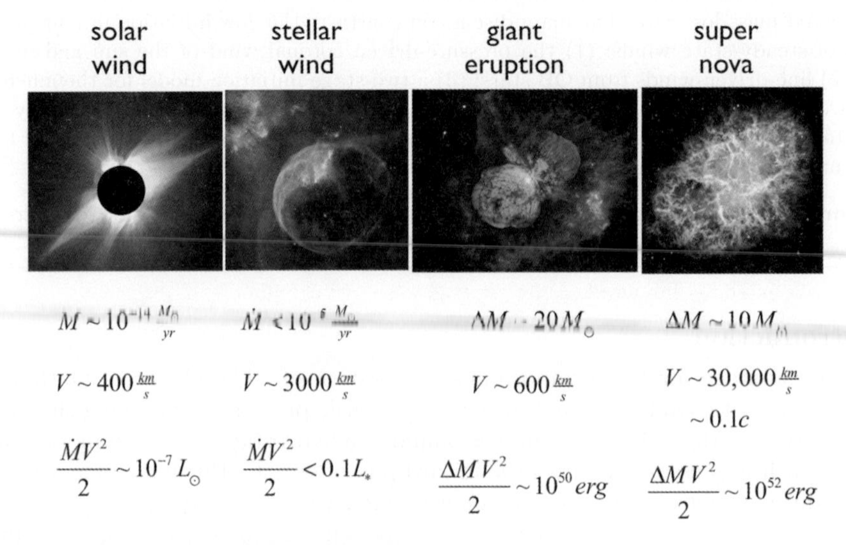

Figure 2. Comparison of the properties of solar and stellar steady wind outflows (left) with the example of eruptive mass loss from η Carinae (right center), and contrasting that with the explosive mass ejection from a supernova (right).

gradient $-(1/\rho)dP/dr$. Using the ideal gas law to write the pressure $P = \rho kT/\mu$ in terms of temperature T, molecular weight μ, and Boltzmann's constant k, we find the pressure drops exponentially with a characteristic pressure scale height,

$$H \equiv \frac{P}{|dP/dr|} = \frac{kT}{\mu g} = \frac{a^2}{v_{orb}^2} R \, . \tag{2.1}$$

Here the last equality casts the value at surface radius $r = R$ in terms of ratio of the (isothermal) sound speed $a \equiv \sqrt{kT/\mu}$ to near-surface orbital speed $v_{orb} \equiv \sqrt{GM/R}$. For the solar photosphere, we find $a \approx 10\,\text{km/s}$ while $v_{orb} \approx 420\,\text{km/s}$, thus implying a scale height that is a tiny fraction $H/R \approx 4 \times 10^{-4}$ of the solar radius. This is the essential reason the edge of the solar disk appears so sharp in white light images, such as shown in the leftmost panel in figure 3.

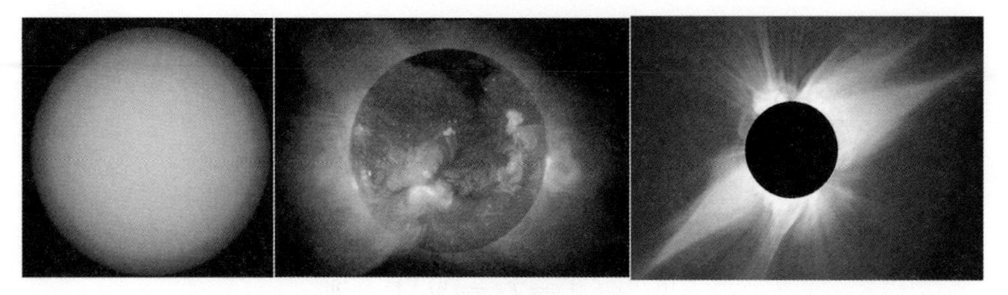

Figure 3. Images of the solar disk. Left: White light showing solar photosphere; Middle: X-rays showing magnetically confined hot coronal loops ; Right: Solar eclipse showing how coronal loops are stretched outward into pointed "helmet" streamers by the expansion into the solar wind.

Indeed, main-sequence stars are all characterized by a similarly small ratio H/R, implying that some other force must kick in to overcome gravity and drive a wind outflow. In the luminous OB and WR stars, this stems from the momentum of scattered radiation, as discussed in §§2.2 and 2.3.

However, for evolved giants, including Red Giants (RG), Red Super-Giants (RSG) and Asymptotic Giant Branch (AGB) stars, the much weaker surface gravity implies a less tiny ratio for $H/R \gtrsim 0.1$, making it easier for internal variations to induce mass loss, as discussed in §2.4.

2.1. *Pressure-driven coronal winds*

The initial prototype for wind mass loss came from the realization by G. Parker that the high (MK) temperature of the solar corona would lead to a supersonic expansion we now know as the solar wind. For a characteristic coronal temperature of 2 MK, we find the scale height ratio is now $H/R \approx 0.14$. The X-ray image in the middle panel of figure 3 shows how the hot corona thus has a much greater extension above the solar surface and beyond the solar limb.

If this coronal temperature is kept high – through extended heating and outward thermal conduction – the radial drop of gravity implies the H/R ratio increases outward; as such, the pressure no longer continues to drop exponentially, but rather asymptotically approaches a finite value, P_∞, at large radii, $r \to \infty$. Relative to the initial pressure P_o at the coronal base, the total drop in pressure for a hydrostatic, isothermal corona is given by

$$\frac{P_o}{P_\infty} \approx e^{R/H} \quad ; \quad \log \frac{P_o}{P_\infty} \approx \frac{6}{T/\mathrm{MK}} \, . \tag{2.2}$$

The latter equality shows the pressure drops by 6 decades for $T = 1\,\mathrm{MK}$, and only 3 decades for $T = 2\,\mathrm{MK}$. By comparison, from the solar transition region (TR) at the coronal base to the interstellar medium (ISM), the inferred pressure drop is actually much greater, $\log(P_{TR}/P_{ISM}) \approx 12$. The upshot is that an extended, hot corona can *not* be maintained in hydrostatic equilibrium; instead, as shown by the outward streamers from the eclipse image in figure 3, it must undergo an outward, supersonic *expansion* known as the solar wind.

As illustrated by the upper left panel of figure 1, this solar wind expansion can be thought of as analogous to the release valve of a pressure cooker, driven fundamentally by mechanical heating generated by magnetic turbulence in the underlying solar atmosphere. As illustrated in figure 4, some of this upward energy flux is lost back to the solar atmosphere through thermal conduction, but the net effect leads to a thermal runaway

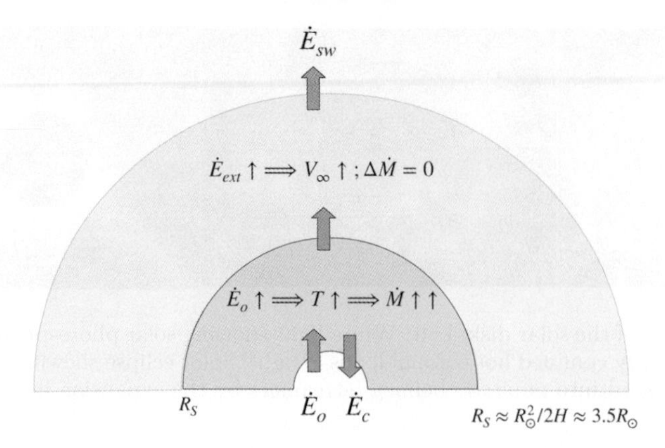

Figure 4. Schematic to illustrate energy input into the nearly hydrostatic, subsonic coronal base vs. that into the supersonic wind above the Parker sonic radius R_s. For the former, the net input vs. loss by conduction back into the underlying atmosphere leads to coronal heating that sets the coronal temperature and location and density at the sonic radius, thus fixing the associated wind mass loss rate. For the latter, any further energy addition increases the wind flow speed, with the wind expansion providing the primary mechanism to carry out the total net amount of coronal heating.

that raises the coronal temperature to millions of Kelvin, several hundreds times higher than the photospheric temperature $T \approx 5800\,\mathrm{K}$.

To account for the advective acceleration $v(dv/dr)$ in a spherically expanding wind, one can use mass continuity to split the pressure gradient force into terms that scale with this acceleration and the sphericity, yielding a steady-state equation of motion,

$$\left(v - \frac{a^2}{v}\right)\frac{dv}{dr} = -\frac{GM}{r^2} + \frac{2a^2}{r}. \tag{2.3}$$

In the subsonic region $v \ll a$, this reduces to the condition for hydrostatic equilibrium; but at a critical ("Parker") radius,

$$R_s = \frac{GM_\odot}{2a^2} = \frac{R_\odot^2}{2H} = 3.5R_\odot\left(\frac{2\mathrm{MK}}{T}\right), \tag{2.4}$$

where the RHS of eq. (2.3) vanishes, the hydrostatic coronal base transitions to a supersonic outflow, driven by the high gas pressure associated with the high coronal temperature.

The density and radius of this sonic point set the wind mass loss rate $\dot{M} = 4\pi\rho_s a R_s^2$, with typical values $\dot{M} \approx 10^{-14} M_\odot/\mathrm{yr}$. Physically, this is set by the level and location of the coronal heating. Heat added within the subsonic coronal base increases the scale height, with less drop in density toward a closer sonic point, and thus a direct increase in $\dot{M}$. In contrast, extended heating into the supersonic wind, where the mass loss rate is already fixed, instead leads to higher energy per unit mass, and thus a higher wind speed.

While the detailed mechanisms for coronal heating remain a subject of much current research, the thermal runaway and associated coronal expansion are thought to be quite robust consequences of the turbulence generated in all cool stars ($T_* < 10,000\,\mathrm{K}$) with convective envelopes associated with the opacity blockage from Hydrogen recombination. The low coronal density needed to avoid strong radiative cooling limits the mass loss rates to values, of order $10^{-14} M_\odot/\mathrm{yr}$, that do not appreciably reduce the stellar mass over evolutionary timescales. But the enhanced angular momentum loss associated with a global magnetic field can lead to an effective spindown of the stellar rotation.

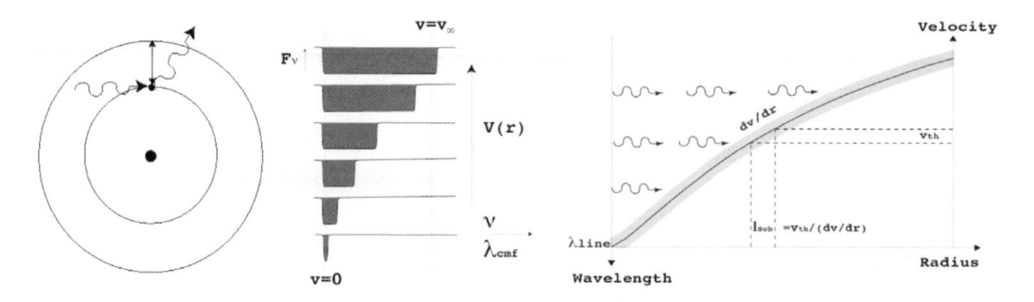

Figure 5. Left: Illustration of resonant scattering between two bound levels of an ion. Middle: The doppler-shifted desaturation that spreads line absorption over a large range of frequencies, ranging up to the frequency associated with the doppler shift from the wind terminal speed, v_∞. Right: Continuum photons propagating to the doppler-shifted line resonance that is thermally broadened by over a Sobolev length $l_{sob} = v_{th}/(dv/dr)$, where v_{th} is the ion thermal speed.

2.2. *Line-driven winds from OB stars*

In more massive, hotter OB stars with surface temperatures $T_* > 10,000\,\mathrm{K}$ hydrogen remains ionized up to the surface; such hot stars thus lack the H-recombination convection zones and associated the magnetic turbulence that heats the hot corona, and associated pressure-driven expansion of cool star winds. However such hot stars have a much higher radiative luminosity, and the outward force from line scattering of this radiation can overcome gravity and so drive a *line-driven* stellar wind outflow. For opacity κ_ν at a frequency ν with radiative flux F_ν, the total radiative acceleration depends on the frequency integral,

$$g_{rad} = \int_0^\infty d\nu \, \frac{\kappa_\nu F_\nu}{c} = \frac{\kappa_e F}{c} \tag{2.5}$$

where the last equality applies in the simplest case of continuum scattering by free electrons, with F the bolometric flux, and c the speed of light; for a fully ionized plasma of standard hydrogen mass fraction $X = 0.72$, the electron scattering opacity is $\kappa_e = 0.2(1+X) = 0.34\,\mathrm{cm^2/g}$.

The ratio of the associated radiative acceleration to gravity is given by the Eddington parameter,

$$\Gamma_e \equiv \frac{\kappa_e F/c}{g} = \frac{\kappa_e L}{4\pi G M c} \approx \frac{M}{200 M_\odot}\,, \tag{2.6}$$

wherein the inverse-square radial dependence of both the radiative flux $F = L/4\pi r^2$ and gravity $g = GM/r^2$ cancels, showing this Eddington parameter depends only on the ratio L/M of luminosity to mass. The last equality applies for the standard radiative envelope scaling $L \sim M^3$, and provides a rationale for the upper limit to stellar mass, which is empirically found to be around $200 M_\odot$ (Figer 2005). Stars that approach or exceed the Eddington limit $\Gamma_e = 1$ can have strong eruptive mass loss, as discussed in §3 for eruptive Luminous Blue Variable stars like η Carinae.

But the steady winds from luminous stars with $\Gamma_e \gtrsim 10^{-3}$ are understood to be driven by the *line* scattering from electrons bound into heavy ions ranging from CNO to Fe and Ni. As illustrated in the leftmost panel of figure 5, the resonance nature of such bound-bound line-scattering greatly enhances the opacity, by a factor $\bar{Q} \gtrsim 10^3$ (Gayley 1995), thus making it possible to overcome gravity and drive a wind outflow for stars with $\Gamma_e \gtrsim 1/\bar{Q} \approx 10^{-3}$.

In practice this maximal line acceleration from *optically thin* scattering is reduced by the saturation of the reduced flux within an optically thick line. But as illustrated in

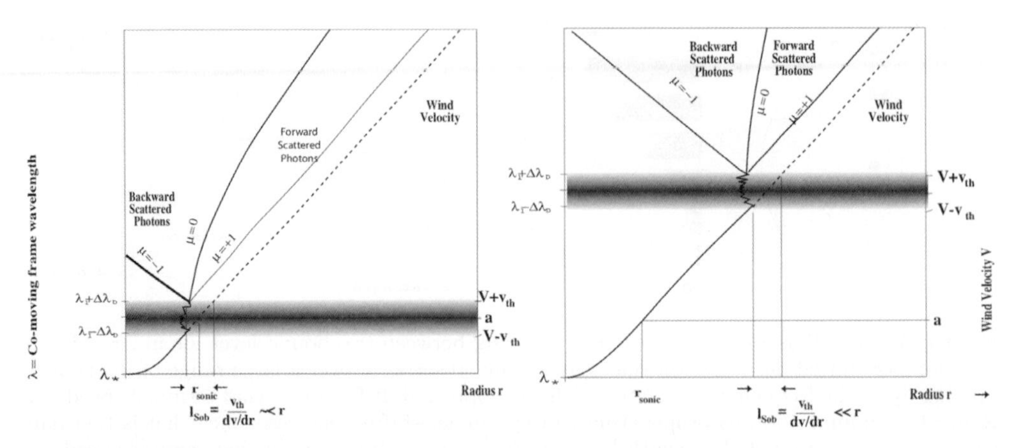

Figure 6. Alternative rendition of doppler shift from wind acceleration, now depicted as redshift of stellar photons until the come into resonance with a thermally broadened line. This allows depiction of the forward and backward scattering of photons escaping this layer. In the outer, supersonic wind of the right panel, this is fore-aft symmetric, canceling any net recoil from the diffuse radiation. But for inner, transonic wind of the left panel, there can be a net asymmetry between the forward/background scattering, leading then to a non-zero diffuse line force. The consequences of this are illustrated in figure 7 and discussed in the text.

the middle panel of figure 5, the doppler shift associated with the wind acceleration acts to *desaturate* this line absorption, effectively sweeping the absorption through a broad frequency band extending out to the frequency associated with the doppler shift from the wind terminal speed, v_∞.

The right panel of figure 5 illustrates how the wind doppler shift of line resonance concentrates the interaction of continuum photons into a narrow resonance layer with width set by the Sobolev length, $\ell \equiv v_{th}/(dv/dr)$ (Sobolev (1960)), associated with acceleration through the ion thermal speed v_{th} that broadens the line profile. In the outer wind where $\ell \ll r$, the line acceleration for optically thick lines is reduced by $1/\tau$, where the Sobolev optical depth $\tau \equiv \bar{Q}\kappa_e \rho \ell$, giving then a line acceleration $\Gamma_{thick} \sim (1/\rho)(dv/dr)$ that itself scales with the wind acceleration. Within this Sobolev approximation, Castor et al. (1975; hereafter CAK) developed a formalism that accounts for the radiative acceleration from an ensemble of thick and thin lines, deriving thereby scalings for the wind speeds and associated wind mass loss rate. Much as in the solar wind, the terminal wind speeds scale with the surface escape speed $v_{esc} \equiv \sqrt{2GM/R}$, with values up to $v_\infty \approx 2000\,\mathrm{km/s}$. However, the mass loss rates can range up to $\dot{M} \sim 10^{-5} M_\odot/\mathrm{yr}$, and so up to a *billion times* that of the solar wind!

A longstanding issue for this Sobolev-based CAK model regards the wind initiation near wind sonic point. The effective desaturation of a Sobolev model requires,

$$\frac{v_{th}}{dv/dr} \equiv \ell \ll H \approx \frac{\rho}{|d\rho/dr|} \approx \frac{v}{dv/dr} \,. \tag{2.7}$$

At the sonic point $v = a$, this requires $v_{th} \ll a$. Fortunately, because line driving is by heavy ions with $v_{th}/a \approx 0.3$, this condition is marginally satisfied; but it does indicate that a more careful treatment is warranted for the radiative transfer and line driving in this transonic region.

For this it is important to recognize that, as illustrated in figure 6, line-transfer actually occurs mainly via *scattering* not pure absorption. The right panel shows that, in the highly supersonic outer winds, the escape of photons from the Sobolev resonance is

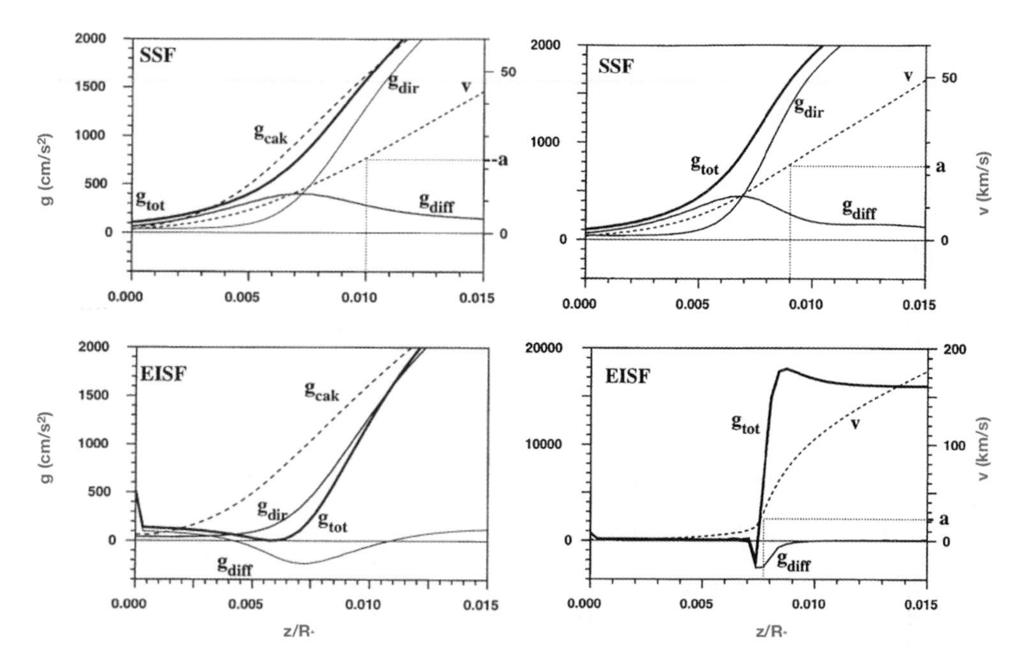

Figure 7. Radiative acceleration with scattering in the transonic regino (left) and the result initiation of wind velocity, for both Smooth Source Function (SSF; top) and Escape Integral Source Function (EISF; bottom) models of the scattering line transfer.

nearly fore-aft symmetric, canceling their recoil, so that the net line-force is the same as if the photon had been purely absorbed. The left panel shows, however, that in the inner, transonic region this scattering can become asymmetric, with a net recoil and so a non-zero *diffuse* line-force.

Figure 7 compares the radiative driving (left) and resulting wind velocity (right) for two different approximations for the transonic scattering line transfer (Owocki & Puls 1996, 1999). The top row shows a *Smooth Source Functon* (SSF) model, which assumes coupling with the continuum can keep the line source function nearly constant through the sonic region; the difference in fore/aft escape leads to a *positive* diffuse line force, which effectively compensates for the reduction in direct line force associated with the incomplete desaturation in this transonic region. The net result is that the wind driving and velocity are very similar to the standard CAK model.

The bottom row shows an *Escape Integral Source Function* (EISF) model, applicable for lower-density cases without sufficient continuum coupling to keep the source function smooth in the transonic region. The greater escape from the increasing velocity gradient leads now to a marked dip in the source function, with a weaker or even inward diffuse line force that further reduces the net line driving the subsonic region, giving thus now a sharp, step-like jump in wind velocity around the sonic point.

An overall point is that in all these models the onset of line-driving near the sonic point represents an effective *line-driven suction*, which draws up mass from the underlying hydrostatic equilibrium of the subsonic region. The reduction in pressure from the outer line-driving induces the underlying subsonic region to expand upward, in much the way that, as illustrated in figure 1, the suction on a straw draws up liquid from a glass. This outside-in suction contrasts with the inside-out thermal expansion of a pressure cooker, and of the analogous gas pressure-driven solar wind.

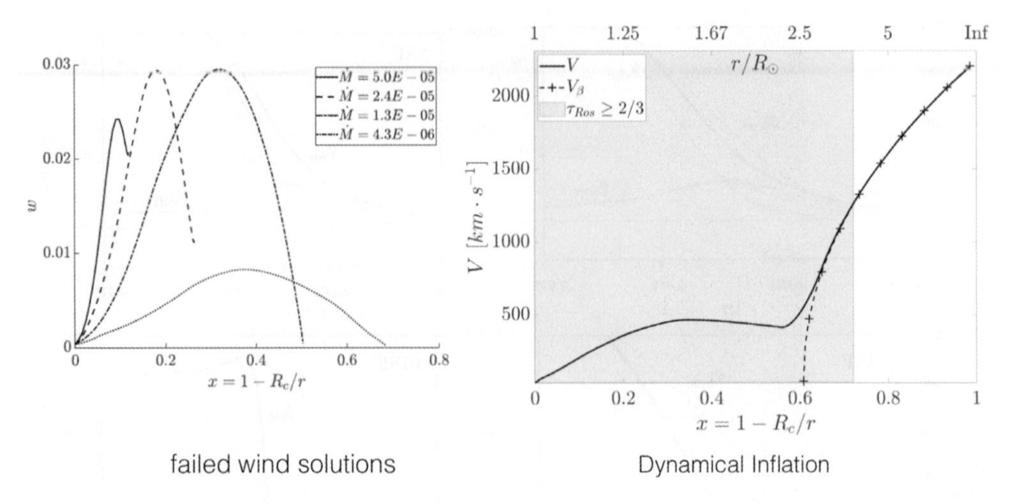

Figure 8. Left: Gravitationally scaled flow energy $w = v^2/v_{esc}^2$ plotted vs. scaled radius $x = 1 - R_c/r$, where R_c is the core radius near the wind sonic point. The various curves illustrate failed outflow solutions initiated at various densities by the iron opacity bump, with thus the various labeled mass loss rates, $\dot{M}$, in $M_\odot$/yr. Right: Flow speed vs. scale radius x for two-stage wind acceleration model. Gray shading shows the optically thick wind region where iron-bump opacity initiates a wind outflow, which is sustained by including velocity desaturation that transitions to a CAK-like line-driven outflow, as fitted in the dashed plus-sign curve by a standard wind velocity law. This two-stage process can be characterized as a kind of "dynamical inflation" of the outer stellar envelope. Figures adopted from Poniatowski et al. (2021).

2.3. *Iron bump initiation of WR winds*

For the much higher mass loss rates of Wolf-Rayet stars, the winds can themselves become optically thick, with the overall wind blanketing increasing the temperature of the transonic region well above the stellar effective temperature, so bringing it closer to the temperature $T \approx 200,000$ K for a strong opacity bump associated with the huge number of overlapping iron lines. In OB stars, this "iron opacity bump" induces a narrow sub-layer of convection that generates gravity waves and the associated surface macro-turbulence inferred from broadening of photospheric spectral lines. But in WR stars close to the Eddington limit, the inefficiency of the more near-surface convection can instead cause the net force to exceed the Eddington limit, leading to either an inflation of the stellar envelope (Gräfener et al. 2012), or even initiation of an outflow (Poniatowski et al. 2021).

As illustrated in the left panel of figure 8, the decline in this iron bump opacity with decreasing temperature means that it alone cannot sustain such outflow driving in the cooler, outer layers, leading then to a gravitational slowing and eventual stagnation of the wind outflow. The right panel shows, however, that including the velocity desaturation of CAK-like line-driving can rekindle the outward acceleration and so sustain the outflow.

The overall two-stage model represents a kind of "dynamical inflation", with associated mass loss rates an order of magnitude higher than obtained in O-stars of comparable luminosity, in extreme cases approaching $\dot{M} \sim 10^{-4} M_\odot$/yr. As illustrated in figure 1, it is somewhat akin to a water bubbler stream that is then taken up by the suction from a drinker. Talks by N. Moens and A. Sander in these proceedings discuss further such WR wind acceleration.

Figure 9. Illustration of large convective plumes of a giant star, which can help initiate overflows mass loss with high mass loss rate $\dot{M}$ but low flow speeds v_∞. Illustration Credit: ESO/L. Calçada

2.4. *Slow overflow mass loss from giant stars*

Finally let us turn to the strong mass loss that can occur from evolved giant stars like Red (Super)Giants (RG and RSG) and those on the Asymptotic Giant Branch (AGB). Like the overflowing glass illustrated in figure 1, this can be thought of a kind spillage, with overflowing material escaping the much weaker gravity of such stars, with flow speeds that barely exceed the reduced escape speed. Even for the lower sound speed associated with a lower surface temperature $T_* \sim 3000\,\mathrm{K}$, the larger radius now can lead to a moderate scale height to radius ratio

$$\frac{H}{R} = 0.2 \frac{T}{3000\mathrm{K}} \frac{R/1000\,R_\odot}{M/M_\odot}\,, \tag{2.8}$$

Figure 9 shows how the associated convection cells in such stars can thus extend over a substantial fraction of a stellar radius. As discussed in the contribution by S. Hoefner in these proceedings, detailed 3D simulations show this can sporadically suspend material to a large enough radius that the temperature becomes low enough ($T \lesssim 1500\,\mathrm{K}$) to initiate dust formation. The high opacity of this dust can lead to a radiative acceleration that exceeds gravity, so driving material to full escape from the star.

For dust grains of radius a, density ρ_d and mass fraction X_d, the dust opacity can be written as

$$\kappa_d = \frac{3X_d}{4a\rho_d} \approx 150 \frac{cm^2}{g} \frac{X_d}{X_{d\odot}} \frac{0.1\mu m}{a} \frac{1\,g/cm^3}{\rho_d}\,, \tag{2.9}$$

where $X_{d\odot} = 0.002$ is the maximum dust fraction for solar abundance of dust-forming elements.

By comparison, the critical (Eddington) opacity to overcome gravity has the scaling (Höfner & Olofsson 2018),

$$\kappa_{crit} = \frac{4\pi GMc}{L} \approx 2.6 \frac{cm^2}{g} \frac{M}{M_\odot} \frac{5000L_\odot}{L}\,. \tag{2.10}$$

Comparison shows that for luminous giant stars even partial dust formation should have an opacity that is sufficient to drive material to escape.

There remains some debate as to whether dust driving is essential to giant-star mass loss, or simply augments it. For example, there are models (e.g., Kee et al. 2021) for steady outflows driven by turbulent pressure, wherein as in pressure-driven coronal models, the mass loss rate is set by the density at the extended sonic point, but with values that greatly exceeds values of the solar wind, as high as $10^{-6} M_{\odot}/\mathrm{yr}$. The presentation by J. Sundqvist in these proceedings provides further discussion of such turbulence-driven wind models. As discussed in the presentation by Iain McDonald, stellar pulsations can also play an important rule in inducing strong mass loss from such giant stars.

3. Eruptive mass loss

3.1. *Energy requirement for eLBV mass ejection*

To complement the above discussion of the initiation of steady winds, let's next review the processes for initiating mass loss in an observational class of luminous, massive stars known as "eruptive Luminous Blue Variables" (eLBV's). Figure 2 summarizes how these have properties that are intermediate between steady solar and stellar winds and the explosive mass ejection of core-collapse supernovae (SNe). In the latter, the energy generated by core collapse deposits sufficient energy into the overlying stellar envelope to blow it completely from the star, over an initial dynamical timescale of seconds, and with ejecta speeds that can approach 0.1c.

Such SNe are thus at the opposite extreme of the gradual, steady mass loss in winds, for which initiation occurs in surface layers when an outward force overcomes the inward gravitational acceleration. Even in the extreme cases of WR winds or mass loss from giant stars, the energy needed to escape is generally a small fraction of the available luminosity emitted by radiation.

The giant eruptions seen from eLBV's have properties intermediate between winds and SNe. While their initiation may be sudden, their evolution can extend over years or even decades, much longer than a dynamical timescale, but generally short compared with a thermal relaxation timescale of the erupting star. The mass fraction ejected can be up to 10-20% of the stellar mass, much less than the full envelope ejection of SNe, but much larger than even the cumulative mass loss of stellar winds. And unlike the fixed terminal speed of winds, eLBV ejecta speeds can have a broad range, extending from a faster, low density leading edge to a bulk of mass that just barely escapes the star's gravity.

A promising paradigm is to consider such eLBV eruptions as arising from a quite sudden addition energy to the star's envelope, which, unlike the explosive addition of SNe, is however only some fraction $f < 1$ of the envelope binding energy.

Figure 10 shows results for 1D hydrodynamical simulations of the response when energy that is a fraction $f = 1/2$ of the stellar binding energy is added to the outer $25\,M_{\odot}$ of a $100\,M_{\odot}$ star, as indicated by the arrow along the left axis of the left panel. In terms of fluid parcels defined by their mass coordinate from the surface, the contours show the time response of the total "Bernouli" specific energy,

$$B \equiv \frac{v^2}{2} - \Phi_{grav} + h\,, \tag{3.1}$$

where v is the flow speed, Φ_{grav} is the gravitational potential, and h is the total specific enthalpy from gas and radiation; in terms of gas density and gas and radiation pressure, this is given by

$$h = \frac{\frac{5}{2}P_{gas} + 4P_{rad}}{\rho} = \frac{5}{2}\frac{kT}{\mu} + \frac{4a_{rad}T^4}{3\rho}\,, \tag{3.2}$$

with a_{rad} is the radiation constant. Note how the initial energy addition induces a pair of direct and reflected shock fronts that propagate toward the surface, heating the gas there so that the increased enthalpy makes the total Bernouli energy positive for the upper $\sim 7 M_{\odot}$ from the surface. The result is an outward expansion of the stellar envelope, with

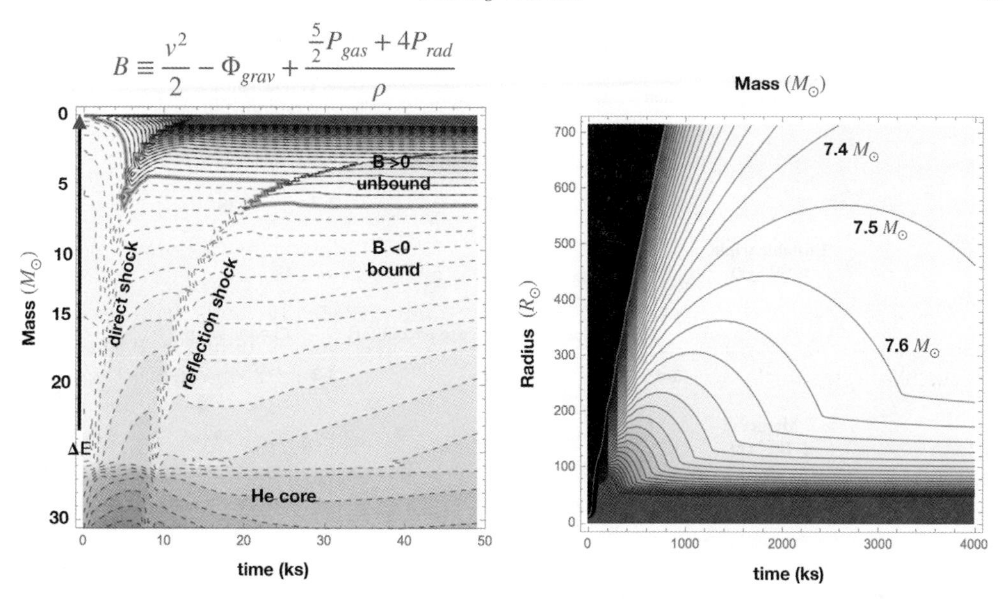

Figure 10. Left: Contours of the Bernouli energy B (defined by the equation at the top of the panel, and by eqn. (3.1)) as a function of the mass below the stellar surface, and the time since the sudden addition of an energy that is half the star's binding energy over the outer $25M_\odot$. The resulting direct and reflected shocks heat the outer layers, making the outer $\sim 7M_\odot$ become unbound with $B > 0$. Right: The resulting outflow of the stellar envelope, with escape of $\sim 7.3M_\odot$, and the rest of the material falling back onto the star. Figures adopted from Owocki et al. (2019).

the positive energy mass parcels with $M < 7.4M_\odot$ escaping completely from the star, while the negative energy mass parcels with $M > 7.5M_\odot$ eventually back onto the star.

The ejecta's variations in time t and radius r for the velocity v, density ρ, and temperature T are quite well fit by similarity forms in the variable $r/t \approx v$. Specifically the scaled density follows a simple exponential decline $\rho t^3 \sim \exp(-r/v_o t)$. This *exponential similarity* leads to analytic scaling relations for total ejecta mass ΔM and kinetic energy ΔK that agree well with the hydrodynamical simulations, with the specific-energy-averaged speed related to the exponential scale speed v_o through $\bar{v} = \sqrt{2\Delta K/\Delta M} = \sqrt{12}v_o$, and a value comparable to the star's surface escape speed, v_{esc}.

Unlike the fixed terminal speed v_∞ of a stellar wind, a small amount of material can be ejected at very high speeds, $> 5000\,\mathrm{km/s}$. But like stellar winds, gravity still plays a central role in controlling the mass and speed of the outflow, through the ratio of the added energy to the gravitational binding energy. This is distinct from standard SNe explosion, for which the added energy essentially overwhelms the gravitational binding of the envelope, leading to explosion speeds of order 0.1c.

3.2. *Merger model for η Carinae*

The 1840's giant eruption of η Carinae is perhaps the most prominent and extreme example of an eLBV, with the estimated $10 - 20\,M_\odot$ ejected mass forming the bipolar 'Homunculus' nebula (shown in the 3rd image in figure 2). Two key challenges are to understand both the energy source powering the eruption, and the causes of the bipolar form. η Carinae is known to have a massive companion in an eccentric ($\epsilon \approx 0.9$ orbit with period $\sim 5.5\,\mathrm{yr}$.

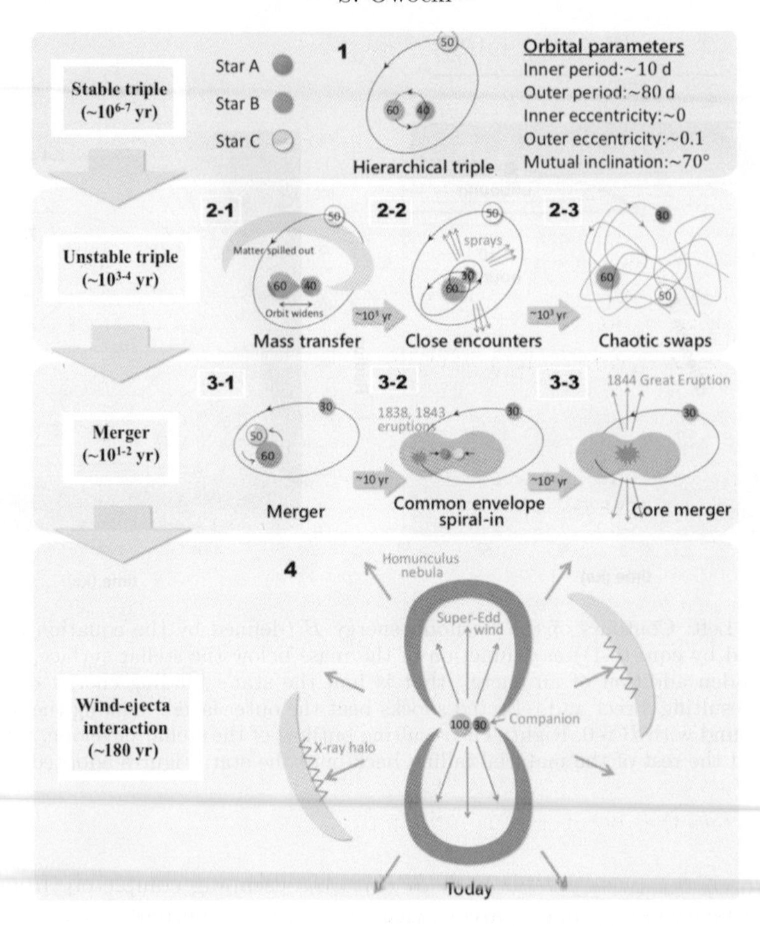

Figure 11. Schematic outline of multi-step process leading to a binary merger that triggers the giant eruption of η Carinae, with the post-eruption, bipolar , super-Eddington wind shaping the Homunculus nebula. Figure adopted from Hirai et al. (2021).

Figure 11 summarizes a recently proposed model by Hirai et al. (2021), in which an original triple system (phase 1) becomes unstable due to mass exchange (phase 2), leading to swaps and close encounters that result in a series of random ejecta that today are observed with source times extending back several centuries. Eventually, orbital decay of the innermost pair leads to a merger (phase 3), powering the 1840's giant eruption. The enhanced luminosity and rapid rotation of the post-merger star (phase 4) drives a strong, bipolar, super-Eddington wind that sculpts the compressed Homonculus nebula seen today.

4. Summary

Within this symposium on "The Origin of Outflows from Evolved Stars", a general overall theme for the above review on the initiation of such outflows is the key role played by gravity, and the need for forces that can overcome its inward pull to start an outflow, but also for this to be sustained to allow escape from the star's gravitational potential.

For coronal winds of the sun and other cool stars with convective envelopes, this is achieved by the high gas pressure associated with turbulent heating and thermal runway to temperatures, for which the gas internal energy becomes comparable to the gravitational binding.

For hot, massive OB stars, the momentum of the high stellar luminosity becomes effectively tapped by the line-scattering for bound-bound transitions of heavy ions, with the resonantly enhanced opacity maintained by desaturation of the doppler shifted outflow; this results in a sudden onset of the line-force near the sonic point, resulting in a line-driven suction of material from the star, for which the details depend on the nature of scattering in this transonic region.

In the much stronger, optically thick winds of Wolf-Rayet stars, the wind blanketing increases the base temperature to be close enough to that of the iron opacity bump, allowing a stronger wind initiation that is then sustained in the outer regions by the line-opacity desaturation.

In giant stars, the reduction in surface gravity and escape speed allow for giant convection cells to suspend material to levels with temperatures $T \lesssim 1500\,\mathrm{K}$, cool enough to initiate dust formation; the coupling of this enhanced dust opacity to the high luminosity can then drive material to full escape from the star. But even without dust formation, the turbulent pressure and pulsations can drive significant mass outflows against the weak gravity.

For eLBV eruptions, a key again is to provide an energy deposition to allow some fraction of the stellar envelope to gain sufficient energy to be ejected from the star. Instead of a steady wind, each mass parcel follows a trajectory tied to its total energy, with now a distribution of escape speeds that are described by a similarity form, with fall back to the star for bulk of material with net negative total energy.

While the outline here presents an idealized view of steady vs. impulsive mass loss, in practice real outflows will exhibit combinations of such traits, e.g. with infall in some WR winds, and with eLBV eruptions punctuated by quasi-steady super-Eddington winds. The further contributions in these proceeding explore in greater detail the many variations on these themes that occur in the outflows from evolved stars.

Acknowledgments

I thank the IAUS366 SOC, and particularly its chair Leen Decin, for the invitation to give this opening review. I also thank L. Poniatowski and R. Hirai for permission to use figures 8 and 11, respectively.

References

Castor, J. I., Abbott, D. C., & Klein, R. I. 1975, *ApJ.*, 195, 157.
Figer, D. F., 2005, *Nature*, 434, 192
Gayley, K. G. 1995, *ApJ*, 454, 410.
Gräfener, G., Owocki, S. P., & Vink, J. S. 2012, *A&A*, 538, A40.
Hirai, R., Podsiadlowski, P., Owocki, S. P., et al. 2021, *MNRAS*, 503, 4276.
Höfner, S. & Olofsson, H. 2018, *A&AR*, 26, 1.
Kee, N. D., Sundqvist, J. O., Decin, L., et al. 2021, *A&A*, 646, A180.
Owocki, S. P. & Puls, J. 1996, *ApJ*. 462, 894.
Owocki, S. P. & Puls, J. 1999, *ApJ*, 510, 355.
Owocki, S. P., Hirai, R., Podsiadlowski, P., et al. 2019, *MNRAS*, 485, 988.
Poniatowski, L. G., Sundqvist, J. O., Kee, N. D., et al. 2021, *A&A*, 647, A151.
Sobolev, V. V. 1960, Cambridge: Harvard University Press, 1960.

Discussion

Decin: In your discussion of eruptive mass loss, what fraction of the stellar mass can be lost in such eruptions?

OWOCKI: It depends on both the location of the energy addition and its fraction of the stellar binding energy. The specific model I discussed ejected about 7% of the mass of the star's $100M_{\odot}$, but a parameter study shows it is possible to eject more or less than this. A large energy added in the stellar core can eject the entire stellar envelope, as in a SNe explosion. Energy addition to a near-surface layer will at most eject the mass in that layer. Energy from a merger seems sufficient to eject $> 10\%$ of a star's mass, as inferred for η Carinae.

MELLAH: I had a question about the bipolar outflow, with the polar outflow being faster and more dilute

OWOCKI: No, it's not more dilute, because in a rapidly rotating radiative envelope, the bright poles can drive both a faster and denser outflow. In this model, that's how the Homunculus shape is formed.

MELLAH: So you would not see the material in the orbital plane.

OWOCKI: Well, there is some material ejected mechanically in the rotational/orbital plane, and this may be the origin of the observed equatorial skirt.

DE MARCO: In a binary, there's of course the straight gravity of the companion to help a star to lose mass. But do you envision a way in which a compansion can trigger mass loss in a way that a star uses its own reservoir of energy?

OWOCKI: Well there are some ideas, including I believe by yourself, for how orbital motion can excite pulsation, with perhaps certain resonances, resulting then in episodic mass ejection. But to get a *giant* eruption, where, as in η Carinae, the star loses 10% or more of its mass, I think you need to add a large source of energy, either externally from a merger, or some kind of ignition of enhanced burning in the core.

SAHAI: In regards to mass loss in binaries, I wanted to point out there's a beautiful example in our Galaxy called the Boomerang nebula, wherein there's a huge amount of mass ejected from an intermediate mass star with relatively high speeds. An other example is V Hydra, where one sees bullet-like ejection triggered from periastron passage of a companion.

OWOCKI: Yes, I didn't intend my talk to be comprehensive, and indeed the main focus was on initiation of wind mass loss. But certainly the topic of eruptive mass loss, and the role of binarity, is a broad area worthy of much further discussion in this symposium.

The Origin of Outflows in Evolved Stars
Proceedings IAU Symposium No. 366, 2022
L. Decin, A. Zijlstra & C. Gielen, eds.
doi:10.1017/S1743921322000230

The first 3D models of evolved hot star outflows

Nicolas Moens and Levin Hennicker

Instituut voor Sterrenkunde, KU Leuven, Celestijnenlaan 200D, 3001 Leuven, Belgium
email: nicolas.moens@kuleuven.be

Abstract. The mechanisms driving mass loss from massive stars in late stages of their evolution is still very much unknown. Stellar evolution models indicate that the last stage before going supernova for many massive stars is the Wolf-Rayet (WR) phase, characterized by a strong, optically thick stellar wind. Stellar models show that these stars exceed the Eddington limit already in deep sub-surface layers around the so-called 'iron-opacity' bump, and so should launch a supersonic outflow from there. However, if the outward force does not suffice to accelerate the gas above the local escape speed, the initiated flow will stagnate and start raining down upon the stellar core. In previous, spherically symmetric, WR wind models, this has been circumvented by artificially increasing either clumping or the line force. Here, we present pioneering 3D time-dependent radiation-hydrodynamic simulations of WR winds. In these models, computed without any ad-hoc force enhancement, the stagnated flow leads to co-existing regions of up- and down-flows, which dynamically interact with each other to form a multi-dimensional and complex outflow. These density structures, and the resulting highly non-monotonic velocity field, can have important consequences for mass-loss rates and the interpretation of observed Wolf-Rayet spectra.

Keywords. Stars: Wolf-Rayet, Radiation: dynamics, Methods: numerical, Hydrodynamics

---

1. Introduction

The Wolf-Rayet (WR) phase is thought to be the last phase in the lives of many stars that are born with an initial mass higher than $\sim 25\,M_\odot$ (Crowther 2007). These stars play an essential role in the chemistry and gas dynamics of their host galaxies, not only because of their strong stellar wind, but also because mass loss in the WR phase determines the type of compact object left after a supernova explosion. WR stars are hot, evolved, compact stars that are observationally characterised by strong, broad emission lines in the optical spectrum. This emission originates in the strong, optically thick stellar wind that surrounds the star.

WR stars have typical masses between $10\,M_\odot$ and $25\,M_\odot$ (Crowther 2007). Their winds have high terminal speeds and high mass-loss rates on the order of $10^{-7} - 10^{-4}\ M_\odot\ \mathrm{yr}^{-1}$ (Hamann *et al.* 2019; Nugis & Lamers 2000). Despite the importance of this final evolutionary phase, the wind launching mechanism is not yet completely understood. Recent theoretical work has put forward a mechanism that combines the effects of line driving with the enhanced radiation force over the Rosseland mean opacity bump due to iron recombination to successfully model the winds of WR stars in a spherically symmetric, 1D setup (Poniatowski et al. 2021).

However, due to convective instabilities in the high Eddington ratio environment close to the star's hydrostatic core, the wind is believed to be prone to structure formation, breaking its spherical symmetry. It is expected that due to this structure formation, the wind of a WR star might be very turbulent. Perhaps not all material that is initially lifted from the stellar surface will be able to completely reach escape velocity before falling back down. Wind inhomogeneities such as described here are believed to be the cause of line profile variations (LPV's) that have been observed for WR stars (Lépine & Moffat 1999).

To understand these effects, WR winds need to be modelled in a time-dependent, multi-dimensional framework, where the radiation force is accurately taken into account by including both the effects of line driving and the iron opacity peak. Here we describe the technique and present the first models that simulate such 3D time-dependent structure formation.

2. Methodology

2.1. *Equations of radiation hydrodynamics*

We present models that are obtained by solving the radiation-hydrodynamics (RHD) equations on a 2D and 3D finite volume mesh in a pseudo-planar, box-in-a-wind approach (see, e.g., Sundqvist et al. 2018). Our numerical domain starts at the hydrostatic core of the star $R_{\rm core}$, is 5 $R_{\rm core}$ long in the radial direction and has a width of 0.5 $R_{\rm core}$ in (both) lateral directions. Gravity is assumed to originate from a point source and we take into account the cooling, heating, and forces due to radiation. The exact equations solved are the conservative equations of hydrodynamics together with the frequency integrated, zeroth moment of the radiative transfer equation (RTE) (see also Moens et al. 2021):

$$\partial_t \rho + \nabla \cdot (\rho \mathbf{v}) = 0, \tag{2.1}$$

$$\partial_t (\mathbf{v}\rho) + \nabla \cdot (\mathbf{v}\rho\mathbf{v} + p) = -\rho \frac{GM}{r^2}\hat{r} + \rho \frac{\kappa \mathbf{F}}{c}, \tag{2.2}$$

$$\partial_t e + \nabla \cdot (e\mathbf{v} + p\mathbf{v}) = -\mathbf{v} \cdot \rho \frac{GM}{r^2}\hat{r} + \mathbf{v} \cdot \rho \frac{\kappa \mathbf{F}}{c} + c\kappa\rho E - 4\pi\kappa\rho B(T_g), \tag{2.3}$$

$$\partial_t E + \nabla \cdot (E\mathbf{v} + \mathbf{F}) = -\nabla \mathbf{v} : P - c\kappa\rho E + 4\pi\kappa\rho B(T_g). \tag{2.4}$$

In the equations above, ρ, $\mathbf{v}$, e, and p are the hydrodynamic gas density, gas velocity, gas total energy density and gas pressure. E, $\mathbf{F}$ and P are the frequency integrated radiation energy density, radiation flux, and radiation pressure. G and M are Newton's gravitational constant and the mass of the stellar object. $B(T_g)$ is the frequency integrated Planck function of a gas with temperature T_g. Finally, κ is the frequency integrated mass absorption coefficient or opacity. This is the interaction constant between the gas and radiation variables and is a crucial quantity in our models. We here assume that κ is isotropic and that the flux mean, Planck mean and energy mean opacities are all equal. On the right hand side of Eq. (2.2), source terms are included that correspond to gravity and the force due to radiation. In Eqs (2.3)–(2.4), source terms include heating and cooling of the gas due to radiation, as well as the terms corresponding to the work provided by the forces of gravity and radiation. The gas pressure is related to the gas energy via the ideal gas law:

$$e = \frac{p}{\gamma - 1} + \frac{\rho v^2}{2}, \tag{2.5}$$

where γ is the adiabatic index. $\mathbf{F}$ and P are related to E via the flux limited diffusion (FLD) closure relation and the Eddington tensor:

$$\mathbf{F} = D\nabla E \tag{2.6}$$

$$P = fE, \tag{2.7}$$

where the diffusion coefficient D and Eddington tensor f are given by an analytic recipe (see, e.g., Moens et al. 2021, Turner & Stone 2001, Levermore & Pomraning 1981).

2.2. *Hybrid opacity formulation*

Opacities in RHD calculations are often obtained via tables such as those provided by the OPAL project (Iglesias & Rogers 1991). These tables provide the Rosseland mean opacity calculated in the static limit as a function of gas density and temperature. However, in the case of WR outflows, such a tabulation assuming a static medium is not correctly describing the interaction between gas and radiation. Indeed, in supersonic media with high velocity gradients, the Doppler shifting of absorption lines significantly increases the total available radiation absorption capacity of the outflowing gas, rendering a more efficient stellar wind.

The winds of WR stars are driven by a combination of the Rosseland mean opacity in the static limit $\kappa^{\rm Ross}$ and a Doppler shift enhanced line opacity $\kappa^{\rm line}$. In our model we have integrated the sum of both in a hybrid opacity model (Poniatowski et al. 2021):

$$\kappa = \kappa^{\rm Ross} + \kappa^{\rm line}. \tag{2.8}$$

The Rosseland mean opacity values $\kappa^{\rm Ross}$ are tabulated by the OPAL project as function of gas density and temperature (Iglesias & Rogers 1991), while for the line opacity $\kappa^{\rm line}$ we rely on calculating the line force in the Sobolev approximation (Sobolev 1960) for a list of lines provided by Pauldrach et al. (1998). This force is then fitted to an analytic formula (Eq. (2.9)) for a range of densities, temperatures and Sobolev optical depths t. The best fitting line force parameters α, $\bar{Q}$ and Q_0 are then tabulated as a function of density and temperature (For a general description of our line force, see Poniatowski et al. 2022, submitted). The line opacity can then be written as:

$$\kappa^{\rm line} = \kappa_0 \frac{\bar{Q}}{1-\alpha} \frac{\left((1+Q_0 t)^{1-\alpha} - 1\right)}{Q_0 t}, \tag{2.9}$$

for Sobolev optical depth:

$$t = c\kappa_0\rho \left|\frac{dv}{dr}\right|^{-1}. \tag{2.10}$$

3. Results

3.1. *General properties of the model*

In Fig. 1 we show the model predictions for a WR star with a mass of $M_* = 10\,M_\odot$, a core radius of $R_{\rm core} = 1\,R_\odot$ and a Luminosity L_* of 0.4 times the classical† Eddington Luminosity $L_{\rm Edd} = 4\pi G M_* c/\kappa_e$. The composition of the gas is hydrogen abundance $X = 0$, helium abundance $Y = 0.98$ and metalicity $Z = 0.02$, where the ratios of different metal abundances is equal to that in the Sun as described by Asplund et al. (2009). The upper and lower panels in Fig. 1 show the relative density and radial velocity of the gas in one of our 2D models. The relative density is computed by dividing the actual density by its time and lateral average. The two panels on the left show the initial conditions, and each consecutive panel to the right is at one dynamical timescale later, where $\tau_{\rm dyn} \coloneqq R_{\rm core}/1000\,{\rm km\,s^{-1}}$. Close to the stellar core, the opacity is dominated by the static Rosseland mean opacity, which is significantly high in these temperature and density regimes due to the recombination of iron. This effect is often called the iron opacity peak. In the early panels of Fig. 1, one can see that the wind is launched from these bottom regions around the iron opacity peak as the radiation force exceeds the

† Here, κ_e is the electron scattering opacity, assumed to have a value of $0.2\,{\rm g^{-1}\,cm^2}$ in the hydrogen free atmosphere of a WR star.

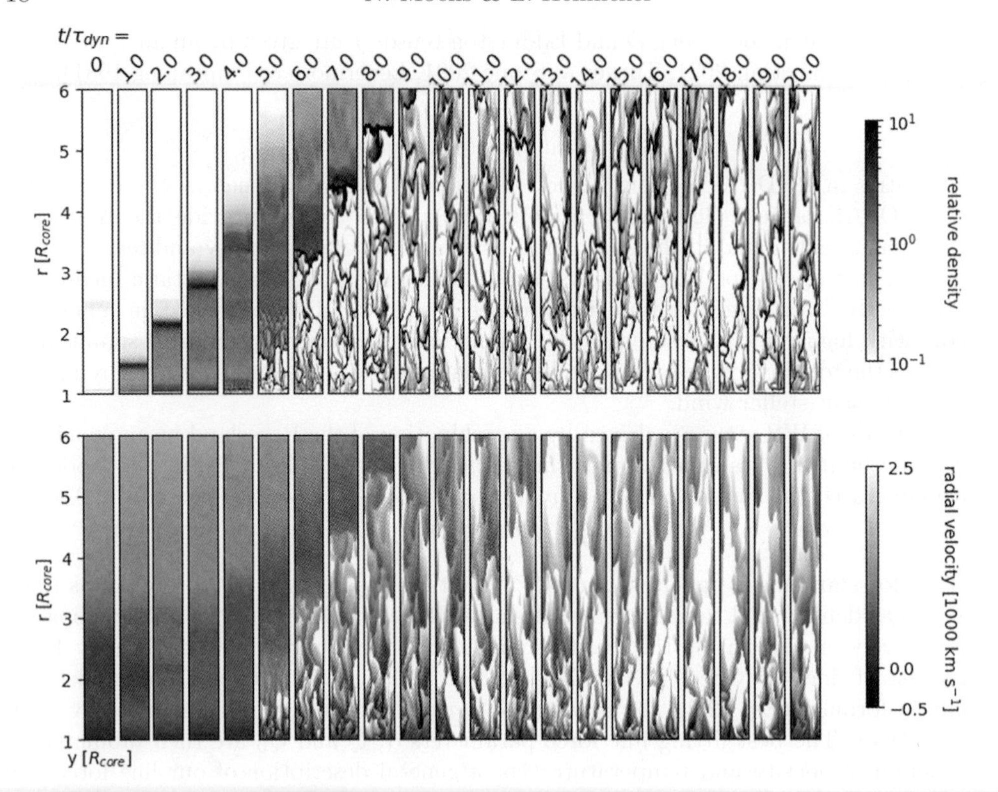

Figure 1. A timeseries, from left to right, displaying the relative density (top) and the radial velocity (bottom) of one of our 2D RHD WR models with a mass of $M_* = 10\,M_\odot$, a core radius of $R_{core} = 1\,R_\odot$ and $L_* = 0.4 L_{Edd}$. The figure shows the breakup from initial conditions and the formation of structure in the regions close to the core which are then carried outward.

force of gravity, and material starts accelerating outwards. Additionally, one can see that material starts breaking up and structure begins to form. This break-up of material is due to a convective like radiative instabillity, which has been described by Castor (2004). Linear perturbation analysis shows that when the stellar atmosphere approaches the Eddington limit and the opacity is a function of density and temperature, it can produce an absolute instability. Due to the structure formation in the low atmosphere, not all gas is able to make it out of the gravitational potential well. Indeed, gas that has been clumped together is driven less efficiently due to self-shadowing and can stagnate or even start accelerating downward back onto the stellar core. This can be seen in the bottom panels of Fig. 1, where in the bottom regions some clumps have low or negative velocities. Gas that does manage to escape will cool down which has a negative impact on the Rosseland mean opacity. However, gas in the outer wind also has a lower density, and this is where line driving takes over. Line driving is very efficient in low density environments where the continuum is optically thin. Indeed, in the optically thin limit, Eq. (2.9) approaches $\kappa^{\rm line} \propto 1/\rho^\alpha$. Further out in the wind, analysis shows that low density material gets accelerated by line acceleration and transfers part of its momentum to the earlier formed clumps via ram pressure thereby driving out the remaining gas in the wind. For a more in depth discussion of the general driving mechanism in our models, we refer to Moens et al. 2022, in prep.

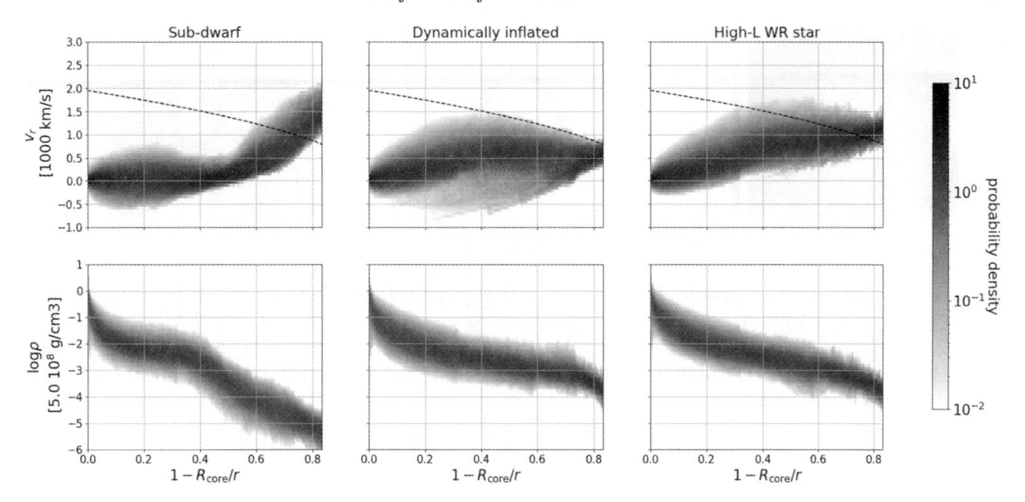

Figure 2. Probability density clouds of radial velocity v_r (top) and gas density ρ (bottom) for three 3D WR wind models with from left to right an increasing Eddington ratio: $L_*/L_{\rm Edd} = 0.33$ for the sub-dwarf, $L_*/L_{\rm Edd} = 0.46$ for the dynamically inflated WR model, and $L_*/L_{\rm Edd} = 0.67$ for the high-L WR star.

3.2. *Transition from sub-dwarf to WR wind*

In addition to the 2D model discussed above, here we discuss a series of three 3D models with the same stellar mass, core radius and gas composition as in the 2D model, but with three different Eddington ratios $L_*/L_{\rm Edd}$. In Fig. 2, we show density probability clouds for radial velocity (top panels) and mass density (bottom panels) for three models increasing in Eddington ratio from left to right. The probability clouds displayed in each panel are constructed by at each radius, calculating the probability density function of the gas densities and velocities present in all the lateral points from 10 3D snapshots. Histograms of these probability density functions are then displayed at each scaled radius in grey scale. The middle column of Fig. 2 corresponds to a star with an Eddington luminosity $L_*/L_{\rm Edd} = 0.46$. The velocity probability distribution of this model correspond to the non-monotonic radial velocity profiles of the dynamically inflated winds such as described by Poniatowski et al. (2021). The wind is initially accelerated due to the iron opacity peak in the Rosseland mean opacity, but afterwards fails and decelerates until line driving takes over at higher radii. One can see a large dispersion in the radial velocities as well as a distribution in density that spans a couple orders of magnitude at any given radial point. The model on the right corresponds to a model of a WR wind with a higher luminosity ($L_*/L_{\rm Edd} = 0.67$). Here, the wind morphology is largely similar to the dynamically inflated WR wind, but due to the high luminosity the wind does not significantly decelerate and the star is able to drive a higher mass-loss rate. Both the middle and right models in Fig. 2 result in a stellar wind that has a relatively high mass-loss rate ($1.5 \cdot 10^{-5}\, M_\odot\, {\rm yr}^{-1}$ and $3.2 \cdot 10^{-5}\, M_\odot\, {\rm yr}^{-1}$) and is optically thick. Finally, in the left model with $L_*/L_{\rm Edd} = 0.33$, the stellar luminosity does not suffice to actually launch a wind via the static Rosseland mean opacity alone. Instead, the launching of the wind gets delayed by expanding the core atmosphere without any significant net radial velocity before $1 - R_{\rm core}/r = 0.5$. Afterwards, the wind is launched and driven by the line driving mechanism. This model corresponds to a hot sub-dwarf, where the expanded atmosphere described here corresponds to sub-surface radiation driven convection. The actual wind is optically thin with a low mass-loss rate ($1.5 \cdot 10^{-6}\, M_\odot\, {\rm yr}^{-1}$).

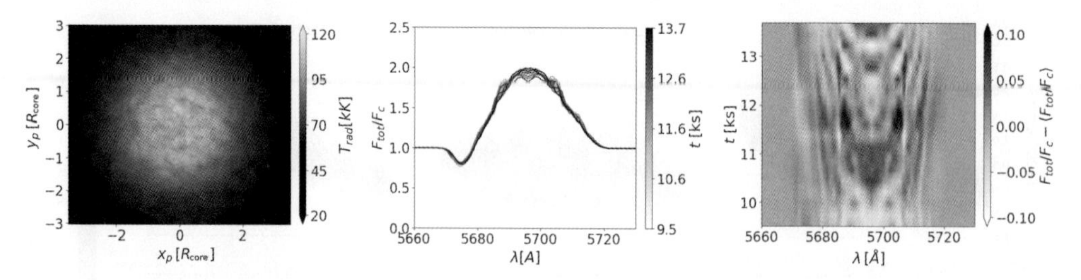

Figure 3. Synthetic observables for the high-L WR model. Left panel: Continuum intensity map at wavelength $\lambda = 5696\ \AA$, expressed in radiation temperature (using $I_\nu =: B_\nu\ (T_{\rm rad})$). Middle panel: Normalized flux profiles at various time snapshots of a generic line transition with central wavelength set to the CIII λ5696 line. Right panel: Mean-subtracted dynamical profile, showing the line profile variation with time. In the labels F_c and $F_{\rm tot}$ are the fluxes due to continuum absorption/emission and total (continuum + line) absorption/emission.

3.3. *Line profile variability*

Finally, we discuss some observable aspects of our work. By converting our 3D box-in-a-wind simulations to full spherical models via a cubed sphere projection we can use a long characteristics radiation transfer solver (Hennicker et al. 2021) to calculate synthetic spectral emission lines for each time snapshot in our 3D models. Due to the structure formation of gas in the wind models, observed spectra show small scale emission and absorption features on top of the main emission line. Temporal evolution shows line profile variation (LPV). LPV's have been observed in WR stars and have been described by e.g. Lépine & Moffat (1999). The left panel in Fig. 3 shows an intensity map of the high-L WR model in the continuum. The two panels on the right show the synthetic emission lines for different snapshots and the relative variation of the emission line as a function of time.

References

Moens, N., Sundqvist, J. O., El Mellah, I., et al. 2022, *A&A*, 657, A81

Moens, N., Poniatowski, L. G., Hennicker, L., Sundqvist, J. O., et. al. 2022, in prep.

Turner, N. J. & Stone, J. M. 2001, *ApJs*, 135, 95.

Levermore, C. D. & Pomraning, G. C. 1981, *ApJ*, 248, 321.

Poniatowski, L. G., Sundqvist, J. O., Kee, N. D., et al. 2021, *A&A*, 647, A151.

Iglesias, C. A. & Rogers, F. J. 1991, Presented at the International Workshop on Radiative Properties of Hot Dense Matter, 22

Poniatowski, L. G., Kee, N. D., Sundqvist, J. O., et al. 2022, submitted to *A&A*

Pauldrach, A. W. A., Lennon, M., et al. 1998, Properties of Hot Luminous Stars, 131, 258

Sundqvist, J. O., Owocki, S. P., & Puls, J. 2018, *A&A*, 611, A17.

Castor, J. I. 2004, Radiation Hydrodynamics, by John I. Castor, pp. 368. ISBN 0521833094. Cambridge, UK: Cambridge University Press, November 2004., 368

Asplund, M., Grevesse, N., Sauval, A. J., et al. 2009, *araa*, 47, 481.

Hennicker, L., Puls, J., Kee, N. D., et al. 2018, *A&A*, 616, A140.

Hennicker, L., Kee, N. D., Shenar, T., et al. 2021, *A&A*, accepted, *arXiv:2111.15345*

Lépine, S. & Moffat, A. F. J. 1999, *ApJ*, 514, 909.

Sobolev, V. V. 1960, Cambridge: Harvard University Press, 1960

Crowther, P. A. 2007, *araa*, 45, 177.

Hamann, W.-R., Gräfener, G., Liermann, A., et al. 2019, *A&A*, 625, A57.

Nugis, T. & Lamers, H. J. G. L. M. 2000, *A&A*, 360, 227

The Origin of Outflows in Evolved Stars
Proceedings IAU Symposium No. 366, 2022
L. Decin, A. Zijlstra & C. Gielen, eds.
doi:10.1017/S1743921322000400

The origin and impact of Wolf-Rayet-type mass loss

Andreas A. C. Sander[1], Jorick S. Vink[2], Erin R. Higgins[2], Tomer Shenar[3], Wolf-Rainer Hamann[4] and Helge Todt[4]

[1]Zentrum für Astronomie der Universität Heidelberg, Astronomisches Rechen-Institut, Mönchhofstr. 12-14, 69120 Heidelberg, Germany
email: andreas.sander@uni-heidelberg.de

[2]Armagh Observatory and Planetarium, College Hill, Armagh BT61 9DG, N. Ireland

[3]Anton Pannekoek Institute for Astronomy, Science Park 904, 1098 XH, Amsterdam, The Netherlands

[4]Institut für Physik und Astronomie, Universität Potsdam, Karl-Liebknecht-Str. 24/25, D-14476 Potsdam, Germany

Abstract. Classical Wolf-Rayet (WR) stars mark an important stage in the late evolution of massive stars. As hydrogen-poor massive stars, these objects have lost their outer layers, while still losing further mass through strong winds indicated by their prominent emission line spectra. Wolf-Rayet stars have been detected in a variety of different galaxies. Their strong winds are a major ingredient of stellar evolution and population synthesis models. Yet, a coherent theoretical picture of their strong mass-loss is only starting to emerge. In particular, the occurrence of WR stars as a function of metallicity (Z) is still far from being understood.

To uncover the nature of the complex and dense winds of Wolf-Rayet stars, we employ a new generation of model atmospheres including a consistent solution of the wind hydrodynamics in an expanding non-LTE situation. With this technique, we can dissect the ingredients driving the wind and predict the resulting mass-loss for hydrogen-depleted massive stars. Our modelling efforts reveal a complex picture with strong, non-linear dependencies on the luminosity-to-mass ratio and Z with a steep, but not totally abrupt onset for WR-type winds in helium stars. With our findings, we provide a theoretical motivation for a population of helium stars at low Z, which cannot be detected via WR-type spectral features. Our study of massive He-star atmosphere models yields the very first mass-loss recipe derived from first principles in this regime. Implementing our first findings in stellar evolution models, we demonstrate how traditional approaches tend to overpredict WR-type mass loss in the young Universe.

Keywords. stars: atmospheres, stars: mass loss, stars: massive, stars: winds, outflows, stars: Wolf-Rayet, stars: evolution, stars: black holes, galaxies: stellar content

---

1. Introduction

The striking appearance of an emission-line dominated optical spectrum has led to the introduction of a new spectral class in the 19th century: the Wolf-Rayet (WR) stars. This purely morphological definition has led to a situation that objects of different evolutionary stages can actually fall into this definition. This includes for example a subgroup of central stars of planetary nebulae showing WR-type spectra. In the massive star regime, the WR phenomenon occurs at multiple evolutionary stages. Among those are the now-called *classical* Wolf-Rayet (cWR) stars, which are evolved, core-He burning objects, partly or completely depleted in hydrogen. Another class are less evolved very massive stars that show WR-type spectra of the WNh subclass which are believed to be core-H burning. The latter essentially form an extension of the main sequence (Crowther & Walborn 2011) for very high initial masses, although the spectral WNh-type alone is not sufficient to infer the evolutionary status (see, e.g., the case of R144 in Shenar et al. 2021).

Their emission-line spectrum turns out to be a challenge for determining the stellar parameters of WR stars, in particular for the radii and the masses. Contrary to OB-type stars, the surface gravity cannot be determined via the wings of absorption lines. For WR stars at about solar metallicity, the whole spectrum is often formed in the wind, making it impossible to deduce a stellar radius (and a corresponding temperature) from spectral fitting due to inherent degeneracies (e.g. Najarro et al. 1997, Hamann & Gräfener 2004). This degeneracy can be arbitrarily broken by invoking a fixed assumption – usually a β-law – about the velocity field $v(r)$, but often results in discrepancies between empirical radii and expectations from stellar structure models (nowadays also termed the "WR radius problem"). These obstacles in pinning down important parameters of WR stars also limit the capabilities of empirical descriptions of WR mass loss as the measurements of the mass-loss rate $\dot{M}$ need to be associated with proper stellar parameters, which ideally should agree with predictions from stellar structure and evolution calculations.

2. Hydrodynamically-consistent atmosphere models

If the spectrum is completely formed in rapidly expanding layers, the location of any stellar radius can only be inferred indirectly. In principle, this could be done by stellar structure calculations (e.g. Grassitelli et al. 2018, Ro 2019, Poniatowski et al. 2021), but would require a treatment of the moving layers where the hydrostatic equation is no longer valid due to a non-zero inertia term (see Grassitelli et al. 2018 for such an approach). Moreover, the presence of a non-negligible velocity leads to a difference between the flux-mean opacity and the Rosseland opacity. While the latter is available in tabulated forms (e.g. the widely used OPAL tables by Iglesias & Rogers 1996, such that a detailed calculation can be avoided, the flux-mean opacity in a medium outside of (local) thermodynamical equilibrium requires a frequency-dependent calculation including the necessary determination of the radiation field and the population numbers.

For these reasons, stellar atmosphere codes provide a promising opportunity. Current codes such as PoWR (Gräfener *et al.* 2002, Hamann & Gräfener 2003) contain the necessary physics for an expanding non-LTE environment, but need to be extended/augmented to solve the hydrodynamic (HD) equation of motion consistently (Gräfener & Hamann 2005, Sander et al. 2017). Studies performed with HD-consistent versions of PoWR have revealed the complex shape of wind velocity fields in classical WR stars (Gräfener & Hamann 2005, Sander et al. 2020, see also Fig. 1), which are a consequence of the various ionization changes in a WR wind. These studies also revealed the fundamental importance of the iron opacities, which are responsible for launching the wind in cWR stars. The fundamental role of iron including its scaling of WR-type mass loss was also found in independent Monte Carlo simulations made by Vink & de Koter (2005).

3. Mass-loss rates for classical WN stars

To investigate the mass-loss of classical WN stars in a more detailed way, we calculated sequences of HD-consistent PoWR models for hydrogen-free stars. Using a fixed stellar temperature of $T_* = 141\,\mathrm{kK}$, each sequence adopted a different metallicity Z, ranging from $2\,Z_\odot$ down to $0.02\,Z_\odot$. To reduce the number of free parameters, we invoked the mass-luminosity relation for hydrogen-free stars from Gräfener et al. (2011). With these assumptions, our model stars approximately correspond to stars on the theoretical He ZAMS. We calculated models with masses up to $400\,M_\odot$ in order to investigate the

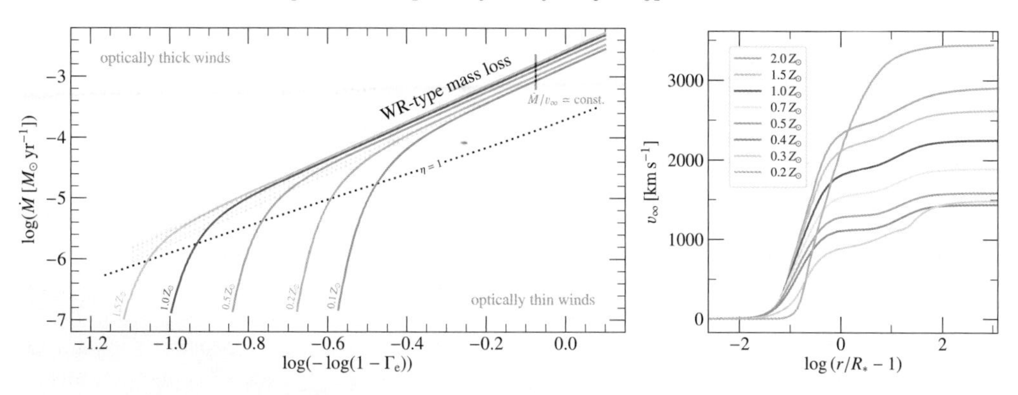

Figure 1. Left panel: Mass-loss $\dot{M}$ rate of He stars for different metallicities based on fitting the results from the hydrodynamic model calculations (full lines). The linear trend in the regime of totally optically thick winds (thin dashed lines) abruptly changes due to the transition to optically thin winds. This coincides with a breakdown of $\dot{M}$. In the optically thick regime, the ratio of $\dot{M}$ over terminal velocity v_∞ is the same for all metallicities. $\Gamma_e \propto L/M$ denotes the Eddington parameter which is proportional to the ratio of luminosity L over mass M.
Right panel: Velocity fields from HD modelling for a 20 $M_\odot$ He star at different metallicities Z.

asymptotic behavior of $\dot{M}$. This turned out to be crucial for extracting a mathematical description which can be extrapolated more realistically than power-law descriptions. Details of the modeling efforts and parameter ranges are given in Sander & Vink (2020).

Our findings are summarized in Fig. 1, where we plot the resulting curves from fitting our data points from the HD model calculations. In the limit of dense winds, WR-type mass loss can be described by a linear relation between $\log \dot{M}$ and $\log\left[-\log\left(1-\Gamma_e\right)\right]$ with different metallicities only causing different offsets. Moreover, all these linear curves align when considering $\log(\dot{M}/v_\infty)$ instead of $\log \dot{M}$ (see also Fig. 2). This means that – in the limit of optically dense winds – the mass-loss scales in the same way as the terminal velocity for WR stars with the same stellar parameters at different metallicities. This is in sharp contrast to OB-star winds and an interesting testable prediction that could potentially provide an important observational mass-loss diagnostic. However, the emission-line dominated spectra will make it challenging to precisely pin down two stars with exactly the same stellar parameters.

Another important result from our modeling efforts is the metallicity-dependent breakdown of $\dot{M}$. As soon as the launching point of the wind approaches the optically thin regime, the linear trend is broken and the mass-loss rate rapidly decreases. At higher metallicity, more line opacities are available and thus a strong mass-loss can be maintained down to lower helium star masses. This rapid breakdown at different regimes qualitatively explains the empirical finding of different minimum luminosities for observed WN stars at different metallicities (see Shenar et al. 2020 for a more detailed discussion and implications on the role of the binary channel to form WR stars).

Incorporating both asymptotic behaviors of our model sequences, we find:

$$\log \frac{\dot{M}}{M_\odot\,\mathrm{yr}^{-1}} = a \cdot \log\left[-\log\left(1-\Gamma_\mathrm{e}\right)\right] - \log(2) \cdot \left(\frac{\Gamma_\mathrm{e,b}}{\Gamma_\mathrm{e}}\right)^{c} + \log \frac{\dot{M}_\mathrm{off}}{M_\odot\,\mathrm{yr}^{-1}} \tag{3.1}$$

$$\text{with} \qquad a = 2.932(\pm 0.016)$$

$$\Gamma_\mathrm{e,b} = -0.324(\pm 0.011) \cdot \log(Z/Z_\odot) + 0.244(\pm 0.010)$$

$$c = -0.44(\pm 1.09) \cdot \log(Z/Z_\odot) + 9.15(\pm 0.96)$$

$$\log \frac{\dot{M}_\mathrm{off}}{M_\odot\,\mathrm{yr}^{-1}} = 0.23(\pm 0.04) \cdot \log(Z/Z_\odot) - 2.61(\pm 0.03)$$

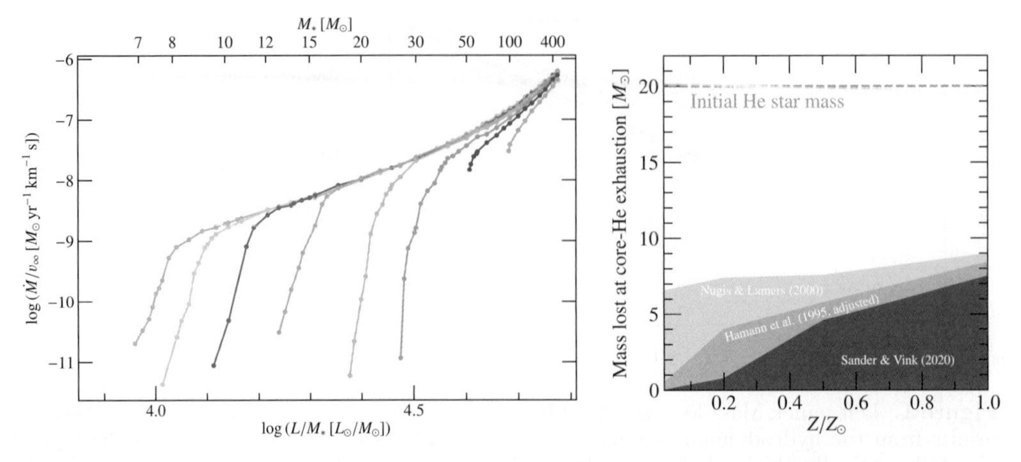

Figure 2. Left panel: Resulting data from dynamically consistent atmosphere models for the ratio of the mass-loss rate $\dot{M}$ over the terminal velocity v_∞ as a function of the luminosity-to-mass ratio L/M: The trends for different metallicities (indicated by different colors) align in the limit of optically thick winds where the wind is launched in deep layers (optical depth $\tau \gg 1$). Right panel: Total mass lost due to stellar winds for a $20\,M_\odot$ He star until core-He exhaustion at different metallicities Z as calculated in the evolutionary models in Higgins et al. (2021). The three colored areas reflect the total mass lost according to the three labeled $\dot{M}$ recipes.

This description marks the first theory-based mass-loss recipe for cWR stars. We do not vary the stellar temperature here and thus we do not expect that our results are valid for much cooler (late-type) WR stars. It is further worth mentioning that we do not account for different chemical abundances in this formula and thus quantitative differences are to be expected for stars with different composition. This is particularly important for stars with leftover hydrogen and for WC stars. However, we expect to see similar trends in follow-up studies for these groups. The results from our pilot study in Sander et al. (2020) where we investigated both WC and WN models revealed a similarity in the trends for WN and WC models (both without hydrogen). However, for the same stellar parameters, the WC mass loss is expected to be lower than the WN mass loss. This is a result of the smaller amount of Thomson opacity in a WC composition, which cannot be compensated by carbon line opacities as the ionization stages of carbon are too high in the deeper layers to notably contribute to the radiative driving. Instead, the carbon line opacities should lead to higher terminal velocities for the same stellar parameters. Our finding of comparably lower mass-loss rates for WC stars is in sharp contrast to the widely used description of Nugis & Lamers (2000) which predicts a higher mass-loss for WC stars as the carbon abundance contributes to their Z-term.

4. Evolution of Wolf-Rayet stars with updated mass-loss treatment

To investigate the impact of our improved mass-loss formalism we calculated a series of MESA helium star models (Higgins et al. 2021). Each of the models was then evolved further with three different mass-loss treatments, namely our new formula from Sander & Vink (2020), the formula from Nugis & Lamers (2000), and the treatment by Hamann et al. (1995) modified as suggested by Yoon et al. (2006), i.e. divided by 10 and scaled with a metallicity-term according to Vink & de Koter (2005). In our evolution models, metallicities between $Z_\odot$ and $0.02\,Z_\odot$ were covered with initial helium star masses ranging from $20\,M_\odot$ to $200\,M_\odot$. Since we were mainly interested deducing overall trends, we used the Sander & Vink (2020) formula also for the WC stage and ignored the slight reduction of $\dot{M}$ predicted in Sander et al. (2020). We ran the models until the

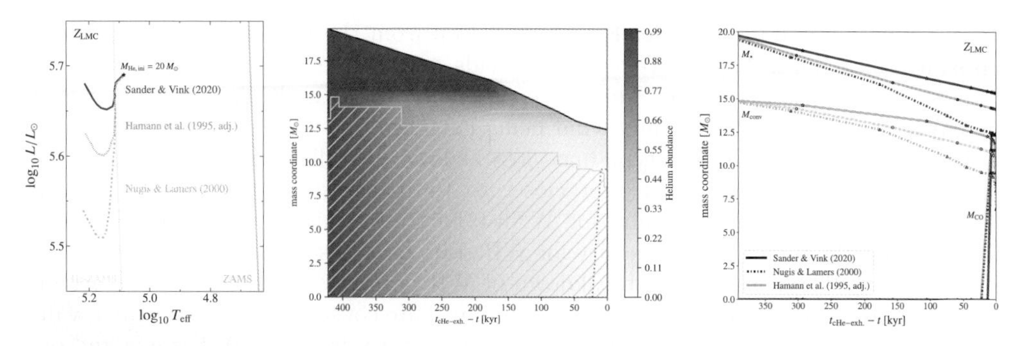

Figure 3. Left panel: Hertzsprung-Russell diagram for the $20\,M_{\odot}$ He star models at $Z_{\rm LMC}$ using different $\dot{M}$ recipes. Middle panel: Kippenhahn diagram for the model using $\dot{M}$ from Nugis & Lamers (2000). Right panel: Time evolution before core-He exhaustion of the total mass (M_{*}), convective core mass ($M_{\rm conv}$), and CO-core mass ($M_{\rm CO}$) for all three $20\,M_{\odot}$ models at $Z_{\rm LMC}$.

exhaustion of core-He burning and investigated the final masses as well as the CO-core masses (see also Fig. 3).

While at $Z = Z_{\odot}$, the change in final masses due to the different recipes is rather mild, the outcome changes drastically at lower metallicities. In particular the treatment of Nugis & Lamers (2000) turns out to be problematic: The incorporation of the carbon and oxygen abundances to the Z-term mimics a "self-enrichment" of the stars, thus leading to unrealistically high mass-loss rates. In fact, the usage of the recipe leads to an effective "mass convergence" that is almost independent of Z (see also Fig. 2). This means that down to $0.1\,Z_{\odot}$ one would always end up with approximately the same final masses if a star manages to reach the cWR stage. This effect is avoided with the other two recipes, as only the initial metallicity Z (or simply the iron abundance) enters here.

Another effect that is clearly visible in our evolution models is caused by the breakdown of WR-type winds predicted in Sander & Vink (2020): While the other two recipes lead to the loss of a few solar masses even at the lowest masses and metallicities, this is not the case in the new formula. This has notable consequences for the range of black hole masses that can be reached in the young Universe. Lower wind mass-loss rates yield higher black hole masses and thus could be an important part of the puzzle found by the black hole mass measurements from gravitational wave events.

For higher-mass He stars, the reduced mass-loss at lower metallicity has another consequence: The CO core at the end of core-He burning can grow larger and thus the regime of pair instability could be reached more often. In fact, our results show that pair instability events could already occur below $0.5\,Z_{\odot}$ if sufficiently massive He stars can be formed. Previous mass-loss treatments would have shifted this limit down to $0.1\,Z_{\odot}$ or below, in particular when using Nugis & Lamers (2000). All the details of our evolution models and a more in-depth discussion of the results can be found in Higgins et al. (2021).

5. Summary & Conclusions

Using sequences of dynamically-consistent atmosphere models, we derived the very first theory-based mass-loss description for classical Wolf-Rayet stars (Sander & Vink 2020). Our models represent stars with a WN-type composition. The sequences reveal a metallicity-dependent breakdown of WR-type mass loss when transiting to thinner winds. In the limit of optically thick winds, we derive a very modest metallicity scaling of $\dot{M} \propto Z^{0.3}$. Moreover, the ratio of mass-loss rate over the terminal velocity seems to be constant in this limit, independent of metallicity. We propose that the empirical relations which predict stronger metallicity-scalings are a consequence of the beginning breakdown of WR-type mass loss, which leads to a deviation from the rather flat $\dot{M}$-slope in the optically thick limit. The derived velocity fields are complex and deviate significantly from simple β-law descriptions. This is a consequence of the multiple ionization changes in a WR-type wind with the inner wind being launched due to iron M-shell opacities.

To reach the regime of WR-type mass loss at lower metallicities and thus lower iron abundances, cWR stars have to get closer to the Eddington limit, similar to what Gräfener & Hamann (2008) found for WNh-type stars. This is in line with empirical findings that the minimum luminosity for WR stars becomes higher at lower metallicity (Shenar et al. 2020). Below the breakdown of WR-type mass loss, our models are also huge sources of He II ionizing flux. This is an immediate consequence of the weaker winds as the radiation in the extreme UV is otherwise used to drive the WR winds and only re-emitted at much longer wavelengths. Using a series of MESA models for helium stars, we demonstrate that implementing the different scaling of WR-type mass loss and its breakdown is crucial for deriving proper black hole masses. The lack of WR-type mass loss at lower metallicity makes it not only easier to form massive black holes, but also allows to reach the regime of pair instability up to $\approx 0.5\, Z_{\odot}$ as long as massive enough He stars can be formed.

Our results presented in these proceedings and the cited papers mark only the first efforts on the way to a full comprehension of WR-type mass loss. Beside expanding to a broader set of stellar parameters and further refinements to account for different chemical compositions, the absolute values of $\dot{M}$ will also depend on a better description of multi-dimensional effects such as a more realistic treatment of clumping and its still quite enigmatic onset in WR winds.

References

Crowther, P. A. & Walborn, N. R. 2011, *MNRAS*, 416, 1311
Gräfener, G. & Hamann, W.-R. 2005, *A&A*, 432, 633
Gräfener, G. & Hamann, W.-R. 2008, *A&A*, 482, 945
Gräfener, G., Vink, J. S., de Koter, A., et al. 2011, *A&A*, 535, A56
Grassitelli, L., Langer, N., Grin, N. J., et al. 2018, *A&A*, 614, A86
Hamann, W.-R., Koesterke, L., & Wessolowski, U. 1995, *A&A*, 299, 151
Hamann, W.-R. & Gräfener, G. 2003, *A&A*, 410, 993
Hamann, W.-R. & Gräfener, G. 2004, *A&A*, 427, 697
Higgins, E. R., Sander, A. A. C., Vink, J. S., et al. 2021, *MNRAS*, 505, 4874
Iglesias, C. A. & Rogers, F. J. 1996, *ApJ*, 464, 943
Najarro F., Hillier D. J., Stahl O. 1997, *A&A*, 326, 1117
Nugis, T. & Lamers, H. J. G. L. M. 2000, *A&A*, 360, 227
Poniatowski, L. G., Sundqvist, J. O., Kee, N. D., et al. 2021, *A&A*, 647, A151
Ro, S. 2019, *ApJ*, 873, 76
Sander, A. A. C., Hamann, W.-R., Todt, H., et al. 2017, *A&A*, 603, A86
Sander, A. A. C., Vink, J. S., & Hamann, W.-R. 2020, *MNRAS*, 491, 4406
Sander, A. A. C. & Vink, J. S. 2020, *MNRAS*, 499, 873
Shenar, T., Gilkis, A., Vink, J. S., et al. 2020, *A&A*, 634, A79
Shenar, T., Sana, H., Marchant, P., et al. 2021, *A&A*, 650, A147
Vink, J. S. & de Koter, A. 2005, *A&A*, 442, 587
Yoon, S.-C., Langer, N., & Norman, C. 2006, *A&A*, 460, 199

The Origin of Outflows in Evolved Stars
Proceedings IAU Symposium No. 366, 2022
L. Decin, A. Zijlstra & C. Gielen, eds.
doi:10.1017/S1743921322000151

Magnetised gas nebulae of evolved massive stars

Dominique M.-A. Meyer

Institut für Physik und Astronomie, Universität Potsdam, Karl-Liebknecht-Strasse 24/25, D-14476 Potsdam, Germany
email: dmameyer.astro@gmail.com

Abstract. Massive stars are amongst the rarest but also most intriguing stars. Their extreme, magnetised stellar winds induce, by wind-ISM interaction, famous multi-wavelengths circumstellar gas nebulae of various morphologies, spanning from large-scale wind bubbles to stellar wind bow shocks, rings and bipolar shapes. We present two- and three-dimensional magneto-hydrodynamical (MHD) simulations of the circumstellar medium of such massive stars at different phase of their evolution. Particularly, we investigate the stability properties of 3D MHD bow shock nebulae around the runaway red supergiant stars IRC-10414 and Betelgeuse. Our results show that their astrospheres are stabilised by an organised, non-parallel ambient magnetic field. These findings suggest that Betelgeuse's bar is of interstellar origin. Last, we explore the circular aspect of the young nebula around the Wolf-Rayet stars. It is found that Wolf-Rayet nebulae are not affected by the ISM gas distribution in which the stellar objects lie, even in the case of fast stellar motion: as testifies the ring-like surroundings of the Milky Way's fastest Wolf-Rayet star, WR124. The morphology of these nebulae is tightly related to their pre-Wolf-Rayet wind geometry and to their phase evolution transition properties, which can generate bipolar shapes. We will further discuss their diffuse projected emission by means of radiative transfer calculations and show that the projected diffuse emission can appear as bipolar structures as in NGC6888.

Keywords. MHD - radiative transfer - circumstellar matter - stars: massive.

1. Introduction

Massive stars are infrequent, seminal and generate ionized superbubbles around stellar clusters (Langer 2012). Throughout their life, their strong winds and explosive death enrich the interstellar medium (ISM), drive away turbulence regulating future star formation and produce cosmic rays. The large variety of scales and physical mechanisms involved therein reflects in their so-called circumstellar medium, shaping as pc-scale bubbles (Weaver *et al.* 1977) or bow shocks (van Buren *et al.* 1988) and finally to remnant nebulae (Meyer *et al.* 2020a; Meyer 2021b).

For decades, magnetic fields have been neglected in the modelling of circumstellar nebulae around evolved massive stars, and only recent works of the past years started investigating the magneto-hydrodynamics (MHD) of the circumstellar medium (van Marle *et al.* 2015). We continue in this direction.

This proceeding is divided as follows. First we will present the 3D MHD models for the circumstellar medium of Betelegeuse and its bar (Section 2), then we will discuss the origin of infrared rings co-moving with some Galactic runaway Wolf-Rayet stars (Section 3). Finally, we will investigate the effects of magnetic fields and wind asymmetries in the

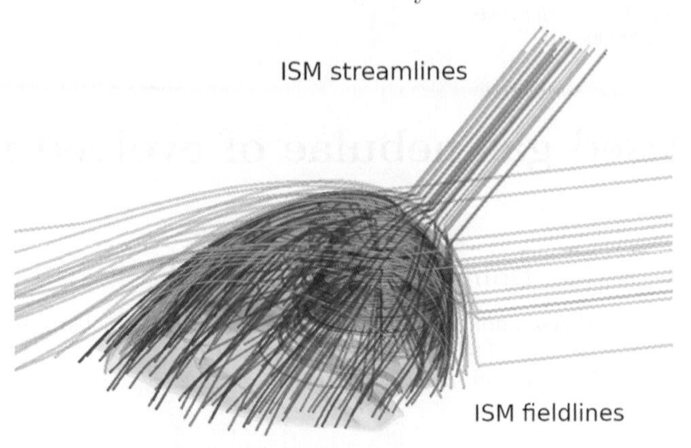

Figure 1. Bow shock nebula of the runaway supergiant star IRC-10414. The transparent surfaces are isodensity contours of the bow shock. The red and yellow lines mark the gas velocity streamline and the magnetic fieldlines, respectively.

shaping of bipolar nebulae around Wolf-Rayet stars (Section 4). Our conclusions are presented in Section 5.

2. 3D MHD stellar wind bow shock around red supergiant stars

In Meyer *et al.* 2021c we used the PLUTO code (Mignone *et al.* 2012) in order to perform full 3D MHD models tailored of the astropsheres of the runaway red supergiant stars IRC-10414 (Gvaramadze *et al.* 2014) and Betelgeuse (Mohamed *et al.* 2012), respectively. The simulations have been carried out on a uniform Cartesian grid, using radiative cooling and heating and we account for special boundary conditions by imposing the runaway star a direction of motion that is not parallel to any of the Cartesian axis, in order to reduce grid effects (Blondin *et al.* 1998).

In Fig. 1 we show a rendering for the surroundings of the runaway supergiant star IRC-10414, with coloured isosurfaces for the astrosphere density structure, as well as gas velocity (in red) streamlines and magnetic fieldlines (in yellow). One sees that the 3D nature of the calculations permits to investigate situations where the star does not move along the same direction as to the local magnetic field direction, while precedent 2D models could not (van Marle *et al.* 2012; Meyer *et al.* 2014b; van Marle *et al.* 2014).

This angle θ between directions of magnetic fields and stellar motion revealed that in the non-parallel case ($\theta \geqslant 5$ degrees), extra magnetic stress and pressure provided by the ISM magnetic field, stabilises the outer region of the astrosphere. It is therefore no more ragged and clumpy as modelled in Blondin *et al.* (1998); Meyer *et al.* (2014b), but adopts a smooth appearance. This profoundly affects the projected emission of the stellar surroundings (see Meyer *et al.* 2021c).

Last, we run a model tailored to the young bow shock of Betelgeuse (Mohamed *et al.* 2012). Radiative transfer calculation for optical Hα emission are shown in Fig. 2 together with HERSCHEL 170 μm infrared emission (ESA Herschel Science Archive Observation ID: 1342242656). It shows that the mysterious bar of Betelgeuse can be explained as an additional interstellar structure to the astrosphere, unlike the pure circumstellar model of Mackey *et al.* 2012.

3. Rings around runaway Wolf-Rayet stars

Wolf-Rayet stars are stellar objects of particularly high initial mass, evolving to a particular phase characterised by mass-loss rates $\geq 10^{-5}\,\mathrm{M}_{\odot}\,\mathrm{yr}^{-1}$ and wind velocities

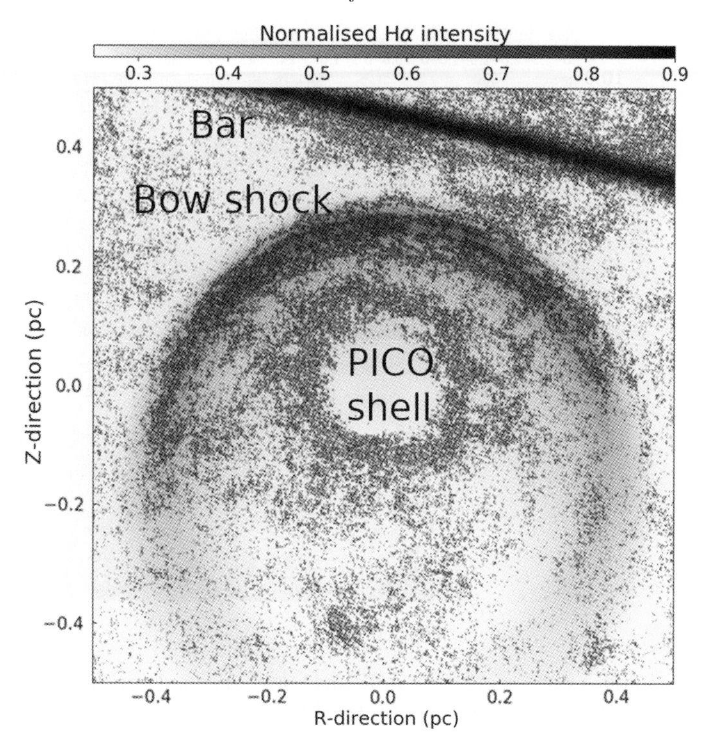

Figure 2. Synthetic Hα image of the circumstellar medium of Betelgeuse (colour), with overplotted Herschel data (black contours). The different elements of this peculiar stellar surroundings are indicated, with the astrosphere (bow shock), the outer bar and the inner photo-ionized-confined shell (PICO, see Meyer *et al.* 2021c), respectively.

3000–5000 km s^{-1} (Hamann *et al.* 2006). These stars display a variety of circumstellar nebulae, both in optical and infrared, generated by the interaction of their wind with their ambient medium. Some of them even run away from their parent cluster and reach the high latitudes of the Milky Way. However, despite of the fact that one could expect such runaway Wolf-Rayet stars to generate bow shocks, no one display one. More intriguingly, some fast-moving Wolf-Rayet stars carry a ring visible in the infrared waveband (Gvaramadze *et al.* 2010).

In Meyer 2021b we performed 2D MHD simulations of the evolution of the circumstellar medium of the 60 M$_\odot$ massive star of Groh *et al.* (2014). This star undergoes a complex pre-supernova evolution, including a main-sequence phase, a B-type phase, several luminous blue variable eruptions and three successive, so-called WN, WC, WO, Wolf-Rayet phases. We showed that an infrared-faint bow shock is produced by the pre-Wolf-Rayet stellar winds, making room for the Wolf-Rayet material to expand into its last precedent wind. Such wind-wind interaction engenders a dense infrared-bright ring which growths as it expands away from the star while conserving its spherical morphology (Fig. 3).

4. Asymmetric nebulae of Wolf-Rayet stars

Amongst the many shapes the surroundings of Wolf-Rayet stars can adopt, several examples reported a bipolar morphology (see Chu *et al.* 1982a), see also the nebula NGC6888 around the star WR 136 (Treffers *et al.* 1982). Although the dust properties of this circumstellar structure suggests that its central star has previously undergone a red supergiant phase (Mesa-Delgado *et al.* 2014), the spherical character of both the red supergiant and Wolf-Rayet winds can not explain the form of NGC6888.

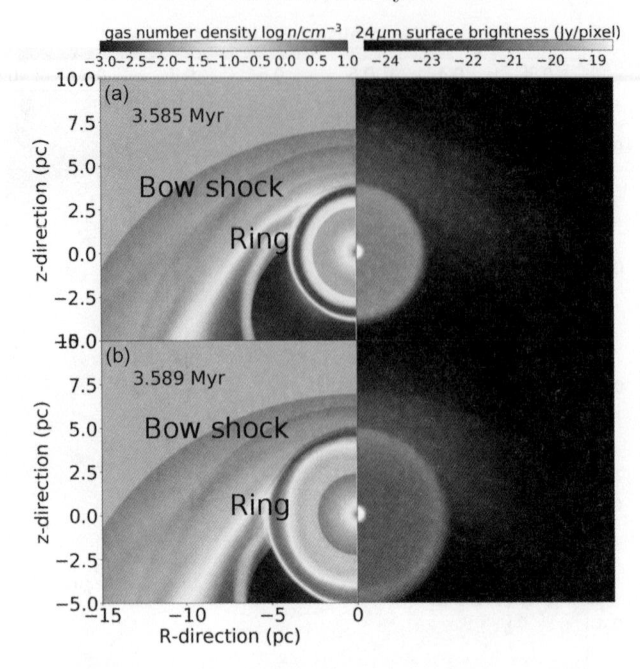

Figure 3. Surroundings of a runaway Wolf-Rayet star, displaying near-infrared bow shock and bright inner ring formed by wind-wind collision (Meyer 2021d).

We explore, by means of 2D simulations, the scenario of a star undergoing a blue supergiant phase before its Wolf-Rayet wind start blowing. As luminous blue variable stars have been observed to expel asymmetric winds into their circumstellar medium, we assume that the blue supergiant wind is blown according to a high polar-to-equatorial density distribution, which we model with the asymmetry function of Raga *et al.* 2008.

Fig. 4 displays two Wolf-Rayet nebulae produced by wind-wind interaction, either assuming an asymmetric blue supergiant wind (left) or a symmetric wind (right). The first generates a peanut-like, bipolar Wolf-Rayet nebula while the latter induces a ring-like spherical bubble. The overplotted contours mark the toroidal component of the wind velocity (solid black line) and velocity (dashed red line), respectively. From this we concluded that the star driving NGC6888 must have undergone a blue supergiant phase in which the wind was launched along the equator.

5. Conclusion

We find that the surface stellar magnetic field of supergiant stars are dynamically unimportant to the shaping of their stellar wind bow shocks, although it makes the simulations more stable computationally. Particularly, when direction of stellar motion and direction of ISM magnetic fields are different, the bow shocks are stabilised and adopt a smoother shape (Meyer *et al.* 2021c). Projection effects, e.g. at Hα, of the 3D MHD bow shocks are important and reveal a diversity of morphologies similar to that in many observed optical and infrared nebulae (van Buren *et al.* 1988). Additionally, we use the 3D MHD models to propose the nature of Betelgeuse's bar as a being interstellar, and not circumstellar as it had been previously suggested (Mackey *et al.* 2012).

Moreover, we explained the formation of rings around the fastest runaway Wolf-Rayet stars such as WR124 by wind-wind collision (Meyer *et al.* 2020b), and showed that an asymmetric pre-Wolf-Rayet wind can drastically change the shape of its subsequent circumstellar medium by forming a bipolar nebula like NGC6888 (Meyer 2021d). These

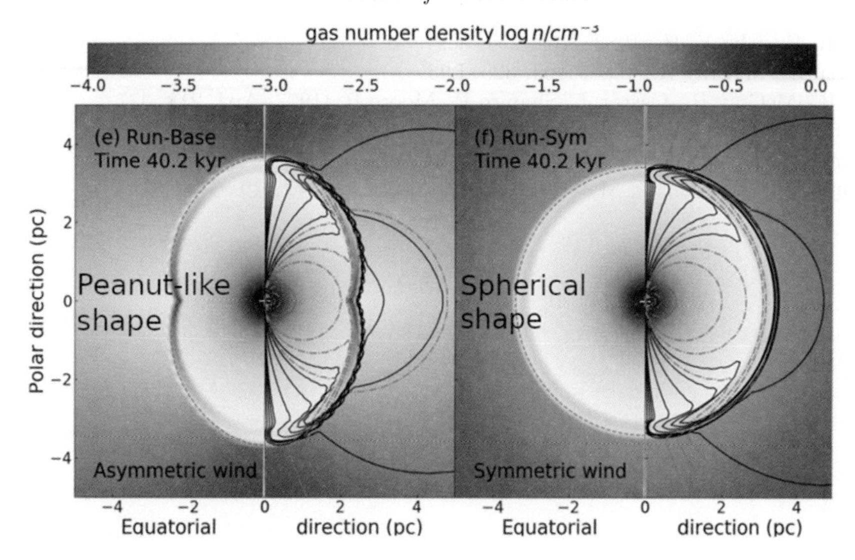

Figure 4. Stellar wind nebulae generated around Wolf-Rayet stars. The circumstellar nebulae are formed by wind-wind interaction, with a blue supergiant wind blown prior to the Wolf-Rayet phase, that is either asymmetric (left) or spherical (right). The subsequent nebulae adopt morphologies which are function of the blue supergiant wind geometry.

surroundings of Wolf-Rayet stars constitute a fantastic laboratory for numerical hydrodynamists and stellar evolution physicists, which should be further investigated by means of high-performance computing simulations.

References

Blondin J. M., Koerwer J. F., 1998, New Astron., 3, 571

Chu Y.-H., 1982a, ApJ, 254, 578

Dullemond C. P., 2012, RADMC-3D: A Multi-Purpose Radiative Transfer Tool, Astrophysics Source Code Library. preprint (ascl:1202.015)

Hamann W. R., Gräfener G., Liermann A., 2006, A&A, 457, 1015

Gvaramadze V. V., Menten K. M., Kniazev A. Y., Langer N., Mackey J., Kraus A., Meyer D. M.-A., Kamínski T., 2014, MNRAS, 437, 843

Groh J. H., Meynet G., Ekstroem S., Georgy C., 2014, A&A, 564, A30

Gvaramadze V. V., Kniazev A. Y., Fabrika S., 2010, MNRAS, 405, 1047

Langer N., 2012, ARA&A, 50, 107

Mackey J., Mohamed S., Neilson H. R., Langer N., Meyer D. M.-A., 2012, ApJ, 751, L10

Mesa-Delgado A., Esteban C., García-Rojas J., Reyes-Pérez J., Morisset C., Bresolin F., 2014, ApJ, 785, 100

Meyer D. M.-A., Mackey J., Langer N., Gvaramadze V. V., Mignone A., Izzard R. G., Kaper L., 2014b, MNRAS, 444, 2754

Meyer D. M.-A., Petrov M., Pohl M., 2020, MNRAS, 493, 3548

Meyer D. M.-A., Oskinova L. M., Pohl M., Petrov M., 2020, MNRAS, 496, 3906

Meyer D. M.-A., Pohl M., Petrov M., Oskinova L. MNRAS, Vol. 502, Issue 4, pp.5340-5355, 2021

Meyer D. M.-A., Mignone A., Petrov M., Scherer K., Velazquez P. F., Boumis P. MNRAS, Vol. 506, 4, 2021

Meyer D. M-A., MNRAS, Volume 507, Issue 4, pp.4697-4714

Mignone A., Zanni C., Tzeferacos P., van Straalen B., Colella P., Bodo G., 2012, ApJS, 198, 7

Mohamed S., Mackey J., Langer N., 2012, A&A, 541, A1

Raga A. C., Cantó J., De Colle F., Esquivel A., Kajdic P., Rodríguez-González A., Velázquez P. F., 2008, ApJ, 680, L45

Treffers R. R., Chu Y.-H., 1982, ApJ, 254, 569
van Buren D., McCray R., 1988, ApJ, 329, L93
Weaver R., McCray R., Castor J., Shapiro P., Moore R., 1977, ApJ, 218, 377
van Marle A. J., Meliani Z., Marcowith A., 2012, A&A, 541, L8
van Marle A. J., Decin L., Meliani Z., 2014, A&A, 561, A152
van Marle A. J., Meliani Z., Marcowith A., 2015, A&A, 584, A49

The Origin of Outflows in Evolved Stars
Proceedings IAU Symposium No. 366, 2022
L. Decin, A. Zijlstra & C. Gielen, eds.
doi:10.1017/S1743921322000722

Can pre-supernova winds from massive stars enrich the interstellar medium with nitrogen at high redshift?

Arpita Roy[1], **Mark R. Krumholz**[2,3], **Michael A. Dopita**[2]†, **Ralph S. Sutherland**[2,3], **Lisa J. Kewley**[2,3] **and Alexander Heger**[3,4]

[1]Scuola Normale Superiore, Piazza dei Cavalieri 7, 56126 Pisa, Italy
email: arpita.roy@sns.it

[2]RSAA, Australian National University, Cotter Road, Weston Creek, ACT 2611, Australia.

[3]ASTRO 3D, Canberra, ACT 2611, Australia

[4]School of Physics and Astronomy, Monash Centre for Astrophysics, 19 Rainforest walk, Monash University, VIC 3800, Australia.

Abstract. Understanding the nucleosynthetic origin of nitrogen and the evolution of the N/O ratio in the interstellar medium is crucial for a comprehensive picture of galaxy chemical evolution at high-redshift because most observational metallicity (O/H) estimates are implicitly dependent on the N/O ratio. The observed N/O at high-redshift shows an overall constancy with O/H, albeit with a large scatter. We show that these heretofore unexplained features can be explained by the pre-supernova wind yields from rotating massive stars ($M \gtrsim 10\,M_\odot$, $v/v_{\rm crit} \gtrsim 0.4$). Our models naturally produce the observed N/O plateau, as well as the scatter at low O/H. We find the scatter to arise from varying star formation efficiency. However, the models that have supernovae dominated yields produce a poor fit to the observed N/O at low O/H. This peculiar abundance pattern at low O/H suggests that dwarf galaxies are most likely to be devoid of SNe yields and are primarily enriched by pre-supernova wind abundances.

Keywords. stars: abundances, (stars:) supernovae: general, stars: winds, ISM: abundances, ISM: evolution, galaxies: dwarf, galaxies: high-redshift, Galaxy: evolution

---

1. Introduction

The origin of nitrogen in the Universe and its cosmic evolution is a question at the forefront of current astrophysical studies. Observationally, the origin of N and its evolution play an important role in optical astronomers' metallicity determinations, which are heavily reliant on the N/O ratio. In fact, one of the best metallicity diagnostics is based on the [NII]/[OII] ratio (Kewley & Dopita 2002). However, at high redshift the OII doublet is often unobservable, and therefore the metallicity estimate is dependent on either [NII]/Hα (Denicoló et al. 2002) or [NII]/[OIII] (Pettini & Pagel 2004) ratio, and in the worst-case scenario when only the red lines of Hα, [NII]λ6584 and [SII]$\lambda\lambda$6717, 6731 doublets are observed, the O/H estimate needs to depend on either [NII]/Hα or [NII]/[SII], or a carefully chosen combination of the two (Dopita et al. 2016).

Our understanding of the N/O ratio is poor because of the complex chemical origin of N. Unlike oxygen, which has a primary origin, N has both primary and secondary origin. An element is defined as a primary element when it is produced entirely via the

†deceased

nucleosynthesis process inside a star without any initial seeds from elements that the star inherited from the interstellar medium (ISM) at birth. By this definition, O is a primary element because it is produced in stars undergoing triple-α reactions that does not rely on pre-existing C, N, or O that was imbibed by the star at birth. On the other hand, N can be produced during the core H burning in massive stars via CNO nucleosynthesis where the C used is inherited from the ISM, and hence this channel of N production is secondary. However, during core He burning in massive stars, shell H burning is catalysed by C produced inside the star itself, and hence this N production is primary. We discuss these two N production scenarios in detail in section 2.

2. Chemical origin of nitrogen and the associated dredge-up mechanisms

A massive star has a large central convective zone where chemical elements are quasi-homogeneously mixed, its immediate outer layer is a radiative zone, and the outermost one is a thin convective zone. In rotating stars, various rotational instability induced diffusions of chemical elements dredge up the heavy elements from the core to the surface, as has been proposed by several authors in the past (Heger et al. 2000; Meynet & Maeder 2000; Maeder & Meynet 2005; Meynet & Maeder 2005). These rotational mixings behave as a bridge between the inner and the outer convective zones that helps in transporting the heavy elements from the core to the surface crossing the radiative barrier. However, in non-rotating stars, dredging up heavy metals from the inner convective zone to the outer convective layer crossing the radiative barrier becomes challenging. Roy et al. (2020) proposes that the mechanism of the exposure of "fossil"-convective cores, where the star exposes regions that are no longer convective but were part of the convective core at an early stage in the evolution for a metal-rich ([Fe/H]$\geqslant -1.0$) massive ($\geqslant 80\,M_{\odot}$) star, can dredge up the heavy elements from the core to the surface. This happens because the star shrinks as it loses mass by main sequence winds. They also find that even the modest amount of mass loss in these stars can expose the "fossil"-convective cores, thereby enhancing the surface abundances of heavy elements.

Given the various dredge-up mechanisms, C produced in the He burning core, undergoing the triple-α process, gets dredged up to the surface and used as a catalyst to N production in the H burning shell. In this method, the H burning shell uses the C that is self-produced by the star for the CNO cycle to convert the C to N, and therefore this N production channel has a primary origin. Whereas, when the core is H burning undergoing CNO process, the C that is converted to N comes from the ISM at birth, and therefore that N has a secondary origin. As O has a primary origin, if N has a primary origin too, then we will expect N/O ratio to be constant with O/H. However, if N has a secondary origin, then we will see N/O to increase linearly with O/H.

3. Observations of N enrichment at high-redshift

Given the challenges and complexities of accurate modelling of N production, many authors aim to probe the origin of N by observing the N/O trend with O/H in metal-poor systems. There are three major approaches frequently undertaken for these observations. Firstly, one can observe metal-poor Galactic halo stars as they carry the footprints of early N production in our Galaxy. Secondly, one can use the HII regions of the Milky-Way and nearby galaxies at various global metallicities for the studies of N production as these systems behave as proxies of galaxies at different redshifts. Thirdly, one can observe high-redshift galaxies directly to study *in situ* N and O production.

Israelian et al. (2004) follows the first approach, and they survey 31 unevolved halo stars within the metallicity range of $-3.1 \leqslant$ [Fe/H] $\leqslant 0$. They find constancy of [N/O] below [O/H]$\leqslant -1.8$ hinting towards the primary origin of N, and a turnover beyond that metallicity with [N/O] $\propto$ [O/H] suggesting the secondary origin of N at high metallicity. In a similar spirit, Spite et al. (2005) also analyse 35 stars with $-4 \leqslant$ [Fe/H] $\leqslant -2$. However, contrary to Israelian et al. (2004)'s observations, they find a broad scatter in [N/O] with no systemic trend in [N/O] with [O/H] at these low metallicities. This suggests a probable more complex origin of N rather than the simple primary origin of N as proposed before.

Observations of HII regions in our Galaxy and local dwarfs also suggest a similar complex origin for N. Izotov et al. (1999) measure N and O yields by observing 50 blue compact galaxies with metallicity $Z_\odot/50 < Z < Z_\odot/7$, and find that N/O is constant with O/H at low metallicity. However, contrary to these observations, several other authors find a large scatter of ~ 1 dex in N/O for a given O/H at low metallicity (Kobulnicky & Skillman 1996; Liang et al. 2006; Pérez-Montero & Díaz 2005; Pérez-Montero & Contini 2009) suggesting no systematic trend in N/O, similar to the observations of metal-poor halo stars.

Combining both these observations from stars and gas in the HII regions, we conclude that there is an overall plateau in N/O ($\log(\mathrm{N/O}) \sim -1.7$) at low metallicity ($12 + \log(\mathrm{O/H}) \leqslant 7.5$), albeit with a large scatter (~ 1 dex).

4. Stellar models and adundance calculation

Having discussed various observational approaches to measure N and O abundances at low metallicities, we discuss our theoretical models to estimate yields from winds and supernovae (SNe) in this section. For stellar models, we use modified MESA Isochrone Stellar Tracks II (MIST-II, Dotter et al., in prep.) that are appropriate for the evolution of massive stars. In this modified version, we include a realistic treatment for diffusion and transport of chemical elements driven by various rotational instabilities and also more accurate implementation of α enhancement to emulate Galactic concordance abundances. For details, we refer readers to Roy et al. (2020, 2021); Grasha et al. (2021). In this proceeding, we show results of our stellar models with masses $10\,\mathrm{M}_\odot \leqslant \mathrm{M} \leqslant 150\,\mathrm{M}_\odot$ for our fiducial rotation rate (see Roy et al. (2021)) of $v/v_{\mathrm{crit}} = 0.4$ for three metallicities, $[\mathrm{Fe/H}] = -2.0$, -3.0, and -4.0. For SN models, we adopt the yields from Limongi & Chieffi (2018).

For the yield estimate, we consider a simple stellar population with the initial mass function (IMF) $\mathrm{d}n/\mathrm{d}m$ at the time of formation $t = 0$. We use the Salpeter IMF with the upper mass limit of $150\,\mathrm{M}_\odot$. The total mass of an element X per unit stellar mass (at formation) that is returned to the ISM gas phase at time t is

$$\psi_{\mathrm{ret}}(X,t) = \psi_w(X,t) + \psi_{\mathrm{SN}}(X,t) = \tag{4.1}$$
$$\frac{1}{\langle m\rangle}\left[\int_0^\infty M_w(X,m,t)\frac{\mathrm{d}n}{\mathrm{d}m}\,\mathrm{d}m + \int_{m_\mathrm{d}}^\infty M_{\mathrm{SN}}(X,m)\frac{\mathrm{d}n}{\mathrm{d}m}\,\mathrm{d}m\right],$$

where ψ_w, ψ_{SN} are wind and SN contribution respectively,

$$\langle m\rangle = \int m\frac{dn}{dm}\,dm \tag{4.2}$$

is the mean stellar mass, m_d is the "death mass" at t, and,

$$M_w(X,m,t) = \int_0^t \dot{M}_w(X,m,t')\,dt' \tag{4.3}$$

is the wind ejected cumulative mass of an element X by a star of initial mass m up to age t.

Now we assume a gas reservoir of mass M_g that converts ϵ_* of its mass instantaneously to stars. If the mass fraction of an element X prior to this star-formation event is $f_0(X)$, then at a time t, after the star formation, the mass of that element in the gas phase is

$$M(X,t) = [(1-\epsilon_*) f_0(X) + \epsilon_* \psi_{\rm ret}(X,t)]\, M_{\rm g}. \tag{4.4}$$

We can therefore write the mass ratio of two elements at time t as,

$$\frac{M(X,t)}{M(Y,t)} = \frac{(1-\epsilon_*) f_0(X) + \epsilon_* \psi_{\rm ret}(X,t)}{(1-\epsilon_*) f_0(Y) + \epsilon_* \psi_{\rm ret}(Y,t)}\,. \tag{4.5}$$

We also consider that subsequent to the star formation, the galaxy accretes an additional primordial pristine gas of mass $M_{\rm p}$ with the fractional abundance of element X as $f_{\rm p}(X)$. In this scenario, the mass ratio of two elements becomes

$$\frac{M(X,t)}{M(Y,t)} = \frac{(1-\epsilon_*) f_0(X) + \epsilon_* \psi_{\rm ret}(X,t) + \epsilon_{\rm p}(t) f_{\rm p}(X)}{(1-\epsilon_*) f_0(Y) + \epsilon_* \psi_{\rm ret}(Y,t) + \epsilon_{\rm p}(t) f_{\rm p}(Y)}\,, \tag{4.6}$$

where $\epsilon_{\rm p}(t) = M_{\rm p}(t)/M_{\rm g}$ is the fractional mass of the added primordial gas. This abundance ratio in terms of number fractions can be rewritten as:

$$\frac{N(X,t)}{N(Y,t)} = \frac{M(X,t)}{M(Y,t)} \frac{m_Y}{m_X}\,, \tag{4.7}$$

where m_X is the atomic mass of element X. We denote $N(X)/N(Y)$ as X/Y.

5. N/O and O/H distributions

Having discussed our model yield calculation in section 4, we now discuss the distribution of N/O versus O/H to predict the observed N/O ratio trend at low metallicity. We show our predicted N/O–O/H distribution for different parameter combinations and for our fiducial rotation rate of $v/v_{\rm crit} = 0.4$ (for justification, see Roy et al. (2021)) for three metallicities, [Fe/H] $= -2$, -3, and -4.0 in Figure 1. We find that when galaxies only retain the wind yields, we see a plateau in N/O with $\log({\rm N/O}) \sim -1.5 - -2.0$ with a scatter of ~ 1 dex at low metallicities. The scatter in N/O readily arises from the variation in metallicities and star formation efficiencies. This prediction of the plateau and scatter strikingly matches with the observed features as discussed in Section 3. This result indicates that the metal-poor dwarf galaxies might have their metal abundances solely from wind yields of massive stars, with high-velocity SNe ejecta escaping them given their shallow potential wells. On the other hand, we also notice that galaxies that can retain both wind and SNe ejecta have higher values of N/O with $\log({\rm N/O}) \sim -1$. This suggests that massive metal-rich galaxies are more suitable candidates that can retain both these low and high-velocity ejecta, and, therefore might have high values of N/O. Also, this high N/O ratio matches well with the observed N/O upturn at high O/H ($12 + \log({\rm O/H}) \gtrsim 8.0$). Thus, this observed upturn in N/O might partially be coming from SNe contribution in metal-rich massive galaxies along with the onset of secondary nitrogen production in AGB stars.

Our modelled yields depend on two major parameters ϵ_* and $\epsilon_{\rm p}$, other than stellar metallicity. The dependence on $\epsilon_{\rm p}$ can be easily comprehended. Changing $\epsilon_{\rm p}$, the parameter that controls the amount of primordial gas added to the galaxy, does not vary N/O ratio, it only changes the O/H ratio, and therefore slides points horizontally in Figure 1. Thus, the more pristine primordial gas is accreted by the high-redshift galaxies, the metal-poor they become. Understanding the dependence of N/O versus O/H on ϵ_* is more subtle. As the star formation efficiency increases, the O/H ratio increases, whereas the N/O ratio decreases. O/H increases because the more the star-formation efficiency, the more the astrated material gets deposited into the ISM thereby enhancing the gas phase O abundance. However, to understand the decrease in the N/O ratio with increasing ϵ_*, we will first have to understand the pure wind-driven case ($\epsilon_* = 1$) without any contribution from the pre-existing ISM abundance. For this case, as the

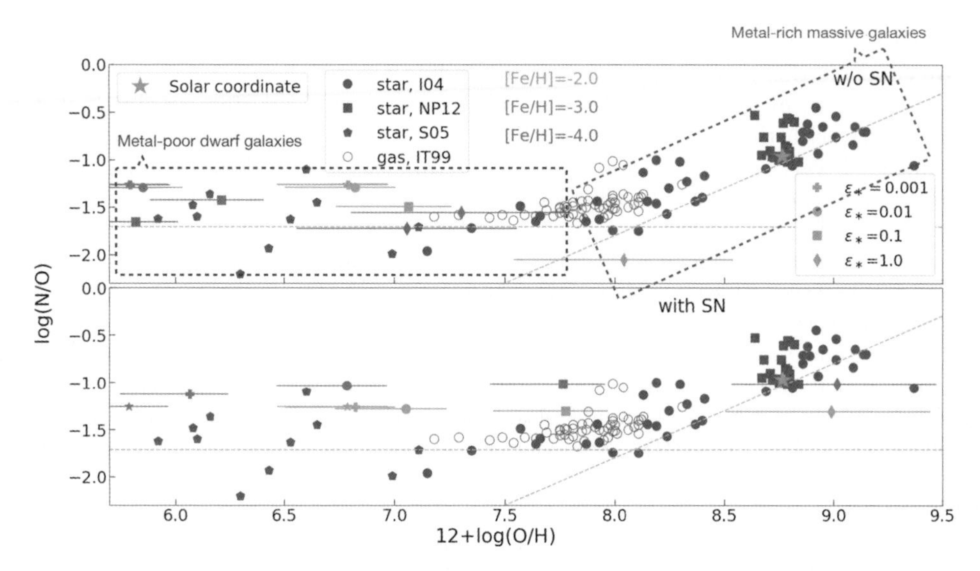

Figure 1. Adapted from Figure 8 of Roy et al. (2021): log(N/O) vs. log(O/H) from our models (coloured data points) and observations (black data points). *Top panel:* Models of galaxies that retain only the low-velocity pre-SN wind yields, *Bottom panel:* galaxies that retain both SNe ejecta and pre-SN winds. Black filled points (representing stellar yields) show the observations of several authors: Israelian et al. (2004) (I04, circles), Spite et al. (2005) (S05, pentagons), Nieva & Przybilla (2012) (NP12, squares), and open circles represent the HII regions (IT99, Izotov et al. (1999)). The orange dashed lines show the primary and secondary N production channels as proposed by Dopita et al. (2016). The red star represents the solar value. The dashed blue and brown boxes divide the N/O versus O/H phase space into two regimes based on our theoretical predictions that metal-poor dwarfs (blue box) retain only wind ejecta, leading to a low N/O ratio with large scatter characteristic of wind yields, while metal-rich massive galaxies (brown box) retain both wind and SN ejecta, leading to a higher N/O ratio characteristic of SNe. Coloured points represent theoretical predictions for different initial metallicity [Fe/H] and star formation efficiency ϵ_* as indicated by the figure legend. For each point, the horizontal bar corresponds to the scenario where a galaxy accretes an additional fraction of primordial ($\epsilon_{\rm p}$) hydrogen and helium followed by the star-formation event. Our assumed range of $\epsilon_{\rm p}$ are $\log \epsilon_{\rm p} = -0.5 - 0.5$. Note that, we omit the $[\mathrm{Fe/H}] = -4$ case for SN yields because our SN yield tables only go down to $[\mathrm{Fe/H}] = -3$ as adopted from Limongi & Chieffi (2018).

majority of massive stars enter the core He-burning phase, primary O is produced more in amount in the triple-α process compared to the primary N production in the shell H-burning. This causes the reduction in the N/O ratio. Having said that, as we decrease ϵ_*, the less amount of stellar astrated yield is deposited into the ISM, and the more of the pre-existing ISM gas abundance start playing a stronger role. We find that the ISM N/O ratio prior to the star formation event is higher compared to the wind ejected N/O. Therefore, as we decrease ϵ_*, the N/O ratio increases.

Combining these findings together, we conclude that we predict a plateau in N/O at low O/H, albeit with a large scatter, and this prediction matches pretty well with observations of metal-poor halo stars and local dwarfs. We also propose that this trend in N/O at low O/H is a signature that metal-poor dwarf galaxies may retain only the low-velocity wind yields, and the high-velocity SNe ejecta will most likely escape their low gravitational potential wells. On the other hand, metal-rich massive galaxies are likely to retain both wind and SNe ejecta, thereby having a significantly higher N/O ratio, and therefore they might be major contributors to the upturn in N/O at high O/H ($12 + \log(\mathrm{O/H}) \gtrsim 8.0$) along with the secondary N from AGB stars.

6. Conclusions

We conclude this paper with three salient findings:

• Winds from metal-poor ([Fe/H] $\leqslant -2.0$) massive ($\geqslant 10\,M_\odot$) stars produce substantial nitrogen with the N/O ratio of $\log(N/O) \sim -2 - -1.5$, independent of total oxygen metallicity. Dwarf galaxies are more likely to retain these low-velocity wind yields, and therefore this predicted N/O ratio is commonly observed in local dwarfs, along with metal-poor halo stars.

• In addition to the mean N/O ratio produced in winds providing a good match to the mean observed values, we also notice the scatter in N/O, similar to observations. In our modelling, this scatter arises from varying metallicity and star formation efficiency.

• SNe produce a higher N/O ratio with $\log(N/O) \sim -1$ compared to winds. Also, the high-velocity SNe ejecta are more likely to be retained by more massive, metal-rich galaxies. Therefore, SNe ejecta might play an important role in the upturn in N/O ratio at higher O/H ($12 + \log(O/H) \gtrsim 8$) along with the secondary N produced in AGB stars.

References

Denicoló, G., Terlevich, R., Terlevich, E. 2002, *MNRAS*, 330, 69
Dopita, M. A., Kewley, L. J., Sutherland, R. S., Nicholls, D. C., 2016, *Ap&SS*, 361, 61
Grasha, K., Roy, A., Kewley, L. J., Sutherland, R. S., 2021, *ApJ*, 908, 241
Heger, A., Langer, N., Woosley, S. E., 2000, *ApJ*, 528, 368
Israelian, G. et al., 2004, *A&A*, 421, 649
Izotov, Y. I., Thuan, T. X., 1999, *ApJ*, 511, 639
Kewley, L. J., Dopita, M. A. 2002, *ApJS*, 142, 35
Kobulnicky, H. A., Skillman, E. D., 1996, *ApJ*, 471, 211
Liang, Y. C., Yin, S. Y., Hammer, F., Deng, L. C., Flores, H., Zhang, B., 2006, *ApJ*, 652, 257
Limongi, M., Chieffi, A., 2018, *ApJS*, 237, 13
Maeder, A., Meynet, G., 2005, *A&A*, 440, 1041
Meynet, G., Maeder, A., 2000, *A&A*, 361, 101
Meynet, G., Maeder, A., 2005, *A&A*, 429, 581
Nieva, M. F., Przybilla, N., 2012, *A&A*, 539, A143
Pérez-Montero, E., Contini, T., 2009, *MNRAS*, 398, 949
Pérez-Montero, E., Díaz, A. I., 2005, *MNRAS*, 361, 1063
Pettini, M., Pagel, B. E. J., 2004, *MNRAS*, 348, L59
Roy, A. et al., 2020, *MNRAS*494, 3861
Roy, A. et al., 2021, *MNRAS*, 502, 4359
Spite, M., et, al., 2005, *A&A*, 430, 655

The Origin of Outflows in Evolved Stars
Proceedings IAU Symposium No. 366, 2022
L. Decin, A. Zijlstra & C. Gielen, eds.
doi:10.1017/S1743921322000783

Effects of wind mass-loss on the observational properties of Type Ib and Ic supernova progenitors

Sung-Chul Yoon[1], Moo-Keon Jung[1], Harim Jin[1,2] and Hyun-Jeong Kim[3]

[1]Department of Physics and Astronomy, Seoul National University, 08826, Seoul, South Korea
email: scyoon@snu.ac.kr

[2]Argelander Institute for Astronomy, University of Bonn, D-53121, Bonn, Germany

[3]Korea Astronomy and Space Science Institute, 34055, Daejeon, South Korea

Abstract. Progenitors of Type Ib and Ic supernovae (SNe) are stripped envelope stars and provide important clues on the mass-loss history of massive stars. Direct observations of the progenitors before the supernova explosion would provide strong constraints on the exact nature of SN Ib/Ic progenitors. Given that stripped envelope massive stars can have an optically thick wind as in the case of Wolf-Rayet stars, the influence of the wind on the observational properties needs to be properly considered to correctly infer progenitor properties from pre-SN observations. Non-LTE stellar atmosphere models indicate that the optical brightness could be greatly enhanced with an optically thick wind because of lifting-up of the photosphere from the stellar surface to the wind matter, and line and free-free emissions. So far, only a limited number of SN Ib/Ic progenitor candidates have been reported, including iPTF13bvn, SN 2017ein and SN 2019yvr. We argue that these three candidates are a biased sample, being unusually bright in the optical compared to what is expected from typical SN Ib/Ic progenitors, and that mass-loss enhancement during the final evolutionary stage can explain their optical properties.

Keywords. Massive Stars, Wolf-Rayet stars, Hydrogen deficient stars, Stellar atmosphere, Stellar evolution, Stellar mass loss, Type Ib supernovae, Type Ic supernovae

---

1. Introduction

The evolution of massive stars is affected by their mass-loss history. Mass loss from massive stars can be caused by various factors like radiation-driven stellar winds, stellar pulsation, rapid rotation, episodic mass eruptions, and binary interactions (e.g., Langer 2012; Smith 2014; Vink 2022). The resulting different final structures of massive stars largely determine the observational diversity of core-collapse supernovae (ccSNe). For example, Type IIP and IIb supernovae would result from progenitors having large (i.e., several solar masses) and small (less than one solar mass) amounts of hydrogen in their envelopes, respectively. Type Ib and Ic supernovae (SNe Ib/Ic) are characterized by absence of hydrogen lines in the SN spectra, implying that their progenitors are helium stars whose hydrogen envelopes were stripped off before the SN explosion.

Direct identifications of ccSN progenitors are crucial for constraining theoretical predictions of massive star evolution, as it would reveal what SN progenitors look like immediately before the SN explosion for different SN types. Many SN II progenitors have been detected in pre-SN images (see Smartt 2015 for a review). By contrast, searches for

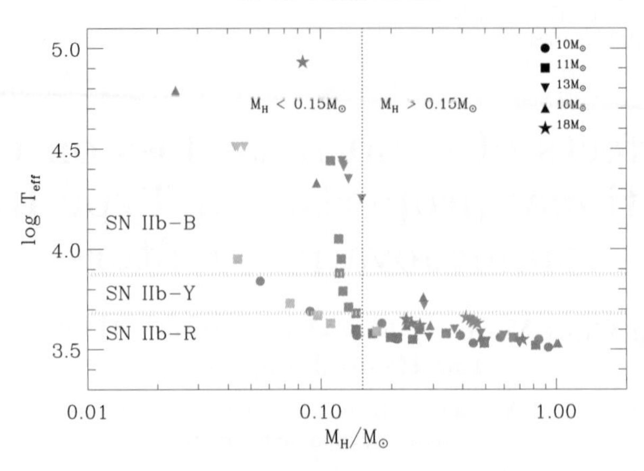

Figure 1. Surface temperatures of massive stars at the pre-SN stage as a function of the remaining hydrogen envelope mass, predicted by binary evolution models where mass loss due to Case B mass transfer from the primary star is considered. The models presented in this figure represent SN IIb progenitors whose hydrogen envelope mass is less than $1.0M_{\odot}$. Different symbols indicate the initial masses of the primary stars of the binary models as indicated by the labels. Purple and Blue denote solar and LMC metalicity models, respectively. Green denotes solar metallicity models with a WR mass loss-rate reduced by a factor of 2 compared to the standard one. The orange dotted lines mark the boundaries for yellow supergiants: $3.681 \leq \log T_{\rm eff}/{\rm K} \leq 3.875$. The back dotted vertical line roughly defines the boundary between red supergiant progenitors and more compact (i.e., yellow/blue supergiant) progenitors. The figure is taken from Yoon *et al.* (2017).

SN Ib/Ic progenitors have been futile (Eldridge *et al.* 2013), except for a few cases including those of iPTF13bvn (Cao *et al.* 2013; Eldridge & Maund 2016; Folatelli *et al.* 2016), SN 2017ein (Kilpatrick *et al.* 2018; Van Dyk *et al.* 2018), and SN 2019yvr (Kilpatrick *et al.* 2021). This is often interpreted as evidence for relatively low-mass helium star progenitors produced in binary systems (e.g., Eldridge *et al.* 2013) rather than Wolf-Rayet (WR) star-like progenitors ($M \gtrsim 8~M_{\odot}$). Studies of SNe Ib/Ic also imply relatively low SN ejecta masses for most cases (i.e., $M_{\rm ej} \approx 1.0 \cdots 4.0 M_{\odot}$) in favor of the binary scenario (e.g., Drout *et al.* 2011; Cano 2013; Lyman *et al.* 2016; Taddia *et al.* 2018; Prentice *et al.* 2019).

However, it should be noted that helium stars are not necessarily brighter in the optical for a higher mass, as pointed out by Yoon *et al.* (2012). This is because helium star spectra depend on the structure of the outermost layers and the presence of optically thick winds as discussed below. A careful study on the details of SN Ib/Ic progenitors is therefore needed for properly guiding future searches of SN Ib/Ic progenitors.

2. SN Ib/Ic progenitor properties

Many SN II progenitors, having an extended hydrogen envelope with a radius of $R >\sim 100~R_{\odot}$, are found to be yellow or red supergiants (e.g., Smartt 2015). Stellar evolution models also predict that most massive stars at the pre-SN stage would be observed as red or yellow supergiants as long as the retained hydrogen envelope mass is higher than about $0.05 \cdots 0.10 M_{\odot}$ (Figure 1; e.g., Yoon *et al.* 2017; Farrell *et al.* 2020), with the exception of blue supergiant progenitors of 1987A-like peculiar SNe IIP which presumably originate from binary mergers (e.g., Podsiadlowski *et al.* 1990; Morris & Podsiadlowski 2007; Menon & Heger 2017; Urushibata *et al.* 2018).

By contrast, hydrogen-deficient SN Ib/Ic progenitors are relatively compact ($R <\sim 10~R_{\odot}$; e.g., Yoon *et al.* 2010). Their surfaces are therefore hotter ($T_{\rm eff} >\sim 10^4$ K) than most SN II progenitors ($T_{\rm eff} < 10^4$ K). The resulting spectra from SN Ib/Ic progenitors would be much harder, leading to fainter optical brightness compared to the case of SN II progenitors for a given bolometric luminosity (e.g., Yoon *et al.* 2012). It is not surprising, therefore, that SN Ib/Ic progenitors have proven more difficult to identify than SN II progenitors in optical images.

The surface properties of SN Ib/Ic progenitors depend on their final mass and chemical composition. In Figure 2, the final luminosity and radius of helium-rich (i.e., $M_{\mathrm{He}} \gtrsim 1.0~M_{\odot}$) SN Ib progenitors predicted by stellar evolution models of Yoon *et al.* (2017) are presented. The bolometric luminosity is proportional to the mass, which is a common feature of all stars in hydrostatic and thermal equilibrium. However, the radius decreases with increasing mass. This is a result of the so-called mirror effect: the carbon-oxygen core becomes more compact for a lower mass while the helium envelope becomes more extended. This has an important consequence for the optical brightness: SN Ib progenitors would be systematically brighter in the optical for a lower final mass as shown in the figure. In other words, relatively low-mass SN Ib progenitors might be easier to detect in pre-SN images despite the fact that more massive progenitors would have higher bolometric luminosities, unless the effects of optically thick winds are significant as discussed below.

On the other hand, SN Ic progenitors would be more compact and hotter than SN Ib progenitors if they were helium-poor carbon-oxygen stars as often assumed in the literature. This could make them even fainter in the optical than SN Ib progenitors as discussed in Yoon *et al.* (2012). In sharp contrast to SN Ib progenitors having a helium-rich envelope which is larger for a lower mass, the radius of helium deficient carbon-oxygen (CO) stars becomes smaller for a lower mass (see Jung *et al.* 2022). Therefore, among SN Ic progenitors, a lower mass progenitor would be fainter.

There also exists the possibility that a large amount of helium is retained in SN Ic progenitors, which can be hidden in the optical spectra. The photosphere temperature of SNe Ib/Ic near the optical peak is not high enough (i.e., $T_{\mathrm{phot}} \sim 7000 - 8000$ K) to make He I absorption lines by thermal processes. Non-thermal processes induced by radioactive ^{56}Ni are therefore invoked to explain He I lines of SNe Ib (Lucy 1991; Swartz 1991). This requires mixing of ^{56}Ni into the helium-rich layer in the SN ejecta, and otherwise He I lines would not be formed (Dessart *et al.* 2012; Hachinger *et al.* 2012; Teffs *et al.* 2020; Williamson *et al.* 2021).

Several studies, however, provide evidence that SN Ic progenitors are distinct from SN Ib progenitors in terms of helium-richness (e.g., Liu *et al.* 2016; Fremling *et al.* 2018; Yoon *et al.* 2019; Dessart *et al.* 2020; Shahbandeh *et al.* 2021). For example, Shahbandeh *et al.* (2021) find dichotomy in the strength of He I $\lambda 2.0581~\mu$m line in near infrared (NIR) early-time spectra between SN Ib and SN Ic: this line is systematically weaker in SNe Ic than in SNe Ib. This can be considered strong evidence for the helium-poor nature of SN Ic progenitors because unlike the case of the He I lines in the optical, helium would be relatively easy to detect with the He I $\lambda 2.0581~\mu$m line (e.g., Dessart *et al.* 2020).

The systematic difference between SNe Ic and Ib is also found in their photometric colors as shown in Figure 3. It is observed that the $B-V$ color of SNe Ic is systematically redder compared to SNe Ib (i.e., $\Delta(B-V) \approx 0.15$) at the V-band peak. A study on radiation-hydrodynamics simulations of SNe Ib/Ic by Jin & Yoon (2022, in preparation) concludes that this color difference can be best explained by helium-rich and helium-poor progenitors for SNe Ib and Ic, respectively. In this case, the photosphere of a SN Ib during the photospheric phase would be located at a relatively deeper and hotter layer than that of a corresponding SN Ic, making SNe Ib systematically bluer than SNe Ic at the optical peak. This is because the outer layer of the SN Ib ejecta where the photosphere is formed is more transparent than SNe Ic.

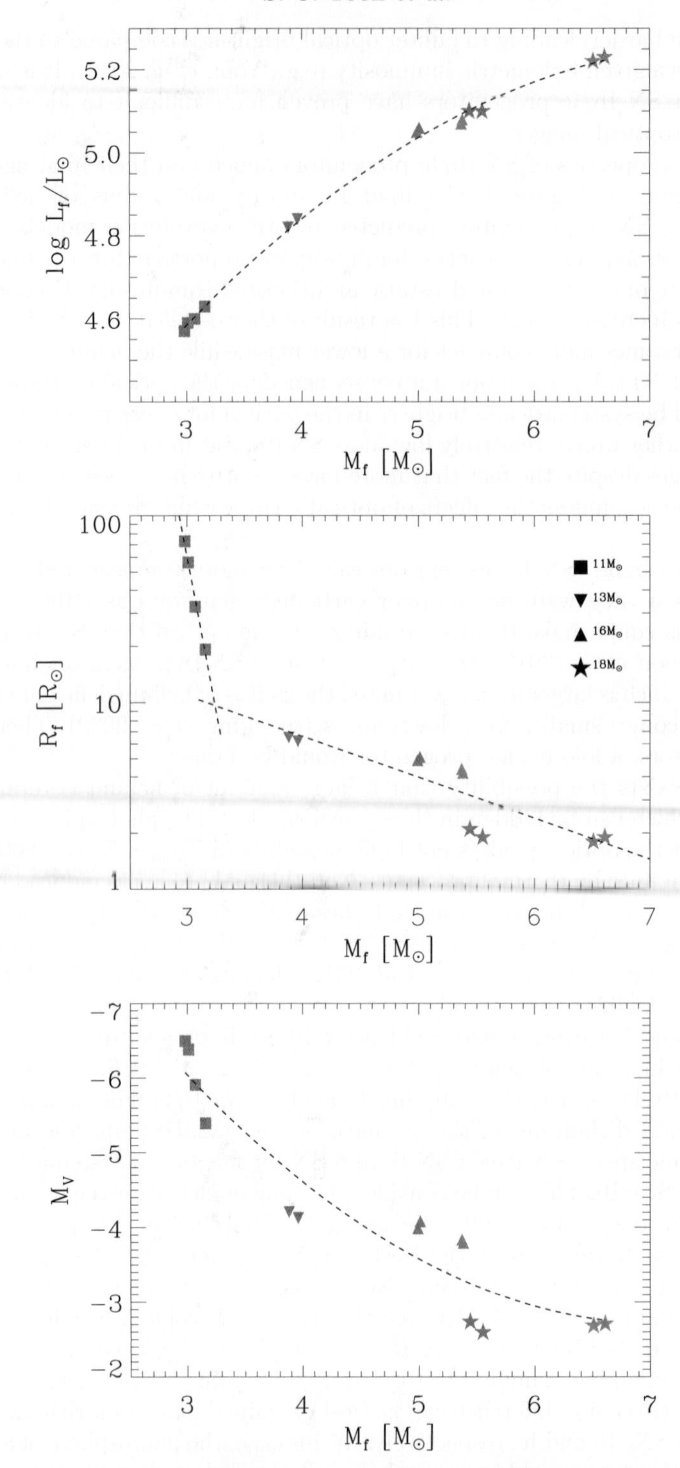

Figure 2. *Top panel:* Final luminosity as a function of the final mass of helium-rich (i.e., $M_{\rm He} \gtrsim 1.0 M_\odot$) SN Ib progenitors at the pre-SN stage predicted by binary stellar evolution models. *Middle panel:* The predicted final radius v.s. final mass relation of the corresponding models. *Bottom panel:* The corresponding V-band magnitude as a function of the final mass under the black-body approximation. The figure is taken from Yoon *et al.* (2017).

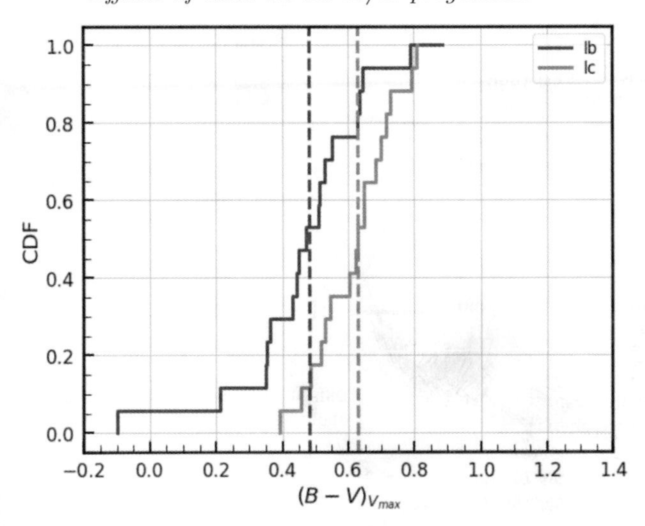

Figure 3. Cumulative distributions of the $B - V$ magnitude difference of SNe Ib and Ic at the peak of V-band magnitude. The SN sample is collected from the literature. See Jin & Yoon (2022, in preparation) for details.

The dichotomy of helium contents in SN Ib/Ic progenitors can also be inferred from the SN color evolution. Yoon *et al.* (2019) show that the color evolution of SNe Ib/Ic during early times is significantly affected by a ^{56}Ni mixing. If ^{56}Ni remains deep inside the SN ejecta, the photosheric color of a SN Ib/Ic would become quickly red initially because of the cooling induced by the ejecta expansion. But it becomes blue several days later when the photosphere recedes to the inner layer heated by ^{56}Ni. By contrast, if ^{56}Ni is fully mixed into the outermost layers of the ejecta, the color evolution becomes monotonic from blue to red color during the photospheric phase. Comparison with observations implies a strong ^{56}Ni mixing in several SNe Ic where monotonic color evolution is found, in which case optical He I lines would be easily produced even with a relatively small amount of helium (Figure 4). This indicates that SNe Ic progenitors are indeed helium-poor.

3. Evolution of massive stars towards SNe Ib/Ic

SN Ib/Ic progenitors lose their hydrogen envelopes during the post-main sequence phase. For single stars at solar metallicity, this can be realized by stellar winds if the initial mass is higher than about $25 \cdots 40\ M_{\odot}$, depending on the uncertain mass-loss rate from supergiant stars (see Yoon 2015 for a review). In close binary systems, mass transfer during the post-main sequence phase (mostly during helium core contraction and core helium burning phase: so-called Case B mass transfer) can remove a large fraction of the hydrogen envelope. The amount of the remaining hydrogen in the envelope after the mass transfer phase depends on the initial parameters of the binary system including the primary mass, mass ratio, and initial period (e.g., Yoon *et al.* 2017). The stripped star would eventually become a pure helium star if all hydrogen is removed from the surface via winds and/or further mass transfer during the post core helium burning phase. Figure 5 provides an example on how the predicted final fates of primary stars in close binary systems depend on the initial mass and orbital period. However, it should be noted that the parameter space for producing pure helium stars would depend on the uncertain mass-loss rate at the post-mass transfer stage (e.g., Vink 2017; Gilkis *et al.* 2019).

The evolution thereafter is largely determined by the mass loss history from helium stars. The dependence of the mass-loss rate of helium stars on the physical conditions

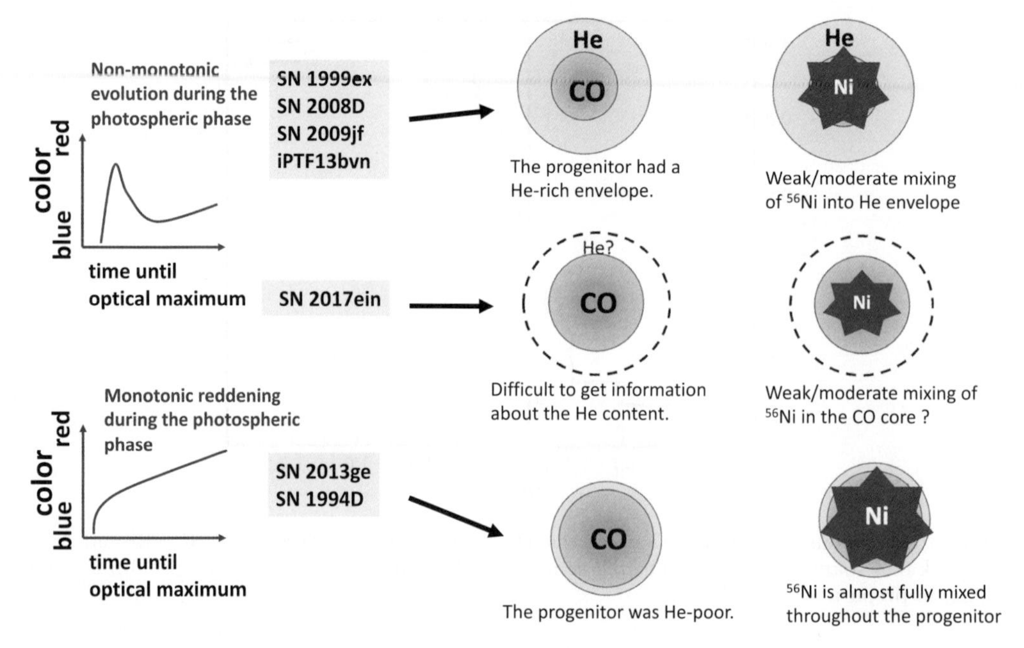

Figure 4. Schematic illustration of the color evolution of SNe Ib and Ic according to the degree of ^{56}Ni mixing in the ejecta (the figures in the left side) according to the numerical result of Yoon *et al.* (2019). Comparison with the observed color evolution implies a relatively weak/moderate ^{56}Ni mixing for SNe Ib and a very strong mixing for some SNe Ic (SN 2013ge and SN 1994D). Given that non-thermal processes due to radioactive ^{56}Ni play an essential role for the formation of He I lines, this result indicates that these SNe Ic had helium-poor progenitors, while SNe Ib had helium-rich progenitors. The figure is taken from Yoon *et al.* (2019).

and different evolutionary stages is not fully understood yet while great progress is being made in this direction (see Vink 2022 for a recent review). Stellar evolution models still cannot provide a robust prediction on whether or not SN Ib and Ic progenitors form a continuous sequence in terms of helium contents.

If the helium star mass is higher than a certain limit (e.g., about $8M_{\odot}$ for solar metallicity; see Shenar *et al.* 2020; Aguilera-Dena *et al.* 2021 for recent related discussions), the mass-loss rate would be high enough for the helium star to appear as a WR star that is characterized by strong emission lines from an optically thick wind. The WR mass-loss rate has been inferred by many observational studies on WR stars in the local Universe. Recently, Yoon (2017) argued, based on the observed WR star sample given by the Potsdam group (Hamann *et al.* 2006; Sander *et al.* 2012; Hainich *et al.* 2014), that the WR mass-loss rate would be enhanced when WR stars undergo a transition from the WN to WC phase (see, however, Sander *et al.* 2020). As shown in Figure 6, a clear dichotomy in the amount of helium is found in the evolutionary models of helium stars using the WR mass-loss rate prescription that includes the effect of the mass-loss enhancement in the WC phase (see also Woosley 2019). This result is consistent with the conclusions of the above-mentioned SN studies on the helium dichotomy between SNe Ib and Ic.

4. Effects of winds on the optical properties of SN Ib/Ic progenitors

As discussed above, SN Ib/Ic progenitors are relatively compact ($R <\sim 10R_{\odot}$) compared to SN II progenitors. Among SN Ib and Ic progenitors, SN Ic progenitors would

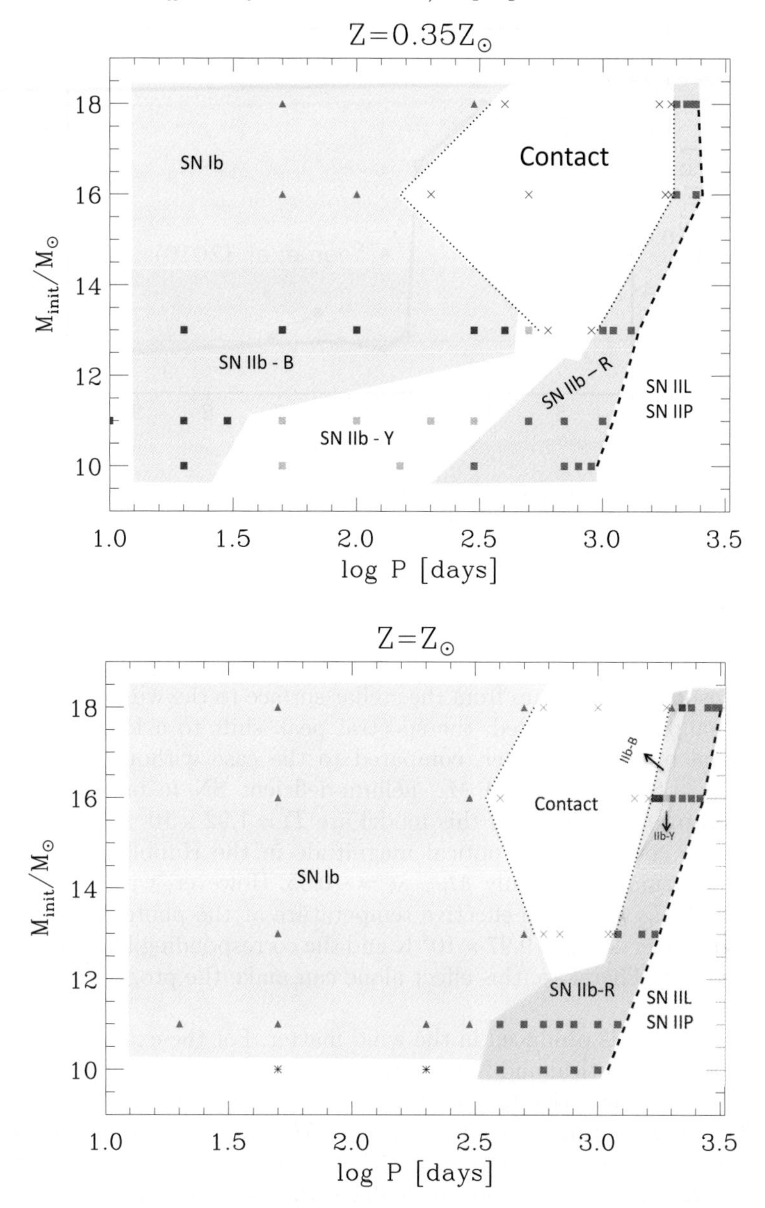

Figure 5. Final fates of primary stars in close binary systems predicted by stellar evolution models for two different metallicities (LMC and solar metallicities for the upper and lower panels, respectively), in the plane spanned by the initial primary star mass and the initial orbital period. In the models, a fixed mass ratio of $q = 0.9$ (the ratio of the secondary to the primary star masses) and mass accretion efficiency of $\beta = 0.2$ (i.e., the ratio of the accreted to the transferred masses) are assumed. Different fates are indicated by different colors as indicated by the labels. SN IIb-B, IIb-Y and II-R denotes SN IIb from blue, yellow and red supergiant progenitors, resepctively. The figure is taken from Yoon *et al.* (2017).

be more compact ($R < 1.0R_{\odot}$) than SN Ib progenitors. Under the black body approximation, therefore, the optical brightness would decrease in the order of SN II, Ib and Ic progenitors for a given bolometric luminosity.

However, massive helium stars like WR stars may have an optically thick wind if the luminosity is sufficiently high. Such a wind can significantly affect the optical brightness and color, because of the following reasons (Kim *et al.* 2015; Jung *et al.* 2022):

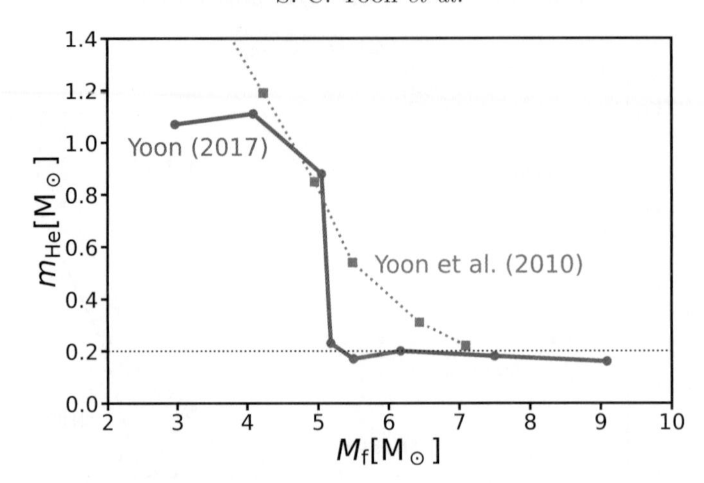

Figure 6. Amounts of helium retained in helium star models at the pre-SN stage as a function of the final mass. The red and green lines give the predictions by Yoon (2017) and Yoon *et al.* (2010), respectively. Note that Yoon (2017) considers mass-loss enhancement of WR stars during the WC phase, while Yoon *et al.* (2010) adopts a mass-loss prescription that only depends on the helium star luminosity. See Yoon (2017) for more details.

• The photosphere is lifted up from the stellar surface to the wind matter. This makes the effective temperature lowered, the spectral peak shift to a longer wavelength and thus the optical brightness higher, compared to the case without wind. In Figure 7, we show an example with a 9.09 $M_\odot$ helium-deficient SN Ic progenitor model. The surface temperature and radius of this model are $T_* = 1.92 \times 10^5$ K and $R_* = 0.52\ R_\odot$, respectively. The corresponding optical magnitude in the Hubble F555W filter under the black-body assumption is only $M_{\rm F555W} = -0.55$. However, with a mass loss rate of $\dot{M} = 1.95 \times 10^{-5}\ M_\odot\ {\rm yr}^{-1}$, the effective temperature at the photosphere formed in the extended wind matter is $T_{\rm eff} = 9.97 \times 10^4$ K and the corresponding black-body magnitude is $M_{\rm F555W} = -2.61$. Therefore, this effect alone can make the progenitor appear 7 times brighter in the optical.

• Free-free emission is produced in the wind matter. For the example in Figure 7, the continuum flux in the optical and NIR range is greatly enhanced compared to the black-body spectrum at the effective temperature of the photosphere because of this free-free emission. The corresponding optical magnitude is $M_{\rm F555W} = -3.34$, which is significantly different from the magnitudes obtained with the black body spectra at the effective temperature ($M_{\rm F555W} = -2.61$) and the stellar surface temperature ($M_{\rm F555W} = -0.55$).

• Strong emission lines are formed in the wind, which can significantly affect magnitudes of specific filters. For the example of Figure 7, the optical magnitude from the full spectrum including the emission lines is $M_{\rm F555W} = -4.39$, compared to $M_{\rm F555W} = -3.34$ from the continuum spectrum only.

Therefore, the presence of an optically thick wind from a SN Ib/Ic progenitor can greatly enhance the optical brightness as shown in Figure 8. This would allow us to constrain the mass-loss rate immediately before the SN explosion as discussed in Jung *et al.* (2022). They also find that the color of SN Ib/Ic progenitors would not become systematically bluer for a higher effective temperature, unlike the prediction with the black-body approximation. This implies that the luminosity and effective temperature of a progenitor candidate cannot be properly inferred by comparing the optical data with standard stellar evolution model predictions, unless the wind effects are properly considered.

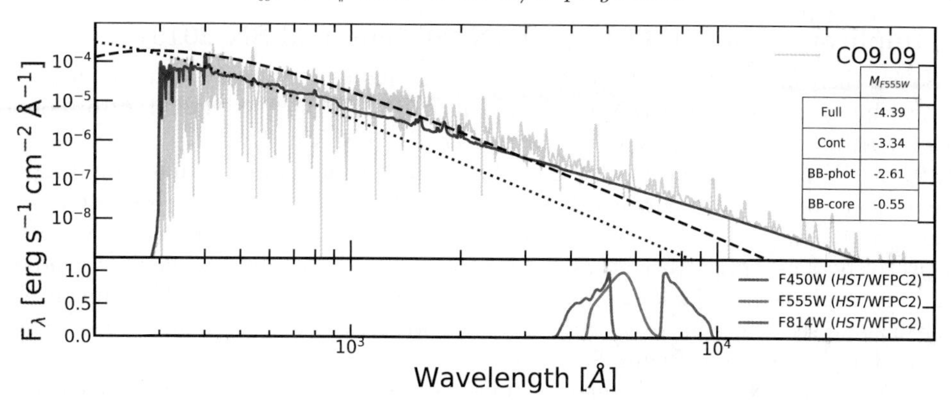

Figure 7. Spectral energy distribution of a helium-deficient SN Ic progenitor model with $M = 9.09\ M_\odot$, $\log L/L_\odot = 5.43$, $T_{\rm surf} = 1.92 \times 10^5$ K, and $R = 0.47 R_\odot$ for a stellar wind mass-loss rate of $1.95 \times 10^{-5}\ M_\odot\ {\rm yr}^{-1}$. The full spectrum was calculated with the non-LTE stellar atmosphere code CMFGEN (Hillier & Miller 1998; Hillier *et al.* 2003). The dotted line gives the black-body spectrum at the surface temperature of the model. The dashed line denotes the black-body spectrum at the effective temperature of the photosphere formed in the wind matter. The purple solid line gives the continuum spectrum. The full spectrum is given by the orange solid line. The figure is taken from Jung *et al.* (2022).

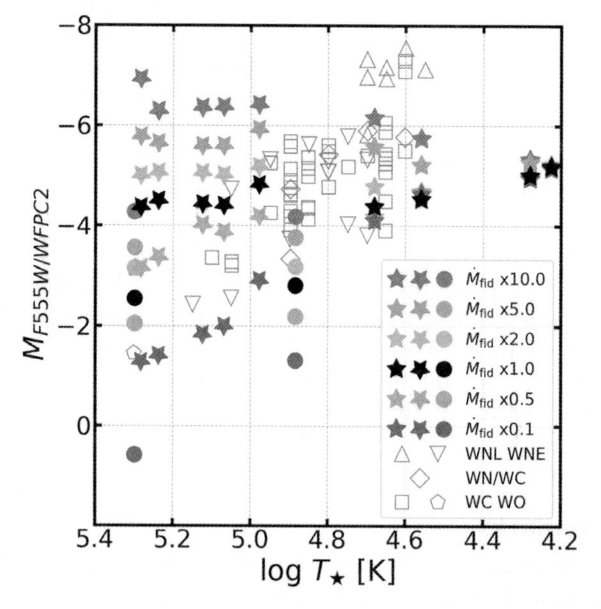

Figure 8. The predicted F555W (HST/WFPC2) filter magnitudes of SNe Ib/Ic progenitors for different mass-loss rates, as a function of the surface stellar temperature. This prediction is made with non-LTE stellar atmosphere models calculated with the CMFGEN code (Hillier & Miller 1998; Hillier *et al.* 2003). The star symbols represent the results from the SN Ib and Ic progenitor models of Yoon (2017) having final masses of 2.91, 2.97, 4.09, 5.05, 5.18, 5.50, 6.16, 7.50, 9.09 $M_\odot$ in the order of decreasing surface temperature. The filled circles denote SN Ic progenitor models of Yoon *et al.* (2019) having final masses of 2.16 and 3.93 $M_\odot$. The black color denotes the result with the fiducial mass-loss rate of helium stars given by Yoon (2017). The results with other mass-loss rates are given by different colors as indicated by the labels in the figure. The open symbols denote Smith v-band filter magnitudes of the Galactic WR star sample of the Potsdam group (Sander *et al.* 2019; Hamann *et al.* 2019). The figure is taken from Jung *et al.* (2022).

5. Implications for iPTF13bvn, SN 2017ein, and SN 2019yvr

So far, there are only three SN Ib/Ic progenitor candidates directly identified in pre-SN optical images: iPTF13bvn (SN Ib; Cao *et al.* 2013; Eldridge & Maund 2016; Folatelli *et al.* 2016), SN 2017ein (SN Ic; Kilpatrick *et al.* 2018; Van Dyk et al. 2018), and SN 2019yvr (SN Ib; Kilpatrick *et al.* 2021). All of these candidates are unusually bright in the optical, compared to our model predictions with the fiducial mass-loss rate given in Figure 8: $M_{\rm F555W} \approx -6.0, -7.5$, and -7.8 for iPTF13bvn, SN 2017ein, and SN 2019yvr, respectively. The optical brightness and color of the iPTF13bvn progenitor candidate might be explained by the lowest mass SN Ib model having an extended helium envelope ($R = 25\ R_{\odot}$; see also Kim *et al.* 2015) but the other two are too bright to be explained by any fiducial model in Figure 8.

Interestingly, we find that enhanced mass loss shortly before the SN explosion can provide a reasonable solution for the optical properties of these progenitor candidates as discussed in Jung *et al.* (2022). The optical magnitudes and color of the iPTF13bvn progenitor candidate can be achieved in SN Ib progenitor models of $M \simeq 4.0 \cdots 5.0\ M_{\odot}$ with a mass-loss rate 10 times higher than the fiducial value (i.e., $\dot{M} \approx 5.5 \cdots 8.0 \times 10^{-5} M_{\odot}\ {\rm yr}^{-1}$; see also Figure 8).

The color of SN 2019yvr progenitor candidate is very red (i.e., $T_{\rm eff} \approx 6800$ K) compared a typical SN Ib progenitor ($T_{\rm eff} = \sim 10^4$ K), but both the color and optical magnitudes of the SN 2019yvr progenitor candidate can be obtained by a SN Ib progenitor model of $\sim 5.0 M_{\odot}$ with a 50 – 100 times higher mass-loss rate than the fiducial value (i.e., $\dot{M} \approx 4.0 \cdots 8.0 \times 10^{-4} M_{\odot}\ {\rm yr}^{-1}$) as shown in Figure 9. Note that such a late time mass-loss enhancement from a helium star is indeed predicted in recent stellar models where heating of the envelope via hydrodynamic waves that transports nuclear energy from the core to the outermost layers is considered (Fuller & Ro 2018; Leung *et al.* 2021). Evidence of mass eruption shortly before the SN explosion is also found in some SNe Ib/Ic (e.g., Jin *et al.* 2021). Alternatively, the SN 2019yvr progenitor itself or its comparison might be a yellow supergiant having an extended tenuous hydrogen-rich envelope (Gilkis & Arcavi 2022; Sun *et al.* 2022).

As for the SN 2017ein progenitor candidate, its optical brightness might be roughly explained by a SN Ic progenitor model with ~ 10 times higher than the fiducial value. However, this object seems to be too blue to be explained by our SN Ic progenitor models or by a massive O-type star companion (see Jung et al. 2022 for more details). Future observations need to confirm if this candidate was a real SN progenitor or not by its disappearance.

6. Conluding remarks

SN Ib/Ic progenitors are stripped-envelope stars and thus provide important constraints for stellar evolution models including mass loss via stellar winds and/or binary interactions (Sections 2 and 3). Although WR stars observed in the nearby universe can also provide useful information on SN Ib/Ic progenitors, most of them are still at core helium burning stage and thus their surface properties would become greatly different from now at the pre-SN stage (Yoon *et al.* 2012; Tramper *et al.* 2015). In addition, observed WR stars seem to be too massive (i.e, $M \gtrsim 10 M_{\odot}$) to explain the observed light curves and spectra of ordinary SNe Ib/Ic.

One of the best ways to understand the exact nature of these progenitors would be direct observation of them before they explode as in the cases of iPTF13bvn, SN 2017ein and SN 2019yvr. However, our recent work indicates that direct comparison of stellar evolution models with the pre-SN observations might be misleading (Jung *et al.* 2022). This is because presence of an optically thick wind can greatly affect optical properties as discussed above. The sensitivity of the optical brightness and color on the wind density

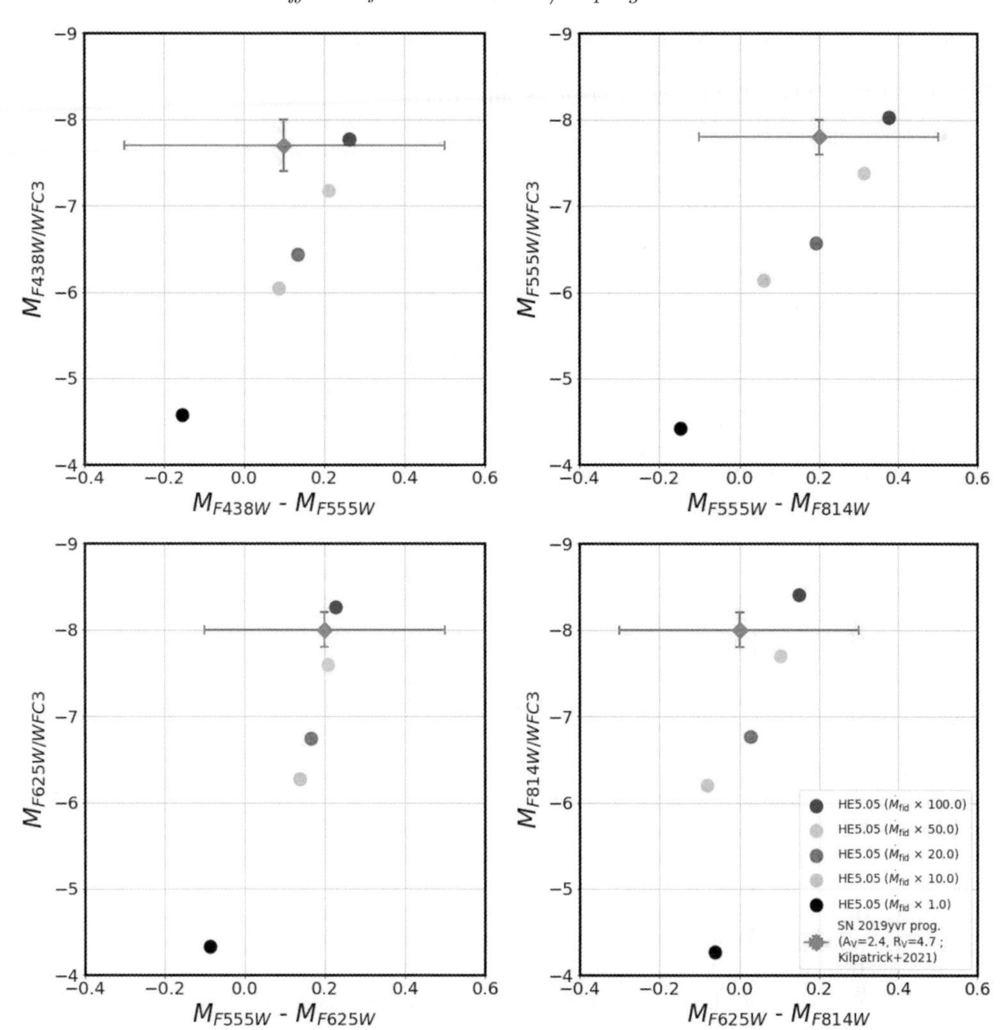

Figure 9. Predicted optical magnitudes and colors for various mass-loss rates (from 1.0 to 100.0 times the fiducial mass-loss rate given by Yoon (2017), as indicated by the labels in the lower right panel) for a 5.05 $M_\odot$ helium-rich SN Ib progenitor, compared to the observation of the SN 2019yvr progenitor candidate. This prediction is obtained with the non-LTE stellar atmosphere code CMFGEN. The figure is taken from Jung *et al.* (2022).

(see Section 4), on the other hand, would allow us to make a good constraint on the mass-loss history during the final evolutionary stages of SN Ib/Ic progenitors. Given that signatures of such mass loss can be found in the early-time SN light curves and spectra, combination of progenitor identification and very early-time SN observation in future studies would be important for making good progress towards a better understanding of SN Ib/Ic progenitors.

References

Aguilera-Dena, D.R., Langer, N., Antoniadis, J., et al. 2021 *arXiv:2112.06948*, submitted to A&A

Cano, Z. 2013 *MNRAS*, 434, 1098

Cao, Y., Kasliwal, M.M., Arcavi, I., et al. 2013 *ApJL*, 775, L7

Dessart, L., Hillier, D.J., Li, C., & Woosley, S.E. 2012 *MNRAS*, 424, 2139

Dessart, L., Yoon, S.-C., Aguilera-Dena, D.R., & Langer, N. 2020 *A&A*, 642, 106

Drout, M.R., Soderberg, A.M., Gal-Yam, A., et al. 2011 *ApJ*, 741, 97
Eldridge, J.J., Fraser, M., & Smartt, S., et al. 2013 *MNRAS*, 436, 774
Eldridge, J.J., & Maund, J.R. 2016 *MNRAS*, 461, L117
Farrell, E.J., Groh, J.H., Meynet, G., Eldridge, J.J. 2020 *MNRAS*, 494, 53
Folatelli, G., Van Dyk, S.D., Kuncarayakti, H., et al. 2016 *ApJL*, 825, L22
Fremling, C., Sollerman, J., Kasliwal, M.M., et al. 2018 *A&A*, 618, 37
Fuller, J., & Ro, S. 2017 *MNRAS*, 476, 1853
Gilkis, A., & Arcavi I. 2022 *MNRAS*, 511, 691
Gilkis, A., Vink, J., Eldridge, J.J., & Tout, C. 2019 *MNRAS*, 486, 4451
Hachinger, S., Mazzali, P.A., Taubenberger, S., et al. 2012 *MNRAS*, 422, 70
Hainich, R., Ruehiling, U., Todt, H., et al. 2014 *A&A*, 565, 27
Hamann, W.-R., Graefener, G., & Liermann, A. 2006 *A&A*, 457, 1015
Hamann, W.R., Graefener, G., Liermann, A., et al. 2019 *A&A*, 625, A57
Hillier, D.J., Lanz, T., Heap, S.R., et al. 2003 *ApJ*, 588, 1039
Hillier, D.J., & Miller, D.L. 1998 *ApJ*, 496, 407
Jin, H., Yoon, S.-C., & Blinnikov, S. 2021 *ApJ*, 910, 68
Jung, M.-K. Yoon, S.-C., & Kim, H.-J. 2022 *ApJ*, 925, 216
Kilpatrick, C.D., Drout, M.R., Auchettl, K., et al. 2021 *MNRAS*, 504, 2073
Kim, H.-J., Yoon, S.-C., & Koo, B.-C. 2015 *ApJ*, 809, 131
Kilpatrick, C.D., Takaro, T., Foley, R.J., et al. 2018 *MNRAS*, 480, 2072
Langer, N. 2012 *ARA&A*, 50, 107
Leung, S.-C., Wu., S., & Fuller, J. 2021 *ApJ*, 923, 41
Liu, Y.-Q., Modjaz, M., Biaco, F., & Graur, O. 2016 *ApJ*, 827, 90
Lyman, J.D., Bersier, D., James, P.A. 2016 *MNRAS*, 457, 328
Lucy, L.B. 1991 *ApJ*, 383, 308
Menon, A., & Heger, A. 2016 *MNRAS*, 469, 46
Morris, Th., & Podsiadlowski 2007 *Science*, 315, 1103
Podsiadlowski, Ph., Joss, P.C., & Rappaport, Sl. 1990 *A&A*, 227, L9
Prentice, S.J., Ashall, C., James, P.A., et al. 2019 *MNRAS*, 485, 1559
Sander, A., Hamann, W.R., & Todt, H 2012 *A&A*, 540, 144
Sander, A., Hamann, W.R., Todt, H., et al. 2019 *A&A*, 621, A92
Sander, A., Vink, J.S., Hamann, W.-R. 2020 *MNRAS*, 491, 440
Shahbandeh, M., Hsiao, E.Y., & Ashall, C. et al. 2021, *ApJ*, 925, 175
Shenar, T., Gilkis, A., Vink, J.S., Sana, H., & Sander, A.A.C. 2020 *A&A*, 634, 79
Sun, N.-C., Maund, J.R., Crowther, P.A., et al. 2022 *MNRAS*, 510, 3701
Teffs, J., Ertl, T., Mazzali, P., et al. 2020 *MNRAS*, 499, 730
Tramper, F., Straal, S., & Sanyal, D., et al. 2015, *A&A*, 581, A110
Smartt, S. 2015, *PASA*, 32, e016
Smith, N. 2014, *ARA&A*, 52, 487
Swartz, D. A. 1991 *ApJ*, 373, 604
Taddia, F., Sollerman, J., Ledoudas, G., et al. 2018, *A&A*, 609, A136
Urushibata, T., Takahashi, K., Umeda, H., Yoshida, T. 2018, *A&A*, 609, A136
Williamson, M., Kerzendorf, W., & Modjaz, M. 2021 *ApJ*, 908, 150
Van Dyk, S. D., Zheng, W, Brink, T. G., Filippenko, A. V. , Milisavljevic, D, Andrews, J. E., Smith, N., Cignoni, M, Fox, O. D., Kelly, P. L., Adamo, A., Yunus, S., Zhang, K., Kumar, S., 2018, *ApJ*, 860, 2
Vink, J.S. 2017, *A&A*, 607, 8
Vink, J.S. 2022, *ARA&A*, in press
Woosley, S.E. 2019 *ApJ*, 878, 49
Yoon, S.-C. 2015, *PASA*, 32, e015
Yoon, S.-C. 2017, *MNRAS*, 470, 3970
Yoon, S.-C., Chun, W., Tostov, A., Blinnikov, S., & Dessart, L. 2019, *ApJ*, 872, 174
Yoon, S.-C., Dessart, L., & Clocchiatti, A. 2017, *ApJ*, 840, 10
Yoon, S.-C., Graefener, G., Vink, J.S., Kozyreva, A., & Izzard, R.G. 2012, *A&A*, 544, L11
Yoon, S.-C., Woosley, S.E., & Langer, N. 2010 *ApJ*, 725, 940

The Origin of Outflows in Evolved Stars
Proceedings IAU Symposium No. 366, 2022
L. Decin, A. Zijlstra & C. Gielen, eds.
doi:10.1017/S1743921322000060

Environments of evolved massive stars: evidence for episodic mass ejections

M. Kraus[1], L. S. Cidale[2,3], M. L. Arias[2,3], A. F. Torres[2,3], I. Kolka[4], G. Maravelias[5,6], D. H. Nickeler[1], W. Glatzel[7] and T. Liimets[1]

[1]Astronomical Institute, Czech Academy of Sciences, Fričova 298, CZ-25165 Ondřejov, Czech Republic, email: michaela.kraus@asu.cas.cz

[2]Departamento de Espectroscopía, Facultad de Ciencias Astronómicas y Geofísicas, UNLP, Paseo del Bosque S/N, La Plata, B1900FWA, Buenos Aires, Argentina

[3]Instituto de Astrofísica de La Plata (CCT La Plata - CONICET, UNLP) Paseo del Bosque S/N, La Plata, B1900FWA, Buenos Aires, Argentina

[4]Tartu Observatory, University of Tartu, 61602 Tõravere, Tartumaa, Estonia

[5]IAASARS, National Observatory of Athens, GR-15236, Penteli, Greece

[6]Institute of Astrophysics, Foundation for Research and Technology-Hellas, GR-70013, Heraklion, Greece

[7]Institut für Astrophysik (IAG), Georg-August-Universität Göttingen, Friedrich-Hund-Platz 1, D-37077 Göttingen, Germany

Abstract. The post-main sequence evolutionary path of massive stars comprises various transition phases, in which the stars shed large amounts of material into their environments. Our studies focus on two of them: B[e] supergiants and yellow hypergiants, for which we investigate the structure and dynamics within their environments. We find that each B[e] supergiant is surrounded by a unique set of rings or arc-like structures. These structures are either stable over time or they display high variability, including expansion and dilution. In contrast, yellow hypergiants are embedded in multiple shells of gas and dust. These objects are famous for their outburst activity. Moreover, the dynamics in their extended atmospheres imply an enhanced pulsation activity prior to outburst. The physical mechanism(s) leading to episodic mass ejections in these two types of stars is still uncertain. We propose that strange-mode instabilities, excited in the inflated envelopes of these objects, play a significant role.

Keywords. supergiants, circumstellar matter, stars: mass loss, stars: winds, outflows

---

1. Introduction

Massive stars ($M > 8\,\mathrm{M}_\odot$) are important cosmic engines. With their life-long mass loss they constantly supply material and energy into their environment, influencing and driving the chemical and dynamical evolution of their host galaxies. The post-main sequence evolution of massive stars bears many unsolved issues, one of them being the question of the triggering mechanism for eruptions and mass ejections occurring in several evolutionary transition phases. As mass loss is crucial for the fate of a massive star, understanding the mechanisms behind mass ejection phases and exploring the amount of mass lost during such events is essential. Our research focuses mainly on two groups of evolved massive stars, the hot B[e] supergiants (B[e]SGs) and the cool yellow hypergiants (YHGs). Both types of objects display clear indication for circumstellar material in form of disks, rings, or shells of material, which must have been released from the star.

2. The environments of B[e] supergiants

B[e] supergiants form a sub-group of the classical B-type supergiants. As such, their temperatures range from 10 000 K to 25 000 K, and they spread over the entire luminosity range of massive stars ($\log L/L_{\odot} \geq 4$). The major characteristics of B[e]SGs is their hybrid appearance (Zickgraf *et al.* 1986). The ultraviolet spectra imply a powerful, high-ionized wind driven by these luminous stars. In contrast to this, pronounced emission-line spectra at optical and infrared wavelengths, a strong infrared excess, and highly polarized emission suggest the co-existence of a dense, disk-like environment composed of low-ionized and neutral gas, along with a considerable amount of warm dust.

To unveil the density structure and dynamics of the circumstellar matter we first focused on molecules, which can be expected to form within the dense, disk-like environments. The most stable and most abundant molecule is CO. We have surveyed B[e]SGs in the K-band and detected intense emission from the CO first-overtone bands in the spectra of about 50% of the objects (Muratore *et al.* 2012; Oksala *et al.* 2013). Modeling of these emission bands provides column density, temperature, and kinematics of the molecular gas (Kraus *et al.* 2000). We found that in all objects the CO emission arises in a dense, rotating ring of gas. The CO temperatures were thereby found to be considerably lower than the CO dissociation temperature (Liermann *et al.* 2010; Cidale *et al.* 2012; Oksala *et al.* 2013), implying that the molecular gas must be localized in detached structures rather than being part of a continuous circumstellar outflowing disk.

Furthermore, in all objects displaying CO band emission we also found emission from the isotopic molecule ^{13}CO. In all cases, the ratio $^{12}\mathrm{C}/^{13}\mathrm{C}$ is significantly smaller than the initial, interstellar value of ~ 90, reinforcing that the circumstellar material has been released from the evolved supergiant star (Kraus 2009). In fact, the finding of circumstellar molecular gas enriched in ^{13}CO in some B[e] stars could even be used to classify these objects as evolved supergiants (Muratore *et al.* 2015; Kraus *et al.* 2020).

The surface of massive stars is oxygen-rich (O/C > 1) and so is the material released into the environment. All excess oxygen that is not bound in CO can thus form other compounds with elements of high abundance and/or high binding energy, and so far we searched for and detected SiO first-overtone band emission in four B[e]SGs (Kraus *et al.* 2015a), and six objects possibly display TiO emission† (Zickgraf *et al.* 1989; Kraus *et al.* 2016; Torres *et al.* 2018). The physical conditions in the SiO band forming region, obtained from modeling of the emission spectra, support the scenario of detached rotating molecular gas ring(s). Modeling of TiO emission is in progress.

Neutral and low-ionized gas provides additional tracers for density and dynamics. These are the lines of [O I]$\lambda\lambda$5577, 6300, 6364 and [Ca II]$\lambda\lambda$7291,7324 (Kraus *et al.* 2010; Aret *et al.* 2012), which can be excited in relatively high-density environments. Their profiles provide further evidence for rotation in a (quasi-)Keplerian manner of the circumstellar matter. Detailed modeling of their profile shapes revealed that the gas is localized in a series of rings, in some cases co-existing with the molecules and with varying, often alternating density (Kraus *et al.* 2016; Torres *et al.* 2018). Moreover, the number of rings and their distribution around the stars is unique for each object (Maravelias *et al.* 2018). In some B[e]SGs, the emission from these rings decreases in strength with time, up to complete disappearance (Torres *et al.* 2012; Liermann *et al.* 2014; Kraus *et al.* 2020), suggesting expansion and dilution of the ejected material. In others, the profiles display asymmetries and time variability implying gaps in the rings or spiral-arm like structures (Maravelias *et al.* 2018; Torres *et al.* 2018), whereas in one

† But we want to caution that not all stars displaying CO band emission have been systematically searched for emission from other molecules yet (for an overview, see Kraus 2019).

object we noted a sudden appearance of CO band emission (Oksala *et al.* 2012), likely due to the accumulation of the steadily decelerating outflowing material (Kraus *et al.* 2010).

Compared to classical B supergiants, B[e]SGs are rare with currently 33 confirmed members and 25 candidates within the Milky Way and the closest Local Group galaxies (Kraus 2019). This could mean that B[e]SGs comprise a short transition phase in the evolution of massive stars, characterized by episodes of enhanced mass loss. In this respect it is noteworthy that a significant number of B[e]SGs display besides the molecular and dusty rings on small scales also nebulae and ejecta on much larger scales speaking for previous mass-loss events (Kraus *et al.* 2021; Liimets et al. in prep.).

3. Yellow Hypergiants and their outburst activity

Yellow hypergiants populate a region in the HR diagram that spreads from 4000 – 8000 K in temperature and from $5.4 \leq \log L/L_{\odot} \leq 5.8$ in luminosity. These stars have started their lives most likely with an initial mass of ~ 20–$40\,M_{\odot}$ and are now on their blue-ward evolution after having passed through the red supergiant (RSG) state. The number of currently known YHGs is small. Just a handful of objects in the Milky Way and its closest neighboring galaxies are confirmed YHGs, and a similar number of objects has the status of a YHG candidate (de Jager 1998; Kourniotis et al., submitted).

YHGs have developed a pronounced core-envelope structure, and their strongly inflated and rarefied envelopes make them susceptible to even tiniest disturbances. It is therefore not surprising that the YHG state is characterized by outbursts during which the stars shed large amounts of material into their environments. Each outburst is identified by a steep drop in stellar brightness accompanied by an apparent decrease in effective temperature so that the star seemingly undergoes an excursion back to the red, cool side of the HR diagram. With the dilution of the ejected material, the star finally re-appears with its real effective temperature. The ejected matter expands and cools, providing an ideal environment for molecule and dust formation. It is noteworthy that, alike the B[e]SGs, about half of the YHG stars and candidates display CO first-overtone bands in emission (e.g., Lambert et al. 1981; Davies *et al.* 2008; Oksala *et al.* 2013) indicating a dense and warm molecular environment. The significantly cooler circumstellar dust may also become observable in form of rings or shells around the star, as in the case of IRAS 17163-3907, which seems to have had three distinct mass-ejection episodes within a 100 year period leading to three shells encompassing the central object (Lagadec et al. 2011; Koumpia et al. 2020; Oudmaijer et al., these proceedings).

A further YHG, famous for its outbursts, is the Galactic object ρ Cas. It is one of the four northern Galactic YHGs that we monitor. The star underwent at least four outbursts since the mid 1940ies, with the most recent one occurring in 2013 (Kraus *et al.* 2019). The strength and intensity of these events is highly variable, and dust formation was recorded so far only in relation to the intense mass-ejection phase in 1945–1947 (Jura & Kleinmann 1990). During the quiescence phases the star undergoes cyclic light variability with quasi-periods between 200 and 400 days.

The temperature sensitive line ratio Fe Iλ6431/Fe IIλ6433 reveals that the atmospheric temperature closely follows the light variability. A temperature increase during stellar brightening phases and decrease during phases of stellar dimming resembles the behavior of radially pulsating stars. Additional support for pulsation activities in the envelope of ρ Cas is provided by the measured radial velocity variations of absorption lines that form in different atmospheric depths. Noteworthy is the enhanced dynamical activity in the outermost layers prior to the 2013 eruption. The same behavior had been recorded before the outburst in 2000 (Lobel *et al.* 2003).

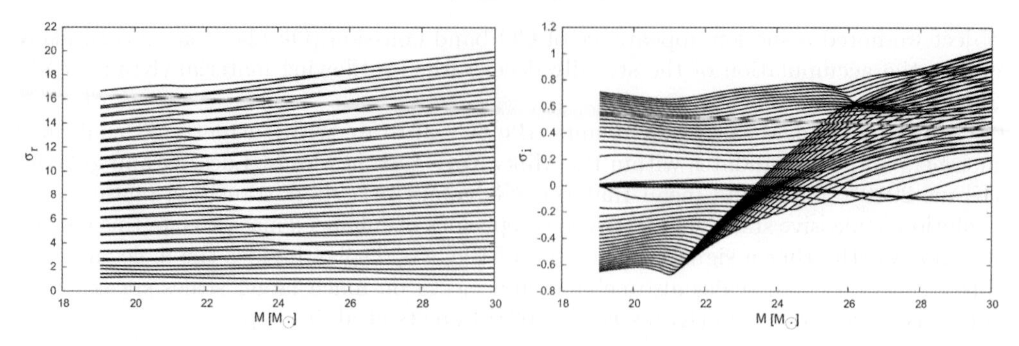

Figure 1. Real parts (= pulsation periods, left panel) and the imaginary parts (right panel) of the eigenfrequencies, which are normalized to the global free-fall time. Positive imaginary parts correspond to damped modes, and negative ones to unstable modes. The computations have been performed for $T_{\rm eff} = 7000\,{\rm K}$ and $\log L/L_{\odot} = 5.7$, matching the observed values of ρ Cas.

Our observations also revealed emission in some Fe I lines (e.g., Fe I λ6359) as well as in the lines of [Ca II]$\lambda\lambda$7291,7324 (Kraus *et al.* 2019). These narrow emission lines are static and are centered on the systemic velocity. They are most prominent during phases of maximum stellar brightness, that is when the star is hottest and hence compact, whereas strong atmospheric absorption conceals the emission during phases of minimum brightness. These findings reinforce the circumstellar nature of the line emission.

Finally, we have identified a trend of decreasing duration and time interval between individual outbursts (Maravelias & Kraus 2021). The increased activity might suggest that ρ Cas is preparing for another major eruption.

4. Pulsation instabilities as trigger for episodic mass ejections

Although B[e]SGs and YHGs populate distinct temperature regimes in the HR diagram, members of both classes display clear signs for episodic mass loss. The observed diversity of rings or arcs around B[e]SGs, and the outburst activity and resolved distinct circumstellar shells around some YHGs require a physical mechanism that is suitable to repeatedly lift matter from the stellar surface into the environments of these objects at detectable rates. Considering their high luminosity over mass ratios, evolved massive stars, particularly post-RSG stars, are ideal candidates for the excitation of strange-mode instabilities (Gautschy & Glatzel 1990; Glatzel 1994). These instabilities have been proposed to drive time-variable mass loss (Glatzel *et al.* 1999) and in fact, strange-mode pulsation and its correlation with phases of enhanced mass loss has been found in the α Cygni variable 55 Cyg from both an observational (Kraus *et al.* 2015b) and a theoretical approach (Yadav & Glatzel 2016). Motivated by these results, we have started to systematically investigate the entire upper HR diagram to identify constellations in stellar parameters that favor the excitation of strange-modes.

We first focused on the temperature and luminosity domain of the YHGs and performed a stability analysis with respect to linear nonadiabatic radial perturbations. A detailed description of these investigations will be published elsewhere. Here, we present our results for parameters matching ρ Cas. We fixed the effective temperature at 7000 K, the stellar luminosity at $\log L/L_{\odot} = 5.7$, and computed models for a range of masses around the value of $24.1 \pm 4.7\,M_{\odot}$ estimated from observations. Each model provides a sequence of complex eigenfrequencies, σ, consisting of the fundamental and overtone modes. These eigenfrequencies, normalized to the global free-fall time, are shown in Figure 1. The left panel displays the real parts of σ, which represent the pulsation frequencies. For each mass we find that many, especially high-overtone modes are excited. Besides the expected homogeneous set of modes we note many modes that are crossing the regular ones. These crossing modes can be identified in the right panel of Figure 1 with negative imaginary parts of σ, meaning that these are all unstable with high growth rates. On the other hand, the homogeneous set of modes have all positive imaginary parts and are consequently stable, that is damped. It is noticeable that the instabilities set in around a stellar mass of $\sim 24\,M_{\odot}$, which is the most likely mass of ρ Cas. In as how much these instabilities are indeed able to drive mass loss from the star needs to be investigated with the help of nonlinear simulations. This work is currently in progress, and first tentative test computations have shown that these instabilities occur also in the non-linear regime.

5. Conclusions

We have presented evidence for episodic mass loss in two types of evolved massive stars: B[e]SGs and YHGs, and we have proposed that strange-mode instabilities might be a suitable mechanism to trigger phases of enhanced mass loss and mass eruptions in these objects. Strange-mode instabilities have been detected in other types of blue supergiants such as the α Cygni variables underlining their suitability and supporting our hypothesis. As a prerequisite for the occurrence of strange mode instabilities a significantly decreased stellar mass (for a fixed luminosity) is favorable. This means that particularly stars in their blue-ward (e.g. post-RSG) evolution are ideal candidates, because massive stars typically lose large amounts of mass during their red-most evolutionary state. While YHGs are definitely post-RSG stars, the situation is less clear for the B[e]SGs. These objects typically display emission-line spectra, rendering it difficult to assess surface abundances and hence to constrain their evolutionary state. Moreover, it was possible to find indication for possible pulsation activity in only one B[e]SG so far (Kraus *et al.* 2016). Some B[e]SGs have been proposed to be indeed post-RSGs based on the strong enrichment of their circumstellar matter in ^{13}CO. In addition, Davies *et al.* (2007) suggested that the mass loss in YHGs might change from initially spherical to axisymmetric at later states of the YHG phase so that YHGs might develop into B[e]SGs. While such a scenario will work only in the specific mass range of YHGs, this is still an interesting idea and needs to be investigated in more detail.

Acknowledgements

M.K., D.H.N, and T.L. acknowledge financial support from the Czech Science Foundation (GA ČR 20-00150S), M.L.A and A.F.T. from the Programa de Incentivos (11/G160) of the Universidad Nacional de La Plata, Argentina, L.S.C. from CONICET (PIP 0177) and the Agencia Nacional de Promoción Científica y Tecnológica (PICT 2016-1971), and G.M. from the European Research Council (ERC) under the European Union's Horizon 2020 research and innovation programme (Grant agreement No. 772086). The Astronomical Institute Ondřejov is supported by RVO:67985815. This work is part of the project "Support of the international collaboration in astronomy (Asu mobility)" with the number: CZ.02.2.69/0.0/0.0/18_053/0016972. Asu mobility is co-financed by the European Union. This project has also received funding from the European Union's Framework Programme for Research and Innovation Horizon 2020 (2014-2020) under the Marie Skłodowska-Curie Grant Agreement No. 823734.

References

Aret, A., Kraus, M., Muratore, M. F., & Borges Fernandes, M. 2012, *MNRAS* 423, 284

Cidale, L. S., Borges Fernandes, M., Andruchow, I., Arias, M. L., Kraus, M., Chesneau, O., Kanaan, S., Curé, M., de Wit, W. J., & Muratore, M. F. 2012, *A&A* 548, 72

Davies, B., Oudmaijer, R. D., & Sahu, K. C. 2007, *ApJ* 671, 2059

Davies, B., Figer, D. F., Law, C. J., Kudritzki, R.-P., Najarro, F., Herrero, A., & MacKenty, J. W. 2008, *ApJ* 676, 1016

de Jager, C. 1998, *A&AR* 8, 145
Gautschy, A., & Glatzel, W. 1990, *MNRAS* 245, 597
Glatzel, W. 1994, *MNRAS* 271, 66
Glatzel, W., Kiriakidis, M., Chernigovskij, S., & Fricke, K. J. 1999, *MNRAS* 303, 116
Jura, M., & Kleinmann, S. G. 1990, *ApJ* 351, 583
Koumpia, E., Oudmaijer, R. D., Graham, V., Banyard, G., Black, J. H., Wichittanakom, C., Ababakr, K. M., de Wit, W. -J., Millour, F., Lagadec, E., Muller, S., Cox, N. L. J., Zijlstra, A., van Winckel, H., Hillen, M., Szczerba, R., Vink, J. S., & Wallström, S. H. J. 2020, *A&A* 635, A183
Kraus, M. 2009, *A&A* 494, 253
Kraus, M. 2019, *Galaxies* 7, 83
Kraus, M., Krügel, E., Thum, C., & Geballe, T. R. 2000, *A&A* 362, 158
Kraus, M., Borges Fernandes, M., & de Araújo, F. X. 2010, *A&A* 517, A30
Kraus, M., Oksala, M., Cidale, L. S., Arias, M. L., Torres, A. F., & Borges Fernandes, M. 2015a, *ApJ* (Letters) 800, L20
Kraus, M., Haucke, M., Cidale, L. S., Venero, R. O. J., Nickeler, D. H., Németh, P., Niemczura, E., Tomić, S., Aret, A., Kubát, J., Kubátová, B., Oksala, M. E., Curé, M., Kamiński, K., Dimitrov, W., Fagas, M., & Polińska, M. 2015b, *A&A* 581, A75
Kraus, M., Cidale, L. S., Arias, M. L., Maravelias, G., Nickeler, D. H., Torres, A. F., Borges Fernandes, M., Aret, A., Curé, M., Vallverdú, R., & Barbá, R. H. 2016, *A&A* 593, A112
Kraus, M., Kolka, I., Aret, A., Nickeler, D. H., Maravelias, G., Eenmäe, T., Lobel, A., & Klochkova, V. G. 2019, *MNRAS* 483, 3792
Kraus, M., Arias, M. L., Cidale, L. S., & Torres, A. F. 2020, *MNRAS* 493, 4308
Kraus, M., Liimets, T., Moiseev, A., Sánchez Arias, J. P., Nickeler, D. H., Cidale, L. S., & Jones, D. 2021, *AJ* 162, 150
Lagadec, E., Zijlstra, A. A., Oudmaijer, R. D., Verhoelst, T., Cox, N. L. J., Szczerba, R., Mékarnia, D., & van Winckel, H. 2011, *A&A* 534, L10
Lambert, D. L., Hinkle, K. H., & Hall, D. N. B. 1981, *ApJ* 248, 638
Liermann, A., Kraus, M., Schnurr, O., & Borges Fernandes, M. 2010 *MNRAS* 408, L6
Liermann, A., Schnurr, O., Kraus, M., Kreplin, A., Arias, M. L., & Cidale, L. S. 2014 *MNRAS* 443, 947
Lobel, A., Dupree, A. K., Stefanik, R. P., Torres, G., Israelian, G., Morrison, N., de Jager, C., Nieuwenhuijzen, H., Ilyin, I., & Musaev, F. 2003 *ApJ* 583, 923
Maravelias, G., & Kraus, M. 2022, *JAAVSO*, in press, arXiv:2112.13158
Maravelias, G., Kraus, M., Cidale, L. S., Borges Fernandes, M., Arias, M. L., Curé, M., & Vasilopoulos, G. 2018, *MNRAS* 480, 320
Muratore, M. F., Kraus, M., & de Wit, W. J. 2012, *BAAA* 55, 123
Muratore, M. F., Kraus, M., Oksala, M. E., Arias, M. L., Cidale, L., Borges Fernandes, M., & Liermann, A. 2015, *AJ* 149, 13
Oksala, M., Kraus, M., Arias, M. L., Borges Fernandes, M., Cidale, L., Muratore, M. F., & Curé, M. 2012, *MNRAS* 426, 56
Oksala, M., Kraus, M., Cidale, L. S., Muratore, M. F., & Borges Fernandes, M. 2013, *A&A* 558, A17
Torres, A. F., Kraus, M., Cidale, L. S., Barbá, R., Borges Fernandes, M., & Brandi, E. 2012, *MNRAS*, 427, L80
Torres, A. F., Cidale, L. S., Kraus, M., Arias, M. L., Barbá, R. H., Maravelias, G., & Borges Fernandes, M. 2018, *A&A* 612, A113
Yadav, A. P., & Glatzel, W. 2016, *MNRAS*, 457, 4330
Zickgraf, F.-J., Wolf, B., Stahl, O., & Humphreys, R. 1989, *A&A* 220, 206
Zickgraf, F.-J., Wolf, B., Stahl, O., Leitherer, C., & Appenzeller, I. 1986, *A&A* 163, 119

The Origin of Outflows in Evolved Stars
Proceedings IAU Symposium No. 366, 2022
L. Decin, A. Zijlstra & C. Gielen, eds.
doi:10.1017/S1743921322000965

How did the Stellar Winds of Massive Stars influence the Surrounding Environment in the Galactic Center?

Mengfei Zhang[1,2] **and Zhiyuan Li**[1,2]

[1]School of Astronomy and Space Science, Nanjing University,
Nanjing 210023, People's Republic of China
email: zmf@nju.edu.cn

[2]Key Laboratory of Modern Astronomy and Astrophysics (Nanjing University),
Ministry of Education,
Nanjing 210023, People's Republic of China

Abstract. In the Galactic center, there are many massive stars blowing strong stellar winds, which will strongly influence the surrounding environment and even the Galactic feedback. The Galactic center is quiescent at present, so the unique continuous energy input source is the massive star, consequently giving rise to many special features, such as the radio bubbles, the X-ray chimneys, the non-thermal filaments and high-metallicity abundance. However, it is difficult to quantify their contributions due to the complex environment in this region, and the past supernovae and Sgr A* activity are also important factors shaping these features. In this work, we discuss some structures possibly related to the stellar winds and perform preliminary simulations to study their evolution. We conclude the stellar winds can obviously influence a large scale $\sim$ 100 pc, and can possibly influence a larger scale environment indirectly.

Keywords. Galaxy: center, stars: winds, ISM: bubbles, supernova remnants

---

1. Introduction

The formation and evolution of galaxies are still mysterious. While the big picture is unambiguous, there are many unresolved confusing details, such as the galactic feedback and supermassive black hole (SMBH) formation. To understand these details, our Milky Way is the best laboratory, which can help us thoroughly check various features.

The most violent galactic feedback process usually happens in the Galactic center, such as the SMBH activity, supernovae and stellar winds, which can reshape the central structures and influence the evolution of the whole galaxy. Fermi bubbles and eROSITA bubbles are both possibly the kpc relics of such a process (Su et al. 2010; Predehl et al. 2020), and they will interact with the intergalactic medium, causing outflow and inflow. In addition, two newly-discovered structures, the radio bubbles and X-ray chimneys with a scale of hundreds parsecs (Heywood et al. 2019, Ponti et al. 2019), can also be taken as a part of the early-stage feedback process. These processes possibly originate from stellar activity or SMBH activity, but it is unclear which one is dominant. The identified supernova remnant (SNR) Sgr A East in the Galactic center implies many supernova explosions here over the past many years, and the X-ray reflection in the central molecular zone (CMZ) also implies the past SMBH X-ray flare (Ponti et al. (2013)). However, these past activities cannot be traced precisely.

The stellar winds of massive stars in this region are also efficient energy input sources, which can be detected directly. Therefore, we can study their influence on the environment in details. There are $\sim$ 200 massive stars (Dong et al. 2012) in the central tens of parsecs, some of which in the central several parsecs are the members of the nuclear star clusters (NSC). The stellar winds of NSC are so strong that they can even prevent the ejecta of SNRs into the central SMBH.

We in this work try to study the possible influence of stellar winds on those interesting structures, and perform a simulation to understand the stellar winds evolution of these massive stars. In Section 2, we describe these structures and their possible relation with the stellar winds. The simulation model and results are presented in Section 3. Section 4 is a summary.

2. Overview

2.1. *The galactic feedback*

The galactic feedback is usually dominated by two mechanisms, AGN or starburst, which push the gas to higher latitude and produce an outflow. In our Milky Way, the Fermi and eROSITA bubbles possibly originate from such an outflow, but it is uncertain whether they are produced by the AGN or starburst activity. Then some gas are accreted back to feed the host galaxy, while some gas just flow away or become diffuse gas surrounding the galaxy. We also detect gas falling into the Galactic disk at radio and X-ray wavelength (Kataoka et al. 2021).

In addition, we usually will neglect the contribution of stellar winds to the feedback, especially for extragalactic sources, because the stellar winds are much weaker than supernovae or SMBH activity. With the development of high-resolution observations, the stellar winds currently can be well studied in our Milky Way, but its relation with the Galactic feedback is still a puzzle. The assembly of many massive stars in the Galactic center implies they can possibly influence the feedback process. Meanwhile, Ressler et al. (2018) simulated the stellar winds of massive stars surrounding Sgr A*, and concluded the stellar winds can feed the SMBH. This can stimulate SMBH activity, then strengthen the outflow.

2.2. *The radio bubbles and the X-ray chimneys*

The relation between stellar winds and large scale feedback is difficult to confirm, but the relation with some smaller structures, such as radio bubbles and the X-ray chimneys, can be discussed quantitatively. We simulated multi-supernovae explosions with strong magnetic field in the central 50 pc of our Galaxy, to simultaneously reproduce the radio bubbles and the X-ray chimneys (Zhang et al. 2021). The explosion frequency is 0.001 yr^{-1}, and the magnetic strength is $\sim 80\ \mu$G. The high explosion frequency make it a limb-brightening bubble, while the strong magnetic field can confine the supernovae ejecta, leading to an elongated shape. After 330 kyr, the simulated results can well match the morphology of the radio bubbles, the X-ray chimneys (see Figure 1). Moreover, even the thermal energy and X-ray luminosity are also roughly consistent with the observation, but a litter lower. Therefore, there are possibly other energy input sources, such as the stellar winds and pulsar wind nebulae.

In addition, in the simulation, we can also reproduce the non-thermal filaments (NTFs), a kind of mysterious structures discovered about 40 years ago, while there are many competitive models. The lifetime of these NTFs is obviously larger than a supernova remnant, so our multi-supernovae model cannot explain NTFs independently. A continuous energy input is necessary to sustain the filamentary shape and accelerate relativistic electrons.

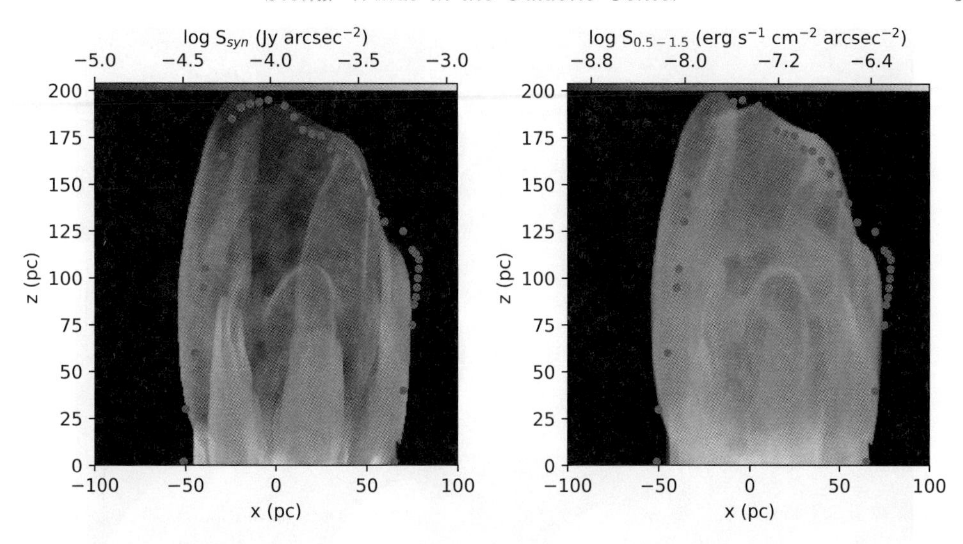

Figure 1. The red dotted line outlines the rim of the northern radio bubble. *Left:* Synchrotron emission, values lower than 10^{-5} Jy arcsec^{-2} are suppressed to enhance visualization of the faint features; *Right:* Synthetic 0.5 − 1.5 keV X-ray intensity distribution in simulation, values lower than 10^{-9} erg s^{-1} cm^{-2} arcsec^{-2} are suppressed to enhance visualization of the faint features.

If the stellar winds are important, it can simultaneously solve the absence of the thermal energy and X-ray luminosity and the NTFs formation in our simulation. Therefore, a study of stellar winds' influence on central hundreds of parsecs is urgent.

2.3. *SNR Sgr A East*

In addition to the importance of stellar winds on the hundreds of parsecs, they are also essential for studying SNR Sgr A East, the unique confirmed SNR in the Galactic center. We simulate the interaction between the shock wave of Sgr A East and the stellar winds of NSC, which can well reproduce the radio morphology of the SNR and an X-ray ridge structure between Sgr A* and the explosion of Sgr A East after $\sim$ 800 years (see Figure 2). However, the X-ray emission inside the SNR is faint, inconsistent with the observation. Because the adiabatic expansion will lead to sharp cooling inside, the bremsstrahlung emission is weak.

The simulation assumes the surrounding environment of the supernova is dominated by the stellar winds of NSC, but there are many molecular clouds or clumps in reality, which can block the shock wave and stimulate the formation of reverse shock. The reverse shock come back to the explosion center and heat the central ejecta, then produce X-ray emission, which is a reasonable explanation for the weak X-ray emission inside. However, the observed clouds are not enough to produce such strong shock wave. We also suggest a stellar wind envelop of the progenitor can be heated directly by the forward shock to produce the X-ray emission, but this is related to non-equilibrium ionization model in the hydrodynamical simulation, which need to be further developed.

The simulation also shows the ejecta cannot propagate into the central 0.4 pc radius due to the strong stellar winds of NSC, which implies the supernova cannot affect the Sgr A* directly.

2.4. *IRS 13E*

In the aforementioned cases, the stellar winds are possibly important, but not dominant. However, next to the Sgr A*, in a complex IRS 13E, the stellar wind is the main

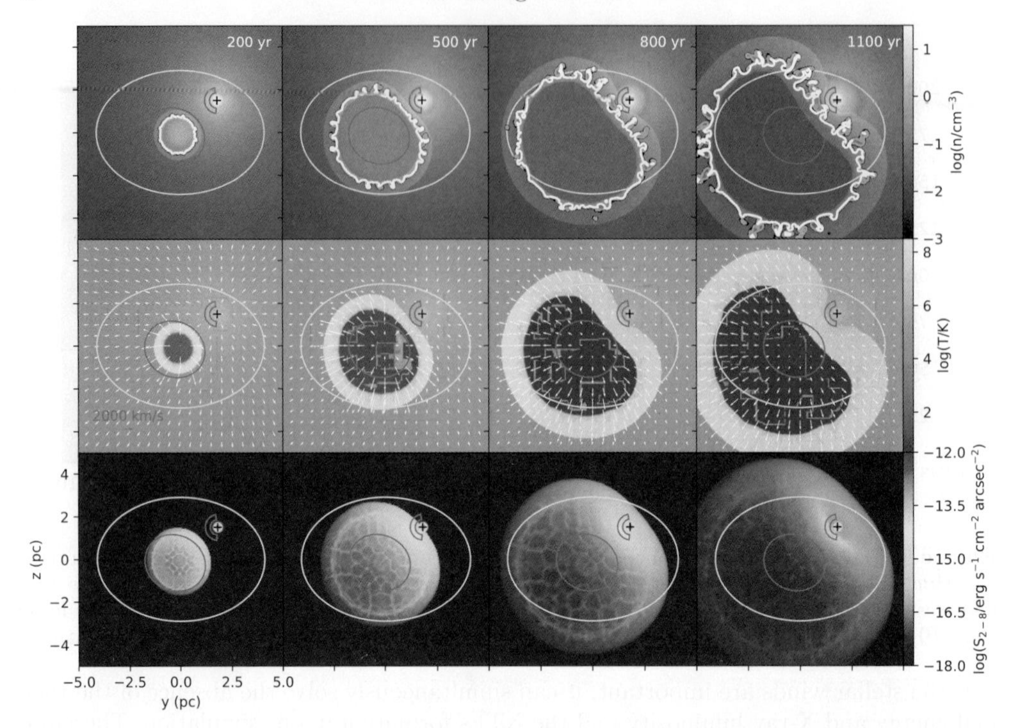

Figure 2. Density (top rows), temperature (middle rows) and X-ray intensity (bottom rows) after 200, 500, 800 and 1100 yr with an explosion energy of 5×10^{50} erg s^{-1} and free hydrogen abundance. The pink and red circle outline the radio and X-ray morphology, while the black plus and red arc indicate the Sgr A* and X-ray ridge, respectively. The ejecta is outlined by the contour, in which the white, red and black indicate 0.9, 0.5, 0.1, respectively. The white arrows indicate the velocity. The density and temperature maps both show slices through the $x = 1$ pc.(From Zhang et al. 2022, to be submitted.)

character. We simulate the collision of stellar winds from massive stars (Zhu et al. 2020), which can match the X-ray luminosity and even the spectrum (see Figure 3).

This work shows the stellar winds can be sufficiently strong to produce high X-ray emission flux. In the Galactic center, the interstellar environment can be largely affected by the stellar winds from massive stars. However, some works related to larger scale structures did not take it serious, which deserve more careful studies.

3. Simulation Model and Results

To quantify the influence of stellar winds, we perform a preliminary simulation for the ~ 200 massive stars in the Galactic center. In the simulation, we assume the mean mass loss rate and wind velocity of these massive stars as 10^{-5} M$_\odot$ yr^{-1} and 2000 km s^{-1}. In the NSC, the mean distance between two stars is so small that we cannot well distinguish them in the simulation, so we take them as one source with a mass loss rate and velocity of 10^{-3} M$_\odot$ yr^{-1} and 100 km s^{-1}. The initial interstellar medium (ISM) density is 0.1 H cm^{-3}. In addition, we take into account the gravitational potential of Sgr A*, NSC and the nuclear disk.

The simulation results after 160 kyr are shown in Figure 4. The stellar winds have blown out the central 100 pc, and are still pushing the outer ISM, which can obviously influence the formation and evolution of the radio bubbles and the X-ray chimneys. Their total energy input during 160 kyr is 2×10^{52} erg, a large value comparable to ten

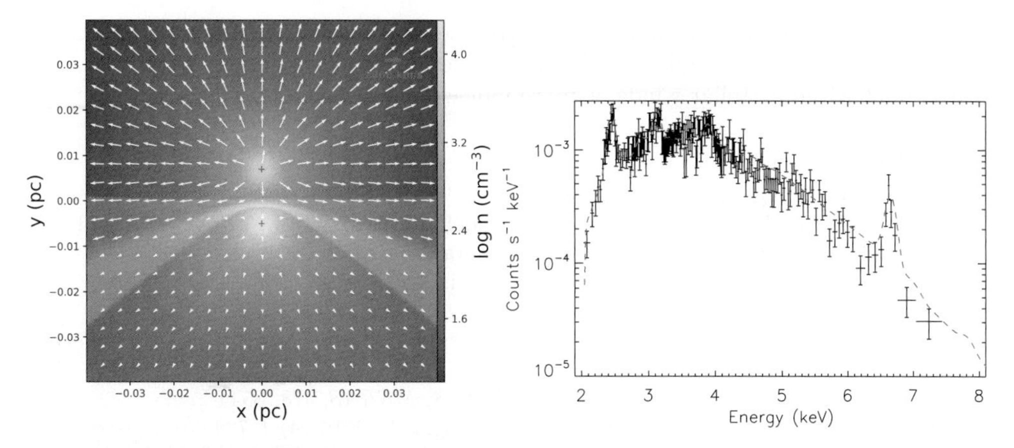

Figure 3. Illustration of the fiducial colliding wind simulation. *Left*: Density distribution in the $z = 0$ plane, overlaid by vectors representing the local velocity. The positions of the two WR stars are marked by '+' signs. *Right*:A comparison between the observed ACIS-S spectrum (black data points) and the simulation-predicted spectrum (red dashed curve). The latter has been multiplied by the energy-dependent foreground absorption and convolved with the instrumental response.

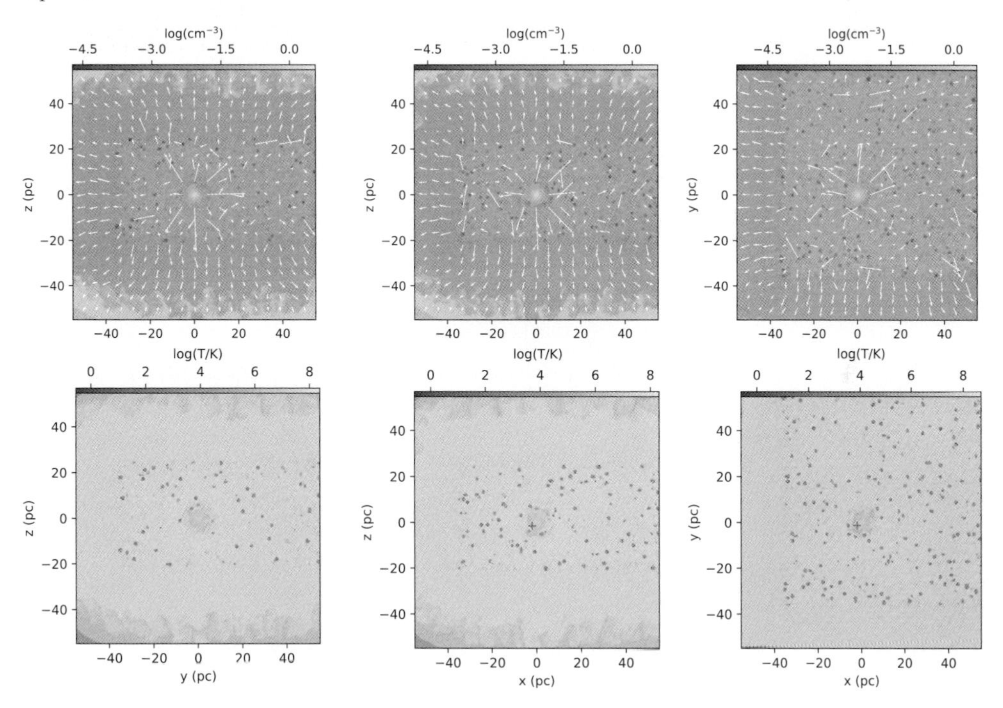

Figure 4. Density (top rows) and temperature (bottom rows) maps after 160 kyr. The white arrows show the velocity. The z-axis runs along the line of sight.

supernovae. However, its relation with the Fermi bubbles and eROSITA bubbles need further investigation.

4. Summary

From large scale to small scale, the influence of the stellar winds on the environment of the Galactic center becomes more important. The X-ray luminosity and thermal energy of X-ray chimney may be contributed by the stellar winds which can also sustain the existence of NTFs. Our simulation also shows the interior X-ray emission of SNR Sgr A East is faint without stellar winds or more molecular components.

References

Dong, H. and Wang, Q. D. and Morris, M. R. 2012, *MNRAS*, 425, 884

Heywood, I., Camilo, F., Cotton, W. D., Yusef-Zadeh, F., et al., 2019, *Nature*, 573, 7773

Kataoka, Jun and Yamamoto, Marino and Nakamura, Yuki et al. 2021, *ApJ*, 908, 14

Predehl, P. and Sunyaev, R. A. and Becker, W. et al. 2020, *Nature*, 588, 227

Ponti, G. and Hofmann, F. and Churazov, E. et al. 2019, *Nature*, 567, 347

Ponti, Gabriele and Morris, Mark R. and Terrier, Regis et al. 2013, *Cosmic Rays in Star-Forming Environments*, 34, 331

Ressler, S. M. and Quataert, E. and Stone, J. M. 2018, *MNRAS*, 478, 3544

Su, Meng and Slatyer, Tracy R. and Finkbeiner, Douglas P. 2010, *ApJ*, 724, 1044

Zhang, Mengfei and Li, Zhiyuan and Morris, Mark R. 2021, *ApJ*, 913, 68

Zhou, Ping and Leung, Shing-Chi and Li, Zhiyuan et al. 2021, *ApJ*, 908, 31

Zhu, Zhenlin and Li, Zhiyuan and Ciurlo, Anna et al. 2020, *ApJ*, 897, 135

The Origin of Outflows in Evolved Stars
Proceedings IAU Symposium No. 366, 2022
L. Decin, A. Zijlstra & C. Gielen, eds.
doi:10.1017/S1743921322001132

The role of mass loss in chemodynamical evolution of galaxies

Chiaki Kobayashi

Centre for Astrophysics Research, Department of Physics, Astronomy and Mathematics
University of Hertfordshire, College Lane, Hatfield AL10 9AB, UK
email: c.kobayashi@herts.ac.uk

Abstract. Thanks to the long-term collaborations between nuclear and astrophysics, we have good understanding on the origin of elements in the universe, except for the elements around Ti and some neutron-capture elements. From the comparison between observations of nearby stars and Galactic chemical evolution models, a rapid neutron-capture process associated with core-collapse supernovae is required. The production of C, N, F and some minor isotopes depends on the rotation of massive stars, and the observations of distant galaxies with ALMA indicate rapid cosmic enrichment. It might be hard to find very metal-poor or Population III (and dust-free) galaxies at very high redshifts even with JWST.

Keywords. nucleosynthesis, stellar abundances, ISM abundances, supernovae, AGB stars, Milky Way Galaxy, galaxies

---

1. Introduction

Explaining the origin of the elements is one of the scientific triumphs linking nuclear physics with astrophysics. As Fred Hoyle predicted, carbon and heavier elements ('metals' in astrophysics) were not produced during the Big Bang but instead created inside stars. So-called α elements (O, Ne, Mg, Si, S, Ar, and Ca) are mainly produced by core-collapse supernovae, while iron-peak elements (Cr, Mn, Fe, and Ni) are more produced by thermonuclear explosions, observed as Type Ia supernovae (SNe Ia; Kobayashi, Karakas & Lugaro 2020, hereafter K20). The production depends on the mass of white dwarf (WD) progenitors, and a large fraction of SNe Ia should come from near-Chandrasekhar (Ch) mass explosions (see Kobayashi, Leung & Nomoto 2020 for constraining the relative contribution between near-Ch and sub-Ch mass SNe Ia). Among core-collapse supernovae, hypernovae ($\gtrsim 10^{52}$ erg) produce a significant amount of Fe as well as Co and Zn, and a significant fraction of massive stars ($\gtrsim 20 M_\odot$) should explode as hypernovae in order to explain the Galactic chemical evolution (GCE; Kobayashi et al. 2006, hereafter K06).

Heavier elements are produced by neutron-capture processes. The slow neutron-capture process (s-process) occurs in asymptotic giant branch (AGB) stars (Karakas & Lugaro 2016), while the astronomical sites of rapid neutron-capture process (r-process) have been debated. The possible sites are neutron-star (NS) mergers (NSM, Wanajo et al. 2014; Just et al. 2015), magneto-rotational supernovae (MRSNe, Nishimura et al. 2015; Reichert et al. 2021), magneto-rotational hypernovae/collapsars (MRHNe, Yong et al. 2021a), and common envelope jets supernovae (Grichener et al. 2022). Light neutron-capture elements (e.g., Sr) are also produced by electron-capture supernovae (ECSNe, Wanajo et al. 2013), ν-driven winds (Arcones Et Al. 2007; Wanajo 2013), and rotating-massive stars (Frischknecht et al. 2016; Limongi & Chieffi 2018).

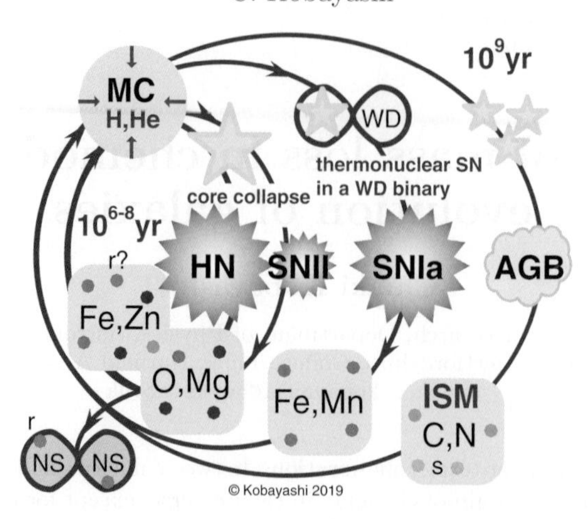

Figure 1. Schematic view of chemical enrichment in galaxies.

The cycles of chemical enrichment are schematically shown in Figure 1, where each cycle produces different elements and isotopes with different timescales. In a galaxy, not only the total amount of metals, i.e. metallicity Z, but also elemental abundance ratios evolve as a function of time. Therefore, we can use all of this information as fossils to study the formation and evolutionary histories of the galaxy. This approach is called Galactic archaeology, and several on-going and future surveys with multi-object spectrographs (e.g., APOGEE, HERMES-GALAH, Gaia-ESO, WEAVE, 4MOST, MOONS, Subaru Prime Focus Spectrograph (PFS), and Maunakea Spectroscopic Explorer (MSE)) are producing a vast amount of observational data of elemental abundances. Moreover, integral field unit (IFU) spectrographs (e.g., SAURON, SINFONI, CALIFA, SAMI, MaNGA, KMOS, MUSE, and HECTOR) allow us to measure metallicity and elemental abundance ratios within galaxies. It is now possible to apply the same approach not only to our own Milky Way but also to other types of galaxies or distant galaxies. Let us call this extra-galactic archaeology.

One of the most important uncertainties in this quest is the input stellar physics, namely, the nucleosynthesis yields, which could directly affect conclusions on the formation and evolutionary histories of galaxies. In this review, I will focus on the impact of stellar mass loss due to stellar rotation in Galactic and extra-galactic archaeology.

2. Galactic chemical evolution

Galactic chemical evolution (GCE) has been calculated analytically and numerically using the following equation:

$$\frac{d(Z_i f_{\rm g})}{dt} = E_{\rm SW} + E_{\rm SNcc} + E_{\rm SNIa} + E_{\rm NSM} - Z_i\phi + Z_{i,\rm inflow}R_{\rm inflow} - Z_i R_{\rm outflow} \quad (2.1)$$

where the mass fraction of each element i in gas-phase ($f_{\rm g}$ denotes the gas fraction) increases via element ejections from stellar winds ($E_{\rm SW}$), core-collapse supernovae ($E_{\rm SNcc}$), Type Ia supernovae ($E_{\rm SNIa}$), and neutron star mergers ($E_{\rm NSM}$). It also decreases by star formation (with a rate ϕ), and modified by inflow (with a rate $R_{\rm inflow}$) and outflow (with a rate $R_{\rm outflow}$) of gas in/from the system considered. It is assumed that the elemental abundance of gas is instantaneously well mixed in the system (called an one-zone model), but the instantaneous recycling approximation is not adopted nowadays. The initial conditions are $f_{{\rm g},0} = 1$ (a closed system) or $f_{{\rm g},0} = 0$ (an open system) with the chemical composition ($Z_{i,0}$) from the Big Bang nucleosynthesis. The first two terms depend only on nucleosynthesis yields, while the third and fourth terms also depend on modelling of the progenitor systems, which is uncertain. The last three terms are galactic terms, and should be determined from galactic dynamics, but are assumed with analytic formula in GCE models.

As in my previous works, star formation rate is assumed to be proportional to gas fraction as $\phi = \frac{1}{\tau_s} f_g$. The inflow rate is assumed to be $R_{\rm inflow} = \frac{t}{\tau_i^2} \exp \frac{-t}{\tau_i}$ (for the solar neighbourhood), or $R_{\rm inflow} = \frac{1}{\tau_i} \exp \frac{-t}{\tau_i}$ (for the rest). The outflow rate is assumed to be proportional to the star formation rate as $R_{\rm outflow} = \frac{1}{\tau_o} f_g \propto \phi$. In addition, star formation is quenched ($\phi = 0$) in case with a galactic wind at an epoch t_w. The timescales are determined to match the observed metallicity distribution function (MDF) in each system (Fig. 2 of K20). The parameter sets that have very similar MDFs give almost identical tracks of elemental abundance ratios (Fig. A1 of Kobayashi, Leung & Nomoto 2020).

Our nucleosynthesis yields of core-collapse supernovae were originally calculated in K06, 3 models of which are replaced in Kobayashi et al. (2011b) (which are also used in Nomoto et al. 2013), and a new set with failed supernovae is used in K20. This is based on the lack of observed progenitors at supernova locations in the HST data (Smartt 2009), and on the lack of successful explosion simulations for massive stars (Janka 2012; Burrows & Vartanyan 2021). The yields for AGB stars were originally calculated in Karakas (2010) and Karakas & Lugaro (2016), but a new set with the s-process is used in K20. The narrow mass range of super-AGB stars is also filled with the yields from Doherty et al. (2014); stars at the massive end are likely to become ECSNe as Crab Nebula exploded in 1054. At the low-mass end, off-centre ignition of C flame moves inward but does not reach the centre, which remains a hybrid C+O+Ne WD. This might become a sub-class of SNe Ia (Kobayashi, Nomoto & Hachisu 2015). For SNe Ia, the nucleosynthesis yields of near-Ch and sub-Ch mass models are newly calculated in Kobayashi, Leung & Nomoto (2020), which used the same code as in Leung & Nomoto (2018) and Leung & Nomoto (2020) but with more realistic, solar-scaled initial composition. The initial composition gives significantly different (Ni, Mn)/Fe ratios. r-process yields are taken from literature (see §2.1 of K20 for more details). The Kroupa initial mass function (IMF) between $0.01 M_\odot$ and $50 M_\odot$ (or $120 M_\odot$ in §2.2, 2.3, and §3) is used throughout this paper. For comparison to observational data, the solar abundances are taken from Asplund et al. (2009), except for $A_\odot(\mathrm{O}) = 8.76$, $A_\odot(\mathrm{Th}) = 0.22$, and $A_\odot(\mathrm{U}) = -0.02$ (§2.2 of K20 for the details).

Using the K20 GCE model for the solar neighbourhood, we summarize the origin of elements in the form of a periodic table. In each box of Figure 2, the contribution from each chemical enrichment source is plotted as a function of time: Big Bang nucleosynthesis (black), AGB stars (green), core-collapse supernovae including SNe II, HNe, ECSNe, and MRSNe (blue), SNe Ia (red), and NSMs (magenta). It is important to note that the amounts returned via stellar mass loss are also included for AGB stars and core-collapse supernovae depending on the progenitor star mass. The x-axis of each box shows time from $t = 0$ (Big Bang) to 13.8 Gyrs, while the y-axis shows the linear abundance relative to the Sun, $X/X_\odot$. The dotted lines indicate the observed solar values, i.e., $X/X_\odot = 1$ and 4.6 Gyr for the age of the Sun. Since the Sun is slightly more metal-rich than the other stars in the solar neighborhood (see Fig. 2 of K20), the fiducial model goes through [O/Fe]=[Fe/H]= 0 slightly later compared with the Sun's age. Thus, a slightly faster star formation timescale ($\tau_s = 4$ Gyr instead of 4.7 Gyr) is adopted in this model. The evolutionary tracks of [X/Fe] are almost identical. The adopted star formation history is similar to the observed cosmic star formation rate history, and thus this figure can also be interpreted as the origin of elements in the universe. Note that Tc and Pm are radioactive. The origin of stable elements can be summarized as follows:

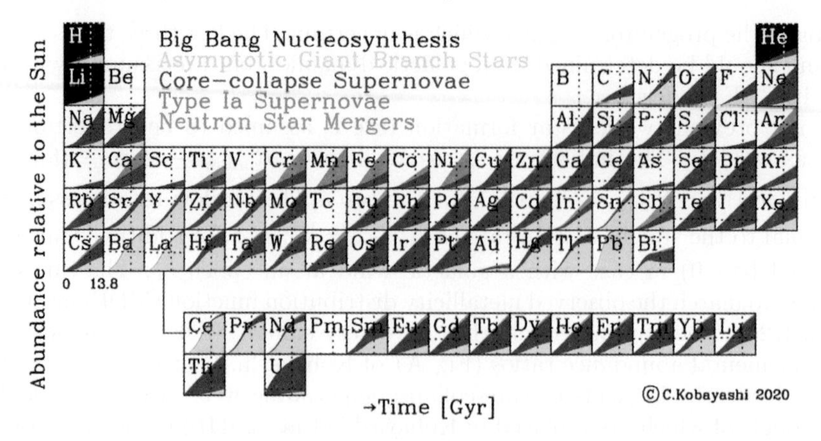

Figure 2. The time evolution (in Gyr) of the origin of elements in the periodic table: Big Bang nucleosynthesis (black), AGB stars (green), core-collapse supernovae including SNe II, HNe, ECSNe, and MRSNe (blue), SNe Ia (red), and NSMs (magenta). The amounts returned via stellar mass loss are also included for AGB stars and core-collapse supernovae depending on the progenitor mass. The dotted lines indicate the observed solar values.

• H and most of He are produced in Big Bang nucleosynthesis. As noted, the green and blue areas also include the amounts returned to the interstellar medium (ISM) via stellar mass loss in addition to He newly synthesized in stars. Be and B are supposed to be produced by cosmic rays, which are not included in the K20 model.

• The Li model is very uncertain because the initial abundance and nucleosynthesis yields are uncertain. Li is supposed to be produced also by cosmic rays and novae, which are not included in the K20 model. The observed Li abundances show an increasing trend from very low metallicities to the solar metallicity, which could be explained by cosmic rays. Then the observation shows a decreasing trend from the solar metallicities to the super-solar metallicities, which might be caused by the reduction of the nova rate (Grisoni et al. 2019); this is also shown in theoretical calculation with binary population synthesis (Kemp et al. 2022), where the nova rate becomes higher due to smaller stellar radii and higher remnant masses at low metallicities.

• 49% of C, 51% of F, and 74% of N are produced by AGB stars (at $t = 9.2$ Gyr). Note that extra production from Wolf-Rayet (WR) stars is not included and may be important for F (see §2.2). For the elements from Ne to Ge, the newly synthesized amounts are very small for AGB stars, and the small green areas are mostly for mass loss.

• α elements (O, Ne, Mg, Si, S, Ar, and Ca) are mainly produced by core-collapse supernovae, but 22% of Si, 29% of S, 34% of Ar, and 39% of Ca come from SNe Ia. These fractions would become higher with sub-Ch-mass SNe Ia (Kobayashi, Leung & Nomoto 2020) instead of 100% Ch-mass SNe Ia adopted in the K20 model.

• A large fraction of Cr, Mn, Fe, and Ni are produced by SNe Ia. In classical works, most of Fe was thought to be produced by SNe Ia, but the fraction is only 60% in our model, and the rest is mainly produced by HNe. The inclusion of HNe is very important as it changes the cooling and star formation histories of the universe significantly (Kobayashi et al. 2007). Co, Cu, Zn, Ga, and Ge are largely produced by HNe. In the K20 model, 50% of stars at $\geq 20 M_\odot$ are assumed to explode as hypernovae, and the rest of stars at $> 30 M_\odot$ become failed supernovae.

• Among neutron-capture elements, as predicted from nucleosynthesis yields, AGB stars are the main enrichment source for the s-process elements at the second (Ba) and third (Pb) peaks.

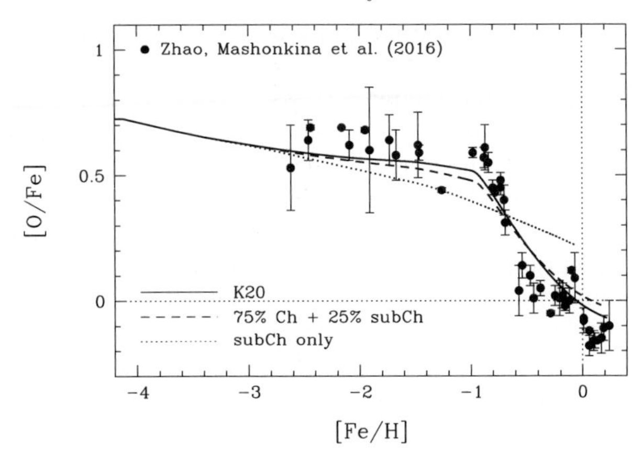

Figure 3. The [O/Fe]–[Fe/H] relations in the solar neighborhood for the models with Ch-mass SNe Ia only (solid line), 75% Ch plus 25% sub-Ch-mass SNe Ia (dashed line), and sub-Ch-mass SNe Ia only (dotted line). The observational data (filled circles) are high-resolution non-local thermodynamic equilibrium (NLTE) abundances.

• 32% of Sr, 22% of Y, and 44% of Zr can be produced from ECSNe, which are included in the blue areas, even with our conservative mass ranges; we take the metallicity-dependent mass ranges from the theoretical calculation of super-AGB stars (Doherty et al. 2015). Combined with the contributions from AGB stars, it is possible to perfectly reproduce the observed trends, and no extra light element primary process (LEPP) is needed. The inclusion of ν-driven winds in GCE simulation results in a strong overproduction of the abundances of the elements from Sr to Sn with respect to the observations.

• For the heavier neutron-capture elements, contributions from both NS-NS/NS-black hole (BH) mergers and MRSNe are necessary, and the latter is included in the blue areas.

In this model, the O and Fe abundances go though the cross of the dotted lines, meaning [O/Fe] = [Fe/H] = 0 at 4.6 Gyr ago. This is also the case for some important elements including N, Ca, Cr, Mn, Ni, Zn, Eu, and Th. Mg is slightly under-produced in the model, although the model gives a $0.2 - 0.3$ dex higher [Mg/Fe] value than observed at low metallicities. This [O/Mg] problem is probably due to uncertain nuclear reaction rates (namely, of $^{12}C(\alpha,\gamma)^{16}O$) and/or the mass loss based by stellar rotation or binary interaction (see also Fig. 9 of K06).

The under-production of the elements around Ti is a long-standing problem, which was shown to be enhanced by multi-dimensional effects (Sneden et al. 2016; see also K15 model in K20). The s-process elements are slightly overproduced even with the updated s-process yields. Notably, Ag is over-produced by a factor of 6, while Au is under-produced by a factor of 5. U is also over-produced. These problems may require revising nuclear physics modelling (see §2.1).

The contributions from SNe Ia depend on the mass of the progenitor WDs, and sub-Ch mass explosions produce less Mn and Ni, and more Si, S, and Ar than near-Ch mass explosions. Figure 3 shows the [O/Fe]–[Fe/H] relations with varying the fraction of sub-Ch-mass SNe Ia. Including up to 25% sub-Ch mass contribution to the GCE (dashed line) gives a similar relation as the K20 model (solid line), while the model with 100% sub-Ch-mass SNe Ia (dotted line) gives too low an [O/Fe] ratio compared with the observational data. For Ch-mass SNe Ia, the progenitor model is based on the single-degenerate scenario with the metallicity effect due to optically thick winds (Kobayashi et al. 1998). For sub-Ch-mass SNe Ia, the observed delay-time distribution is used since the progenitors are the combination of mergers of two WDs in double degenerate systems and low accretion in single degenerate systems; Kobayashi, Nomoto & Hachisu (2015)'s formula are for those in single degenerate systems only.

The [α/Fe]–[Fe/H] relation is probably the most important diagram in GCE. In the beginning of the universe, the first stars form and die, but the properties such as mass and rotation are uncertain and have been studied using the second generation, extremely metal-poor (EMP) stars. Secondly, core-collapse supernovae occur, and their yields are imprinted in the Population II stars in the Galactic halo. The [α/Fe] ratio is high and stays roughly constant with a small scatter. This plateau value does not depend on the star formation history but does on the IMF. Finally, SNe Ia occur, which produce more Fe than O, and thus the [α/Fe] ratio decreases toward higher metallicities; this decreasing trend is seen for the Population I stars in the Galactic disk. Because of this [α/Fe]–[Fe/H] relation, high-α and low-α are often used as a proxy of old and young age of stars in galaxies, respectively. Note that, however, that this relation is not linear but is a plateau and decreasing trend (called a 'knee'). The location of the knee depends on the star formation timescale and is at a high metallicity in the Galactic bulge, followed by the Galactic thick disk, thin disk, and satellite galaxies show at lower metallicities.

2.1. *Mystery of gold*

Figure 4 shows the evolutions of neutron-capture elements as [X/Fe]–[Fe/H] relations. As known, AGB stars can produce the first (Sr, Y, Zr), second (Ba), and third (Pb) peak s-process elements, but no heaver elements (navy long-dashed lines). It is surprising that ECSNe from a narrow mass range ($\Delta M \sim 0.15 - 0.2 M_\odot$) can produce enough of the first peak elements; with the combination of AGB stars, it is possible to reproduce the observational data very well (cyan short-dashed lines). This means that no other light element primary process (LEPP), such as rotating massive stars, is required. However, the elements from Mo to Ag seem to be overproduced, which could be tested with the UV spectrograph proposed for VLT, CUBES. Additional production from ν-driven winds leads to further over-production of these elements in the model (green dot-long-dashed lines), but this should be studied with more self-consistent calculations of supernova explosions.

Neutron star mergers can produce lanthanides and actinides, but not enough (olive dotted lines); the rate is too low and the timescale is too long, according to binary population synthesis. This is not improved enough even if neutron-star black-hole mergers are included (orange dot-short-dashed lines). An r-process associated with core-collapse supernovae, such as MRSNe, is required. The same conclusion is obtained with other GCE models and more sophisticated chemodynamical simulations (e.g., Haynes & Kobayashi 2019), as well as from the observational constraints of radioactive nuclei in the solar system (Wallner et al. 2021).

In the GCE model with MRSNe (magenta solid lines), it is possible to reproduce a plateau at low metallicities for Eu, Pt, and Th, relative to Fe. However, even with including both MRSNe and neutron star mergers, the predicted Au abundance is more than ten times lower than observed. This underproduction is seen not only for the solar abundance but also for low metallicity stars although the observational data are very limited. UV spectroscopy is needed for investigating this problem further.

In the GCE models, we adopt the best available nucleosynthesis yields in order to explain the universal r-process pattern: $25 M_\odot$ "b11tw1.00" 3D yields from Nishimura et al. (2015) and $1.3 M_\odot + 1.3 M_\odot$ 3D/GR yields from Wanajo et al. (2014). These yields are sensitive to the electron fraction, which depends on hydrodynamics

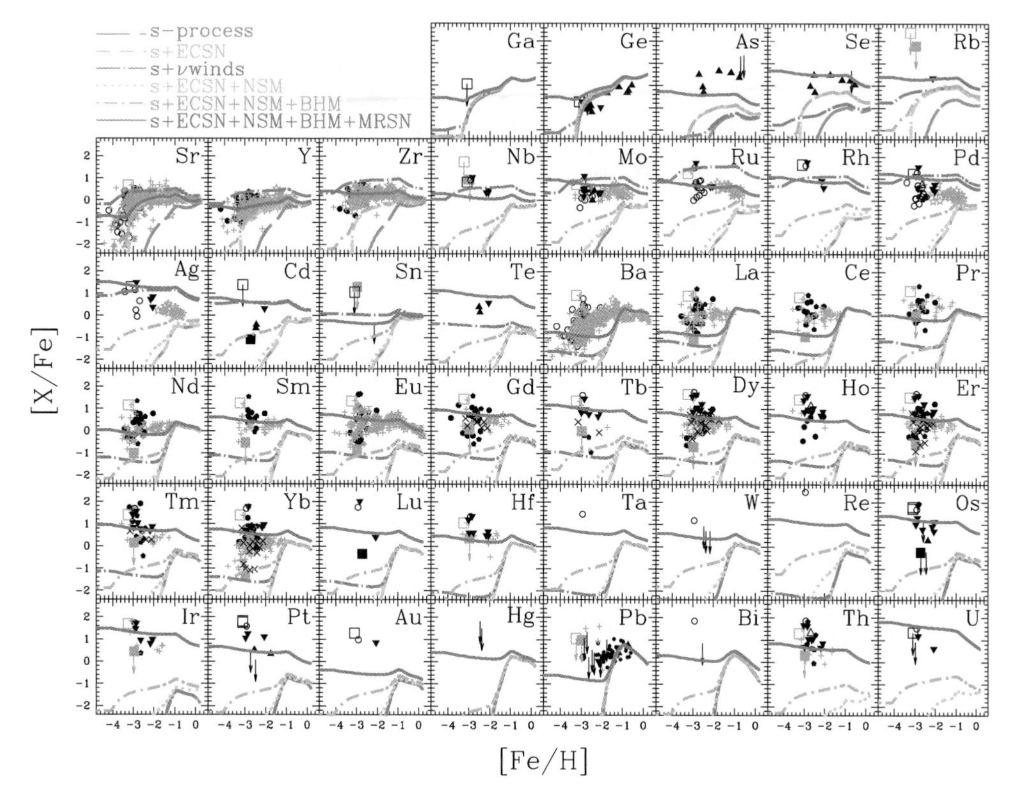

Figure 4. The [X/Fe]–[Fe/H] relations for neutron capture elements, comparing to the models in the solar neighborhood, with s-process from AGB stars only (blue long-dashed lines), with s-process and ECSNe (light-blue short-dashed lines), with s-process, ECSNe, and ν-driven winds (green dotted-long-dashed lines), with s-process, ECSNe, and NS-NS mergers (olive dotted lines), with s-process, ECSNe, and NS-NS/NS-BH mergers (orange dotted-short-dashed lines), with s-process, ECSNe, NS-NS/NS-BH mergers, and MRSNe (red solid lines). Observational data are updated from K20.

and ν-processes during explosions. Post-process nucleosynthesis calculations of successful explosion simulations of massive stars are required. However, it would not be easy to increase Au yields only since Pt is already in good agreement with the current model. There are uncertainties in some nuclear reaction rates and in the modelling of fission, which might be able to increase Au yields only, without increasing Pt or Ag. The predicted Th and U abundances are after the long-term decay, to be compared with observations of metal-poor stars, and the current model does not reproduce the Th/U ratio either.

It seems necessary to have the r-process associated with core-collapse supernovae, such as MRSNe. Is there any direct evidence for such events? There are a few magneto-hydrodynamical simulations and post-process nucleosynthesis that successfully showed enough neutron-rich ejecta to produce the 3rd peak r-process elements (Winteler et al. 2012; Mösta et al. 2018) as in the Sun. However, there is no observational evidence that support magneto-rotational supernovae. The predicted iron mass is rather small (Nishimura et al. 2015; Reichert et al. 2021), and the astronomical object could be faint.

Extremely metal-poor (EMP) stars have been an extremely useful tool in the Galactic archeology. At the beginning of galaxy formation, stars form from a gas cloud that was enriched only by one or very small number of supernovae, and hence the elemental abundances of EMP stars have offered observational evidences of supernovae in the past. As a result, it is found that quite a large fraction of massive stars became faint supernovae that

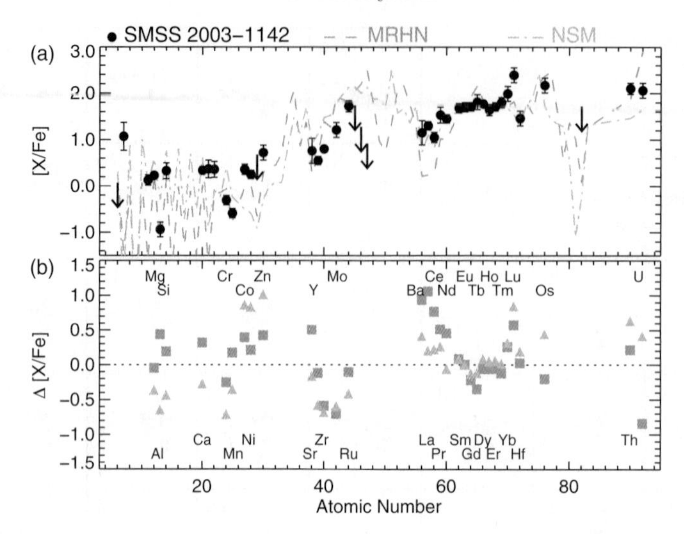

Figure 5. The elemental abundance of a extremely metal-poor star SMSS J200322.54-114203.3, which has [Fe/H]$= -3.5$, comparing with mono-enrichment from a magneto-rotational hypernovae (blue dashed line and squares) and multi-enrichment including a neutron star merger (orange dot-dashed line and triangles). The lower panel shows the differences, i.e., the observed values minus model values.

give a high C/Fe ratio leaving a relatively large black hole ($\sim 5M_\odot$, Nomoto et al. 2013). It might even possible to form a black hole from $10 - 20M_\odot$ stars (Kobayashi, Ishigaki et al. 2014). It is challenging to find EMP stars due to the expected small number, and it is also challenging to measure elemental abundances in detail. Thus only a limited number of EMP stars have been analysed in previous works. The Australian team has been using a strategic approach to increase EMP data; ~ 26000 candidates are found from photometric data on the SkyMapper telescope, 2618 of which have spectroscopic observations at ANU's 2.3m telescope (Da Costa et al. 2019), and for ~ 500 stars detailed elemental abundances are measured with higher-resolution spectra taken at larger telescopes such as Magellan, Keck, and VLT (Yong et al. 2021b). SMSS J200322.54-114203.3 was discovered in the SkyMapper EMP survey and reported in Yong et al. (2021a).

Figure 5 shows the elemental abundance of SMSS J200322.54-114203.3, which has [Fe/H]$= -3.5$, and is 2.3 kpc away with the Galactic halo orbit. This star has a very clear detection of uranium and thorium, and thus it is a so-called actinides boost star. The stars also clearly showed the solar r-process pattern from $Z \sim 60$ to 70. Surprisingly, the observed Zn abundance is very high ([Zn/Fe]$= 0.72$), which indicates that the enrichment source was an energetic explosion ($\gtrsim 10^{52}$ erg). It also showed normal α enhancement as for the majority of Galactic halo stars, although a normal supernova ($\sim 10^{51}$ erg) gives a range of α enhancement depending on the progenitor mass. The star is not carbon enhanced ([C/Fe]< 0.07) but shows nitrogen enhancement ([N/Fe]$= 1.07$), which indicates that the progenitor stars were rotating. All of these features at $Z \leqslant 30$ are consistent with the enrichment from a single hypernova.

The observed abundance pattern is compared with two theoretical models. The blue dashed line is for mono-enrichment from a single, magneto-rotational hypernova (MRHN), where the nucleosynthesis yields are obtained combining the iron-core of a magneto-rotational supernovae (Nishimura et al. 2015) to the envelope of a $25M_\odot$ hypernova. The orange dot-dashed line shows a model of multi-enrichment where a neutron star merger occur in the interstellar matter chemically enriched by previous generations of core-collapse supernovae. The MRHN model gives better fit at and below Zn ($Z \leqslant 30$).

Both models show overproduction around the 1st peak ($Z \sim 40$) and around Ba ($Z = 56$), which may be due to the uncertainties in nuclear astrophysics. The MRSN scenario is also consistent with the short timescale of the EMP star formation; due to the required long timescales of neutron star mergers, it is unlikely that this EMP star is enriched by a neutron star merger.

Future prospects: This star provided the first observational evidence of the r-process associated with core-collapse supernovae. The r-process site could be in the jets as in the MRHN model, or could also be in the accretion disk around black hole as in the collapsar model (Siegel et al. 2019). According to the chemical evolution models, this event is rare, only one in 1000 supernovae. MRHNe should be observable in future astronomical transient surveys such as with LSST/Rubin, since they eject a significant amount of iron; the ejected iron mass is $0.017 M_\odot$ in the $25 M_\odot$ model, while $0.33 M_\odot$ in the $40 M_\odot$ model (Yong et al. 2021a). It might be possible to detect the signature of Au and Pt production directly in the supernova spectra.

2.2. *Ramp-up of fluorine at high-redshifts*

Fluorine is an intriguing element. The major production site has been debated for 30 years among low-mass AGB stars, stellar winds in massive stars, and the ν-process in core-collapse supernovae (Kobayashi et al. 2011a). The observational constraints have been obtained for stars, but accessible lines are only in infrared, and thus the sample number was limited. Moreover, there was confusion in the excitation energies and transition probabilities for the HF lines (Jönsson et al. 2014). The vibrational-rotational lines get too weak at low metallicities, and the available sample of EMP stars is only for carbon enhanced stars (Mura-Guzmán et al. 2020), which may not be representative of galactic chemical evolution. Franco et al. (2021) opened a new window for constraining fluorine production in the early universe by directly measuring HF abundance in a galaxy at redshift $z = 4.4$.

The galaxy NGP-190387 was discovered by the H-ATLAS survey with the Herschel satellite, with the redshift confirmed by the Northern Extended Millimetre Array (NOEMA). It is a gravitationally lensed galaxy with a magnification factor $\mu \simeq 5$. The lowest rotational transitions from the HF molecule appeared as an absorption line in the Atacama Large Millimeter/submillimeter Array (ALMA) data. This molecule is very stable and the dominant gas-phase form of fluorine in the ISM.

Figure 6 shows the fluorine abundance of NGP-190387 (yellow point), comparing with galactic chemical evolution models with various star formation timescales. In the models, it is assumed that star formation took place soon after the reionization, which was boosted probably due to galaxy merger 100 Myr before the observed redshift $z = 4.4$. This assumption is based on the properties of other submillimetre galaxies (SMGs) at similar redshifts. The models show rapid chemical enrichment including fluorine, but in order to explain the observed fluorine abundance, the contribution from Wolf-Rayet stars (green lines) is required. The nucleosynthesis yields from Limongi & Chieffi (2018) are adopted in the models, assuming the rotational velocity of 300 km s^{-1} for $13 - 120 M_\odot$ stars.

Future prospects: This galaxy provided the first observational evidence for rotating massive stars with initial masses $\gtrsim 30 M_\odot$ producing a significant amount of fluorine in the early universe. Since it is not easy to explode these stars with ν-driven mechanism, these stars are likely to remain black holes with masses $\gtrsim 20 M_\odot$, which may be detected with gravitational wave missions. It would be possible to constrain the formation history further if CNO abundances and/or ^{13}C isotopic ratio are measured, because the yields depend on the mass of progenitor stars.

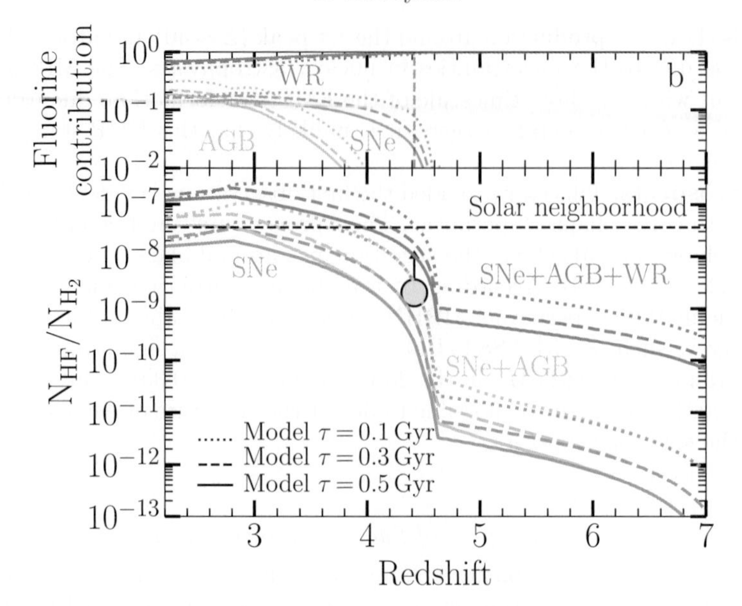

Figure 6. The fluorine abundance in the galaxy NGP-190387 at redshift $z = 4.4$, comparing with galactic chemical evolution models with only supernovae (blue lines), plus AGB stars (orange lines), plus WR stars (green lines).

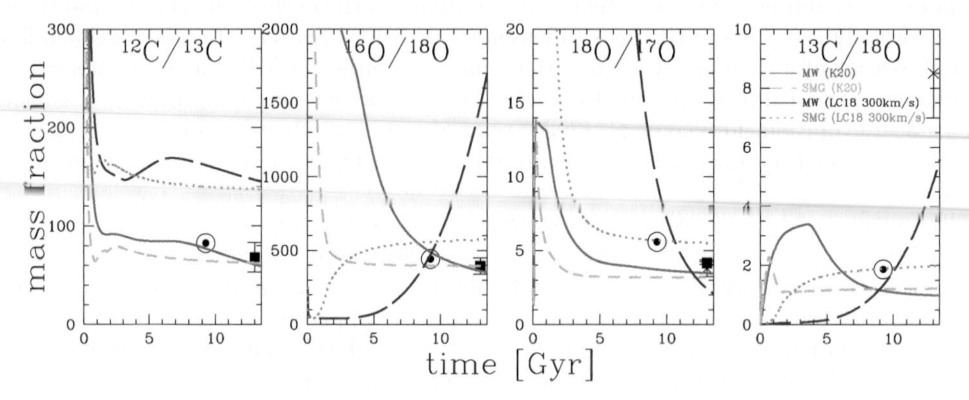

Figure 7. Evolution of isotopic ratios of GCE models for the Milky Way (red solid and blue long-dashed lines) and for a submillimetre galaxy (green short-dashed and magenta dotted lines), which are calculated with the K20 yields (red solid and green short-dashed lines) or the yields from Limongi & Chieffi (2018) with the rotational velocity of 300 km s^{-1} (blue long-dashed and magenta dotted lines). The solar symbol indicates the solar ratios (Asplund et al. 2009). See Romano et al. (2019) for the observational data sources of the ISM (filled squares and crosses).

2.3. *GCE with rotating massive stars*

The evolution of isotopic ratios is shown in Figs. 17-19 of Kobayashi et al. (2011b) and Fig. 31 of K20 (see also Romano et al. 2019). ^{13}C and ^{25,26}Mg are produced from AGB stars (^{17}O might be overproduced in our AGB models), while other minor isotopes are more produced from metal-rich massive stars, and thus the ratios between major and minor isotopes (e.g., ^{12}C/^{13}C, ^{16}O/^{17,18}O) generally decrease as a function of time/metallicity. But this evolutionary trend is completely different if rotating massive stars are included.

Figure 7 shows the evolution of isotopic ratios in the GCE models with and without rotating massive stars, comparing to observational data. For the Milky Way (red solid

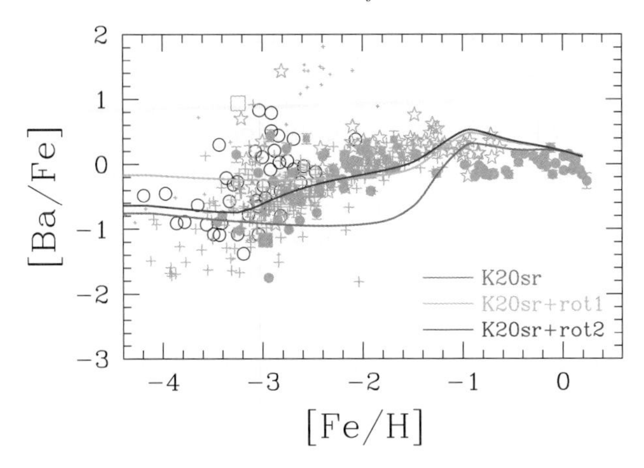

Figure 8. The [Ba/Fe]–[Fe/H] relation, comparing with K20 model (red line) and models with rotating massive stars with different rotational velocities (green and blue lines). See K20 for the other data points and model details.

and blue long-dashed lines), the adopted star formation history is the same as the solar neighbourhood model in §2. For the SMG (green short-dashed and magenta dotted lines), rapid gas accretion and star formation with the timescales of 1 Gyr are assumed, and after ~ 1.5 Gyr the star formation is self-regulated with a constant gas fraction of 50% (see M. Doherty et al., in prep. for more details). The K20 model (red solid lines) is not bad, well reproducing C and O isotopic ratios in the Sun (the solar symbol) and in the present-day ISM (filled squares), but not the $^{13}C/^{18}O$ ratio (cross).

The model with rotating massive stars (blue long-dashed lines) gives a better evolutionary track of the $^{13}C/^{18}O$ ratio, showing a rapid increase in the past 5 Gyrs in the Milky Way. However, ^{12}C is overproduced at all metallicities, and ^{16}O is overproduced at high metallicities in this model. Here the yields from Limongi & Chieffi (2018) with the rotational velocity of 300 km s^{-1} are used. With 150 km s^{-1}, the $^{12}C/^{13}C$ ratio becomes even larger, and the $^{13}C/^{18}O$ ratio becomes slightly smaller. The model with 150 km s^{-1} is worse than the model with 300 km s^{-1} in order to reproduce the observations, and using an averaged rotational velocity as in Prantzos et al. (2018) does not solve this problem. With these yields set, even excluding stellar rotation, the $^{12}C/^{13}C$ ratio is systematically higher than in the K20 yields, and the difference may be caused by their choice of nuclear reaction rates, mass-loss, and/or convection. It is possible that these effects could change the evolution of the $^{18}O/^{17}O$ ratio. The overproduction of ^{16}O is probably due to their choice of remnant masses, and in fact these yields do not reproduce the observed [O/Fe]–[Fe/H] relation.

The impact of stellar rotation should be studied from s-process elements, since the weak-s process forms s-process elements from the existing seeds at a much shorter timescale than AGB stars (Frischknecht et al. 2016; Limongi & Chieffi 2018). For s-process elements, there is some underproduction from [Fe/H] ~ -3 to ~ -1 in the K20 model, which could be improved in the inhomogeneous enrichment effect in chemodynamical simulations (§3.1). Alternatively, the models with rotating massive stars can explain the observed average trend (Fig. 8); in the first model (green line, also plotted in Yong et al. 2021b) the same metallicity-dependent distribution of rotational velocity as in Prantzos et al. (2018) is used, while in the second and better model (blue line) 150 and 300 km s^{-1} is assumed for HNe and MRHNe, respectively. As already noted, the nucleosynthesis yields from Limongi & Chieffi (2018) cannot reproduce the observations of many elements, and thus we use the contributions from stellar envelopes and winds only in addition to the K20 yields (see Kobayashi et al., in prep. for more details).

3. Chemodynamical evolution of galaxies

Thanks to the development of super computers and numerical techniques, it is now possible to simulate the formation and evolution of galaxies from cosmological initial conditions, i.e., initial perturbation based on Λ cold dark matter (CDM) cosmology. Star formation, gas inflow, and outflow in Eq.(2.1) are not assumed but are, in principle, given by dynamics. Due to the limited resolution, star forming regions in galaxy simulations, supernova ejecta, and active galactic nuclei (AGN) cannot be resolved, and thus it is necessary to model star formation and the subsequent effects – feedback – introducing a few parameters. Fortunately, there are many observational constraints, from which it is usually possible to choose a best set of parameters. To ensure this, it is also necessary to run the simulation until the present-day, $z = 0$, and reproduce a number of observed relations at various redshifts. A schematic view of the chemodynamical evolution of galaxies is given in Figure 9.

Although hydrodynamics can be calculated with publicly available codes such as Gadget, RAMSES, and AREPO, modelling of baryon physics is the key and is different depending on the simulation teams/runs, such as EAGLE, Illustris, Horizon-AGN, Magneticum, and SIMBA simulations. These are simulations of a cosmological box with periodic boundary conditions, and contain galaxies with a wide mass range (e.g., $10^{9-12} M_\odot$ in stellar mass) at $z = 0$. In order to study massive galaxies it is necessary to increase the size of the simulation box (e.g., $\gtrsim 100$ Mpc), while in order to study internal structures it is necessary to improve the spacial resolution (e.g., $\lesssim 0.5$ kpc). The box size and resolution are chosen depending on available computational resources. On the other hand, zoom-in techniques allow us to increase the resolution focusing a particular galaxy, but this also requires tuning the parameters with the same resolution, comparing to a number of observations in the Milky Way.

For the baryon physics, the first process to calculate is **radiative cooling**, and photo-heating by a uniform and evolving UV background radiation. We use a metallicity-dependent cooling function computed with the MAPPINGS III software (see Kobayashi et al. 2007 for the details), and find candidate gas particles that can form stars in a given timestep. The effect of α enhancement relative to Fe is taken into account using the observed relation in the solar neighborhood.

Our **star formation** criteria are: (1) converging flow, $(\nabla \cdot \boldsymbol{v})_i < 0$; (2) rapid cooling, $t_{\rm cool} < t_{\rm dyn}$; and (3) Jeans unstable gas, $t_{\rm dyn} < t_{\rm sound}$. The star formation timescale is taken to be proportional to the dynamical timescale ($t_{\rm sf} \equiv \frac{1}{c_*} t_{\rm dyn}$), where c_* is a star formation timescale parameter. The parameter c_* basically determines when to form stars following cosmological accretion, and has a great impact on the size–mass relation of galaxies (Fig. 4 of Kobayashi 2005). We found that it is better not to include a fixed density criterion. Note that based on smaller-scale simulations, more sophisticated analytic formula are proposed including the effects of turbulence and magnetic fields.

A fractional part of the mass of the gas particle turns into a new star particle (see Kobayashi et al. 2007, for the details). Note that an individual star particle has a typical mass of $\sim 10^{5-7} M_\odot$, i.e. it does not represent a single star but an association of many stars. The masses of the stars associated with each star particle are distributed according to an IMF. We adopt Kroupa IMF, which is assumed to be independent of time and metallicity, and limit the IMF to the mass range $0.01 M_\odot \leqslant m \leqslant 120 M_\odot$. This assumption is probably valid for the resolution down to $\sim 10^4 M_\odot$.

We then follow the evolution of the star particle, as a simple stellar population, which is defined as a single generation of coeval and chemically homogeneous stars of various

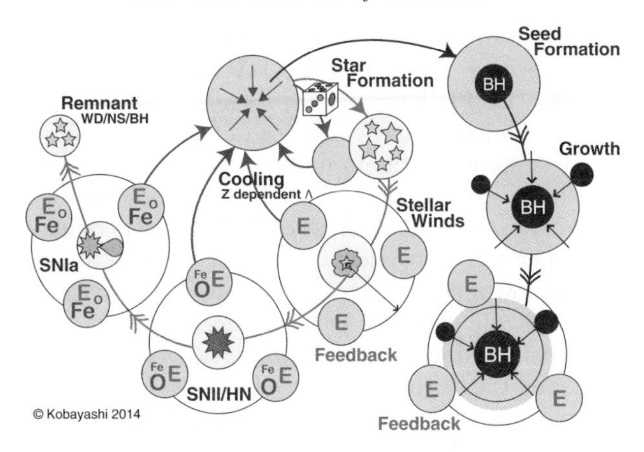

Figure 9. Schematic view of chemodynamical evolution of galaxies.

masses, i.e. it consists of a number of stars with various masses but the same age and chemical composition (Kobayashi 2004). The production of each element i from the star particle (with an initial mass of m_*^0) is calculated using a very similar equation as Eq.(2.1):

$$E_Z(t) = m_*^0 \left[E_{\rm SW} + E_{\rm SNcc} + E_{\rm SNIa} + E_{\rm NSM}\right]. \tag{3.1}$$

Similarly, the energy production from the star particle is:

$$E_e(t) = m_*^0 \left[e_{e,{\rm SW}}\mathcal{R}_{\rm SW}(t) + e_{e,{\rm SNcc}}\mathcal{R}_{\rm SNcc}(t) + e_{e,{\rm SNIa}}\mathcal{R}_{\rm SNIa}(t)\right] \tag{3.2}$$

which includes the event rates of SWs ($\mathcal{R}_{\rm SW}$), core-collapse supernovae ($\mathcal{R}_{\rm SNcc}$), and SNe Ia ($\mathcal{R}_{\rm SNIa}$), and the energy per event (in erg) for SWs: $e_{e,{\rm SW}} = 0.2 \times 10^{51}(Z/Z_\odot)^{0.8}$ for $> 8M_\odot$ stars, Type II supernovae: $e_{e,{\rm SNII}} = 1 \times 10^{51}$, hypernovae: $e_{e,{\rm HN}} = 10, 10, 20, 30 \times 10^{51}$ for $20, 25, 30, 40M_\odot$ stars, respectively, and for SNe Ia: $e_{e,{\rm SNII}} = 1.3 \times 10^{51}$.

We distribute this **feedback** metal mass and energy from a star particle j to a fixed number of neighbour gas particles ℓ, $N_{\rm FB}$, weighted by the smoothing kernel as:

$$\frac{d(Z_{i,k}\, m_{{\rm g},k})}{dt} = \sum_j \left[W_{kj} E_{Z,j}(t) / \sum_\ell^{N_{\rm FB}} W_{\ell j}\right] \tag{3.3}$$

and

$$\frac{du_k}{dt} = (1 - f_{\rm kin}) \sum_j \left[W_{kj} E_{e,j}(t) / \sum_\ell^{N_{\rm FB}} W_{\ell j}\right] \tag{3.4}$$

where u_k denotes the internal energy of a gas particle k, and $f_{\rm kin}$ denotes the kinetic energy fraction, i.e., the fraction of energy that is distributed as a random velocity kick to the gas particle. Note that in the first sum the number of neighbour star particles is not fixed to $N_{\rm FB}$. To calculate these equations, the feedback neighbour search needs to be done twice in order to ensure proper mass and energy conservation; first to compute the sum of weights for the normalization, and a second time for the actual distribution.

The parameter $N_{\rm FB}$ determines the average energy distributed to gas particles; a large value of $N_{\rm FB}$ leads to inefficient feedback as the ejected energy radiatively cools away shortly, while a small value of $N_{\rm FB}$ results in a only small fraction of matter affected by feedback. Alternatively, with good resolution in zoom-in simulations, feedback neighbors could be chosen within a fixed radius ($r_{\rm FB}$), which affects a number of observations including the MDF (Fig. 12 of Kobayashi & Nakasato 2011). The parameter $f_{\rm kin}$ has a great impact on radial metallicity gradients in galaxies, and $f_{\rm kin} = 0$, i.e., purely thermal feedback, gives the best match to the observations (Fig. 14 of Kobayashi

2004). Note that various feedback methods are proposed such as the stochastic feedback (Dalla Vecchia & Schaye 2008, used for EAGLE) and the mechanical feedback (Hopkins et al. 2018, used for FIRE).

Eq.(3.3) gives positive feedback that enhances radiative cooling, while Eq.(3.4) gives negative feedback that suppresses star formation. The mass ejection in Eq.(3.3) never becomes zero, while the energy production in Eq.(3.4) becomes small after 35 Myrs. Therefore, it is not easy to control star formation histories depending on the mass/size of galaxies within this framework. In particular, it is not possible to quench star formation in massive galaxies, since low-mass stars keep returning their envelope mass into the ISM for a long timescale, which will cool and keep forming stars.

Therefore, in order to reproduce observed properties of massive galaxies, additional feedback was required, and the discovery of the co-evolution of super-massive black holes and host galaxies provided a solution – **AGN feedback**. Modelling of AGN feedback consists of (1) seed formation, (2) growth by mergers and gas accretion, and (3) thermal and/or kinetic feedback (see Taylor & Kobayashi 2014 for the details). In summary, we introduced a seeding model where seed black holes originate from the formation of the first stars, which is different from the 'standard' model by Springel et al. (2005) and from most large-scale hydrodynamical simulations. In our AGN model, seed black holes are generated if the metallicity of the gas cloud is zero ($Z=0$) and the density is higher than a threshold density ($\rho_g > \rho_c$). The growth of black holes is the same as in other cosmological simulations, and is calculated with Bondi-Hoyle accretion, swallowing of ambient gas particles, and merging with other black holes. Since we start from relatively small seeds, the black hole growth is driven by mergers at $z \gtrsim 3$ (Fig. 2 of Taylor & Kobayashi 2014). On a very small scale, it is not easy to merge two black holes, and an additional time-delay is applied in more recent simulations (P. Taylor et al. in prep.).

Proportional to the accretion rate, thermal energy is distributed to the surrounding gas, which is also the same as in many other simulations. In more recent simulations, non-isotropic distribution of feedback area is used to mimic the small-scale jet (e.g., Davé et al. 2019). There are a few parameters in our chemodynamical simulation code, but Taylor & Kobayashi (2014) constrained the model parameters from observations, and determined the best parameter set: $\alpha = 1$, $\epsilon_f = 0.25$, $M_{seed} = h^{-1} 10^3 M_\odot$, and $\rho_c = 0.1 h^2 m_H \, cm^{-3}$. Our black holes seeds are indeed the debris of the first stars although we explored a parameter space of $10^{1-5} M_\odot$. This is not the only one channel for seeding, and the direct collapse of primordial gas and/or the collapse of dense stellar clusters ($\sim 10^5 M_\odot$, Woods et al. 2019) are rarer but should be included in larger volume simulations. Nonetheless, our model can successfully drive large-scale galactic winds from massive galaxies (Taylor & Kobayashi 2015b) and can reproduce many observations of galaxies with stellar masses of $\sim 10^{9-12} M_\odot$ (Taylor & Kobayashi 2015a, 2016, 2017). The movie of our fiducial run is available at `https://www.youtube.com/watch?v=jk5bLrVI8Tw`

3.1. *Extra-galactic archaeology*

Elemental abundances and isotopic ratios provide additional constrains on the timescales of formation and evolution of galaxies. This extra-galactic archaeology will be possible in the near future with the James Webb Space Telescope (JWST), the wavelength coverage of which will allow us to measure CNO abundance simultaneously (Vincenzo & Kobayashi 2018a). At high redshifts, metallicities have been measured mainly with emission lines in star-forming galaxies, where Fe abundance is not accessible. Instead of α/Fe ratios in Galactic archaeology, CNO abundances can be used for extra-galactic archaeology.

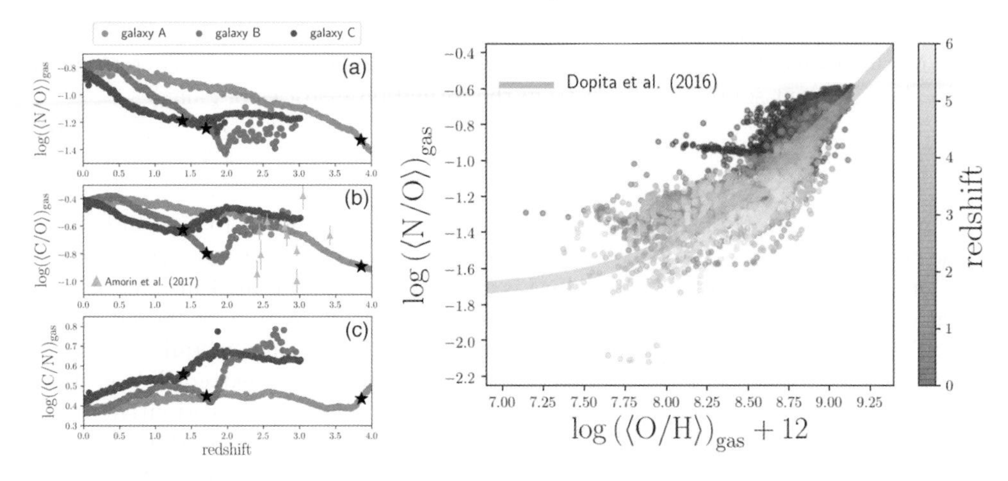

Figure 10. (Left) Evolution of CNO abundance ratios in simulated disk galaxies with different formation timescales, taken in a cosmological simulation. (Right) Evolution of the N/O–O/H relation of simulated galaxies in a cosmological simulation. The gray bar indicates the compilation of observational data.

The left panel of Figure 10 shows theoretical predictions of CNO abundance ratios for simulated disk galaxies that have different formation timescales. These galaxies are chosen from a cosmological simulation, which is run by a Gadget-3 based code that includes detailed chemical enrichment (Kobayashi et al. 2007). In the simulation, C is mainly produced from low-mass stars ($\lesssim 4M_\odot$), N is from intermediate-mass stars ($\gtrsim 4M_\odot$) as a primary process (Kobayashi et al. 2011b), and O is from massive stars ($\gtrsim 13M_\odot$). C and N are also produced by massive stars; the N yield depends on the metallicity as a secondary process, and can be greatly enhanced by stellar rotation (as for F in §2.2).

In the nearby universe, the N/O–O/H relation is known for stellar and ISM abundances, which shows a plateau (~ -1.6) at low metallicities and a rapid increase toward higher metallicities. This was interpreted as the necessity of rotating massive stars by Chiappini et al. (2006). However, this should be studied with hydrodynamical simulations including detailed chemical enrichment, and Kobayashi (2014) first showed the N/O–O/H relation in a chemodynamical simulation. Vincenzo & Kobayashi (2018b) showed that both the global relation, which is obtained for average abundances of the entire galaxies, and the local relation, which is obtained for spatially resolved abundances from IFU data, can be reproduced by the inhomogeneous enrichment from AGB stars. Since N yield increases at higher metallicities, the global relation originates from the mass–metallicity relation of galaxies, while the local relation is caused by radial metallicity gradients within galaxies. Moreover, the right panel of Figure 10 shows a theoretical prediction on the time evolution of the N/O–O/H relation, where galaxies evolve along the relation. Recent observation with KMOS (KLEVER survey) confirmed a near redshift-invariant N/O-O/H relation (Hayden-Pawson et al. 2022).

Future prospects: It is possible to reproduce the observed N/O–O/H relation only with AGB stars and supernovae in our chemodynamical simulations. If we include N production from WR stars, the N/O plateau value at low metallicities becomes too high compared with observations. The enrichment from the first stars is not included either, which can be significantly different because very massive stars ($\geq 140M_\odot$) explode as pair-instability supernovae without leaving any remnants, or massive stars ($\sim 25M_\odot$) explode as failed supernovae that leave relatively more massive black holes ($\sim 5M_\odot$). The former is based on theoretical prediction (but no observational evidence yet) and the latter is inferred from observational results (Nomoto et al. 2013). It would be very useful if other elemental abundances (e.g., Ne, S) are measured in order to determine the chemical enrichment sources and constrain the formation histories of galaxies.

3.2. *Galactic archaeology*

Inhomogeneous enrichment in chemodynamical simulations leads to a paradigm shift on the chemical evolution of galaxies. As in a real galaxy, i) the star formation history is not a simple function of radius, ii) the ISM is not homogeneous at any time, and iii) stars migrate (Vincenzo & Kobayashi 2020), which are fundamentally different from one-zone or multi-zones GCE models. As a consequence, (1) there is no age–metallicity relation, namely for stars formed in merging galaxies. It is possible to form extremely metal-poor stars from accretion of nearly primordial gas, or in isolated chemically-primitive regions, at a later time. (2) Enrichment sources with long time-delays such as AGB stars, SNe Ia, and NSMs can appear at low metallicities. This is the reason why the N/O plateau is caused by AGB stars in §3.1. However, this effect is not enough to reproduce the observed [Eu/(O,Fe)] ratios only with NSMs (Haynes & Kobayashi 2019). (3) There is a significant scatter in elemental abundance ratios at a given time/metallicity, as shown in Figs. 16-18 of Kobayashi & Nakasato (2011).

Figure 11 shows the frequency distribution of elemental abundance ratios in our chemodynamical zoom-in simulation of a Milky Way-type galaxy ($r \leqslant 100$ kpc). The colour indicates the number of stars in logarithmic scale. In order to compare the scatter, the range of y-axis is set to the same, from -1.5 to $+0.5$ for the elements except for He and Li. In the simulation we know the ages and kinematics of star particles as well as the locations within the galaxy, which are used in the following summary:

- The Big Bang nucleosynthesis gives $Y = 0.2449$. He is also synthesized during stellar evolution and the He abundance increases up to $Y \sim 0.33$ showing a tight function of metallicity (see also Vincenzo et al. 2019).
- The initial Li abundance is set to be a theoretical value of the Big Bang nucleosynthesis, $A(\mathrm{Li}) = 2.75$, although it is higher than the observed value of EMP stars. The Li abundance increases by the production from AGB stars, but decreases at super-solar metallicities due to the destruction in AGB stars. There is a large scatter from $A(\mathrm{Li}) \sim 2.6$ to 0.3 at $[\mathrm{Fe/H}] \gtrsim -1$.
- The average [C/Fe] ratio is around ~ 0. There is only a small number of C-enhanced EMP stars with $[\mathrm{C/Fe}] \geq 0.7$ due to the inhomogeneous enrichment from AGB stars in our model. This number is much smaller than observed because faint supernovae are not included.
- The [N/Fe] ratio shows a large scatter of ± 0.5 dex even at $[\mathrm{Fe/H}] \sim -3$ due to the inhomogeneous enrichment, and it is possible to reproduce the observational data only with AGB stars, without rotating massive stars.
- The α elements, O, Mg, Si, S, and Ca, show similar trends, a plateau with a small scatter at $[\mathrm{Fe/H}] \lesssim -1$, and then bimodal decreasing trends both caused by the delayed Fe enrichment from SNe Ia. The high-α stars tend to have high v/σ and old ages, forming the thick disk. On the other hand, the low α stars are rotationally supported and relatively younger, belonging to the thin disk. The thick disk reaches only $[\mathrm{Fe/H}] \sim 0$, while thin disk expends to $[\mathrm{Fe/H}] \sim 0.2$. The stars with $[\mathrm{Fe/H}] > 0.2$ in this figure are located in the bulge region ($r \leqslant 1$ kpc in this paper).
- The odd-Z elements, Na, Al, and Cu, show bimodal increasing trends from $[\mathrm{Fe/H}] \sim -2$ to higher metallicities, which are predominantly caused by the metallicity dependent yields of core-collapse supernovae. The higher [(Na,Al,Cu)/Fe] trend is composed of the high α stars in the thick disk, while the lower [(Na,Al,Cu)/Fe] trend is made of the low α

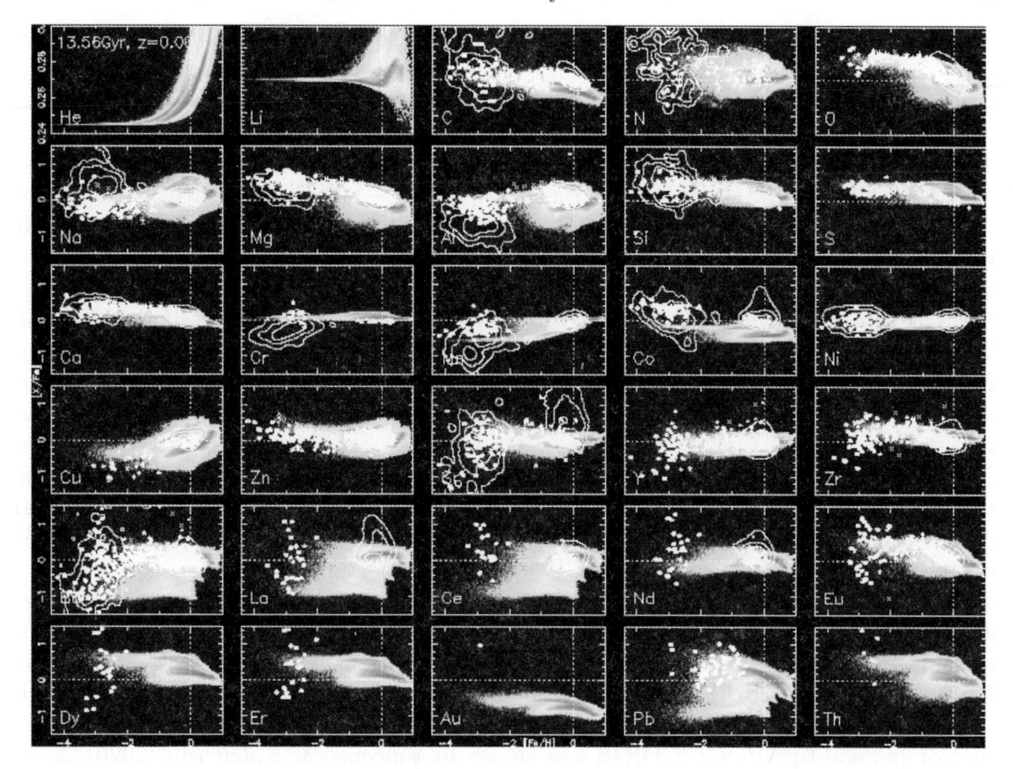

Figure 11. Frequency distribution of elemental abundance ratios in a chemodynamical zoom-in simulation of a Milky Way-type galaxy. The filled circles are observational data from high-resolution spectra, and the metal-rich and metal-poor contours show the HERMES-GALAH survey DR3 (Buder et al. 2021) and the SkyMapper EMP survey (Yong et al. 2021b), respectively.

stars in the thin disk; these differences are caused by SNe Ia. Moreover, [Na/Fe] shows a greater increase with a steeper slope for the thick disk than that for the thin disk, which is due to an additional contribution from AGB stars. Such a difference in the slopes is not seen for Al and Cu.

• The [Cr/Fe] ratio is almost constant at ~ 0 over all ranges of metallicities. In observations, EMP stars tend to show [Cr/Fe] < 0 values affected by the NLTE effect, but Cr II observations show [Cr/Fe] ~ 0 (K06). No decrease of [Cr/Fe] is caused by SNe Ia either if we use the latest SN Ia yields (Kobayashi, Leung & Nomoto 2020).

• In contrast to the [α/Fe] ratios, [Mn/Fe] shows a plateau value of ~ -0.5 at [Fe/H] $\lesssim -1$, and then bimodal increasing trends because Ch-mass SNe Ia produce more Mn than Fe (K06). Unlike one-zone models, [Mn/Fe] does not reach a very high value ($\sim +0.5$) in chemodynamical simulations because not all SN Ia progenitors have high metallicities due to the inhomogeneity. The inclusion of sub-Ch mass SNe Ia would worsen the situation.

• Co and Zn can be enhanced by higher energy explosions (K06). There are two plateaus with [Zn/Fe] ~ 0.2 for the thick disk and [Zn/Fe] ~ 0 for the thin disk. At [Fe/H] > 0.2, [Zn/Fe] shows a rapid increase, which is seen only for the bulge stars in our model. Very similar features are seen also for Co, although Co is under-produced overall, which is due to the lack of multi-dimensional effect in the nucleosynthesis yields (Sneden et al. 2016).

• The first-peak s-process elements (Sr, Y, and Zr) are produced by ECSNe and AGB stars. Sr and Y do not show a bimodality, while Zr shows a very similar trend as for the α elements.

• The second-peak s-process elements (Ba, La, and Ce) show a very large scatter of 1.5 dex especially at low metallicities. These elements are first produced by MRSNe/MRHNe in our model with a floor value of [(Ba,La,Ce)/Fe] ~ -1, and are greatly enhanced by AGB stars. The contribution from AGB stars appears from [Fe/H] ~ -2; [(Ba,La,Ce)/Fe] ratios reach ~ 0 at [Fe/H] ~ -1, peaks at [Fe/H] ~ 0, then decrease due to SNe Ia. The decrease is significant in the bulge stars at [Fe/H] > 0.2.

• The r-process elements (Nd, Eu, Dy, Er, Au, and Th in the figure) do show a bimodality in the model with MRSNe/MRHNe. The average [Nd/Fe] is about ~ 0, while Eu, Dy, Er, and Th behave very similar to the α elements. As discussed in §2.1, Au is underabundant overall, but the thick disk stars are systematically more gold-rich than for the thin disk stars!

• Pb is one of the third-peak s-process elements, and the distribution at low metallicities are similar to the second-peak s-process elements. However, [Pb/Fe] shows a steeper decrease from [Fe/H] ~ -1 to higher metallicities, due to the metallicity dependence; Pb is produced more from metal-poor AGB stars.

Compared with observational data, the scatter of s-process elements seem too large around [Fe/H] ~ -1, which also supports the weak-s process rotating massive stars. The elemental abundance distributions depend on the locations within galaxies, as shown in Fig. 1 of Kobayashi (2016) for a different chemodynamical simulation. These distributions should be compared with non-biased, homogeneous dataset of observational data and the comparison to APOGEE DR16 was shown in Vincenzo & Kobayashi (2020). More analysis will be shown in Kobayashi (2022, in prep). The movie of our fiducial run is available at `https://star.herts.ac.uk/~chiaki/works/Aq-C-5-kro2.mpg`

4. Conclusions and Discussion

Thanks to the long-term collaborations between nuclear and astrophysics, we have good understanding on the origin of elements (except for the elements around Ti and a few neutron-capture elements such as Au). Inhomogeneous enrichment is extremely important for interpreting the elemental abundance trends (§3.2). It can reproduce the observed N/O–O/H relation only with AGB stars and supernovae (§3.1), but not the observed r-process abundances only with NSMs; an r-process associated with core-collapse supernovae such as magneto-rotational hypernovae is required, although the explosion mechanism is unknown (§2.1). It is necessary to run chemodynamical simulations from cosmological initial conditions, including detailed chemical enrichment. Theoretical predictions depend on input stellar physics, and the effects of stellar mass loss due to rotation and/or binary interaction should be investigated further. The importance of WR stars is indicated by the high F abundance in $z \sim 4.4$ galaxy (§2.2), as well as for some isotopic ratios and Ba and Pb abundances around [Fe/H] ~ -1 in the Milky Way (§2.3). If WR stars are producing heavy elements on a very short timescale, it might be hard to find very metal-poor or Population III (and dust-free) galaxies at very high redshifts even with JWST. Finally, mass loss from these stars will change the dust production as well, and thus it is also important to calculate element-by-element dust formation, growth, and destruction, as well as the detailed chemical enrichment.

Galactic archaeology is a powerful approach for reconstructing the formation history of the Milky Way and its satellite galaxies. APOGEE and HERMES-GALAH surveys have provided homogeneous datasets of many elemental abundances that can be statistically compared with chemodynamical simulations. Future surveys with WEAVE and 4MOST will provide more. Having said that, the number of EMP stars will not be increased so much in these surveys, and a target survey such as the SkyMapper EMP survey is also needed, in particular for constraining the early chemical enrichment from the first stars.

Reflecting the difference in the formation timescale, elemental abundances depend on the location within galaxies. Although this dependence has been explored toward the Galactic bulge by APOGEE, the dependence at the outer disk is still unknown, which requires 8m-class telescopes such as the PFS on Subaru telescope. Despite the limited spectral resolution of the PFS, α/Fe and a small number of elements will be available. The PFS will also be able to explore the α/Fe bimodality in M31; it is not yet known if M31 has a similar α/Fe dichotomy or not.

The next step will be to apply the Galactic archaeology approach to external or distant galaxies. Although it became possible to map metallicity, some elemental abundances, and kinematics within galaxies with IFU, the sample and spacial resolution are still limited even with JWST. Integrated physical quantities over galaxies, or stacked quantities at a given mass bin, will also be useful, which can be done with the same instruments developed for Galactic archaeology (although optimal spectral resolutions and wavelength coverages are different). In addition, ALMA has opened a new window for elemental abundances and isotopic ratios in high-redshift galaxies.

It is very important to understand stellar physics when these observational data are translated into the formation timescales or physical processes of galaxies. For example, the famous [α/Fe] ratio is not a perfect clock. Analysing low-α EMP stars, Kobayashi, Ishigaki et al. (2014) summarized various reasons that cause low α/Fe ratios: (1) SNe Ia, (2) less-massive core-collapse supernovae ($\lesssim 20 M_\odot$), which become more important with a low star formation rate (3) hypernovae, although the majority of hypernovae are expected to give normal [α/Fe] ratios (~ 0.4), and (4) pair-instability supernovae, which could be very important in the early universe. Therefore, low-α stars are not necessarily enriched by SNe Ia in a system with a long formation timescale. It is necessary to also use other elemental abundances (namely, Mn and neutron-capture elements) or isotopic ratios, with higher resolution ($> 40,000$) multi-object spectroscopy on 8m class telescopes (e.g., cancelled WFMOS or planned MSE).

Acknowledgements

I thank my collaborators, D. Yong, M. Franco, F. Vincenzo, P. Taylor, A. Karakas, M. Lugaro, N. Tominaga, S.-C. Leung, M. Ishigaki, and K. Nomoto, and V. Springel for providing Gadget-3. I acknowledge funding from the UK Science and Technology Facility Council through grant ST/M000958/1, ST/R000905/1, ST/V000632/1. The work was also funded by a Leverhulme Trust Research Project Grant on "Birth of Elements".

References

Arcones, A., Janka, H.-Th., & Scheck, L. 2007, *A&A*, 467, 1227
Asplund, M., Grevesse, N., Sauval, A. J., & Scott, P. 2009, *ARAA*, 47, 481
Buder, S., Sharma, S., Kos, J., et al. 2021, *MNRAS*, 506, 150
Burrows, A. & Vartanyan, D. 2021, *Nature*, 589, 29
Chiappini, C., Hirschi, R., Meynet, G., et al. 2006, *A&A*, 449, L27
Da Costa, G. S. et al. 2019, *MNRAS*, 489, 5900
Dalla Vecchia, C., Schaye, J., 2008, *MNRAS*, 387, 1431
Daveé, R., Anglés-Alcázar, D., Narayanan, D., et al. 2019, *MNRAS*, 486, 2827
Doherty, C. L., Gil-Pons, P., Lau, H. H. B., et al. 2014a, *MNRAS*, 437, 195
Doherty, C. L., Gil-Pons, P., Siess, L., Lattanzio, J. C., Lau, H. H. B., 2015, *MNRAS*, 446, 2599
Franco, M., Coppin, K. E. K., Geach, J. E., et al. 2021, *Nature Astronomy*, 5, 1240
Frischknecht, U., Hirschi, R., Pignatari, M., et al. 2016, *MNRAS*, 456, 1803
Haynes, C. J. & Kobayashi, C. 2019, *MNRAS*, 483, 5123

Janka, H.-T., 2012, *Annual Review of Nuclear and Particle Science*, 62, 407
Karakas, A. I. 2010, *MNRAS*, 403, 1413
Karakas, A. I., Lugaro, M., 2016, *ApJ*, 825, 26
Kemp, A. J., Karakas, A. I., Casey, A. R., et al. 2022, *MNRAS*, 509, 1175
Kobayashi, C., 2004, *MNRAS*, 347, 740
Kobayashi, C., 2005, *MNRAS*, 361, 1216
Kobayashi, C. 2014, *IAU Symposium*, 298, 167
Kobayashi, C. 2016, *IAU Symposium*, 317, 57
Kobayashi, C., Ishigaki, M. N., Tominaga, N., & Nomoto, K. 2014, *ApJ*, 5, L5
Kobayashi, C., Izutani, N., Karakas, A. I. et al, 2011a, *ApJ*, 739, L57
Kobayashi, C., Karakas, I. A., & Umeda, H. 2011b, *MNRAS*, 414, 3231
Kobayashi, C., Karakas, I. A., & Lugaro, M. 2020a, *ApJ*, 900, 179 (K20)
Kobayashi, C., Leung, S.-C. & Nomoto, K. 2020b, *ApJ*, 895, 138
Kobayashi, C. & Nakasato, N. 2011, *ApJ*, 729, 16
Kobayashi, C., Nomoto, K., & Hachisu, I. 2015, *ApJ*, Letter, 804, 24
Kobayashi, C., Springel, V, & White, S. D. M. 2007, *MNRAS*, 376, 1465
Kobayashi, C., Tsujimoto, T., Nomoto, K., Hachisu, I, & Kato, M. 1998, *ApJ*, 503, L155
Kobayashi, C., Umeda, H., Nomoto, K., et al. 2006, *ApJ*, 653, 1145 (K06)
Grichener, A., Kobayashi, C., & Soker, N. 2022, *ApJ*, Letter, in press
Grisoni, V., Matteucci, F., Romano, D., et al. 2019, *MNRAS*, 489, 3539
Hayden-Pawson, C., Curti, M., Maiolino, R., et al. 2022, *MNRAS*, submitted
Hopkins, P. F. et al. 2018, *MNRAS*, 477, 1578
Jönsson, H., Ryde, N., Harper, G. M., et al. 2014, *A&A*, 564, A122
Just, O., Bauswein, A., Ardevol Pulpillo, R., et al. 2015, *MNRAS*, 448, 541
Leung, S.-C. & Nomoto, K. 2018, *ApJ*, 861, 143
Leung, S.-C. & Nomoto, K. 2020, *ApJ*, 888, 80
Limongi, M., & Chieffi, A. 2018, *ApJS*, 237, 13
Mösta, P., Roberts, L. F., Halevi, G., et al. 2018, *ApJ*, 864, 171
Mura-Guzmán, A., Yong, D., Abate, C., et al. 2020, *MNRAS*, 498, 3549
Nishimura, N., Takiwaki, T., Thielemann, F. K. 2015, *ApJ*, 810, 109
Nomoto, K., Kobayashi, C., & Tominaga, N. 2013, *ARAA*, 51, 457
Prantzos, N., Abia, C., Limongi, M., Chieffi, A., & Cristallo, S. 2018, *MNRAS*, 476, 3432
Reichert, M., Obergaulinger, M., Eichler, M., Aloy, M. A. & Arcones 2021, *MNRAS*, 501, 5733
Romano, D., Matteucci, F., Zhang, Z.-Y., et al. 2019, *MNRAS*, 490, 2838
Smartt, S. J. 2009, *ARAA*, 47, 63
Sneden, C., Cowan, J. J., Kobayashi, C., et al. 2016, *ApJ*, .817, 53
Siegel, D. M., Barnes, J. & Metzger, B. D. *Nature*, 569, 241
Springel, V., Di Matteo, T., & Hernquist, L. 2005, *MNRAS*, 361, 776
Taylor, P. & Kobayashi, C. 2014, *MNRAS*, 442, 2751
Taylor, P. & Kobayashi, C. 2015a, *MNRAS*, 448, 1835
Taylor, P. & Kobayashi, C. 2015b, *MNRAS*, 452, L59
Taylor, P. & Kobayashi, C. 2016, *MNRAS*, 463, 2465
Taylor, P. & Kobayashi, C. 2017, *MNRAS*, 471, 3856
Vincenzo, F. & Kobayashi, C. 2018a, *A&A*, 610, L16
Vincenzo, F. & Kobayashi, C. 2018b, *MNRAS*, 478, 155
Vincenzo, F. & Kobayashi, C. 2020, *MNRAS*, 496, 80
Vincenzo, F., Miglio, A., Kobayashi, C., et al. 2019, *A&A*, 630, A125
Wallner, A., Froehlich, M. B., Hotchkis, M. A. C., et al. 2021, *Science*, 372, 742
Wanajo, S., 2013, *ApJ*, 770, L22
Wanajo, S., Janka, H.-T., Müller, B., 2013, *ApJ*, 767, L26
Wanajo, S., Sekiguchi, Y., Nishimura, N., et al. 2014, *ApJ*, 789, L39
Winteler, C., Käppeli, R., Perego, A., et al. 2012, *ApJ*, Letter, 750, 22
Woods, T. E., Agarwal, B., Bromm, V., et al. 2019, *PASA*, 36, 27
Yong, D., Da Costa, G. S., Bessell, M. S., et al. 2021, *MNRAS*, 507, 4102
Yong, D., Kobayashi, C., Da Costa, G. S., et al. 2021, *Nature*, 595, 223

The Origin of Outflows in Evolved Stars
Proceedings IAU Symposium No. 366, 2022
L. Decin, A. Zijlstra & C. Gielen, eds.
doi:10.1017/S1743921322001090

The interplay between mass-loss and binarity

Hugues Sana

Institute of Astrophysics, KU Leuven, Celestijnlaan 200D, 3001 Leuven, Belgium
email: hugues.sana@kuleuven.be

Abstract. Most stars with birth masses larger than that of our Sun belong to binary or higher order multiple systems. Similarly, most stars have stellar winds. Radiation pressure and multiplicity create outflows of material that remove mass from the primary star and inject it into the interstellar medium or transfer it to a companion. Both have strong impact on the subsequent evolution of the stars, yet they are often studied separately. In this short review, I will sketch part of the landscape of the interplay between stellar winds and binarity. I will present several examples where binarity shapes the stellar outflows, providing new opportunities to understand and measure mass loss properties. Stellar winds spectral signatures often help clearly identifying key stages of stellar evolution. The multiplicity properties of these stages then shed a new light onto evolutionary connections between the different categories of evolved stars.

1. Introduction

Stars with birth masses larger than about twice the mass of our Sun (spectral types OBA on the main sequence) are fundamental cosmic engines. These intermediate- and high-mass stars heat and enrich the interstellar medium, they drive the chemical evolution of galaxies and, at the high-mass end, their end-of-life explosions are bright enough that they can be seen throughout cosmic times. Given their decisive role in a wide range of astrophysical phenomena, it is of paramount importance to understand how these stars live and die.

The lifetimes of intermediate and massive stars spans three orders of magnitudes: from about 2 Gyr for a 2 $M_\odot$ star, to just over 2 Myr for stars with inital masses of 100 $M_\odot$ or more. While vastly different, these times scales are significantly shorter than the age of the universe. As a consequence, their full evolutionary cycle, from their formation in dense molecular clouds to their final stages where (most of) these stars shed their envelopes through strong stellar winds, can be directly observed, providing us with direct insight into their evolution. Yet, many uncertainties remain that affect predictions of stellar evolution models. Among these, the issues related to their internal structure, rotation and mixing properties, the strengths and structure of their outflows and their stellar multiplicity are some of the most problematic, preventing proper astrophysical understanding of stellar evolution.

In this paper, we start from the postulate that stellar outflows and multiplicity are not only ubiquitous phenomena among intermediate and high-mass stars, but that much is to be gained to consider them as just but two sides of the same coin: that of the physics of the objects that host them. This paper is organised as follows. In Sect. 2, I will present several examples where binarity shapes the stellar outflows, providing new opportunities to understand and measure mass loss properties. In Sect. 3, we will use the spectral signatures resulting from stellar outflows to help identifying key stages of stellar evolution. The multiplicity properties at each of these stages then shed a new light onto stellar evolution, possibly raising new questions on the nature and roles of different categories of evolved stars.

2. Stellar outflows shaped by binarity

In binary systems where each of the stars in the system has a strong stellar wind, typically O+O and WR+O binaries, the winds from the two stars collide at supersonic velocities, creating a wind collision region the shape of which is defined by the ram pressure equilibrium surface between the two winds (Usov (1992); Stevens et al. (1992)). For WR+O system, a simplified but useful 3D geometrical model of such collision zone is provided by the cone-shock model (Lührs (1997)) where the star with the weaker wind is carving a cavity in the wind of the primary star. Both theoretical considerations (e.g., Usov (1992)) and hydrodynamical simulations (e.g., Pittard (2009); Parkin & Gosset (2011)) however reveal that the reality is more complex: there are two shock fronts, one for the collision of each wind with the wind-interaction region, i.e., the denser and hotter downstream region in which the gas has been compressed and heated up by the aforementioned shocks. These simulations further reveal the presence of various instabilities (shear, Rayleigh-Taylor, Kelvin-Helmholtz, thin-shell, ...) that can dramatically modify the appearance and the properties of the collision (Pittard (2009); Madura et al. (2013)), from a relatively stable configuration in the adiabatic case (e.g., WR 22, Parkin & Gosset (2011), HD93403, Rauw et al. (2002) to a chaotic wind collision structure in the isothermal case (e.g., HD152248, Sana et al. (2004)). The main parameter delineating these two extremes depends on the speed at which the shocked gas cools down with respect to how fast it escapes the collision zone (Stevens et al. (1992)). This in turn strongly depends on the separation between the two stars, hence on the orbital period (P). In highly eccentric systems, the wind collision region can actually transition for an adiabatic to a radiative collision zone as the two stars near the periastron passage (Lamberts et al. (2012), Clementel et al. (2015a), Clementel et al. (2015b)).

Typical observational signatures of these wind-wind collision include thermal X-ray emission, as the shocked gas is heated up to temperatures of 10 million Kelvin, as well as synchrotron (in the radio domain) or inverse Compton (in the hard X-rays domain) emission resulting from the presence of populations of electrons that have been accelerated to relativistic speed in the vicinity of the shock fronts. The shocked regions have also been proposed as the locus of acceleration of cosmic rays. If the cooling in the collision region is quick enough, the ionised gas might recombine, creating measurable emission in the optical, infrared and UV recombination lines. In the following, I will outline a small selection of recent results illustrating the vast diversity of behaviour and observational signatures resulting from wind-wind collisions. The focus will be on colliding wind systems hosting a Wolf-Rayet star.

2.1. *Episodic dust producers*

In colliding wind systems harbouring a carbon-rich Wolf-Rayet star (WC) and an O star, the enhanced density in the wind-wind collision produces favourable conditions for the nucleation of dust. Combined with the Coriolis force due to the orbital motion, this gives rise to so-called pinwheel nebulae which are characterised by a dust plume following an Archimedian spiral pattern. Because the conditions are more easily reached at periastron passage, the dust production might be episodic only.

While the pinwheel nebulae have been know for over two decades and have been directly imaged by e.g., Tuthill et al. (1999), some recent works are shedding more light to their large range of properties (see also Fig. 1):

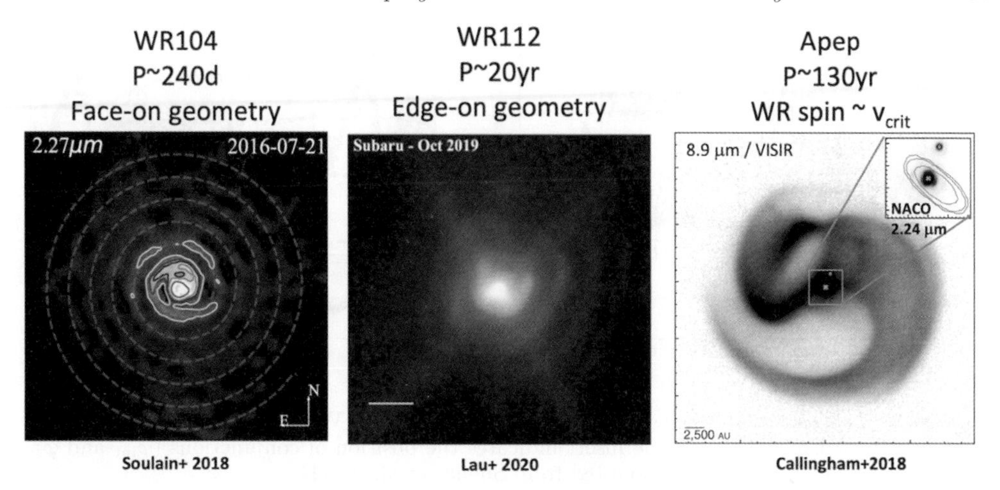

Figure 1. High-angular resolution images of the three WC dust producers WR 104, WR 112 and Apep, illustrating the wide diversity of morphologies and orbital periods. Figure adapted from Soulain et al. (2018) (left), Lau et al. (2020) (middle) and Callingham et al. (2019) (right).

- WR 104 ($P \sim 241$d, Tuthill et al. (2008)) is the prototypical pinwheel nebula where the spiral structure is clearly visible thanks to the face-on orientation of the system with respect to our line of sight. First imaged in the near-infrared thanks to sparse aperture masking (Tuthill et al. (1999)), the system has recently been observed with VLT/SPHERE and VISIR (Soulain et al. (2018)) providing a more complete view of its spiral structure produced over the last 20 years, constraining the dust to mass ratio in the spiral arms in the range of 1 to 10% and confirming that WR 104 is a hierarchical triple with the third companion at a projected distance of about 2600 au.
- WR 140 is an eccentric WC7+O5 system on an $P \sim 8$ yr orbit (Thomas et al. (2021)), and an episodic dust maker. It is one of the best characterised colliding wind binaries, with an almost continuously X-ray light curve illustrating the clear $1/D$ dependance of the X-ray thermal emission, with D the separation between the two stars. In the radio domain, time-resolved images of the synchrotron emission from the wind interaction region shows how its position and shape change as the O star plunge into the radiosphere of the WC star while nearing periastron (Dougherty et al. (2005)).
- WR 112 ($P \sim 20$ yr) offers very different though no less spectacular images than WR 104. This results from its edge-on geometry. Lau et al. (2020) recently constrained the dust mass-loss rate of the WC star to be about 3×10^{-6} M$_\odot$ yr^{-1}, making it one of the largest dust-producing pinwheel nebula known.
- Apep ($P \sim 130$ yr) has a very peculiar plume. Recent work by Callingham et al. (2019) indicates that the plume can only be reproduced by assuming that the WC object is (near-)critically rotating, making Apep a candidate gamma-ray-burst progenitor.

These few examples not only illustrate the large diversity of orbital periods involved (over two orders of magnitudes) but also how the structure and shape of the wind-wind collision allow astronomers to constrain the mass-loss rates of evolved massive stars (a key ingredient in stellar evolution) as well as some of the stellar properties that are needed to properly predict their end-of-life scenarios.

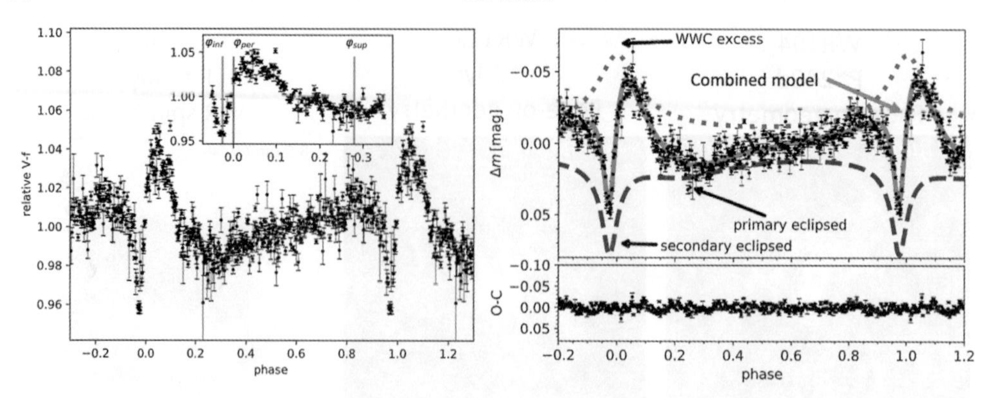

Figure 2. *Left panel:* OGLE light curve of the R144 WNh5+WNh6 binary system revealing a clear heartbeat-like signature. The insert indicates the position of conjunctions ($\varphi_{\rm inf}$ and $\varphi_{\rm sup}$) and of periastron ($\varphi_{\rm per}$). Figure adapted from Shenar et al. (2021).

2.2. *A heartbeat light curve produced by wind-wind collision*

While wind-wind collision (WWC) is known to contribute to the optical emission lines, whether in O+OB or in WR+O binary systems, its direct impact on the optical light-curve of a system is usually rather limited. Here we highlight new results from the R144 WR+WR binary in the Large Magellanic Cloud showing that this is not necessarily the case (Shenar et al. (2021)).

WR 144 is a hydrogen rich Wolf-Rayet object in the Tarantula nebula. It is also the brightest X-ray source in the region and, given its predicted bolometric luminosity, has been suspected for a long time to be a binary system. One had to wait the advent of the X-Shooter spectrograph at ESO/VLT to detect a clear binary signature, with significant radial velocity (RV) variations over time scales of months and the presence of two hydrogen rich WNh stars (Sana et al. (2013b)), suggesting that WR 144 is one of the most massive binary system known so far.

Using additional data from the Tarantula Massive Binary Monitoring program (Almeida et al. (2017)) from the FLAMES and XSHOOTER instruments at the VLT, Shenar et al. (2021) obtained the first orbital solution, with $P \sim 74$ d and a significant eccentricity of $e \approx 0.5$. Spectral disentangling allowed for a more precise spectral typing and atmosphere modelling, revealing a WNh5 and WNh6 components and measured mass-loss rates of the order of 4×10^{-5} M$_\odot$ yr^{-1}.

Given the properties of the system. it was clear that wind-wind collision was to be expected, and indeed, strong phase-locked variation of He and N lines was observed. More intriguing though, the OGLE light curve of R144 displayed the characteristic of heartbeat systems where a strong dip in the light curve over a very short period of time is followed by a sharp rise before returning to its average level (Fig. 2). Heartbeat systems are usually interpreted as tidal deformation near periastron passage in eccentric binaries and are typically lower mass stars (Thompson et al. (2012)).

A first attempt to model the heartbeat light curve with tidal deformation provided a good fit but required stellar radii that were about 10 times larger than expected for WNh stars. The second attempt turned out more successful. Indeed Shenar et al. (2021) showed that the light curve could be modelled by a combination of double wind eclipses and a phase-locked contribution of the order of a few tenths of magnitude from the emission lines to the optical magnitude (Fig. 2). The obtained model is very sensitive to inclination and the best fit yields $i = 60 \pm 2^{\rm o}$ and $\dot{M} \approx 4$ to 5×10^{-5} M$_\odot$ yr^{-1} (in excellent agreement with estimates from the spectroscopic analysis), resulting in absolute (dynamical) masses of 74 and 69 (± 4) M$_\odot$ for the WNh stars in R144. This makes R144 one of the most massive binaries known and one of the few for which direct mass measurements can be obtained thanks to a careful analysis of its wind-wind collision.

2.3. *A unified scheme to explain the geometry of the outflows of AGB stars*

Asymptotic giant branch stars (AGBs) are post-core-He-burning low and intermediate mass stars that are characterised by strong, chemically enriched winds and extended circumstellar enveloppes. Their strong winds strip the AGB stars of their enveloppes, leading them to the post-AGB and planetary nebula (PN) phases. In the latter phase, the hot PN central star emits copious amount of UV radiation that ionizes the ejecta, giving rise to beautiful planetary nebulae. Aside from their certain aesthetical qualities, one of the fascinating aspects of these planetary nebulae is the wide diversity of shapes encountered.

Similarly spectacular, ALMA observations are now able to reveal similar diversities of morphologies around AGB stars (Decin et al. (2020)), suggesting that a common physical mechanism is shaping the outflows of both AGBs ad PNe. The mass-loss rates of the AGB outflows can be well measured with modern ALMA data. Such mass-loss rates provide a reference clock with which, one can show, the AGB morphologies are correlated (Decin et al. (2020)). This key result suggests that there is an evolutionary connection between the different morphologies of AGB winds. A “young” AGB in a binary system will likely produce an equatorial density enhancement. As the system ages and the mass loss increases, the orbit will widen due to angular momentum loss leading first to a bipolar nebula, then a spiral-like structure.

This novel results is yet another example of the evolutionary diagnostic power that can be obtained by combining stellar outflow and orbital properties. We will come back to another such case for higher mass stars in Sect. 3.

2.4. *Wind-wind-collision and the habitability of Earth-like planets*

While this review focuses on intermediate and high-mass stars, the following example shows that wind-wind collision is actually of broader relevance. Johnstone et al. (2015) studied the impact of the wind-wind collision, between the winds of two solar-type stars separated by 0.5 au, on the habitability of a circumbinary earth-like planet. The authors show that, at radial distances of about 1 to 2 au, the wind-wind collision region has density and temperature enhancements of about a factor 3 and 5, respectively, compared to the material outside the wind-wind region. A circumbinary planet orbiting the system will be swept by the collision region multiple times during its orbit, so that it spends about a quarter of its orbital period within the density- and temperature-enhanced wind-wind collision region. The main effect is to compress the magnetosphere by about 20%. While this effect in principle remains limited for a fully formed earth-like planet, a “young” earth with a denser and more expanded atmosphere might actually be partially stripped by these effects.

3. Evolutionary connections in the upper HRD

In this second part of this paper, we will focus on evolutionary connections between various populations of evolved stars. We will focus exclusively on massive stars, that is stars massive enough to become red supergiants and to end their life in core-collapse events. These evolved populations of massive stars are mostly identified through atmospheric and wind-related diagnostics. Specifically, we will consider five broad categories: OB stars, red supergiants (RSGs), classical Wolf-Rayet stars (WR), luminous blue variables (LBVs) and Be stars.

In a single-star context, the evolution of massive stars is generally described by the Conti scenario (Conti (1976)), which makes evolutionary connections between different categories of massive stars. A simplified version of it is given below. A more detailed version of it can be found in, e.g., Lamers & Levesque (2017).

$$40M_\odot \lesssim M_{\rm ini} \lesssim 90M_\odot :$$
$$\text{early O/Of/WNh} \rightarrow LBV \rightarrow WN \rightarrow WC(\rightarrow WO),$$

$$25M_\odot \lesssim M_{\rm ini} \lesssim 40M_\odot :$$
$$\text{mid O} \rightarrow YSG \rightarrow RSG \rightarrow WN \rightarrow WC,$$

$$8M_\odot \lesssim M_{\rm ini} \lesssim 25M_\odot :$$
$$\text{late O/early B} \rightarrow YSG \rightarrow RSG(\rightarrow blue\ loop/YSG? \rightarrow RSG).$$

The main ingredient that defines a given evolutionary pathway is the star's initial mass. The amount of mass lost during the evolved phases is also important as it controls whether a post-hydrogen burning objects (RSG, LBV) will be stripped of its envelope, hence returning to the hotter part of the Hertszprung-Russell diagram (HRD).

While the Conti scenario looks at the connections between different evolutionary stages through a single-star, wind-focused angle, one can repeat the same exercise from different perspectives. For example, Smith & Tombleson (2015), Humphreys et al. (2016) and Smith (2019) used the relative isolation of stars in each category. Here, we propose to use the multiplicity properties of these different categories of evolved massive stars to investigate their evolutionary connections.

3.1. *main-sequence OB stars*

In this review, we will use the "OB star" denomination to broadly designate hot main-sequence massive stars (Walborn et al. (2014)), i.e., hydrogen-burning stars with an initial mass larger than about 8 $M_\odot$. This category thus encompasses most O-type stars, B supergiants, and B dwarfs with spectral type earlier than B3 on the zero-age main-sequence. Without loss of generality, we will include the very WNh massive stars in this category despite the fact that their spectral type is actually that of an hydrogen-rich Wolf-Rayet (WNh, de Koter et al. (1997)). From an evolutionarily perspective though, these WNh stars are indeed very massive hydrogen-burning stars and thus fit our definition of "OB stars".

In the Milky Way, the multiplicity fraction of stars is know to increase with stellar mass (Duchêne & Kraus (2013), Moe & Di Stefano (2017)), from slightly less that 50% for solar type stars, to >60% for early B-stars (Abt et al. (1990); Dunstall et al. (2015); Banyard et al. (2022)) and $>70\%$ for O-type stars (Sana et al. (2012)). These lower limits on the multiplicity fractions increase when considering wider binaries, observed e.g. through high-angular resolution imaging. For example, Sana et al. (2014) derived 90% for all O-type stars and 100% for O dwarfs, a result confirmed by more recent work on the Orion and M17 star forming region by Grellmann et al. (2013), Karl et al. (2018) and Bordier et al. (2022), respectively.

To decide whether a binary system will interact and when this interaction will happen (e.g., within or beyond the main-sequence phase), the most important property is probably its orbital period. Of interest, the derived period distributions of O and early-B stars derived from a variety of environments are strikingly similar (Sana et al. (2013a); Villaseñor et al. (2021); Banyard et al. (2022)), pointing towards a common, metallicity independent mechanism responsible to the pairing of massive binaries.

While these period distributions will be used as initial conditions in the following, a word of caution is needed. It is indeed possible that these distributions do not directly result from the star formation process itself. They may have been altered by early dynamical processing, through e.g., migration as suggested by recent results on the dearth of short-period systems in M17, one of the youngest massive star forming regions studied in depth (Ramírez-Tannus et al. (2017); Sana et al. (2017)). The early dynamical processing of the orbital properties is further supported by an observed correlation of the dynamical dispersion as a function of cluster age as put forward by Ramírez-Tannus et al. (2021). The observed orbital distributions of OB stars might not be the exact end product of star formation; yet, they probably provide good enough starting points to compute their subsequent evolution and future fate. One major uncertainty, perhaps, remain the role of triple and higher order multiples and whether dynamical effects might impact the frequency, timing, and nature of binary interactions.

3.2. *Red supergiants*

Red Supergiants (RSGs) are typically cooler than 5000 K and characterised by a strong wind outflow, the launching mechanism of which remains poorly understood and may require a combination of various processes, such as pulsation, magnetic pressure and radiation pressure on molecular lines (e.g., Kee et al. (2021)). Red supergiants represent the He-burning phase of stars with initial masses typically between 8 and 25 $M_{\odot}$. RSGs may directly collapse and explode as supernovae or loose their envelope and turn into classical Wolf-Rayet stars.

The RSGs multiplicity is a much more difficult affair than that of their unevolved counterparts. This results not only from their fewer numbers, making statistically meaningful results difficult to reach, but also because their much larger radii. As a consequence, the orbital periods are the order of years to decades, so that corresponding radial velocity variations become of a similar magnitude as the stochastic noise expected from the large convection cells at the surface of these stars.

Figure 3 summarises some of the recent results, mostly obtained in the Magellanic Clouds. These studies consistently obtained observed multiplicity fractions of the order of 10 to 20%, and bias-corrected fractions of 20 to 30% (e.g., Patrick et al. (2019), Patrick et al. (2020), Neugent et al. (2020), Dorda & Patrick (2021)). These fractions are significant smaller than the > 60 to 70% claimed for their progenitors, the main-sequence OB stars with initial mass lower than $\sim 25 - 40$ $M_{\odot}$.

The lower RSG binary fraction can be understood, at least qualitatively, by invoking binary interaction. Indeed, most binaries with orbital separation smaller than the typical size of a RSG will actually interact either during the main sequence (for $P < 6$ d, approx.) or when crossing the Hertzsprung gap ($P < 2\,000$ d, approx). A simple integration of the (bias-corrected) initial period distributions presented in Fig. 5 suggest that about 2/3 of the systems will interact. Depending on the mass ratio and nature of the interaction, we would face either a merging (resulting in a single star), a stable mass-transfer (stripping the primary of (most of) its envelope, thus preventing the primary star to reach the RSG phase) or a common envelope ejection (also resulting in a non-RSG product). The net result is to effectively remove the close binary systems from the RSG sample, so that the multiplicity fraction of RSGs is indeed expected to be a factor 3 to 5 lower than that of their main-sequence counterparts.

While this exercise shows that there is no fundamental tension between the multiplicity fraction of OB main-sequence stars and RSGs, there is however a strong corollary: a significant number of post-interaction products produced by the interaction of binaries born as OB systems. The number of post-interaction systems known, while growing, remains

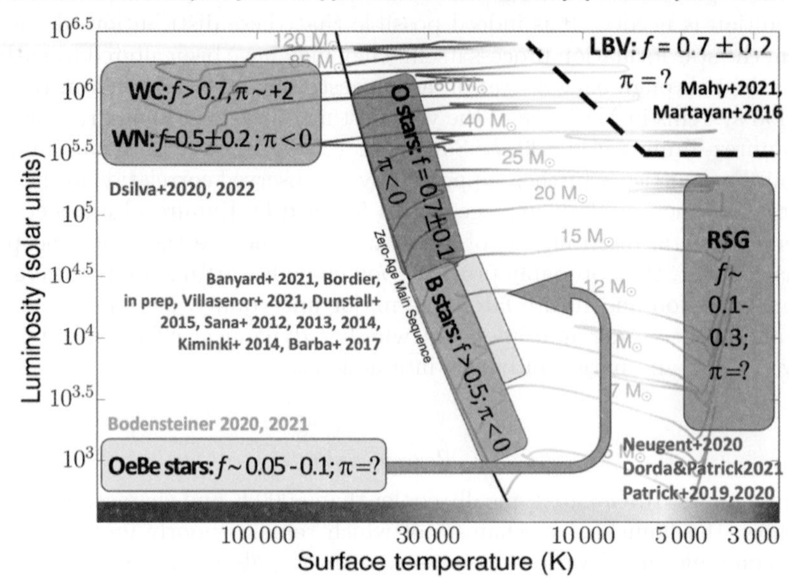

Figure 3. Overview of our current knowledge of the multiplicity in the upper Hertszprung-Russell diagram (HRD) based on spectroscopic results. f is the multiplicity fraction, π is the index of the period distribution ($f_{\log P} \propto (\log P)^{\pi}$). The multiplicity of OB stars nears $f = 1$ once the interferometric data are included. Brott et al. (2011) evolutionary tracks have been overlaid.

consuspiciously small. In the next section we consider the proposition that classical Be stars are an important population of interaction products.

3.3. *Classical Be stars*

Be stars are B-type stars with hydrogen line emission in their spectra. While there are various types of Be stars, we will focus here on classical Be stars, i.e., Be stars whose emission is believed to be associated with a decretion "disk". The presence of a disk is expected to result from an equatorial density enhancement in (near)critical, rapidly rotating stars where the centrifugal forces at the equator become equal or larger than the gravitational attraction. There are three main scenarios to explain the origin of the classical Be phenomenon, all linked to how Be stars acquire their (near-)critical spins: star formation (high spin at birth), single-star evolution (angular momentum transfer from the core to the envelope when the star nears the terminal age main sequence (TAMS)), angular momentum accretion due to mass-transfer in a binary system. In the following, we will only consider the latter two.

One of the expected consequences of stable Roche lobe overflow is the transfer of angular momentum from the initially more massive donor star to the initially least massive accreting star. It has been shown that even a small accretion efficiency would quickly spin up the accretor to critical rotation velocities. Candidate post-interaction systems could thus host a main-sequence rapidly rotating star and classical Be stars offer a naturally match to this description.

Let us now investigate the predictions for the multiplicity of Be stars depending on the formation mechanism:

Scenario 1 - Binary Star Channel: producing a Be star by mass transfer and accretion will strip the primary star from (most of) its envelope, resulting in a so-called stripped-star, i.e. an (almost) naked He core. In general, these are not very luminous and relatively low mass objects, so they do not induce large RV variation of their Be companion. Besides, obtaining accurate RV measurements of Be stars is notoriously difficult so that chances are that Be+He systems would likely remain undetected and appear as presumably single Be stars in RV surveys.

Scenario 2 - Single Star Channel: At least half the early B-type stars are found in B+BA spectroscopic binary systems. If the Be star phenomenon is produced by angular momentum transfer from core to envelope when the star is nearing the TAMS, then this is expected to be a widespread phenomenon affecting all B-type binaries with period larger than about 10 days. The latter requirement on the orbital period ensures that the orbital separation is large enough so that no interaction happens during the main-sequence lifetime of the primary. This is indeed needed so that the B-type primary has the opportunity to become a Be star via the single-star channel before any interaction occurs. Within periods ranging from 10 days to a few 100s of days, these Be+BA binaries would appear as spectroscopic binaries that are reasonably easy to spot either through double line profiles or through large RV variations of the Be star (orbital velocity $v_{\rm orb} > 20$ km s^{-1}).

Qualitatively, the predictions of this discussion is counter-intuitive: in the single star channel for the formation of Be stars, one expects a fair fraction of them ($>25\%$ given some bias estimates, Bodensteiner et al. (2020)) of them to be detected as spectroscopic binaries with a main-sequence, unevolved lower mass companion. Alternatively, if the binary channel is responsible for the formation of Be stars, one should observe Be stars as single stars or in binaries with an evolved (stripped) companion.

In that context, Bodensteiner et al. (2020) performed a literature study of about 300 Be stars with spectral-type earlier than B1.5 and periods shorter than 5000 days. They noted that 91% of the sample is reported to be single, 5% have an evolved companion and 4% have a companion of an unidentified nature, i.e. they could be evolved companions or main-sequence stars. Even if all these companions are main-sequence companions, this is significantly lower than the 25% expected from the initial multiplicity properties of B-type stars and strongly suggests that the binary channel is a dominant channel to produce classical Be stars.

3.4. *Luminous Blue Variables*

Luminous Blue Variables (LBVs) are evolved massive stars that are observationally described by their name: luminous, blue and variable. They are found in the upper right part of the HRD. The exact nature of LBVs and their evolutionary connection with other categories of stars remain debated. In the main stream view, they are believed to be the evolved counterpart of very massive stars that have crossed the Hertzsprung gap after the end of hydrogen burning. By doing so, they have encountered the Humphreys-Davidson limit, a limit in the HRD where the radiative pressure at the surface becomes of the same order of magnitude as the gravitational attraction. This situation is expected to be unstable, hence the expected variabilities, and possibly leads to massive outbursts.

Using interferometry of a sample of LBV and LBV candidates, Mahy et al. (2022) constrained the LBV multiplicity fractions of about 60% to 80%, a value compatible with that derived from spectroscopy alone (Fig. 4). Such a high binary fraction disfavours a scenario in which LBVs are predominantly runaway stars that have gained mass in a previous binary interaction. Mahy et al. (2022) thus suggest that LBV should come from one of the three following channels: (i) single-star evolution in wide binary, i.e. wide

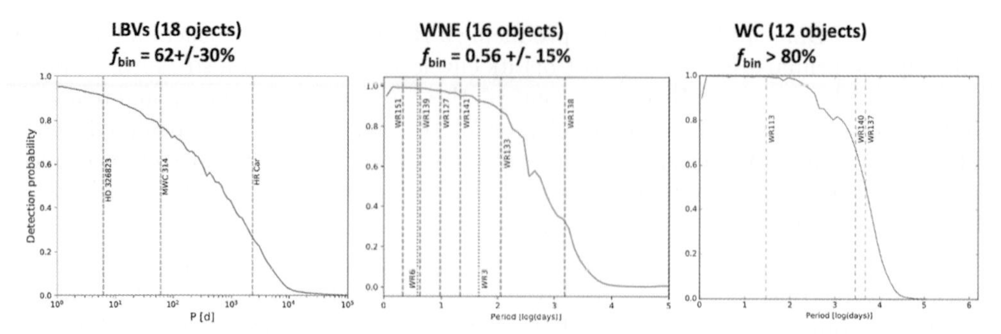

Figure 4. Detection probability curves of the spectroscopic campaigns of Mahy et al. (2022) (LBV, left), Dsilva et al. (2022) (WNE, center) and Dsilva et al. (2020) (WC, right). The location of the objects with known orbital periods is indicated by dashed lines. Sample sizes and bias-corrected binary fractions are indicated on top of their respective plot.

enough to not interact; (ii) the result of a merger in an initially triple system where one does not detect the outer companion, or (iii) the descendants of short-period binaries that have widen a lot due to non-conservative mass transfer. While the multiplicity fraction of LBVs is similar to that of OB stars, one has to note that most of the detected companions around LBVs are on much wider orbits given the larger radii of LBV stars.

3.5. *classical Wolf-Rayet stars*

Classical Wolf-Rayet stars (cWRs) are (post)-Helium burning massive stars that have been stripped of their hydrogen envelope, revealing their inner core that has been enriched in nucleosynthesis products. Depending on the chemical compositions of their outer atmosphere, one distinguishes three main categories of cWRs: the nitrogen-, carbon-, and oxygen-rich classical Wolf-Rayet (WNs, WCs, and WOs, respectively). Because of their luminosities, temperatures and small sizes, the radiation pressure on the atomic lines in the far-ultraviolet domain drives a very strong, optically thick stellar wind. Observationally, the wind signatures dominate their spectral appearance, with strong and often broad emission lines of He, C, N and O.

In single stars, the stripping of the envelope is expected to be due to stellar winds, successively revealing deeper and deeper layers, offering an evolutionary progression from WN to WC to, in a small number of cases, WO. In the Milky Way, one estimates a minimum initial mass of 20 to 30 $M_{\odot}$ (Shenar et al. (2020a)) for a single star to be able to strip itself to the cWR stage through stellar winds. In a close binary system, stripping might also occur through mass transfer once the initially more massive star fills its Roche lobe. In principle, binary interaction decreases the minimum initial mass needed to produce a Wolf-Rayet object, possibly contributing to populate a lower luminosity region in the HRD than that occupied by wind-stripped cWRs (Shenar et al. (2020a)).

Following the Conti scenario, early-type Wolf-Rayet stars of the Nitrogen sequence (WNE) will evolve into WC stars if they shed enough mass before exploding as SNe. Dsilva et al. (2020) investigated the multiplicity properties of two similar sized samples of WNE and WC stars. They found that the bias-corrected binary fractions of WC and WNE are respectively $> 80\%$ and $56 \pm 15\%$ respectively. While error bars are large, the results remain intriguing. On the one hand, how can WNE stars evolved into WC stars if their binary fraction is smaller? A possible explanation is that the RV sensitivity of WNE observations is limited by their wind-induced line profile variations, preventing a reliable probing at large orbital periods. On the other hand, even the distributions of the orbital periods are puzzling. Looking at the Galactic Wolf-Rayet Star catalog

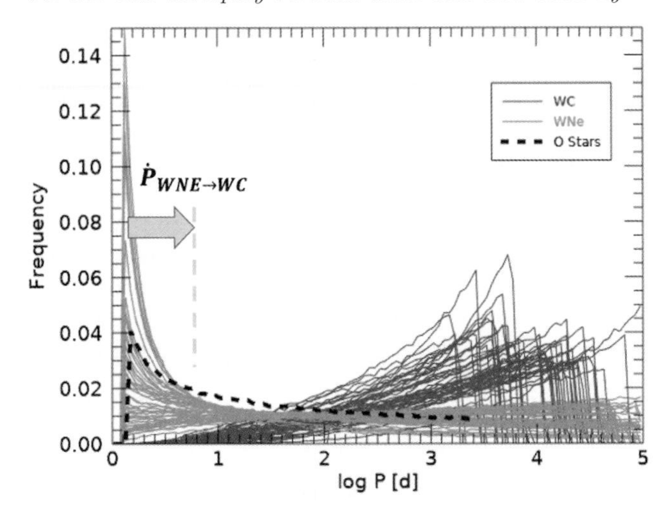

Figure 5. Preliminary measured distributions of the WNE (red) and WC (blue) compared to that of O-type stars (black). Each curve follows $f_{\log P} \propto (\log P)^{\pi}$ and is characterised by a maximum cutoff period $P_{\max}$. The WNE and WC curves provide 50 realisations drawn from the π and $P_{\max}$ posterior distributions of Dsilva et al. (2022). The yellow arrow indicates the expected widening of the orbits from the WNE to the WC phases resulting from angular momentum loss due to stellar winds. See Dsilva et al. (2022) for final results.

(Rosslowe & Crowther (2015)), Dsilva et al. (2022) reported that most of the know WNE binaries have period of the order of 100 days or less. Most of the known WC binaries have periods of 1000 days or more (Dsilva et al. (2020)). An initial attempt to estimate the true orbital period distribution of each sample from Dsilva et al. is displayed in Fig. 5. We refer to Dsilva et al. (2022) for updated results, but the conclusions remain. WNE and WC stars seem to have different period distributions, with the first one being heavily tweaked towards short period systems and the latter is flatter and favours long-period systems. Mass losses from the WNE to the WC phase will inevitably take away angular momentum and broaden the orbit, but the impact on the period values is expected to be limited to about a factor of few, i.e., insufficient to reconcile the WNE and WC period distributions (Fig. 5). These initial results require further investigations but illustrate the diagnostic power of the information contained in the multiplicity properties of these evolved populations.

4. Conclusions

Stellar winds and binarity are ubiquitous phenomena among stars heavier than our Sun. Taken separately, they are both responsible for interesting physical processes, e.g., wind instabilities, clumping or mass transfer. Both phenomena have strong impact on the mass of the stars, it is therefore not surprising that both are critical ingredients to properly compute the evolution of stars.

In this short review, we outlined two avenues that jointly consider stellar winds and binarity. On the one hand, we considered stars with strong outflows in binary systems, creating new physical processes. Their studies offer new diagnostics for stellar winds (e.g. enabling new ways to measure mass-loss rates) but also for the binary systems and their components. In the second avenue, we focused on the massive stars and discussed evolutionary connections between various categories of young and evolved massive stars in the light of their multiplicity properties. We show that the multiplicity properties of RSGs are compatible to first order with that of OB main-sequence stars under the caveat that binary interaction strongly impact most binaries with separation smaller than the typical RSG radius. This implies that a large fraction of systems should interact before any if the stars reach the RSG phase, suggesting the presence of a significant population of binary products. We report on new results by Bodensteiner et al. (2020) suggesting that Be stars could potentially offer such a population, by showing that binary interaction formation channel for Be star matches the lack of main-sequence + Be binaries.

We also looked at the properties of the evolved counterpart of the most massive stars, WRs and LBVs. We reported on recent results on LBVs and WRs which, though based on small samples, raised sufficiently intriguing results to deserve future investigations. While more data are certainly needed to offer full scale conclusions, this short review has shown that much insight is to be gained by considering both outflows and multiplicity as complementary diagnostics tools to better understand stellar evolution in general.

References

Abt, H. A., Gomez, A. E., & Levy, S. G. 1990, The Astrophysical Journal Supplement Series, 74, 551

Almeida, L. A., Sana, H., Taylor, W., et al. 2017, Astronomy & Astrophysics, 598, A84

Banyard, G., Sana, H., Mahy, L., et al. 2022, Astronomy & Astrophysics, 658, A69

Bodensteiner, J., Shenar, T., & Sana, H. 2020, Astronomy & Astrophysics, 641, A42

Bordier, E., Frost, A. J., Sana, H., et al. 2022, Astronomy & Astrophysics, in press (arXiv:2203.05036)

Brott, I., de Mink, S. E., Cantiello, M., et al. 2011, Astronomy & Astrophysics, 513, A115

Callingham, J. R., Tuthill, P. G., Pope, B. J. S., et al. 2019, Nature Astronomy, 3, 82

Clementel, N., Madura, T. I., Kruip, C. J. H., Paardekooper, J. P., & Gull, T. R. 2015b, Monthly Notices of the Royal Astronomical Society, 447, 2445

Clementel, N., Madura, T. I., Kruip, C. J. H., & Paardekooper, J. P. 2015a, Monthly Notices of the Royal Astronomical Society, 450, 1388

Conti, P. 1976, Mem. Soc. Roy. Sciences Liège, IX, 193

de Koter, A., Heap, S. R., & Hubeny, I. 1997, The Astrophysical Journal, 477, 792

Decin, L., Montargès, M., Richards, A. M. S., et al. 2020, Science, 369, 1497

Dorda, R. & Patrick, L. R. 2021, Monthly Notices of the Royal Astronomical Society, 502, 4890

Dougherty, S. M., Beasley, A. J., Claussen, M. J., Zauderer, B. A., & Bolingbroke, N. J. 2005, The Astrophysical Journal, 623, 447

Dsilva, K., Shenar, T., Sana, H., & Marchant, P. 2020, Astronomy & Astrophysics, 641, A26

Dsilva, K., Shenar, T., Sana, H., & Marchant, P. 2022, Astronomy & Astrophysics, in press

Duchêne, G. & Kraus, A. 2013, Ann. R. Astron. Astroph., 51, 269

Dunstall, P. R., Dufton, P. L., Sana, H., et al. 2015, Astronomy & Astrophysics, 580, A93

Grellmann, R., Preibisch, T., Ratzka, T., et al. 2013, Astronomy & Astrophysics, 550, A82

Humphreys, R. M., Weis, K., Davidson, K., & Gordon, M. S. 2016, The Astrophysical Journal, 825, 64

Johnstone, C. P., Zhilkin, A., Pilat-Lohinger, E., et al. 2015, Astronomy & Astrophysics, Volume 577, A122

Karl, M., Pfuhl, O., Eisenhauer, F., et al. 2018, Astronomy & Astrophysics,strik 620, A116

Kee, N. D., Sundqvist, J. O., Decin, L., de Koter, A., & Sana, H. 2021, Astronomy & Astrophysics, 646, A180

Lamberts, A., Dubus, G., Lesur, G., & Fromang, S. 2012, Astronomy & Astrophysics, 546, A60

Lamers, H. & Levesque, E. 2017, Understanding Stellar Evolution (IOP Publishing Ltd)

Lau, R. M., Hankins, M. J., Han, Y., et al. 2020, The Astrophysical Journal, 900, 190

Lührs, S. 1997, Publ. Astron. So. Pac., 109, 504

Madura, T. I., Gull, T. R., Okazaki, A. T., et al. 2013, Monthly Notices of the Royal Astronomical Society, 436, 3820

Mahy, L., Lanthermann, C., Hutsemékers, D., et al. 2022, Astronomy & Astrophysics, 657, A4

Moe, M. & Di Stefano, R. 2017, The Astrophysical Journal Supplement Series, 230, 15

Neugent, K. F., Levesque, E. M., Massey, P., Morrell, N. I., & Drout, M. R. 2020, The Astrophysical Journal, 900, 118
Parkin, E. R. & Gosset, E. 2011, Astronomy and Astrophysics, 530, A119
Patrick, L. R., Lennon, D. J., Britavskiy, N., et al. 2019, Astronomy & Astrophysics, 624, A129
Patrick, L. R., Lennon, D. J., Evans, C. J., et al. 2020, Astronomy & Astrophysics, 635, A29
Pittard, J. M. 2009, Monthly Notices of the Royal Astronomical Society, 396, 1743
Ramírez-Tannus, M. C., Backs, F., de Koter, A., et al. 2021, Astronomy & Astrophysics, 645, L10
Ramírez-Tannus, M. C., Kaper, L., de Koter, A., et al. 2017, Astronomy & Astrophysics, 604, A78
Rauw, G., Vreux, J.-M., Stevens, I. R., et al. 2002, Astronomy & Astrophysics, 388, 552
Rosslowe, C. K. & Crowther, P. A. 2015, Monthly Notices of the Royal Astronomical Society, 447, 2322
Sana, H., Stevens, I. R., Gosset, E., Rauw, G., & Vreux, J.-M. 2004, Monthly Notices of the Royal Astronomical Society, 350, 809
Sana, H., de Mink, S. E., de Koter, A., et al. 2012, Science, 337, 444
Sana, H., de Koter, A., de Mink, S. E., et al. 2013, Astronomy & Astrophysics, 550, A107
Sana, H., van Boeckel, T., Tramper, F., et al. 2013, Monthly Notices of the Royal Astronomical Society, 432, L26
Sana, H., Le Bouquin, J.-B., Lacour, S., et al. 2014, The Astrophysical Journal Supplement Series, 215, 15
Sana, H., Ramírez-Tannus, M. C., de Koter, A., et al. 2017, Astronomy & Astrophysics, 599, L9
Shenar, T., Gilkis, A., Vink, J. S., Sana, H., & Sander, A. A. C. 2020a, Astronomy & Astrophysics, 634, A79
Shenar, T., Sablowski, D. P., Hainich, R., et al. 2020b, Astronomy & Astrophysics, 641, C2
Shenar, T., Sana, H., Marchant, P., et al. 2021, Astronomy & Astrophysics, 650, A147
Smith, N. 2019, Monthly Notices of the Royal Astronomical Society, 489, 4378
Smith, N. & Tombleson, R. 2015, Monthly Notices of the Royal Astronomical Society, 447, 598
Soulain, A., Millour, F., Lopez, B., et al. 2018, Astronomy & Astrophysics, 618, A108
Stevens, I. R., Blondin, J. M., & Pollock, A. M. T. 1992, The Astrophysical Journal, 386, 265
Thomas, J. D., Richardson, N. D., Eldridge, J. J., et al. 2021, Monthly Notices of the Royal Astronomical Society, 504, 5221
Thompson, S. E., Everett, M., Mullally, F., et al. 2012, The Astrophysical Journal, 753, 86
Tuthill, P. G., Monnier, J. D., & Danchi, W. C. 1999, Nature, 398, 487
Tuthill, P. G., Monnier, J. D., Lawrance, N., et al. 2008, The Astrophysical Journal, 675, 698
Usov, V. V. 1992, The Astrophysical Journal, 389, 635
Villaseñor, J. I., Taylor, W. D., Evans, C. J, et al. 2021, Monthly Notices of the Royal Astronomical Society, 507, 5348
Walborn, N. R., Sana, H., Simón-Díaz, S., et al. 2014, Astronomy & Astrophysics, 564, A40

The Origin of Outflows in Evolved Stars
Proceedings IAU Symposium No. 366, 2022
L. Decin, A. Zijlstra & C. Gielen, eds.
doi:10.1017/S1743921322000631

The Link between Hot and Cool Outflows

Jorick S. Vink, A.A.C. Sander, E.R. Higgins and G.N. Sabhahit

Armagh Observatory and Planetarium, College Hill, BT61 9DG, Armagh, Northern Ireland
email: jorick.vink@armagh.ac.uk

Abstract. The link between hot and cool stellar outflows is shown to be critical for correctly predicting the masses of the most massive black holes (BHs) below the so-called pair-instability supernova (PISN) mass gap. Gravitational Wave (GW) event 190521 allegedly hosted an "impossibly" heavy BH of 85 $M_\odot$. Here we show how our increased knowledge of both metallicity Z *and* temperature dependent mass loss is critical for our evolutionary scenario of a low-Z blue supergiant (BSG) progenitor of an initially approx 100 $M_\odot$ star to work. We show using MESA stellar evolution modelling experiments that as long as we can keep such stars above 8000 K such low-Z BSGs can avoid strong winds, and keep a very large envelope mass intact before core collapse. This naturally leads to the Cosmic Time dependent maximum BH function below the PISN gap.

Keywords. Winds, mass loss, black holes, massive stars, stellar evolution

---

1. Introduction

Accurate mass-loss rates – as a function of effective temperature $\dot{M} = f(T_{\rm eff})$ – are needed for making reliable predictions for the evolution of the most massive stars, including the black hole (BH) mass function with respect to metallicity Z (see Sander et al. these proceedings).

Over the last five years gravitational wave (GW) observatories have shown the existence of very heavy black holes. The current record holder is the primary object in the GW event 190521 with 85 $M_\odot$. Due to the fact that this BH mass is almost twice as large as the generally accepted lower boundary of the pair-instability (PI) supernova (SN) mass gap at approximately 50 $M_\odot$ (Farmer et al. 2019; Woosley & Heger 2021), the GW event discoverers argued the 85$M_\odot$ BH is most likely a second generation BH, as it would be *impossible* for a progenitor star to have directly collapsed into a BH within the PISN mass gap spanning 50 and 130 $M_\odot$ (Abbott et al. 2020).

In this contribution we show that this conclusion could be premature, as we have constructed a robust blue supergiant (BSG) scenario for the collapse of a very massive star (VMS) of order 100 $M_\odot$ at low Z (Vink et al. 2021). The key physiscs involves both the Z-dependence and the effective temperature dependence of the mass-loss rate of evolved supergiants.

2. Overview of hot and cool mass-loss rates

When a massive star burns hydrogen (H) in the core it traditionally evolves from the hot blue side to the cool red part of the stellar Hertzsprung-Russell diagram (HRD). When this takes place at approximately constant luminosity L the key physical parameters are (i) the amount of mixing by processes such as core overshooting, as these set the duration of the wind mass-loss phase during H-burning and the total mass being

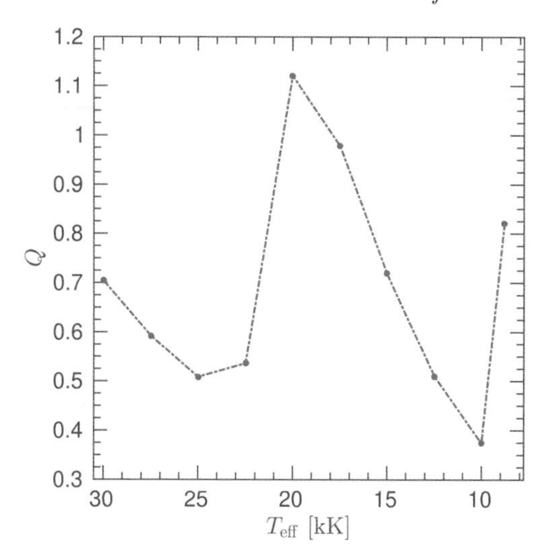

Figure 1. The first and second bi-stability jump in terms of the global wind parameter Q (see Petrov et al. 2016). Note that the exact location of the second jump is below the lower temperature boundary, i.e. with $T_{\rm eff}$ lower than 8800 K.

lost on the main sequence, as well as (ii) the absolute rate of mass-loss (dependent on the host galaxy Z) and (iii) how this mass loss varies from the hot to the cool side of the HRD. For hot stars above 10 kK the winds are driven by gas opacity (see Vink 2022 for a recent review) and while the absolute mass-loss values are still under debate (e.g. Björklund et al. 2021) the implication that mass-loss rates of hot-star winds are Z-dependent is undisputed. The exact Z dependence still needs to be established however (Vink & Sander 2021).

When $T_{\rm eff}$ drops during stellar evolution – starting from approx. 40 kK – the mass-loss rate is first expected to drop (see Fig. 1). The reason for this behaviour is that the line acceleration is set by the product of the stellar flux and the opacity. When $T_{\rm eff}$ drops, the stellar flux gradually moves from the ultraviolet (UV) part of the spectral energy distribution (SED) to the optical, while the opacity is still predominately 'left behind' in the UV part of the SED. In other words, there is a growing mismatch between the flux and the opacity, implying the flux-weighted opacity drops, and so does $\dot{M}$ (Vink et al. 1999).

This situation changes abruptly when the dominant line driving element iron (Fe) recombines, causing a bi-stability jump (BSJ) in the wind parameters. The first recombination is that from Fe IV to III at approx. 21 kK, the second recombination from Fe III to II takes place below 8800 K (Vink et al. 1999; Petrov et al. 2016). The exact location of this second BSJ has not yet been determined, as the current generation of sophisticated co-moving frame (CMF) radiative transfer model atmospheres has not yet been able to converge below the recombination of H at approx 8000 K. This uncertainty of mass-loss in the yellow supergiant / hypergiant phase is of key relevance for setting the Humphreys-Davidson (HD) limit (Gilkis et al. 2021; Sabhahit et al. 2021) and YSG mass-loss should therefore play an important role empirically (e.g. Koumpia et al. 2020; Oudmaijer & Koumpia these proceedings). Accurate mass-loss rates of $T_{\rm eff}$ dependent $\dot{M}$ are also critical for constructing the next generation of hydrodynamical stellar evolution models for both luminous blue variable and YSG phases (Grassitelli et al. 2021).

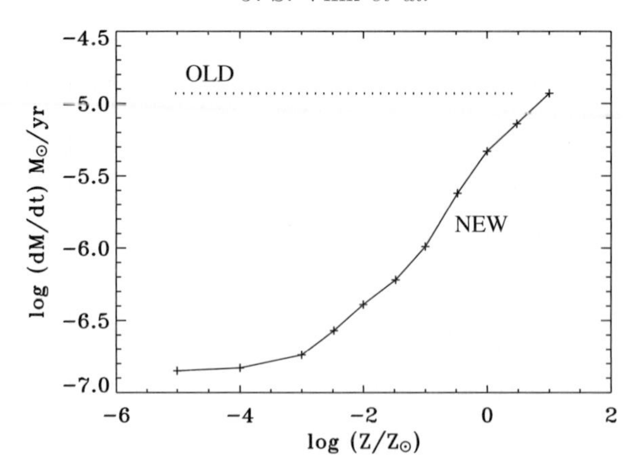

Figure 2. The Z dependence of WR stars. In stellar models before 2005 the mass-loss rate of WR stars was assumed to be independent of the host galaxy Z ("OLD") as the high abundance of self-enriched elements such as carbon (C) was thought to be dominant. Vink & de Koter (2005) showed that despite the huge amount of C in WR atmospheres the winds are nonetheless predominately driven by Fe and thus strongly Fe-dependent, indicated by "NEW". See Sander et al. these proceedings for new mass-loss predictions.

3. Current mass-loss recipes in use - and the link to wind physics

One of the most used mass-loss recipes currently in use in massive star models is the "Dutch" wind loss recipe in MESA Paxton et al. (2013). In this collection of mass-loss prescriptions, massive stars undergo modest mass loss on the main-sequence and enhanced mass loss below the first BSJ according to Vink et al. (1999, 2001). The second BSJ is not directly covered in the Dutch recipe, but it follows a similar approach as Brott et al. (2011) where the second BSJ is *in*directly included by switching from the Vink et al. theoretical recipe to the empirical cool star recipe of de Jager et al. (1988) at approx. 10 kK. This prescription yields relatively large mass-loss rates, although the physics of cool red supergiant (RSG) winds is still under debate, and this also means that whether or not RSG winds have a Z-dependence is presently unclear.

For stars evolving back to the hotter part of the HRD above 10 kK and with enriched atmospheres with a helium mass fraction Y larger than approx. 0.6, the models generally assume the total Z – including all elements heavier than He – empirical Wolf-Rayet (WR) recipe of Nugis & Lamers (2000). Note that a scaling with this 'total' Z is unlikely to be physically correct, as Vink & de Koter (2005) showed the host galaxy Fe to be the main wind driver despite the larger abundances of self-enriched elements (see Fig. 2).

In order to account for Fe-dependent winds as well as the knowledge that the second BSJ is located below 10 kK and even below 8800 K, Vink et al. (2021) and Sabhahit et al. (2021) provided an updated version of the Dutch wind mass-loss recipe in MESA. With this improved treatment, stars typically have lower mass-loss rates in the BSG phase, which prevents excessive mass loss at low Z, and helps stars maintain sufficient envelope mass to form very heavy BHs as long as they remain hotter than $\sim$8000 K (see Fig. 3).

4. Implications for impossible black holes over Cosmic Time

Vink et al. (2021) realized that in order to enable the collapse of a VMS to an 85 $M_\odot$ BH this needs 2 key ingredients. The first one is an intrinsically low Z in order for Z-dependent mass loss to not evaporate the initial stellar mass. The second requirement involves a relatively low amount of core overshooting, equivalent to a step overshooting parameter $\alpha_{\rm ov}$ of 0.1 or below. The reason for this second ingredient is 3-fold.

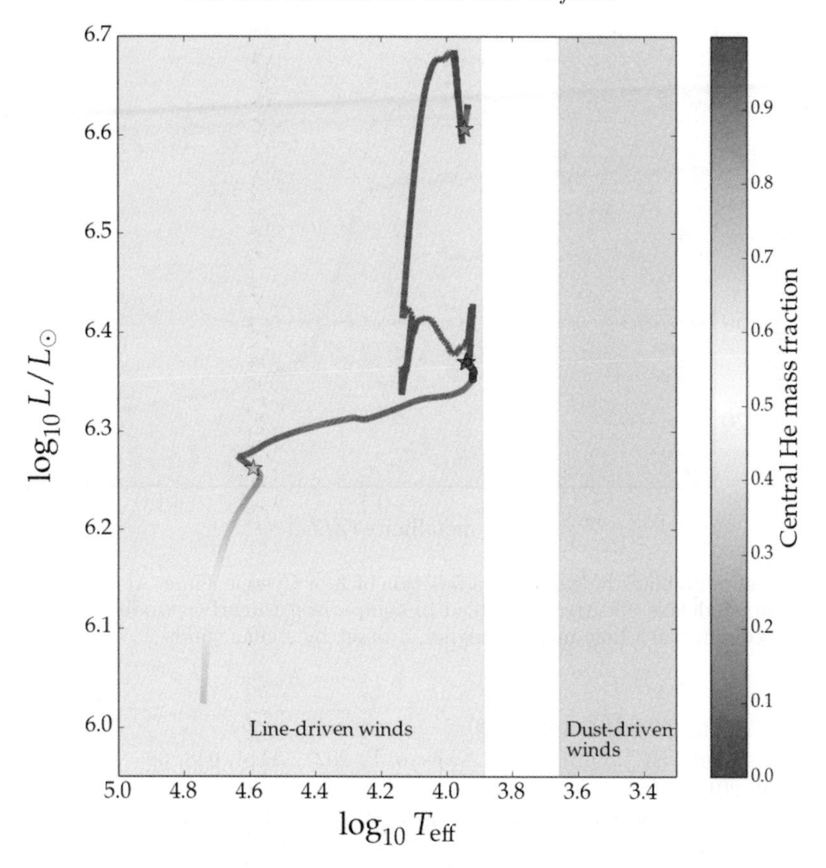

Figure 3. Evolution of our BSG progenitor model in a Hertzsprung-Russell diagram (HRD). The colour bar represents the core He abundance, with a yellow star showing the TAMS position, a blue star illustrating the end of core He-burning, and a red star marking the end of core O-burning. Blue dots (near the blue star) show time-steps of 50,000 years after core H exhaustion, where time is spent as a BSG (i.e. above 8 kK). Shaded regions highlight the area in the HRD where RSGs (red) evolve with dust-driven winds (as generally assumed, but the physical mechanism is still debated) or BSGs (blue) evolve with line-driven winds.

Firstly, low overshooting keeps the star more compact, and the collapse of the entire envelope of a very massive BSG is easier to accomplish than that for a RSG (e.g. Fernandez et al. 2018). Secondly, a low overshooting keeps the core mass below the PISN limit, and enforces a larger envelope mass (e.g. Higgins & Vink 2019). Thirdly, if the star remains above the effective temperature of the second BSJ, the regime of high mass-loss rates at low $T_{\rm eff}$ can be avoided (see Figs. 1 & 3).

Vink et al. (2021) showed that at Z values below approx. 10% of the solar metallicity 90-100 $M_\odot$ stars initially could have core masses below the critical 37 $M_\odot$ limit, and collapse into 80-90 $M_\odot$ BHs. Such impossibly heavy BHs are firmly within the canonical PISN mass gap, which should therefore be adjusted. A schematic maximum BH mass with Z is shown in Fig.4. The figure shows an almost twice as large maximum upper BH mass below the PISN gap at early Cosmic times at low Z, while for larger Z the maximum BH mass is directly set by stellar wind mass loss.

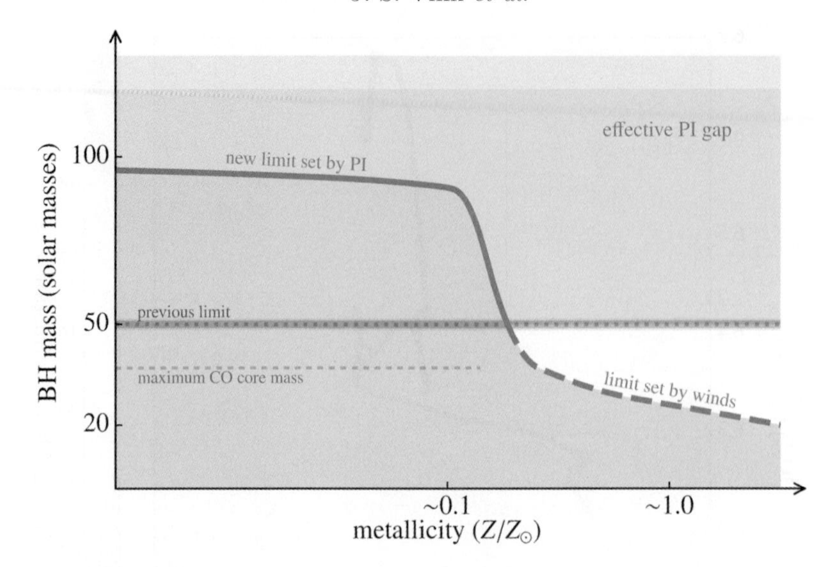

Figure 4. Maximum black hole mass as a function of Z or Cosmic Time. At low Z the maximum mass from our models is effectively doubled in comparison to earlier models and assumptions, while the maximum black hole mass at higher Z is set by stellar winds.

References

Abbott, R. et al. 2020, *PhRvL*, 125, 1102
Björklund, R., Sundqvist, J.O., Puls, J., Najarro, F. 2021, *A&A*, 648, 36
Brott, I., et al. 2011, *A&A*, 530, 115
Farmer, R., Renzo, M., de Mink, S.E., Marchant, P., Justham, S. 2019, *ApJ*, 887, 53
Fernandez, R., Quataert, E., Kashiyama, K., Coughlin, E.R. 2018, *MNRAS*, 476, 2366
Gilkis, A., et al. 2021, *MNRAS*, 503, 1004
Grassitelli, L., et al. 2021, *A&A*, 647, 99
Higgins, E.R., & Vink, J.S. 2019, *A&A*, 622, 50
de Jager, C., Nieuwenhuijzen, H., van der Hucht, K.A. 1988, *A&AS*, 72, 259
Koumpia, E., et al. 2020, *A&A*, 635, 183
Nugis, T., & Lamers, H.J.G.L.M. 2000, *A&A*, 360, 227
Paxton, B., et al. 2013, *ApJS*, 208, 4
Petrov, B., Vink, J.S., & Grafener, G. 2016, *MNRAS*, 458, 1999
Sabhahit, G.N., Vink, J.S., Higgins, E.R., Sander, A.A.C. 2021, *MNRAS*, 506, 4473
Vink, J.S., de Koter, A., & Lamers, H.J.G.L.M. 1999, *A&A*, 350, 181
Vink, J.S., de Koter, A., & Lamers, H.J.G.L.M. 2001, *A&A*, 369, 574
Vink, J.S., & de Koter, A. 2005, *A&A*, 442, 587
Vink, J.S., & Sander, A.A.C. 2021, *MNRAS*, 504, 2051
Vink, J.S., Higgins, E.R., Sander, A.A.C., Sabhahit, G.N. 2021, *MNRAS*, 504, 146
Vink, J.S. 2022, *ARAA*, in press. ArXiv 2109.08164
Woosley, S., & Heger, A. 2021, *ApJ*, 912, 31

The Origin of Outflows in Evolved Stars
Proceedings IAU Symposium No. 366, 2022
L. Decin, A. Zijlstra & C. Gielen, eds.
doi:10.1017/S174392132200014X

Atmospheric structure and dynamics of evolved massive stars. Thanks to 3D radiative hydrodynamical simulations of stellar convection

A. Chiavassa[1,2]

[1]Université Côte d'Azur, Observatoire de la Côte d'Azur, CNRS, Lagrange, CS 34229, Nice, France
email: andrea.chiavassa@oca.eu

[2]Max-Planck-Institut für Astrophysik, Karl-Schwarzschild-Straße 1, 85741 Garching, Germany

Abstract.
Evolved massive stars are major cosmic engines, providing strong mechanical and radiative feedback on their host environment. They contribute to the enrichment of their environment through a strong stellar winds, still poorly understood. Wind physics across the life cycle of these stars is the key ingredient to accomplish a complete understanding of their evolution in the near and distant Universe. Nowadays, the development of the observational instruments is so advanced that the observations became very sensitive to the details of the stellar surface making possible to quantitatively study what happens on their surfaces and above where the stellar winds become dominant. Three-dimensional radiative hydrodynamics simulations of evolved stars are essential to a proper and quantitative analysis of these observations. This work presents how these simulations have been (and will be) crucial to prepare and interpret a multitude of observations and how they are important to achieve the knowledge of the mass-loss mechanism.

Keywords. Massive evolved stars, hydrodynamics, radiative transfer

1. Introduction

Evolved and massive cool stars are major cosmic engines, providing strong mechanical and radiative feedback on their host environment (Langer 2012). Through strong stellar winds and supernova ejections, they enrich the interstellar medium with chemical elements, which are the building blocks for the next generation of stars. In particular, these objects are known to propel strong winds and stellar evolution models are not able to reproduce these winds without ad hoc physics. Therefore, a complete understanding of stellar evolution in the near and distant Universe and its impact on the cosmic environment cannot be achieved without a detailed knowledge of wind physics. This requires to trace the total mass ejected as well as its nature, the velocity of the winds, and the behaviour of the circumstellar envelope.

Massive ($M \geq 8\ M_{\odot}$, the exact value of the upper limit depends on the treatment of convection; Höfner & Olofsson 2018) evolved cool stars are objects that have reached the late phases of their evolution when the nuclear fuel in the interior is almost exhausted. These stars grow dramatically in size and become Red Supergiant (RSG) stars. RSGs are precursors of core-collapse supernovae and bear high luminosity ($L > 1000\ L_{\odot}$) with effective temperatures between 3450 and 4100 K and stellar radii up to several hundreds of $R_\odot$, or even more than 1000 $R_\odot$ (Levesque et al. 2005). Several mechanisms triggering mass-loss have been discussed, including magneto-hydrodynamic waves (Cranmer & Saar, 2011) and radiation pressure on molecules and dust (Josselin & Plez 2007), but still there is no realistic quantitative wind model (Meynet et al. 2015) that can explain the observed broad mass-loss rate range ($\dot{M} = 10^{-7} - 10^{-4}$ $M_\odot$/yr; De Beck et al. 2010). A whole picture of all the physical processes that simultaneously trigger and shape the strong winds is still missing. As underlined in Höfner & Olofsson (2018) and Decin (2021), the mass-loss mechanism is hard to discern because it involves a range of interacting, time-dependent physical processes on microscopic and macroscopic scales coupled with dynamical phenomena such as convection and pulsation in sub-photospheric layers, strong radiating shocks in the atmosphere, and dust condensation as well as radiative acceleration in the wind forming regions. In addition to this, it should also be noted that the situation is even more complex in the presence of (sub)stellar companions that are known to shape the outflow of cool evolved stars (Decin et al. 2020).

In this context, two physical processes play an important role in initiating and feeding up the strong mass-loss. In the first place the evolution of these objects is impacted by stellar convection. The convection process is non-local, three dimensional, and involves non-linear interactions over many disparate scale lengths. Moreover, it is often responsible for transporting heat up to the visible surface (Nordlund et al. 2009). In RSG atmospheres, convection is inferred from a few giant structures observed at the stellar surface with sizes comparable to the stellar radius and evolving on weekly or yearly time scales (Montargès et al. 2021, 2018; Chiavassa et al. 2010a, 2011a). These result into more extreme atmospheric conditions than in the Sun: very large variations in velocity, density and temperature produce strong radiative shocks in their extended atmosphere that can cause the gas to levitate and thus contribute to mass-loss (Höfner & Freytag 2019; Freytag et al. 2017; Chiavassa et al. 2011b).
The second ingredient is the magnetic field. Cranmer & Saar, (2011) presented a predictive description of mass loss, based on Alfvén-wave-driven wind that require open flux tubes, radially directed away from the star, in order for the gas to be accelerated and escape (Höfner & Olofsson 2018). Several authors have detected and monitored over years low intensity integrated magnetic field of the order of 1-10 Gauss (Mathias et al. 2018; Aurière et al. 2010), but its origin is still under debate and it would most likely be very different from the dynamo at work in solar-type stars due to both their slow rotation and the fact that only a few convection cells are present at their surface at any given time (Aurière et al. 2010; Freytag et al. 2002).

2. 3D radiation-hydrodynamics simulations of stellar convection of massive evolved stars

In recent years, with increased computational power, it has been possible to compute grid of 3D radiation-hydrodynamics (RHD) simulations of the whole stellar envelope that are used to predict reliable synthetic spectra and images for several stellar types. The red supergiant star simulations are computed with CO5BOLD (Freytag et al. 2012). The code solves the coupled equations of compressible hydrodynamics and non-local radiative energy transport in the presence of a fixed external spherically symmetric gravitational field on a 3D cartesian grid. No artificially pulsations are added to the simulations (e.g., by a piston) but they are self-excited. The code uses a óstar-in-a-boxó configuration where the computational domain is a cubic grid equidistant in all directions; the same open boundary condition is employed for all sides of the box. The 3D simulations are characterized by realistic input physics and reproduce the effects of convection and non-radial waves. Currently they do not include a radiative-driven wind. The important input parameters for the simulation are (Chiavassa et al. 2011b): the stellar mass (contributing

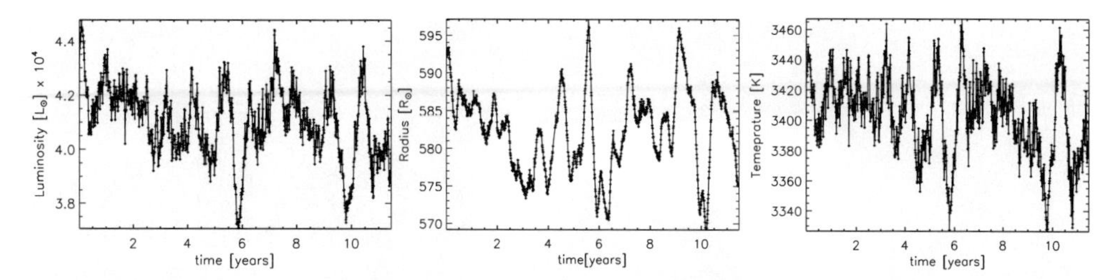

Figure 1. Spherical averages of the luminosity (left), the radius (center), and the effective temperature (right) as a function of time. See Table 1 for more details.

Table 1. Stellar parameters of the RHD simulation used in this work. The first column shows the simulation name, then the next 5 columns the stellar parameters such as the total mass, the average luminosity, the radius, the effective temperature, and the surface gravity. The different quantities are averaged over spherical shells (as in Chiavassa et al. 2009) and epochs (7th column, $t_{\rm avg}$). Errors are one standard-deviation fluctuations with respect to the time average. The solar metallicity is assumed.

Simulation	$M_\star$ $[M_\odot]$	$L_\star$ $[L_\odot]$	$R_\star$ $[R_\odot]$	$T_{\rm eff}$ [K]	log g [cgs]	$t_{\rm avg}$ [yr]	**Grid points**	$x_{\rm box}$ $[R_\odot]$
st35gm04n38	5	41517.3±1074.4	582.03±4.7	3414.2±16.8	−0.40± 0.01	11.46	401^3	1631

to the gravitational potential), the input luminosity in the core, and the abundances that were used to construct the equation-of-state and the opacity tables. The latter are gray or use a frequency-binning scheme (3 to 5 bins). In the end, average values of stellar radius, effective temperature, and surface gravity have to be derived from a relaxed model (Fig. 1). Once the RHD simulation is done, the snapshots are used for detailed post-processing treatment to extract interferometric, spectrophotometric, astrometric, and imaging observables that in the end are compared to the observations. For this purpose, we use the 3D pure-LTE radiative transfer code Optim3D (Chiavassa et al. 2009) to compute synthetic spectra and intensity maps. Optim3D takes into account the Doppler shifts caused by the convective motions. The radiative transfer is calculated using pre-tabulated extinction coefficients generated with the MARCS code (Gustafsson et al. 2008) and by adopting the solar composition of, e.g., Asplund et al. (2009).

3. What 3D simulations predict

In this Section, we show different properties of convection-related structures using a particular 3D RHD simulation. Table 1 display the temporal and spherical averaged stellar parameters and the numerical box details of this simulation, that has been compared to interferometric (Climent et al. 2020) and spectroscopic observations (Kravchenko et al. 2019). Fig. 1 shows that even if the simulation has reached a stable state, the spherical averaged quantities still varies as a function of time as a consequence of the turbulent medium. In the end, RHD simulations of massive evolved stars show a very heterogeneous photospheric patter evolving on timescales of weeks to years (Chiavassa et al. 2011a). In the simulations, the radiation is of primary importance for many aspects of convection and the envelope structure in a RSG simulations. It does cool the surface to provide a somewhat unsharp outer boundary for the convective heat transport and it also contributes significantly to the energy transport in the interior. Below the photospheric visible layers (i.c., optical depth $\tau_{\rm Rosselend} > 1$, the opacity has its peak causing a very steep temperature jump which is very prominent on top of upflow regions. At the same time a density inversion appears, which is a sufficient condition of convective instability (Chiavassa et al. 2011b).

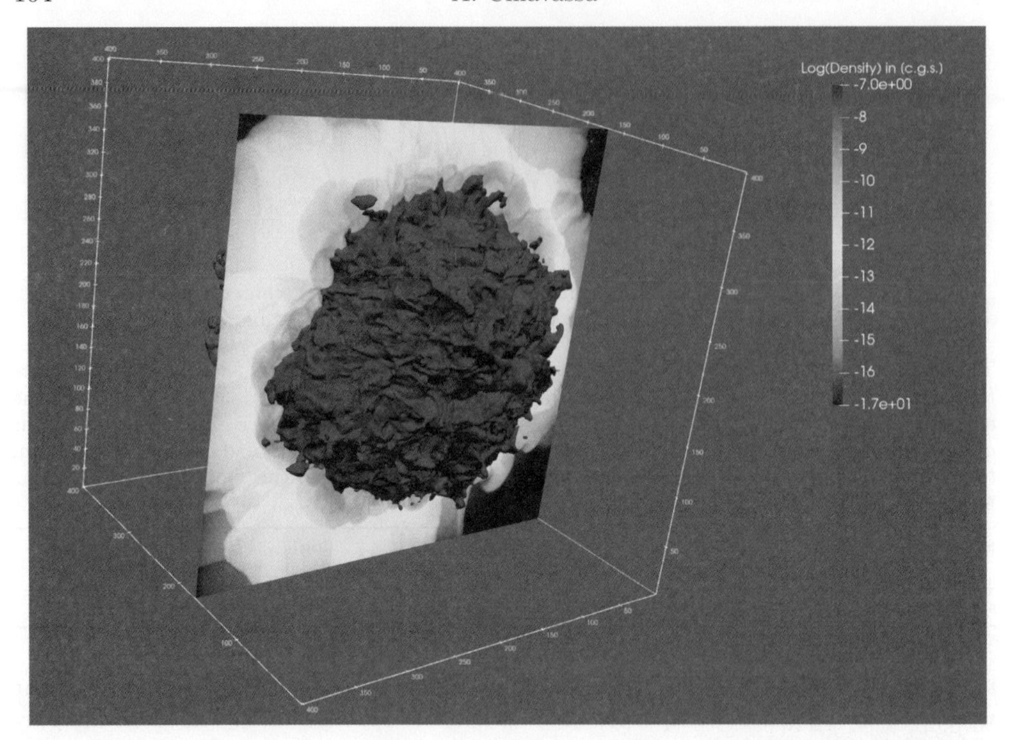

Figure 2. Logarithm of the density (from red to dark blue) of the 3D RHD simulation of Table 1 overplotted to the isosurface (amaranth color) of the Rosseland optical depth equal to one, where approximatively the continuum flux is formed.

The rising material originates in the deep convective zone (defined as the region below the Rosseland radius) and develops as an atmospheric shock when it reaches higher values in radius. This is explained in Freytag et al. (2017) and Liljegren et al. (2018): the sound waves produced by non-stationary convection (e.g., merging down-drafts or other localized events) travel through the stellar interior ($\tau_{\rm Rosselend} > 1$) to the outer layers ($\tau_{\rm Rosselend} << 1$) where the waves are slowed down and compressed because of the temperature drop. Moreover, in the outer layers (i.e., above the Rosseland radius) the density drops several orders of magnitude (Fig. 2) and the turbulent pressure dominates over the gas pressure ($P_{turbulent}/P_{gas}$ is larger than 10, Chiavassa et al. 2011b) increasing the amplitude of the rising sound wave.
Eventually, the wave becomes a shock which propagates all the way from the stellar surface to the outer atmospheric layers with significant Mach numbers (up to 8, or even larger, Fig. 3). In these layers, the density (and the temperature) shows irregular structures with convection cells in the interior and a network of shocks in the atmosphere (Fig. 2). Local fluctuations in high Mach numbers and small-scale heights due to shocks pose high demands on the stability for the hydrodynamics. A side effect of the steep and significant temperature jump is the increase in pressure scale height from small photospheric values to values that are a considerable fraction of the radius in layers just below the photosphere.

Figure 4 displays the maps of the radial velocities. The fluffy layers (dark red, 20 km/s) correspond to the continuum forming region at Rosseland optical depth equal to one. Above, the high and heterogeneous velocities (up to ~30 km/s) are accompanied by energetic pressure fluctuations, which in turn have a strong influence on shock waves. Following Freytag et al. (2017), who did this analysis for Asymptotic Giant Branch

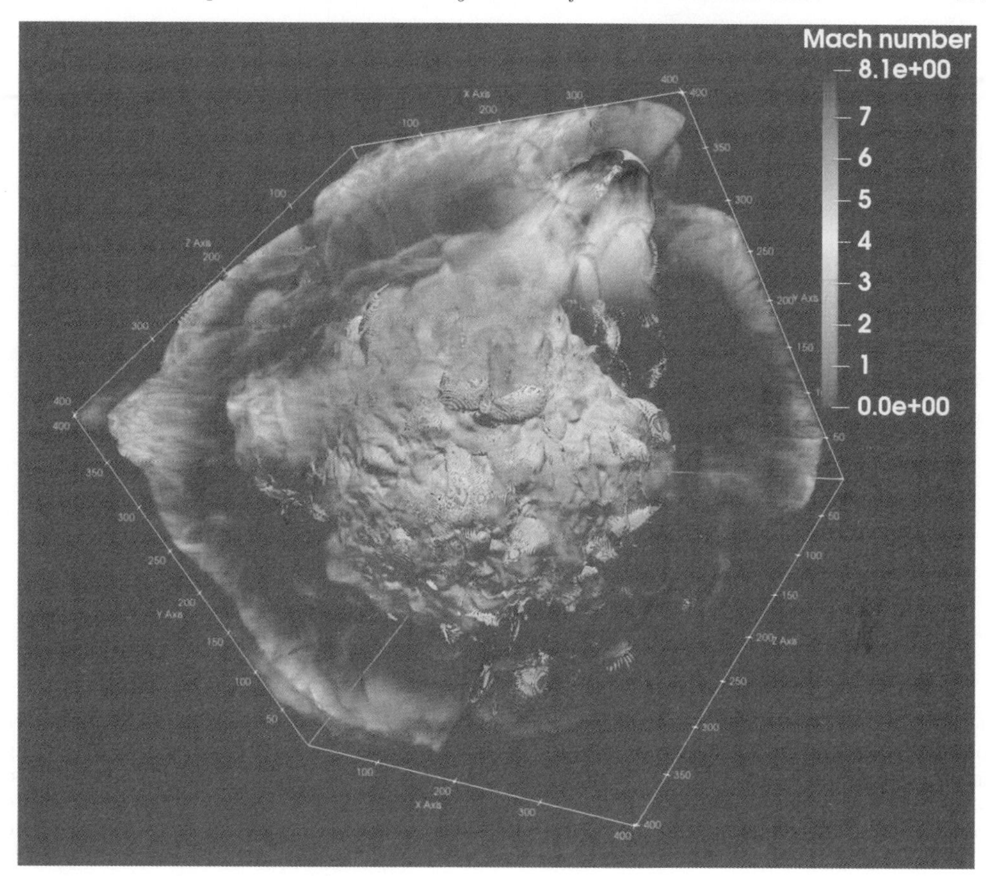

Figure 3. Radial Mach numbers computed for the 3D RHD simulation of Table 1 (blue to red colors) overplotted to the isosurface of the temperature equal to 3500 K (white). This temperature corresponds approximatively to the expected effective temperature of 3D simulations (Table 1). The simulations shows that the outer boundaries are either hit at some angle by an outgoing shock wave or let material fall back (mostly with supersonic velocities larger than Mach ∼3, Fig. 4). In the end, the shocks pass through the boundaries with a simple and stable prescription in the code based on filling typically two layers of ghost cells where the velocity components and the internal energy are kept constant Freytag et al. (2012).

(AGB) simulations, we investigated the radial motions in the photosphere using averages over spherical shells of the radial velocities for each snapshot (Fig. 5). As for AGB stars, the behavior of the inner part of the model differs from that of the outer layers: below $\sim600\,R_\odot$ (the nominal radius is $582\,R_\odot$, Table 1) the velocity field is rather regular and coherent over all layers, close to a standing wave. The differents slopes visible in the outer layers (above $\sim600\,R_\odot$) are clearly indicating the presence of propagating shock waves but in a much less regular and smooth way than in the AGB case (as a matter of comparison, see Fig. 5 of Freytag et al. 2017).

4. Two examples of applications for 3D simulations: convection cycles and spatially resolved surfaces

To provide quantitative constraints to the physics of massive evolved stars, observational techniques have reached such a level of excellence that it is now possible to reconstruct spatially and temporal resolved images of the stellar surface in the near IR and in the optical with interferometric or imagery techniques (e.g., Montargès et al.

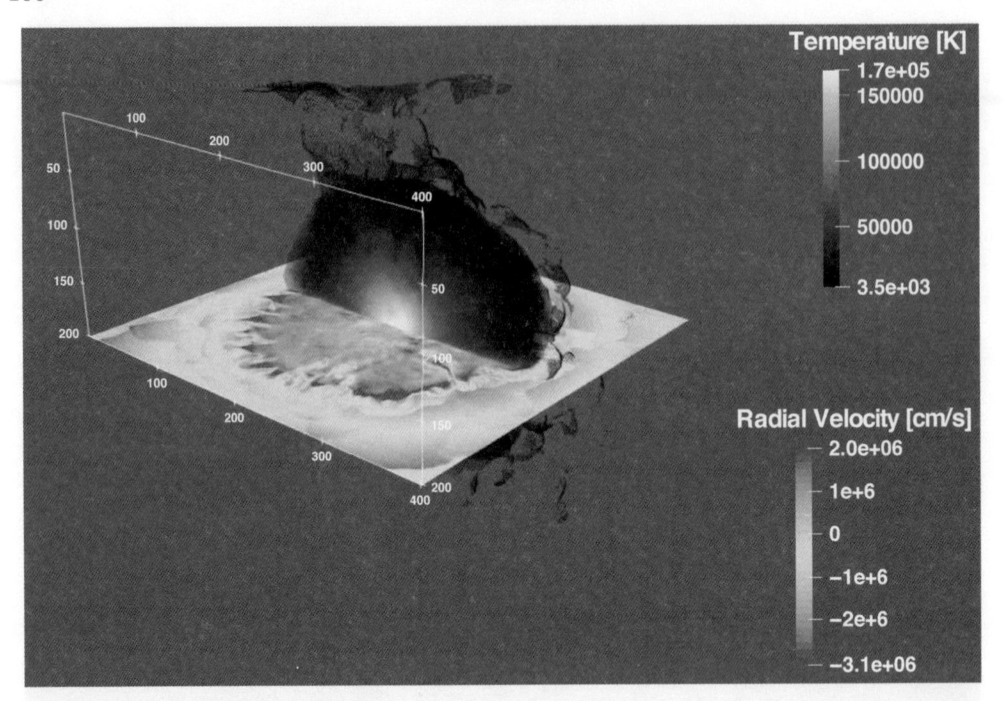

Figure 4. Central slice across the box of Fig. 3 showing the radial velocity: blue indicates outward and red inward flow. In addition to the velocities, the temperature volume rendering is also displayed (yellow to black colors). The temperature values arbitrary stops at 3500 K to show the approximative position of the $\tau_{\rm Rosselend}=1$, while in the simulation the temperature range covered is between ∼1000 and 170 000 K. The large shocks can be up to ∼250 $R_\odot$ wide (each grid point is about 4 $R_\odot$) with local temperature of ∼2500 K and log(density in [c.g.s.]) of ∼-13.

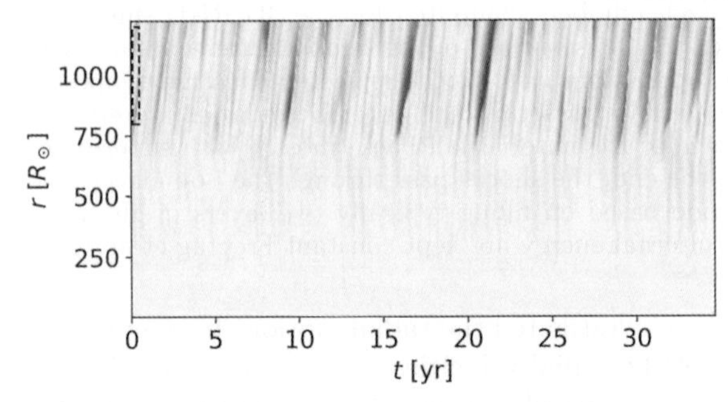

Figure 5. Spherically averaged radial velocities for the full run time and radial distance of the simulation of Table 1. The different colors show the average vertical velocity at that time and radial distance. The velocity range and color is the same as in Fig. 4.

2021; Norris et al. 2021; Cannon et al. 2021; Climent et al. 2020; Montargès et al. 2018; Ohnaka et al. 2017; Kervella et al. 2016; Chiavassa et al. 2010b; Haubois et al. 2009). In addition to this, long term spectro-photometric surveys are also available for observing RSG stars (e.g., Kravchenko et al. 2021, 2019; Lebzelter et al. 2019; Mathias et al. 2018; Kiss et al. 2006). The interpretation of these observations requires realistic modelling that takes into account most of the processes at work in the atmosphere (i.e., convection, shocks, pulsation, radiative transfer, ionization, molecules and dust formation, magnetic field). We present in this Section two examples based on the 3D RHD simulations done with CO5BOLD code and post-processed with the radiative transfer code Optim3D.

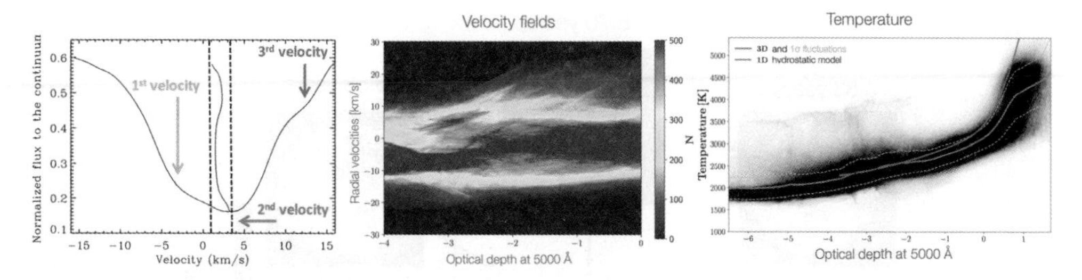

Figure 6. *Left panel:* Synthetic spectral line of the Ti I at 6261.11 Å for one snapshot of a 3D RHD simulation of an RSG star (Chiavassa & Freytag 2015). The vertical dashed line shows the spanned velocities of the line bisector. The different arrows and colors display the positions of different velocity components which contribute to the shape of the line. *Central panel:* The distribution of vertical velocities extracted from a 3D simulation as a function of the optical depth at 5000 Å. The color code shows areas with high (red) or low (blue) density of points (Kravchenko et al. 2018). *Right panel:* 3D simulation of the thermal structure. Darker areas correspond to more frequent temperature values. The red line is the mean 3D temperature profile. The orange dashed lines correspond to the one σ values around the average. The blue line refers to a 1D hydrostatic model.

4.1. *Convection cycles*

RHD simulations provide a self-consistent ab-initio description of the non-thermal velocity field generated by convection, shock waves, and overshoot that manifests itself in spectral line shifts and changes in the equivalent width. They combine important properties such as velocity amplitudes and velocity-intensity correlations, which affect the line shape, shift, and asymmetries. Figure 6 (left panel) shows an example for the optical Ti I line at 6261.11 Å. The line shape constitutes of more than one velocity component that contributes through the different atmospheric layers where the line forms. As a consequence, the line bisector† is not straight and span values up to 5 km/s on a temporal scale of few weeks (as already pointed out by Gray 2008). As the vigorous convection is prominent in the emerging flux, the radial velocity measurements for evolved stars are very complex and need a sufficiently high spectral resolution to possibly disentangle all the sources of macro-turbulence. In addition to the velocity field (Fig. 6, central panel), other elements affect the line formation: (i) the strength of the transition depends on the mean thermal gradient in the outer layers ($\tau_{5000} < 1$ in right panel of Fig. 6), for instance a shallow mean thermal gradient weakens the contrast between the continuum and line forming regions; (ii) the temperature (and density) inhomogeneities that affect the opacity run through the photospheric layers where the line forms.

In this context, we used the tomographic method to recover the distribution of the component of the velocity field projected on the line of sight at different optical depths in the stellar atmosphere (Fig. 6, central panel). This method was proposed for the first time by Alvarez et al. (2000), Alvarez et al. (2001a), Alvarez et al. (2001b) for AGB stars and then adapted and implemented in Optim3D for RSGs by Kravchenko et al. (2018). The authors successfully managed to show that in 3D simulations, the spectral lines do not form in the same limited number of layers as in 1D hydrostatic models, but they spread over different optical depths due to the non-radial convective muvements. Additionally, this method allows to recover the dependence of the velocity field across the atmosphere.

The tomography of the stellar photosphere tomographic opens a new doorway for the study of stellar dynamical cycles in evolved stars, and in particular RSGs.

† It is the locus of the midpoints of the line. A symmetric profile has a straight vertical bisector, while the "C"-shaped line bisector reveals asymmetries.

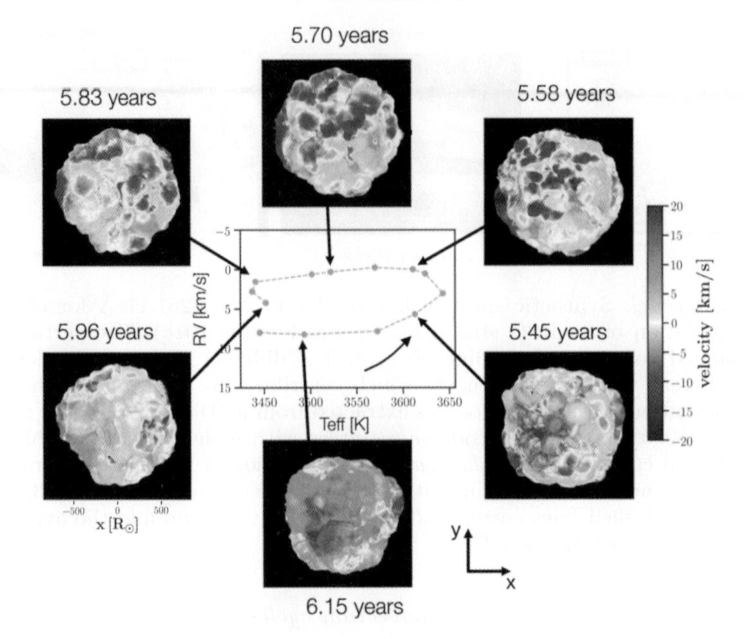

Figure 7. Velocity maps for different snapshots of a RSG simulation of Table 1 during a convection cycle (central part of the panel). The velocity is weighted with the contribution function, which is a useful diagnostics for studying spectral line formation in stellar atmospheres (e.g., Kravchenko et a. 2018). The red/blue colors correspond to inward/outward moving material, respectively. The central panel shows the behaviour of the radial velocity (RV). The arrow indicates the direction of the evolution along the hysteresis loop (Kravchenko et al. 2019, 2021).

Kravchenko et al. (2019) were able to interpret the long-term (almost 7 years of high-resolution spectra observed with the HERMES spectrograph) photometric variability of the RSG star μ Cep. The authors denoted, in the observations, the presence of an hysteresis loop for convection. The hysteresis loop illustrates the convective turn-over of the material in the stellar atmosphere: first, the rising hot matter reaches upper atmospheric layers, then temperature drops as the matter moves horizontally and finally matter falls and cools down (Gray 2008). Kravchenko et al. (2019) showed that 3D RHD simulations explain this observed hysteresis behaviour and are useful to interpret time-dependent signatures, detectable in the observations, that relay on convection. As a matter of example, the velocity maps in Fig. 7 reveal upward and downward motions of matter extending over large portions of the stellar surface. The relative fraction of upward and downward motions is what distinguishes the upper from the lower part of the hysteresis loop (central panel of Fig. 7), its top part (zero velocity) being characterized by equal surfaces of rising and falling material. The bottom part of the hysteresis loop occurs, as expected, when the stellar surface is covered mostly by downfalling material.

Another example concerns the Great dimming episode of the RSG star Betelgeuse, when the brightness decreased dramatically to about 35% of its typical brightness in December 2019 (Guinan et al. 2020) before swiftly recovering over the next few months. Using the tomography and long-term HERMES data, Kravchenko et al. (2021) revealed the presence of two subsequent shocks in February 2018 and January 2019, the second one amplifying the effect of the first one. This produced a rapid expansion of a portion of the atmosphere of Betelgeuse and an outflow between October 2019 and February 2020. The final result was a sudden increase in molecular opacity in the cooler upper photosphere and, as a consequence, an unusual plumbing of the stellar brightness. This phenomenon has been described in the literature as òmolecular plumesó rising from the

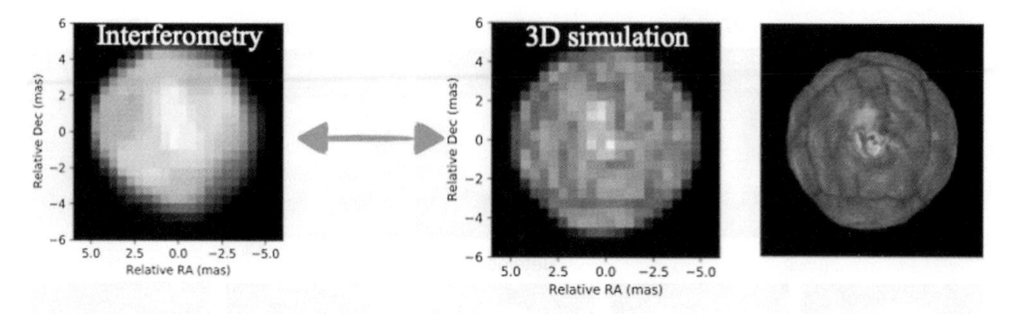

Figure 8. Representation of convective pattern size on a RSG star in observations and simulations. *Left panel:* Image of the stellar surface of the RSG CE Tau ($M_\star \approx 15\,M_\odot$, $L_\star \approx 6.6 \times 10^4\,L_\odot$, $T_{\rm eff} \approx 3820\,{\rm K}$, $R_\star \approx 587\,R_\odot$, $\log g \approx 0.05$) reconstructed from interferometric data collected with PIONIER@VLTI (Montargès et al. 2018). *Center and right panel:* Synthetic intensity map calculated from a 3D simulation ($M_\star \approx 12\,M_\odot$, $L_\star \approx 8.9 \times 10^4\,L_\odot$, $T_{\rm eff} \approx 3430\,{\rm K}$, $R_\star \approx 846\,R_\odot$, $\log g \approx -0.3$, rightmost panel) and degraded to the observation spatial resolution (central panel).

photosphere of supergiants (Kervella et al. 2016) or òmolecular reservoirsó (Harper et al. 2020). In the literature, there are also other explanations of this Betelgeuse's dimming: Levesque & Massey 2020; Cotton et al. 2020; Safonov et al. 2020; Dupree et al. 2020; Dharmawardena et al. 2020; Montargès et al. 2021; Davies & Plez 2021.

4.2. *Spatially resolved surfaces: stellar surface details explained by simulations*

Spatially resolved stellar surface observations, among which interferometry contributes substantially, are of great importance for evolved stars for two reasons: (i) they afford the direct detection and characterization of the convective pattern related to the surface dynamics, and (ii) they allow to determine the stellar parameters.

Two main observables are used in interferometry: the visibility and the closure phases. Visibilities measure the surface contrast of the source and are primarily used to determine fundamental stellar parameters and limb-darkening. Closure phases combine the phase information from three (or more) telescopes and provide direct information on the morphology of the source (Monnier 2003). The wise combination of both contributes to the image reconstruction of the observed targets. For a correct interpretation of the observations, it is necessary to simultaneously explain both observables with the same model as well as the intensity contrast and shape as a function of wavelengths. This is outlined in Fig. 8 where the reconstructed image (left panel) is compared to the synthetic image obtained from a 3D simulation (rightmost panel) convolved with the instrumental beam (central panel).

During the last decade, several observational works (Fig. 9) used 3D RHD simulations to explain the interferometric data of massive evolved stars. For instance, the first works concerned the RSG star Betelgeuse for which Chiavassa et al. (2009) and Chiavassa et al. (2010a) detected and measured the characteristic sizes of convective cells using measurements in the infrared and in the optical. Chiavassa et al. (2010b) and Chiavassa et al. (2021) reconstructed the images of another RSG, VX Sgr, with different instruments from the H to the N band to probe the presence of large convective cells on its surface. Montargès et al. (2014), Montargès et al. (2016), and Montargès et al. (2017) reported a series of reconstructed images, interpreted with 3D simulations, for several RSGs and different instruments (AMBER and PIONIER at VLTI).

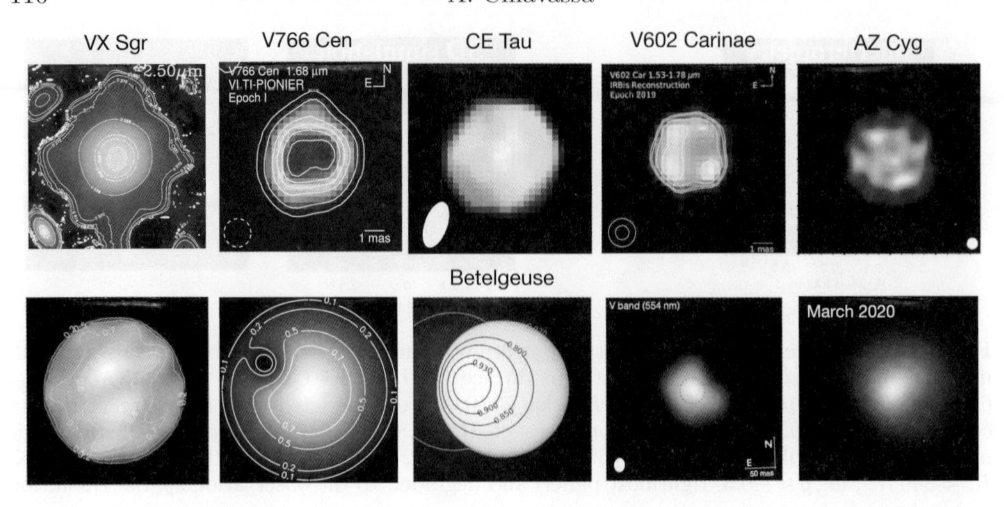

Figure 9. Few examples of images obtained from observations of different RSG stars. *Top row:* VX Sgr with AMBER@VLTI (Chiavassa et al. 2010b), V766 Cen with PIONIER@VLTI (Wittkowski et al. 2017), CE Tau with PIONIER@VLTI (Montargès et al. 2018), V602 Carinae with PIONIER@VLTI (Climent et al. 2020), AZ Cyg with MIRC@CHARA (Norris et al. 2021). *Bottom row:* Betelgeuse with IOTA and COAST interferometers (first two images, Chiavassa et al. 2010a), with PIONIER@VLTI (Montargès et al. 2016), with SPHERE (Kervella et al. 2016; Montargès et al. 2021).

However also the temporal evolution (at different wavelengths) is a key point in the understanding of stellar dynamics. For instance, Wittkowski et al. (2017), Montargès et al. (2018), Climent et al. (2020), Norris et al. (2021), and Montargès et al. (2021) showed the importance of temporal variability in the observations. To tackle all the different astrophysical problems related to evolved stars, recent and future interferometers have to challenge the combination of high spectral and spatial resolution as well as the time monitoring on relatively short timescales (weeks/month) of these objects (Chiavassa et al. 2011a; Montargès et al. 2021).

The direct measurement of stellar angular diameters has been the principal goal of most attempts with astronomical interferometers since the pioneering work of Michelson & Pease (1921). Nowadays with the advent of Gaia, for stars of known distance the angular diameter becomes of paramount importance to yield the stellar radius and eventually to the absolute magnitude. These quantities are essential links between the observed properties of stars and the results of theoretical calculations on stellar structure and evolution. Few survey works (Cruzalèbes et al. 2013; Arroyo-Torres et al. 2014, 2015; Wittkowski et al. 2017) characterized the fundamental parameters and atmospheric extensions of evolved stars in our neighbourhood using AMBER instrument (now decommissioned) at VLTI. In particular, the last two papers observed a linear correlation between the visibility ratios of observed RSGs and the luminosity and surface gravity, indicating an increasing atmospheric extension with increasing luminosity and decreasing surface gravity, indirectly supporting a mass-loss scenario of a radiatively driven extension caused by radiation pressure on Doppler-shifted molecular lines. These results are confirmed for AGB stars (Wittkowski et al. 2016) where the atmospheric extension is detected and explained by the RHD simulations for a sample of interferometric observations, supporting the mass-loss scenario of pulsation- and shock-induced dynamics that can levitate the molecular atmospheres of Mira/AGB variables to extensions that are consistent with observations.

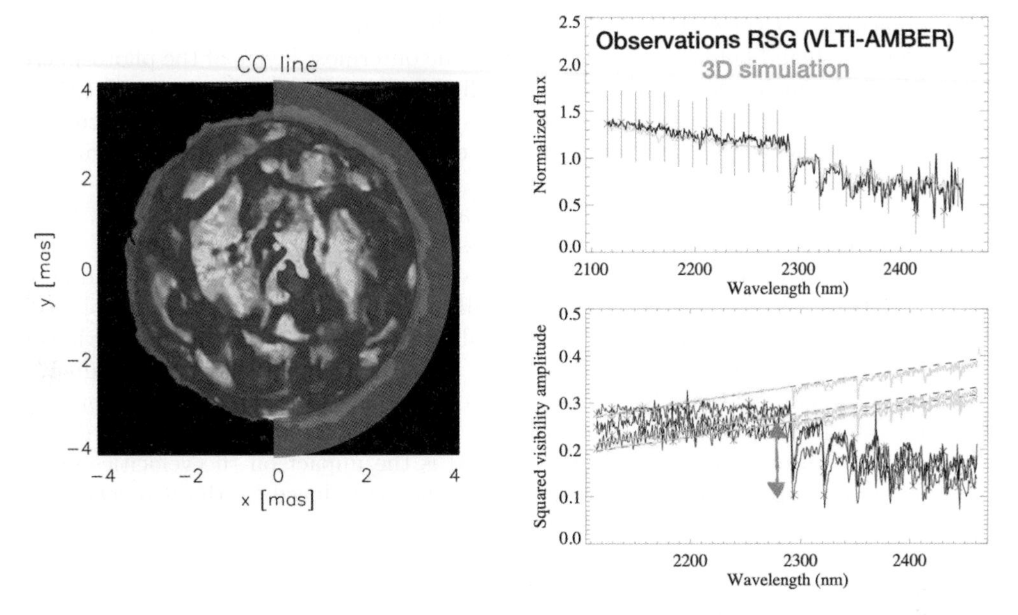

Figure 10. *Left panel:* Synthetic image in the Carbon Monoxide (CO) lines at about 2.3 μm. The semi-circle in red displays the expected atmospheric extension needed to explain the drop in the squared visibility (right panels). *Right panels:* Interferometric observations of red supergiants with AMBER (black) compared to 3D simulation predictions (green). While the flux adjustment (top panel) is in a good agreement, the synthetic visibilities cannot reproduce the observed atmospheric extension (bottom panel, red arrow). See Arroyo-Torres et al. (2015) for more details.

5. Improving 3D simulations: towards the solution of the mass-loss mechanism?

Recent advances in instrumental techniques in interferometry, imaging, and spectroscopy have achieved an astonishing level of accuracy. Despite the very satisfactory comparisons of the 3D simulations with observations, a number of studies have highlighted the current limitations of 3D simulations that need to be solved to provide a quantitative response to the problem of mass-loss in red supergiants. These four points are the cornerstone for future developments in the field and are listed below in order of importance.

(1) *The radiative pressure.* A comparison of 3D simulations to interferometric observations with AMBER at VLTI have shown that the extension of the observed red supergiant atmospheres is not interpretable with current models (Arroyo-Torres et al. 2015). Figure 10 shows that while the synthetic flux is adequate to match the observed AMBER flux (top right panel), the squared visibility are completely off (bottom right panel). The interferometric visibilities are linked to the surface brightness contrast of the observed object. The observed visibility amplitudes show strong visibility drops in the molecular bands (emphasizing a major extension of the photosphere, red arrow in bottom right panel) that cannot be explained by the simulations. This is also visible in the synthetic intensity map (left panel) where the missing photospheric extension is highlighted in red.
This finding was confirmed by the optical data obtained with SPHERE, where also here the extension of the 3D photosphere is too limited compared to the

observations (Kervella et al. 2016). The inclusion of radiative pressure in the simulations should help the gas to levitate in the outermost layers of the photosphere, where the molecular opacity is not negligible (e.g., TiO molecules) and explain (at least in part) the mechanism of mass loss: radiative pressure on molecules as suggested by Josselin & Plez (2007), based on an order of magnitude calculation by Gustafsson & Plez (1992).

(2) *The magnetic field.* The presence of a magnetic field in stars is intimately linked to the convection across the stellar photosphere. A typically magnetic field results in the increase of atmospheric velocities and higher temperatures in the chromosphere. As a consequence, the overall structure of the stellar atmosphere is affected. In the case of evolved stars, local dynamos are expected to appear in correspondence to the large convective cells (Freytag et al. 2002). The introduction of a magnetic field into 3D simulations is under development and should produce large-scale local dynamos, that grow for decades and saturate with only moderate fluctuations. What has to be explored is the impact on the velocities in the lower photosphere. Here, the velocities should be reduced by the magnetic field. Meanwhile, in the outer layers (where the boundary conditions are crucial), the velocities should increase and sustain the stellar winds.

(3) *The numerical resolution.* Resolving the òturbulentó character of an RSG photosphere is a complex task that has an impact on several aspects: model stratification, numerical viscosity, Doppler shift in the spectral lines. The spatial resolution of small-scale structures close to the grid box need large box sides (e.g., 1000^3 or even larger depending on the spectral type) which is extremely computer-time intensive. Refining the computational box means resolving better the turbulent medium (Nordlund et al. 2009). The latter point is precisely what is missing in 3D RHD simulation. Figure 11 shows that the velocities in the simulations are at maximum 20-30 km/s (red curve), which is far too low to reach the escape velocity (light blue curve) even though the velocities are supersonic (the sound speed is plotted in violet). The under-estimation of the simulation's velocities has a direct consequence on the gas levitation by the vigorous convection in these stars. The current solution is to increase the number of points to at least 1000^3.

(4) *The stellar rotation.* Observations of Betelgeuse indicate that RSG stars may rotate: Uitenbroek et al. (1998) found with HST an angular rotational velocity between 2.0 and 2.5 km/s (i.e., a projected equatorial velocity of 5.0 km/s); Kervella et al. (2018) reported a projected equatorial velocity of about 5.47 km/s using ALMA. The observed rotational velocity is about six times lower than the turbulent velocity due to the convection-related surface structures, but it could also take part in the mass-loss mechanism. Freytag et al. (2017) described the effect of rotation on AGB simulations carried out with CO5BOLD and demonstrated that the temperature stratification shows hardly any effect, while the average density in the atmosphere increases with shorter rotation period. The authors insisted on the actual shortcomings in their simulations where the approximation of the smoothed stellar core plays a larger role than for purely convective flows that are not rotating. However, in the context of RSG stellar winds, it remains valuable to question how the angular momentum is advected in a very slowly rotating convective envelope and the role (if any) of the magnetic fields/convection coupling across the stellar photosphere and above.

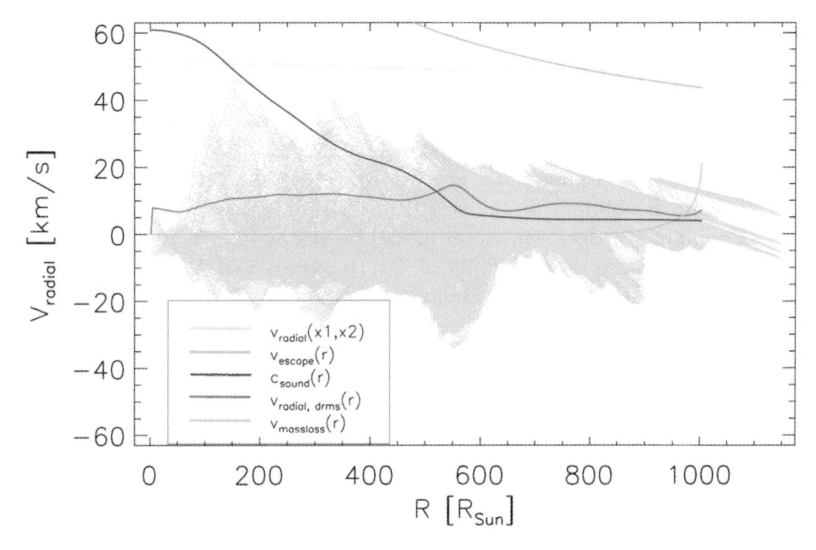

Figure 11. Scatter plot of the radial velocities (green points) for one particular snapshot of the 3D simulation of Table 1 and for all the grid points in the numerical box. The red line displays the velocities' spherical average. The violet line shows the sound speed. The light blue displays the escape velocities for a 5 $M_\odot$ star. The orange line is an example of 1D velocity derived from mass conservation for a mass-loss rate set to 10^{-6} $M_\odot$/yr. The nominal radius of the simulation is 582 $R_\odot$ (Table 1).

6. Conclusions

We presented 3D radiation-hydrodynamics simulations for massive evolved RSG stars. These simulations are computed with the CO5BOLD code that takes into account the full convective envelope and are characterized by realistic input physics. The simulations reproduce the effects of convection and non-radial waves. RHD simulations predict photospheric structures with extremely inhomogeneous temperatures and densities. The rising material originates in the deep convective zone and develops as an atmospheric shock with supersonic velocities when it reaches the outer layers. Additionally, the important and natural temporal variability of the convective muvements affects the stellar parameters and, more generally, all the observables of RSG stars. More details will appear in a future paper (Chiavassa & Kravchenko, 2022, Living Reviews in Computational Astrophysics).

The stellar variability, heterogeneity, and dynamics demonstrated by 3D RHD simulations of RSGs may have an impact on the environement of core-collapse supernovae (Giacinti et al. 2019); on the quantitative studies of metallicity in our Galaxy (Levesque 2018) and in nearby galaxies (Davies et al. 2017); and on the mass-loss mechanism and thus in the physics implemented in stellar evolution codes (Meynet et al. 2015).

These stars are among the largest stars in the Universe and their luminosities place them among the brightest stars, visible to very long distances. RSGs are major cosmic engines with strong stellar winds whose origin is still under debate. Ongoing and future developments of 3D RHD simulations will help to lead towards the solution of the mass-loss problem.

References

Alvarez, R., Jorissen, A., Plez, B., et al. 2000, *A&A*, 362, 655
Alvarez, R., Jorissen, A., Plez, B., et al. 2001, *A&A*, 379, 305
Alvarez, R., Jorissen, A., Plez, B., et al. 2001, *A&A*, 379, 288
Arroyo-Torres, B., Martí-Vidal, I., Marcaide, J. M., et al. 2014, *A&A*, 566

Arroyo-Torres, B., Wittkowski, M., Chiavassa, A., et al. 2015, *A&A*, 575, A50
Asplund, M., Grevesse, N., Sauval, A. J., et al. 2009, *ARAA*, 47, 481
Aurière, M., Donati, J.-F., Konstantinova-Antova, R., et al. 2010, *A&A*, 516, L2
Cannon, E., Montargès, M., de Koter, A., et al. 2021, *MNRAS*, 502, 369
Chiavassa, A., Plez, B., Josselin, E., et al. 2009, *A&A*, 506, 1351
Chiavassa, A., Haubois, X., Young, J. S., et al. 2010, *A&A*, 515, A12
Chiavassa, A., Lacour, S., Millour, F., et al. 2010, *A&A*, 511, A51
Chiavassa, A., Pasquato, E., Jorissen, A., et al. 2011a, *A&A*, 528, A120
Chiavassa, A., Freytag, B., Masseron, T., et al. 2011b, *A&A*, 535, A22
Chiavassa, A. & Freytag, B. 2015, Why Galaxies Care about AGB Stars III: A Closer Look in Space and Time, 497, 11
Chiavassa, A., Kravchenko, K., Millour, F., et al. 2020, *A&A*, 640, A23
Chiavassa, A., Kravchenko, K., Montargès, M., et al. 2021, *A&A* in press, arXiv:2112.10695
Climent, J. B., Wittkowski, M., Chiavassa, A., et al. 2020, *A&A*, 635, A160
Cotton, D. V., Bailey, J., De Horta, A. Y., et al. 2020, Research Notes of the American Astronomical Society, 4, 39
Cranmer, S. R. & Saar, S. H. 2011, *ApJ*, 741, 54
Cruzalèbes, P., Jorissen, A., Rabbia, Y., et al. 2013, *MNRAS*, 434, 437
Davies, B., Kudritzki, R.-P., Lardo, C., et al. 2017, *ApJ*, 847, 112
Davies, B. & Plez, B. 2021, *MNRAS*, 508, 5757
De Beck, E., Decin, L., de Koter, A., et al. 2010, *A&A*, 523, A18
Decin, L., Montargès, M., Richards, A. M. S., et al. 2020, *Science*, 369, 1497
Decin, L. 2021, *ARAA*, 59, 337
Dharmawardena, T. E., Mairs, S., Scicluna, P., et al. 2020, *ApJ*, 897, L9
Dupree, A. K., Strassmeier, K. G., Matthews, L. D., et al. 2020, *ApJ*, 899, 68
Freytag, B., Steffen, M., & Dorch, B. 2002, Astronomische Nachrichten, 323, 213
Freytag, B., Steffen, M., Ludwig, H.-G., et al. 2012, Journal of Computational Physics, 231, 919
Freytag, B., Liljegren, S., & Höfner, S. 2017, *A&A*, 600, A137
Giacinti, G., Dwarkadas, V., Marcowith, A., et al. 2019, 36th International Cosmic Ray Conference (ICRC2019), 36, 74
Gray, D. F. 2008, *ApJ*, 135, 1450
Guinan, E., Wasatonic, R., Calderwood, T., et al. 2020, The Astronomer's Telegram, 13512
Gustafsson, B. & Plez, B. 1992, Instabilities in Evolved Super- and Hypergiants, 86
Gustafsson, B., Edvardsson, B., Eriksson, K., et al. 2008, *A&A*, 486, 951
Harper, G. M., Guinan, E. F., Wasatonic, R., et al. 2020, *ApJ*, 905, 34
Haubois, X., Perrin, G., Lacour, S., et al. 2009, *A&A*, 508, 923
Höfner, S. & Freytag, B. 2019, *A&A*, 623, A158
Höfner, S. & Olofsson, H. 2018, *ARAA*, 26, 1
Josselin, E. & Plez, B. 2007, *A&A*, 469, 671
Kervella, P., Lagadec, E., Montargès, M., et al. 2016, *A&A*, 585, A28
Kervella, P., Decin, L., Richards, A. M. S., et al. 2018, textitA&A, 609, A67
Kiss, L. L., Szabó, G. M., & Bedding, T. R. 2006, *MNRAS*, 372, 1721
Kravchenko, K., Van Eck, S., Chiavassa, A., et al. 2018, *A&A*, 610, A29
Kravchenko, K., Chiavassa, A., Van Eck, S., et al. 2019, *A&A*, 632, A28
Kravchenko, K., Jorissen, A., Van Eck, S., et al. 2021, *A&A*, 650, L17
Langer, N. 2012, *ARAA*, 50, 107
Lebzelter, T., Trabucchi, M., Mowlavi, N., et al. 2019, *A&A*, 631, A24
Levesque, E. M., Massey, P., Olsen, K. A. G., et al. 2005, *ApJ*, 628, 973
Levesque, E. M. 2018, *ApJ*, 867, 155
Levesque, E. M. & Massey, P. 2020, *ApJ*, 891, L37
Liljegren, S., Höfner, S., Freytag, B., et al. 2018, *A&A*, 619, A47
Mathias, P., Aurière, M., López Ariste, A., et al. 2018, *A&A*, 615, A116
Meynet, G., Chomienne, V., Ekström, S., et al. 2015, *A&A*, 575, A60

Michelson, A. A. & Pease, F. G. 1921, *ApJ*, 53, 249. doi:10.1086/142603
Monnier, J. D. 2003, Reports on Progress in Physics, 66, 789
Montargès, M., Kervella, P., Perrin, G., et al. 2014, *A&A*, 572, A17
Montargès, M., Kervella, P., Perrin, G., et al. 2016, *A&A*, 588, A130
Montargès, M., Chiavassa, A., Kervella, P., et al. 2017, *A&A*, 605, A108
Montargès, M., Norris, R., Chiavassa, A., et al. 2018, *A&A*, 614, A12
Montargès, M., Cannon, E., Lagadec, E., et al. 2021, *Nature*, 594, 365
Nordlund, Å., Stein, R. F., & Asplund, M. 2009, Living Reviews in Solar Physics, 6, 2
Norris, R. P., Baron, F. R., Monnier, J. D., et al. 2021, *ApJ*, 919, 124
Ohnaka, K., Weigelt, G., & Hofmann, K.-H. 2017, *Nature*, 548, 310
Safonov, B., Dodin, A., Burlak, M., et al. 2020, arXiv:2005.05215
Uitenbroek, H., Dupree, A. K., & Gilliland, R. L. 1998, *ApJ*, 116, 2501
Wittkowski, M., Chiavassa, A., Freytag, B., et al. 2016, *A&A*, 587, A12
Wittkowski, M., Arroyo-Torres, B., Marcaide, J. M., et al. 2017, *A&A*, 597, A9
Wittkowski, M., Abellán, F. J., Arroyo-Torres, B., et al. 2017, *A&A*, 606, L1

Discussion

ORSOLA DE MARCO: There is a large literature on stabilising large stars like RSG and AGB for common envelope simulations. I do not understand how you overcome some numerical challenges in particular when computing the luminosity of the star. What do you do when you set up your star?

ANDREA CHIAVASSA: In a cartesian equidistant box, the initial model is produced starting from a sphere in hydrostatic equilibrium with a weak velocity field inherited from a previous model with different stellar parameters. The input luminosity enters into few central grid cells of the box and the envelope stellar mass in the equation for the gravitational potential. After some time, the limb-darkened surface without any convective signature appears but with some regular patterns due to the numerical grid. After several years of stellar time, a regular pattern of small-scale convection cells develops and, after cells merge the average structures, it becomes big and the regularity (due to the Cartesian grid) is lost. The intensity contrast grows with time.

ILEYK EL MELLAH: Did you try to inject any molecular network? Do you have an idea of the distribution of abundances of molecular species as a function of distance? Is this coherent with the olivine detection you see in VX Sgr?

ANDREA CHIAVASSA: No, we did not try so far to include any molecular network dependence. The approach presented is related to a tentative model fitting with RADMC3D code and there is not, so far, a link with the expected abundances of molecules in the extended photospehre. However, this is something we have in mind to do in the future.

RAGHVENDRA SAHAI: In your very sophisticated simulations, the word chromosphere did not come out. Do you produce something like a chromosphere in these models? Is there a reasonable amount of ionised gas around these stars?

ANDREA CHIAVASSA: Indeed, in our actual simulations, the temperature and density drops go farther from the star and thus we can say that the chromosphere is not included so far.

LEEN DECIN: The formation of carbon Monoxide is calculated using a thermal equilibrium scheme? and so what is the abundances used?

ANDREA CHIAVASSA: Yes, it was at thermal equilibrium with solar metallicity. Maybe I was a bit quick on this point during the talk. The spectro-interferometric observations I showed concerned the flux and the visiblity curves. While the synthetic flux had a good agreement with observations, the visibility (ie, the spatially resolved intensity brightness) was completely off. This means that we cannot resolve the extension of the atmosphere of those stars while we actually reproduce the integrated flux of the object.

LEEN DECIN: Is it enough to approximate the contribution to the photospheric flux using two different hydrostatic MARCS models? In the case of Betelgeuse?

ANDREA CHIAVASSA: In the paper we did with Ben Davies in 2013, we used two MARCS components to mimic the surface brightness of the 3D simulated. However, I think that we would need many more 1D structures for this purpose.

BEN DAVIES: Comment. We recently added a Betelgeuse wind to the MARCS structure, including temperature inversion in the upper wind and we obtain a very good fit of the entire SED.

The Origin of Outflows in Evolved Stars
Proceedings IAU Symposium No. 366, 2022
L. Decin, A. Zijlstra & C. Gielen, eds
doi:10.1017/S1743921322000667

The Great Dimming of Betelgeuse from the VLT/VLTI

M. Montargès[1], E. Cannon[2], E. Lagadec[3], A. de Koter[4,2], P. Kervella[1], J. Sanchez-Bermudez[5,6], C. Paladini[7], F. Cantalloube[8], L. Decin[2,9], P. Scicluna[7], K. Kravchenko[10], A. K. Dupree[11] S. Ridgway[12], M. Wittkowski[13], N. Anugu[14,15], R. Norris[16], G. Rau[17,18], G. Perrin[1], A. Chiavassa[3], S. Kraus[15], J. D. Monnier[19], F. Millour[3], J.-B. Le Bouquin[19,20], X. Haubois[7], B. Lopez[3], P. Stee[3] and W. Danchi[17].

[1]LESIA, Observatoire de Paris, Université PSL, CNRS, Sorbonne Université, Université de Paris, 5 place Jules Janssen, 92195 Meudon, France
email: miguel.montarges@observatoiredeparis.psl.eu

[2]Institute of Astronomy, KU Leuven, Celestijnenlaan 200D B2401, 3001 Leuven, Belgium

[3]Université Côte d'Azur, Observatoire de la Côte d'Azur, CNRS, Laboratoire Lagrange, Bd de l'Observatoire, CS 34229, 06304 Nice cedex 4, France

[4]Anton Pannenkoek Institute for Astronomy, University of Amsterdam, 1090 GE, Amsterdam, The Netherlands

[5]Max Planck Institute for Astronomy, Königstuhl 17, 69117, Heidelberg, Germany

[6]Instituto de Astronomía, Universidad Nacional Autónoma de México, Apdo. Postal 70264, Ciudad de México, 04510, México

[7]European Southern Observatory, Alonso de Cordova 3107, Vitacura, Santiago, Chile

[8]Aix Marseille Université, CNRS, LAM (Laboratoire d'Astrophysique de Marseille) UMR 7326, 13388, Marseille, France

[9]School of Chemistry, University of Leeds, Leeds LS2 9JT, UK

[10]Max Planck Institute for extraterrestrial Physics, Giessenbachstraße 1, D-85748 Garching, Germany

[11]Center for Astrophysics, Harvard & Smithsonian, 60 Garden Street, Cambridge, MA 02138, USA

[12]NSF's National Optical-Infrared Astronomy Research Laboratory, PO Box 26732, Tucson, AZ 85726-6732, USA

[13]European Southern Observatory, Karl-Schwarzschild-Str. 2, 85748, Garching bei München, Germany

[14]Steward Observatory, 933 N. Cherry Avenue, University of Arizona, Tucson, AZ, 85721, USA

[15]University of Exeter, School of Physics and Astronomy, Stocker Road, Exeter, EX4 4QL, UK

[16]Physics Department, New Mexico Institute of Mining and Technology, 801 Leroy Place, Socorro, NM 87801, USA

[17]NASA Goddard Space Flight Center, Exoplanets & Stellar Astrophysics Laboratory, Code 667, Greenbelt, MD 20771, USA

[18]Department of Physics, Catholic University of America, Washington, DC 20064, USA

[19]Department of Astronomy, University of Michigan, Ann Arbor, MI, 48109, USA

[20]Univ. Grenoble Alpes, CNRS, IPAG, 38000 Grenoble, France

Abstract. From November 2019 to April 2020, the prototypical red supergiant Betelgeuse experienced an unexpected and historic dimming. This event was observed worldwide by astrophysicists, and also by the general public with the naked eye. We present here the results of our observing campaign with ESO's VLT and VLTI in the visible and infrared domains. The observations with VLT/SPHERE-ZIMPOL, VLT/SPHERE-IRDIS, VLTI/GRAVITY and VLTI/MATISSE provide spatially resolved diagnostics of this event. Using PHOENIX atmosphere models and RADMC3D dust radiative transfer simulations, we built a consistent model reproducing the images and the photometry.

Keywords. Stars, Stellar evolution, Time-domain astronomy, Transient astrophysical phenomena, Stars: individual: Betelgeuse, supergiants, Stars: mass-loss, Stars: imaging, Techniques: high angular resolution.

1. Introduction

Mass loss of red supergiants (RSG) is far from being understood. In particular, the triggering mechanism that allows the material to escape the photosphere remains unknown. Two main scenarios have been proposed to explain the origin of the outflow: (1) the combined action of a lowering of the effective gravity through turbulent velocities initiated by giant convective cells (Kee et al. 2021), possibly supported by radiative pressure on molecular lines (Josselin & Plez 2007), and (2) Alfvén waves dissipation through the chromosphere (Airapetian et al. 2000). Polarimetric adaptive-optics imaging in the visible revealed dust clumps at a few stellar radii from the photosphere (Kervella et al. 2016 and Cannon et al. 2021), pointing to an episodic and non-homogeneous mass loss, further complicating the picture.

High angular resolution observations represent the best tool to find the missing mechanism by observing material ejection close to the photosphere. Being the second closest RSG (222^{+48}_{-34} pc; see Harper et al. 2017 and Joyce et al. 2020), Betelgeuse (α Ori) is among the preferred targets for such observations, and often considered to be the prototypical star of this luminosity class. Its visual light curve shows two main periods of $\sim 400-420$, and ~ 2100 days respectively (Stothers 2010). At the end of 2019, while it was on its way to its anticipated minimum, its brightness decreased faster than usual (Guinan et al. 2019). We now know that this was the beginning of what has been called 'the Great Dimming of Betelgeuse' with the star reaching its historic visible minimum on 7-13 February 2020 (Guinan et al. 2020 and Fig. 1), few days after the expected minimum. Its V band magnitude was then $V = 1.614 \pm 0.008$ mag.

Here we provide a concise summary of the main findings of the refereed article Montargès et al. (2021), that have been presented during the conference.

2. Observations

We obtained observations of Betelgeuse one year before the 'Great Dimming'. By chance these have been scheduled exactly at the previous light minimum. The VLT/SPHERE (Beuzit et al. 2019) data were obtained on the night of January 1st 2019, both with the Zurich Imaging Polarimeter (ZIMPOL, Schmid et al. 2018) in visible polarimetric imaging, and with the infra-red dual imaging and spectrograph (IRDIS, Dohlen et al. 2008) for infrared sparse aperture masking (SAM, Cheetham et al. 2016). VLTI/GRAVITY (Gravity Collaboration et al. 2017) data were secured in the compact configuration of the auxiliary telescopes (A0-B2-D0-C1) on 20th January 2019. Following successful Director Discretionary Time proposals, observations were executed during the 'Great Dimming' on 14th February 2020 with VLTI/GRAVITY, 27th December 2019

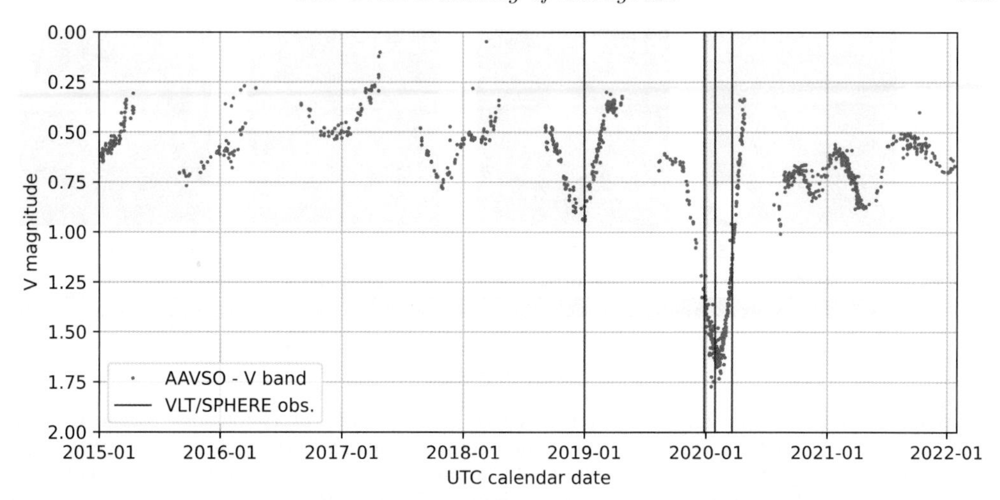

Figure 1. V band light curve of Betelgeuse between 2015 and 2022 from the American Association of Variable Star Observers (AAVSO) measurements. The vertical lines correspond to the dates of the VLT/SPHERE-ZIMPOL images.

with VLT/SPHERE-IRDIS, and 27th December 2019 (before the light minimum), 28th January 2020 (at the light minimum), and 18th, 19th and 21st March 2020 (after the light minimum) with VLT/SPHERE-ZIMPOL.

The GRAVITY data, combined with the short baseline IRDIS SAM observations, allow to precisely measure the angular diameter in the near infrared (NIR). The visible ZIMPOL direct imaging spatially resolves the photosphere of the star, owing to its large angular diameter (Sect. 59).

The data reduction and calibration are described in details in Montargès et al. (2021).

3. Measuring the angular diameter with optical interferometry

The 'Great Dimming' could have been caused by a change of size of Betelgeuse. The angular diameter is also an important input of any numerical model. We used the combination of interferometric observables from VLT/SPHERE-IRDIS (Cnt_K2 filter) in SAM mode and VLTI/GRAVITY (in the K band continuum at $2.22 - 2.28$ μm) to measure it. The angular diameters values obtained are $\theta_{\rm UD} = 42.61 \pm 0.05$ mas (reduced $\chi^2 = 26.5$) before the Dimming, and $\theta_{\rm UD} = 42.11 \pm 0.05$ mas (reduced $\chi^2 = 46.3$) during the Dimming. An attempt to fit the data with a power-law limb-darkened disk does not change significantly the reduced χ^2 values, nor the angular diameter.

The angular diameter variation is far from the 30% decrease that would be necessary to fully reproduce the 'Great Dimming'. It is consistent with the variations registered over the previous decades (Ohnaka et al. 2009).

4. Spatially resolved imaging of the photosphere

The spatially resolved images of Betelgeuse in the visible (Fig. 2) reveal an almost spherical photosphere before (January 2019) the 'Great Dimming', slightly elongated in the North-East to South-West direction. During the 'Great Dimming', the Southern hemisphere appears much dimmer than the rest of the star. This is modeled in the following sections.

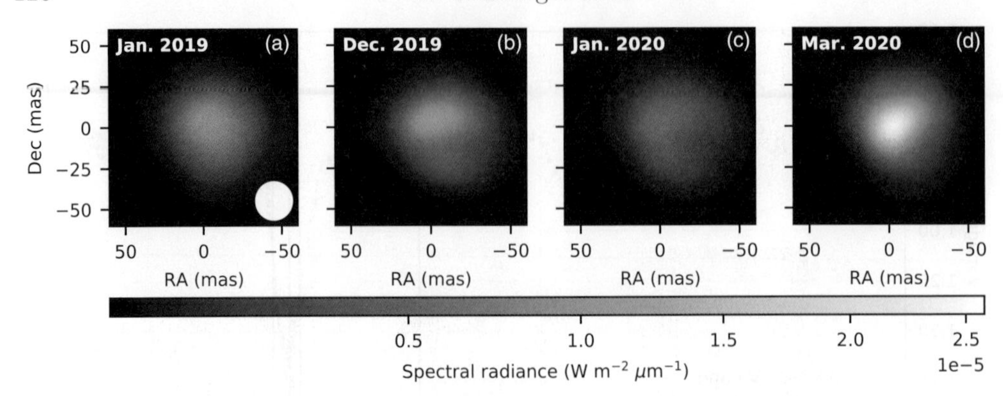

Figure 2. Deconvolved VLT/SPHERE-ZIMPOL observations of Betelgeuse in the Cnt_Hα filter (644.9 nm). North is up; east is left. The beam size of ZIMPOL is indicated by the white disk in panel a. A power-law scale intensity with an index of 0.65 is used to enhance the contrast. Panel a: January 2019. b: December 2019. c: January 2020. d: March 2020.

Table 1. Best matching parameters of the synthetic cool patch PHOENIX models with the ZIMPOL observations. $\log \mathcal{L}$ is the log-likelihood. See Sect. 33 for the definition of the parameters.

Parameter	**December 2019**	**January 2020**	**March 2020**
x_p (mas)	−7.1	−2.4	−28.4
y_p (mas)	−14.2	−2.4	−35.6
r_p (mas)	23.7	19.0	45.0
T_{hot} (K)	3,700	3,700	3,700
T_{cool} (K)	3,400	3,400	3,200
$\log \mathcal{L}$	-8.8×10^6	-5.5×10^7	-4.0×10^7

4.1. *A constant dust component in the line of sight*

It has been known for many decades that Betelgeuse is permanently surrounded by a relatively thin dusty envelope (see for example Verhoelst et al. 2009 and Kervella et al. 2011 and references therein). In order to determine the dust contribution to the 'Great Dimming' (if any), it is important to determine the dust component already detectable in the January 2019 images.

In order to estimate the contribution of the dust in the pre-dimming images, we derived the photometry from the SPHERE-ZIMPOL images, and collected the contemporaneous data from the American Association of Variable Stars Observers (AAVSO). We modeled these photometric measurements using a Cardelli extinction law (Cardelli et al. 1989) applied to a PHOENIX atmosphere of a RSG (Lançon et al. 2007) at 3600 K. We derive $R_V = 4.2$ and $A_V = 0.65$. Hereafter we consider that the 'Great Dimming' event happens in addition to this pre-existing extinction.

4.2. *Photospheric cool spot model*

We first attempt to reproduce the 'Great Dimming' by considering a cool patch on the photosphere of Betelgeuse. We model this by creating a composite stellar disk from PHOENIX atmosphere models (Lançon et al. 2007) at different temperatures.

Three parameters fully define the cool patch in addition to its temperature: its center position on the stellar disk (x_p, y_p), and its size r_p. The modeling details are available in Montargès et al. (2021). Table 1 shows the best match parameters, and the corresponding images are shown in Fig. 3. The photometry is plotted on Fig. 4.

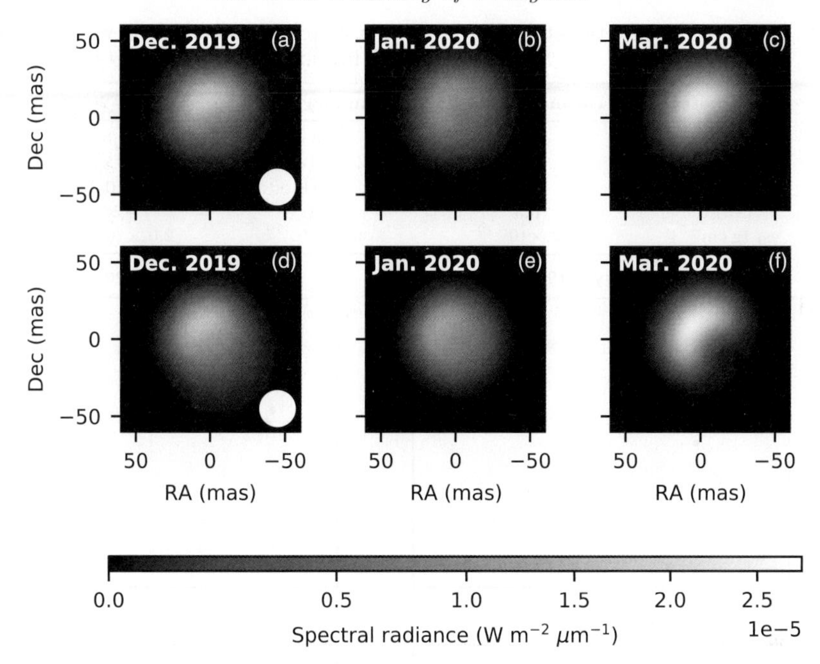

Figure 3. Best matching models for the ZIMPOL images in the Cnt_Hα filter (644.9 nm). The upper images correspond to the cool spot PHOENIX model, the lower images to the dusty clump RADMC3D simulations.

Here we have only assessed the compatibility of a cool spot model with the 'Great Dimming'. Precisely constraining the temperatures of the photosphere and the spot would require multi-spectral spatially resolved imaging,

4.3. *Dusty clump model*

Another scenario to explain the 'Great Dimming' of Betelgeuse could be the formation of a cloud of dust in the line of sight. We explore this hypothesis through RADMC3D (Dullemond 2012) radiative transfer simulations.

In our models we consider the x axis to be oriented West to East, the y axis South to North, and the z axis towards the observer. The spherical dust clump is centred at $(x_c\ , y_c\ , z_c)$, r_c being its radius, and possess a constant dust density of ρ_0. We consider the dust to be made of $MgFeSiO_4$ (Jaeger et al. 1994; Dorschner et al. 1995). We set the grain size to have maximum absorption by the dust clump in the visible, which is achieved adopting grain radii in the range 0.18 to 0.24 μm, following a gaussian size distribution.

We produced a grid of simulations to be matched with the SPHERE-ZIMPOL images and their photometry. The procedure did not converge for the images taken in January and March 2020. For these two epochs, we used best guesses obtained by manually matching the observables. The results are summarized in Table 2, and the images are shown in Fig. 3. The corresponding photometry is represented on Fig. 4.

5. Conclusion

Both types of models (cool patch and dusty clump) are able to reproduce the images of the 'Great Dimming' obtained by VLT/SPHERE-ZIMPOL. However, both are not fully reproducing the SED, particularly in the NIR (Fig. 4). Comparison with the other observations of the 'Great Dimming' are discussed in Montargès et al. (2021). In order

Table 2. Best matching parameters for the RADMC3D clump simulations with the ZIMPOL observations. See Sect. 34 for the definition of the parameters.

Parameter	December 2019	January 2020	March 2020
x_c (au)	−1.9	−0.8	−1.9
y_c (au)	−3.0	−0.6	−1.8
z_c (au)	12.5	20.0	20.0
r_c (au)	6.5	5.0	4.5
ρ_0 (g cm^{-3})	3.2×10^{-19}	4.0×10^{-19}	2.0×10^{-18}

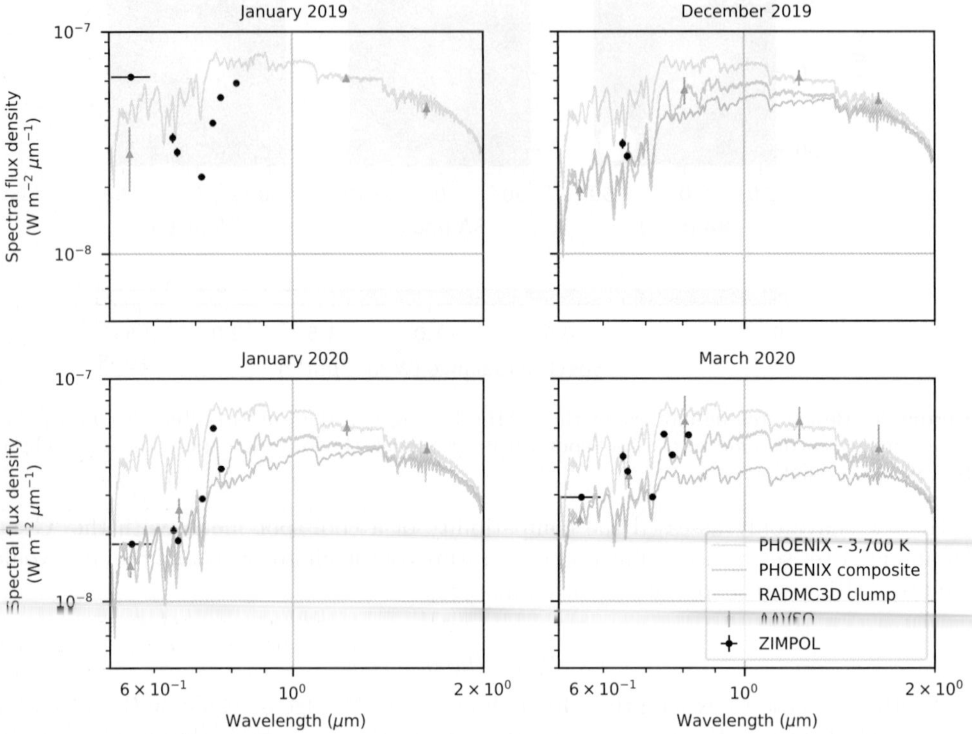

Figure 4. Observed photometry and modeled SEDs for Betelgeuse before (a), and during the 'Great Dimming' (b, c, and d). The black dots correspond to the ZIMPOL photometry, the gray triangles to the AAVSO measurements. The orange curve represents the SED of a PHOENIX 15 $M_\odot$ RSG with $T_{\rm eff} = 3700$ K. The blue curve corresponds to the best matching PHOENIX cool patch model, and the purple curve to the best dusty clump RADMC3D simulation.

to attempt a full explanation of the event, and to frame it within the general pulsation pattern of Betelgeuse (the exceptional light minimum happend 4 days later than the expected 2020 minimum), we conclude that both scenarios happened consecutively. A gas cloud had been expelled by the star months or years before the event. In late 2019 a cool patch appeared on the visible hemisphere of Betelgeuse, which triggered dust nucleation in the dust cloud located above the photosphere. The 'Great Dimming' is the combination of the cool spot and the dusty clump, which explains why each hypothesis cannot fully explain the observations individually.

References

Airapetian, V. S., Ofman, L., Robinson, R. D., Carpenter, K., & Davila, J. 2000, *ApJ* 528, 965
Beuzit, J. L., Vigan, A., Mouillet, D., et al. 2019, *A&A* 631, A155
Cannon, E., Montargès, M., de Koter, A., et al. 2021, *MNRAS* 502(1), 369
Cardelli, J. A., Clayton, G. C., & Mathis, J. S. 1989, *ApJ* 345, 245

Cheetham, A. C., Girard, J., Lacour, S., et al. 2016, in F. Malbet, M. J. Creech-Eakman, & P. G. Tuthill (eds.), *Optical and Infrared Interferometry and Imaging V*, Vol. 9907 of *Society of Photo-Optical Instrumentation Engineers (SPIE) Conference Series*, p. 99072T

Dohlen, K., Langlois, M., Saisse, M., et al. 2008, in I. S. McLean & M. M. Casali (eds.), *Ground-based and Airborne Instrumentation for Astronomy II*, Vol. 7014 of *Society of Photo-Optical Instrumentation Engineers (SPIE) Conference Series*, p. 70143L

Dorschner, J., Begemann, B., Henning, T., Jaeger, C., & Mutschke, H. 1995, *A&A* 300, 503

Dullemond, C. P. 2012, *RADMC-3D: A multi-purpose radiative transfer tool*, Astrophysics Source Code Library

Gravity Collaboration, Abuter, R., Accardo, M., et al. 2017, *A&A* 602, A94

Guinan, E., Wasatonic, R., Calderwood, T., & Carona, D. 2020, *The Astronomer's Telegram* 13512, 1

Guinan, E. F., Wasatonic, R. J., & Calderwood, T. J. 2019, *The Astronomer's Telegram* 13341, 1

Harper, G. M., Brown, A., Guinan, E. F., et al. 2017, *AJ* 154, 11

Jaeger, C., Mutschke, H., Begemann, B., Dorschner, J., & Henning, T. 1994, *A&A* 292, 641

Josselin, E. & Plez, B. 2007, *A&A* 469, 671

Joyce, M., Leung, S.-C., Molnár, L., et al. 2020, *ApJ* 902(1), 63

Kee, N. D., Sundqvist, J. O., Decin, L., de Koter, A., & Sana, H. 2021, *A&A* 646, A180

Kervella, P., Lagadec, E., Montargès, M., et al. 2016, *A&A* 585, A28

Kervella, P., Perrin, G., Chiavassa, A., et al. 2011, *A&A* 531, A117

Lançon, A., Hauschildt, P. H., Ladjal, D., & Mouhcine, M. 2007, *A&A* 468, 205

Montargès, M., Cannon, E., Lagadec, E., et al. 2021, *Nature* 594(7863), 365

Ohnaka, K., Hofmann, K.-H., Benisty, M., et al. 2009, *A&A* 503, 183

Schmid, H. M., Bazzon, A., Roelfsema, R., et al. 2018, *A&A* 619, A9

Stothers, R. B. 2010, *ApJ* 725, 1170

Verhoelst, T., van der Zypen, N., Hony, S., et al. 2009, *A&A* 498, 127

The Origin of Outflows in Evolved Stars
Proceedings IAU Symposium No. 366, 2022
L. Decin, A. Zijlstra & C. Gielen, eds
doi:10.1017/S1743921322001065

The inner dust shell of Betelgeuse seen with polarimetry

Xavier Haubois

European Organisation for Astronomical Research in the Southern Hemisphere, Casilla 19001, Santiago 19, Chile
email: xhaubois@eso.org

Abstract. The origin of red supergiant mass loss still remains to be understood. Characterizing the formation zone and the dust distribution within a few stellar radii above the surface is key to understanding the mass loss phenomenon. With its angular diameter of about 42 mas in the optical, Betelgeuse makes an ideal target to resolve the inner structures that represent potential signatures of dust formation. Past polarimetric observations reveal a dust environment in the first stellar radii. Depending on their characteristics and composition, dust grains could interact with the stellar radiation, trigger mass loss by momentum transfer from photons to dust to gas. Using spatially-resolved polarimetric observations of Betelgeuse, we detect a quasi-symmetric inner dust shell centered at ~0.5 stellar radii above the photosphere and attempt at constraining its dust population.

1. Introduction

The recent findings on the 2019-2020 Betelgeuse dimming highlighted further the importance of understanding dust formation to account for photometric variability that is observed in evolved stars (e.g., Dupree et al. 2020; Montargès et al. 2021). High angular-resolution polarimetric observations in these objects enable us to detect and characterize their dusty environments. In the case of stars belonging to the Asymptotic Giant Branch (AGBs), the combination of interferometry with polarimetry allows us to estimate the angular extent and flux ratio of dust environments with respect to the central photosphere (Ireland et al. 2005; Norris et al. 2012). Under assumptions of Mie scattering, dust grain sizes can be estimated. This technique was used for Betelgeuse where a polarizing structure was interpreted as a thin dust shell located only 0.5 stellar radius above the photosphere. With grain sizes of about 300 nm, the dust composition was found to be compatible with $MgSiO_3$ (enstatite) and Mg_2SiO_4 (forsterite) (Haubois et al. 2019). As these datasets didn't allow for reliable interferometric imaging, simple geometries made of spherical thin shells were assumed. However, depending on the scenario, mass loss mechanisms can leave characteristic signatures in the morphology of the dust environments. We here present preliminary results of an imaging campaign of Betelgeuse with the VLT/SPHERE instrument (Beuzit et al. 2019).

2. Observations with SPHERE

Observations in January 2019 were taken in 10 filters but we discarded the broad V-band filter as it showed evidence for instrumental polarization crosstalk. We present typical set of Stokes parameters and linear polarized intensity images in Fig. 1. The linear polarized intensity ($PI = \sqrt{Q^2 + U^2}$) and angle of linear polarization ($AoLP =$

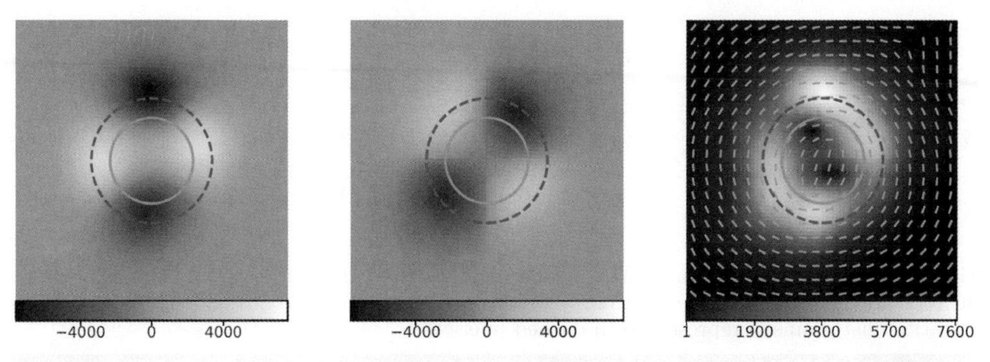

Figure 1. Stokes Q (left), U (middle) and polarized intensity (right) at 644nm. Solid red and dashed blue circles represent the photospheric radius (22 mas) and the position of the inner dust shell previously found in Haubois et al. (2019) (32 mas), respectively. On the polarized intensity map, angles of polarization are marked as orange segments. Units are W.m^{-2}.sr^{-1}.

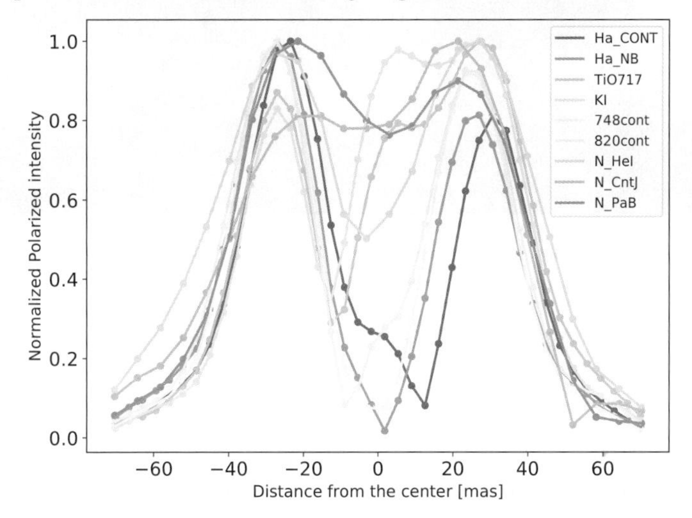

Figure 2. Normalized vertical cut of Polarized Intensities in each filter.

$\frac{1}{2}\arctan\left(\frac{U}{Q}\right)$) are consistent with a quasi-azimuthal polarizing structure centered around 30 mas of distance (see vertical cuts as a function of the filters in Fig. 2). The variation of mean polarized intensity along a ring of 30-mas radius is about 50%.

As a first modelling step, such ring-like structures can be reproduced with spherical dust shells. In order to fit the images, their external radius should be of about 9.5 AU (about 2 stellar radii). The spectral variation of the polarized intensity varies depending on the azimuthal sector but can be relatively well approximated using a silicate composition for the dust shell (Fig. 3). More details of the modeling will be presented in a forthcoming publication. The nature of the dark central features seen on the right part of Fig. 1 is ambiguous. In principle, they could be the results of a lesser scattering efficiency, e.g. zones of multi-scattering , under-densities, or effects of the radiation field illuminating this inner dust shell, like large surface convective cells. It is then useful to make use of complementary techniques such as spectro-polarimetry where brillance maps can be reconstructed from linear polarization spectral profiles (López Ariste et al. 2018). Comparison of the two datasets are on-going.

Finally, the temporal variation of polarized intensities can give us precious hints on the origin and mechanisms affecting the inner dust shell of Betelgeuse. Fig. 4 shows three epochs in two filters. Over a few years, we can see that the external radius doesn't change

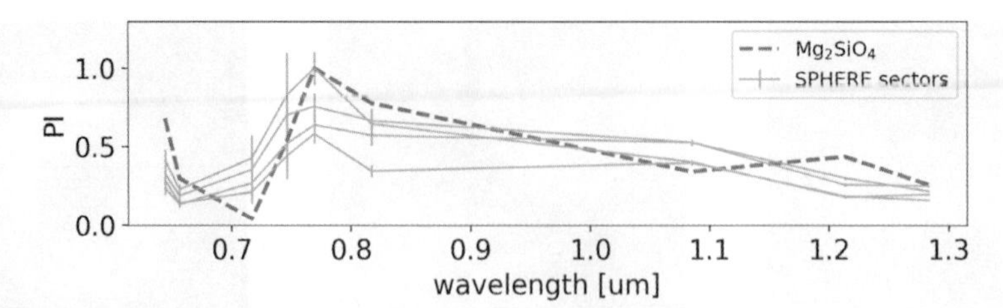

Figure 3. Spectral variation of the normalized polarized intensities for 4 azimuthal sectors (sectors are defined by large departure to the average value). A dust shell model made of enstatite dust grains is represented in dashed line.

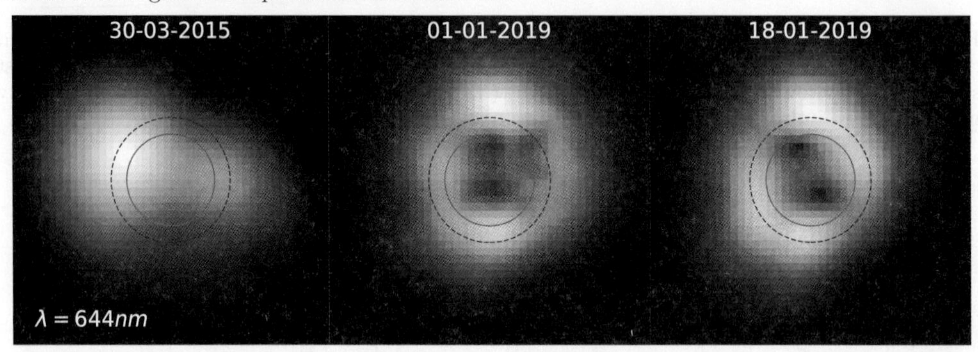

Figure 4. Evolution of Betelgeuse polarized intensities at 644 nm. The meaning of circles is the same as in Fig. 1.

dramatically but that the morphology does with a timescale that is probably beyond 1–2 months. This is consistent with Safonov et al. (2020) who shows a particularly well time-sampled variation of images obtained with speckle polarimetry during Betelgeuse's dimming. It is not surprising that so close to the photosphere, timescales are similar to those of surface convection in red supergiants. However, the way a convective surface would shape the inner dust shell remains to be established.

3. Conclusion

High angular-resolution polarimetry reveals that the inner dust shell of Betelgeuse is encircled in the first 2 stellar radii and its morphology and density distribution vary with a timescale of about 1–2 months. A characterization of the dust population can be attempted using radiative transfer methods. To connect this dusty structure to other mass loss components, it appears critical to perform a polarimetric monitoring at a sufficient cadence and combine it with other techniques such as spectro-polarimetry.

References

Beuzit, J. L., Vigan, A., Mouillet, D., et al.,2019,*A&A*, 631, A155
Dupree, A. K., Strassmeier, K. G., Matthews, L. D., et al. 2020, *ApJ*, 899, 68
Haubois, X., Norris, B., Tuthill, P. G., et al. 2019, *A&A*, 628, A101
Ireland, M. J., Tuthill, P. G., Davis, J., et al. 2005, *MNRAS*, 361, 337
Kervella, P., Perrin, G., Chiavassa, A., et al. 2011, *A&A*, 531, A117
López Ariste, A., Mathias, P., Tessore, B., et al. 2018,*A&A*, 620, A199
Montargès, M., Cannon, E., Lagadec, E., et al. 2021, *Nature*, 594, 365
Norris, B. R. M., Tuthill, P. G., Ireland, M. J., et al. 2012, *Nature*, 484, 220
Safonov, B., Dodin, A., Burlak, M., et al., 2020, arXiv e-prints, arXiv:2005.05215

The Origin of Outflows in Evolved Stars
Proceedings IAU Symposium No. 366, 2022
L. Decin, A. Zijlstra & C. Gielen, eds
doi:10.1017/S1743921322000989

Analytic, Turbulent Pressure Driven Mass Loss Rates from Red Supergiants

J.O. Sundqvist[1] and **N.D. Kee**[1,2]

[1]Institute of Astronomy, KU Leuven, Celestijnenlaan 200D, B-3001 Leuven, Belgium,
email: jon.sundqvist@kuleuven.be

[2]National Solar Observatory, 22 Ohi'a Ku St, Makawao, HI 96768, USA
email: dkee@nso.edu

Abstract. Although red supergiants (RSGs) are observed to be undergoing vigorous mass loss, explaining the mechanism launching their winds has been a long-standing problem. Given the importance of mass loss to stellar evolution in this phase, this is a key uncertainty. In this contribution we present a recently published model (Kee et al. 2021) showing that turbulent pressure alone can extend the stellar atmosphere of an RSG to the degree that a wind is launched. This provides a fully analytic mass-loss prescription for RSGs. Moreover, utilising observationally inferred turbulent velocities for these objects, we find that this wind can carry an appropriate amount of mass to overall match observations. Intriguingly, when coupled to stellar evolution models the predicted mass-loss rates show that stars with initial masses above $M_{\rm ini} \sim 17 M_\odot$ may naturally evolve back to the blue and as such not end their lives as RSGs; this is also in overall good agreement with observations, here of Type II-P/L supernova progenitors. Moreover, since the proposed wind launching mechanism is not necessarily sensitive to metallicity, this could have important implications for stellar evolution predictions in low-metallicity environments.

1. Introduction

The lack of a satisfactory theory explaining the strong, $> 10^{-7}$ M$_\odot$ yr^{-1}, mass loss for evolved massive stars on the red supergiant (RSG) branch has been a long standing problem in our understanding of these objects (see Levesque 2017, for a recent review). Namely, while for lower-mass asymptotic giant branch (AGB) stars it is generally assumed that strong pulsations lift gas up to radii where radiation pressure on dust grains can drive it out of the stellar potential (see, e.g., contribution by S. Höffner in these proceedings), in comparison the dust-condensation radius of RSGs is believed to (on average) be located much further away from the stellar surface. Indeed, modeling attempts have been generally unsuccessful in generating the atmospheric extensions of RSGs necessary to put enough material at the dust sublimation front (e.g., Arroyo-Torres *et al.* 2015).

An alternative suggestion has been that pulsational motions might be accompanied or replaced by significant atmospheric turbulence (Gustafsson & Plez 1992; Josselin & Plez 2007), and that this turbulence might be seeded by the vigorous convection expected in the atmospheres of RSGs (Freytag *et al.* 2012); indeed, observations of red supergiants do indicate that the outer layers of these stars are very turbulent (e.g., Josselin & Plez 2007; Ohnaka et al. 2017). Inspired by the work of Gustafsson & Plez (1992) and Josselin & Plez (2007), we have recently derived analytic mass-loss rates that focus on these large observed turbulent velocities present in RSGs (Kee et al. 2021).

2. The model

As outlined in detail by Kee et al. (2021), for a constant mass-loss rate $\dot{M} = 4\pi \rho v r^2$ we write the 1D, stationary equation of motion as

$$v \left(1 - \frac{a^2 + v_{\rm turb}^2}{v^2}\right) \frac{dv}{dr} = \frac{2\left(a^2 + v_{\rm turb}^2\right)}{r} - \frac{GM_*\left(1-\Gamma\right)}{r^2}, \tag{2.1}$$

where a is the isothermal sound speed, $v_{\rm turb} = \sqrt{P_{\rm turb}/\rho}$ is the turbulent velocity with associated turbulent pressure $P_{\rm turb}$, and $\Gamma \equiv \kappa L_*/(4\pi G M_* c)$ is the Eddington factor expressing the ratio of radiative to gravitational acceleration for an opacity κ, stellar luminosity L_*, and stellar mass M_*.

The location of the modified Parker (1958) radius, defined here as the point at which the flow velocity equals an 'effective' sound speed $a_{\rm eff} \equiv \sqrt{a^2 + v_{\rm turb}^2}$, is

$$R_{\rm p,mod} = \frac{G\,M_*\left(1-\Gamma\right)}{2\left(a^2 + v_{\rm turb}^2\right)}, \tag{2.2}$$

yielding the generic mass-loss rate

$$\dot{M} = 4\,\pi\,\rho(R_{\rm p,mod})\,a_{\rm eff}(R_{\rm p,mod})\,R_{\rm p,mod}^2\,. \tag{2.3}$$

For a given effective sound speed $a_{\rm eff}$, the problem in hand thus boils down to estimating the density ρ at this modified Parker radius $R_{\rm p,mod}$.

Assuming first an isothermal atmosphere with temperature $T = T_{\rm eff}$ and constant opacity κ, this density can be analytically estimated by computing the optical depth τ from an assumed stellar radius at $R_* \equiv r(\tau = 2/3)$ to $R_{\rm p,mod}$ (see Kee et al. 2021, their Sect. 2., for details). This yields a fully analytic expression for $\dot{M}$ as function of the input stellar parameters L_*, M_*, R_*, and (an assumed constant) $v_{\rm turb}$. Relaxing the isothermal assumption, we next compute a temperature structure following Lucy (1971) (see also eqns. 16-17 in Kee et al. 2021), numerically solve the equation of motion, and iterate toward an internally consistent mass-loss rate; comparing this then to the fully analytic isothermal result, we derive a non-isothermal correction factor to the analytic model. This yields a final mass-loss rate as predicted by our model:

$$\dot{M} = \dot{M}_{\rm an} \left(\frac{v_{\rm turb}/(17\,{\rm km\,s^{-1}})}{v_{\rm esc}/(60\,{\rm km\,s^{-1}})}\right)^{1.30}, \tag{2.4}$$

where $v_{\rm esc}$ is the escape speed from the stellar surface R_*, and the analytic mass-loss rate $\dot{M}_{\rm an}$ is given by equations (5),(7),(8),(11), and (13) in Kee et al. (2021).

As demonstrated, the above essentially is a modified Parker-like wind model, where the potential for initiating a large RSG mass loss simply lies in the very loosely bound envelopes of these stars. This can be seen more directly by using the effective (i.e., the one reduced by $1-\Gamma$) escape speed from the stellar surface to re-write the modified Parker radius as

$$\frac{R_{\rm p,mod}}{R_*} = \frac{1}{4}\frac{v_{\rm esc,eff}^2}{a_{\rm eff}^2}\,. \tag{2.5}$$

For a sun-like star the escape speed from the stellar surface ($v_{\rm esc,\odot} \sim 600\,{\rm km/s}$) is very much larger than the effective photospheric sound speed ($a_{\rm eff} \sim 8{\rm km/s}$). This means that a very hot corona with $T \sim 10^6$K is required to lift material up to a Parker point located only a few radii above the stellar surface. On the other hand, for RSGs the effective escape speed is about an order of magnitude lower than for sun-like stars, so that only a modest amount of turbulent velocity is required to shift the location of the Parker point to regions reasonably close to R_*. This is the essential point as to why such atmospheric turbulence can play a key role in initiating significant mass loss from the very extended RSG atmospheres, while it will be a very ineffective mechanism for high-gravity stars on the main-sequence.

3. Some first analysis and implications of new mass-loss rates

Because of the essentially exponential dependency of density on the effective atmospheric scale-height, the predicted mass-loss rates in our model are extremely sensitive to the quantitative input value of $v_{\rm vturb}$. However, from the RSG samples compiled by Josselin & Plez (2007) and Ohnaka et al. (2017) a high mean observed velocity dispersion $v_{\rm disp} = 20.3\,{\rm km/s}$ can be inferred (Kee et al. 2021). In these studies, the characteristic values of $v_{\rm disp}$ have been obtained from analysing line-of-sight velocity shifts in spectral lines using a tomography technique (Josselin & Plez 2007) and by means of direct mapping of the projected velocity across the stellar surface as observed in some strategic molecular lines (Ohnaka et al. 2017); on the other hand, reproducing the corresponding observationally inferred mass-loss rates for the same stars within our model only requires $v_{\rm turb} = 18.2\,{\rm km/s}$ (Kee et al. 2021). As such, to the extent that we may identify these inferred velocity dispersions with the turbulent velocity entering our model, the predicted mass-loss rates indeed lie in the correct range. This illustrates the large potential of turbulent pressure for levitating RSG atmospheres, and lends some first support to the proposed mass-loss model. Nonetheless, we emphasise that these characteristic values should be interpreted only in this kind of average manner; when inspecting individual RSGs, there is large scatter both regarding inferred velocity dispersions and empirically derived mass-loss rates.

The latter is also reflected in the large discrepancies present in the various empirical mass-loss recipes for RSGs present on the market. Indeed, even for a given luminosity these empirical mass loss scalings can differ by huge amounts, up to several orders of magnitude depending on the chosen recipe (see Kee et al. 2021, their Fig. 8, for a comparison of different recipes). Given these large uncertainties in current empirical calibrations, the model proposed here may also be taken as a reasonable option for various applications where RSG mass loss is important.

As just a first example of this, we here compute stellar evolution tracks using i) the (quite standard) empirical mass loss calibration by de Jager et al. (1988) and ii) our new predicted rates. Specifically, while the applied mass-loss rates are assumed to be equivalent for hot stars ($T_{\rm eff} > 10$ kK), for cool stars ($T_{\rm eff} < 10$ kK) we do two separate simulation sets. The first of these retains the standard de Jager et al. (1988) mass-loss rates as a baseline. The other preferentially uses our new Kee et al. (2021) prescription with the suggested default $v_{\rm turb} = 18.2$ km s^{-1} from that paper and above. However, turbulent pressure initiated mass loss is developed for application on the RSG branch itself, and as such is not (yet) well calibrated for yellow supergiants. We therefore take the maximum between the de Jager et al. (1988) and the Kee et al. (2021) mass-loss rate whenever 10 kK $> T_{\rm eff} >$ 5 kK. This has the effect of using the de Jager et al. (1988) rates on the first crossing of the Hertzsprung gap, before the star inflates on the RSG branch, and instead using the Kee et al. (2021) rates for post-RSG objects. Finally, for $T_{\rm eff} < 5$ kK, this new scheme always uses Kee et al. (2021). The implementation of this "Leuven-modified Dutch mass loss scheme" in the stellar evolution code MESA (Paxton et al. 2011; Paxton et al. 2013) is available at https://doi.org/10.5281/zenodo.4333564.

Further specifications for these MESA calculations regard possible C/O enhancements in opacities as discussed in Paxton et al. (2011), mixing length theory applied according to the Ledoux criterion with a semiconvective mixing efficiency 0.01, and the MLT++ prescription as described in Paxton et al. (2013), their Section 7.2. In order to simplify

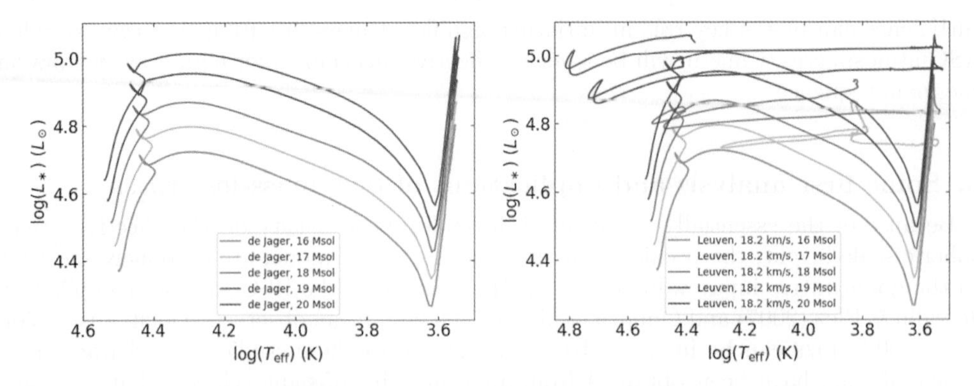

Figure 1. Comparison of stellar evolution tracks beginning from zero-age main sequence masses 16 to 20 $M_\odot$ in 1 $M_\odot$ increments. Stars in the left panel have been evolved with the de Jager et al. (1988) RSG mass-loss rates, while stars in the right panel were evolved using our new 'Leuven' mass-loss rates as described in the text.

the current models, we further omit convective overshooting in the simulations. The inlist files used for these simulations are available at https://doi.org/10.5281/zenodo.4333564.

Figure 1 shows simulated evolution tracks of stars with initial masses from 16 to 20 $M_\odot$ using the de Jager et al. 1988 mass-loss rates in the left panel and the simulations with our new ('Leuven') rates in the right panel. All stars are evolved up to carbon core depletion. The difference between these simulations is strikingly evident as all simulations using the de Jager rates die on the RSG branch while stars with initial mass $M_{\rm ini} \geq 17\ M_\odot$ using the Leuven mass-loss rates do not. Indeed, this is in general good agreement with the observationally inferred upper limit to the initial mass for Type II-P/L SNe ($16 M_\odot \lesssim M_{\rm ini} \lesssim 23 M_\odot$ (Smartt et al. 2009). These results are different than what was found in the recent study by Beasor *et al.* (2021), where the authors used their own new empirical RSG mass loss scaling in similar evolution models and found that then stars with initial masses below $30 M_\odot$ do not evolve back to the blue.

This difference in behaviour of the evolution models arises from the strong dependence of the Leuven mass-loss rates on stellar surface gravity ($\propto M_*/R_*^2$). Namely, as the star climbs up the RSG branch and loses mass, its surface gravity decreases further, thereby increasing the mass-loss rate in a positive feedback loop. This feedback of increased mass loss with RSG evolution effectively generates a competition of time scales between mass-loss induced stripping of the star's hydrogen envelope and the core nuclear burning timescale. Below the critical transition mass, here $\sim 17\ M_\odot$, the star runs out of nuclear fuel before losing its hydrogen envelope and ends its life as a Type II-P/L supernova. At that transition mass and above, mass loss wins out, the star loses almost its entire Hydrogen envelope, and in reaction the star contracts off the RSG branch back toward hotter effective temperatures.

Actually, also according to our models the stars lose mass at quite moderate rates during most of their time as RSGs. However, as the star evolves toward ever lower masses it eventually enters a short-lived RSG phase with strongly enhanced mass loss, which ultimately allows the star to lose most of its hydrogen envelope. In the evolution models displayed in the right panel of Fig. 1 here, the $M_{\rm ini} = 17 M_\odot$ ($M_{\rm ini} = 20 M_\odot$) model spends 3 % (14 %) of its RSG life-time having $\dot{M} > 10^{-4}\ {\rm M_\odot/yr}$. The de Jager et al. 1988 prescription misses this as their mass-loss rates do not scale with stellar mass, and it is further also unclear how well the new empirical scalings by Beasor *et al.* (2021) are able to capture these short-lived phases associated with strongly enhanced RSG mass loss.

Finally, the value of the maximum initial mass below which stars are predicted to die on the RSG branch indeed also depends on the choice of $v_{\rm turb}$ for the Kee et al. (2021)

mass-loss rates. Here we have taken an average value (see above) as being characteristic for the complete RSG phase, but it would certainly not be unreasonable to suspect that this might also vary with the RSG evolution. Nonetheless, it is interesting to note that the simple average $v_{\rm turb} = 18.2\,{\rm km/s}$ applied here, and obtained directly from comparison to empirical studies, immediately yields an upper limit to the initial mass for Type II-P/L SNe that seems to agree rather well with observations.

4. Origin of the turbulent velocity?

The turbulent velocity enters our model as an essentially free input parameter, albeit adjusted according to the observations that clearly indicate its presence. Naturally, however, a fully consistent theoretical model for RSG mass loss must also be able to predict $v_{\rm turb}$. As mentioned in the introduction, a natural candidate for this regards the vigorous convective motions expected to occur in the surface and sub-surface layers of RSG stars. Although such convective simulations typically have shown turbulent velocities that are smaller than suggested by observations (e.g., Arroyo-Torres *et al.* 2015), we note that the characteristic velocities observed in the recent radiation-hydrodynamic simulations by Goldberg et al. (2021) seem to be significantly higher. Moreover, these 3D simulations (as well as 1D evolution models such as those presented above) also show that RSG atmospheres breech the Eddington limit (defined by $\Gamma = 1$) already in deep sub-surface atmospheric layers. That is, just like for hotter stars (see contributions by S. Owocki, N. Moens), an approach accounting carefully for also the radiative acceleration around sub-surface (atomic) "opacity bumps" might be necessary when modelling the turbulent RSG surface and wind initiation. Moreover, if this wind launching mechanism ultimately is connected to hydrogen (or helium) recombination, this might have far-reaching consequences for massive-star evolution at low metallicity; indeed, assuming a constant $v_{\rm turb}$ the Kee et al. rates do not contain any direct dependency on the stellar metallicity.

References

Arroyo-Torres, B., Wittkowski, M., Chiavassa, A., et al. 2015, *A&A*, 575, A50

E.R. Beasor, B. Davies, N. Smith, 2021, *ApJ*, 922, 55

de Jager, C., Nieuwenhuijzen, H., & van der Hucht, K. A. 1988, *A&AS*,72, 259

Freytag, B., Steffen, M., Ludwig, H. G., et al. 2012, *J. of Comp. Ph.*, 231, 919

Goldberg, J. A., Jiang, Y.-F., & Bildsten, L. 2021, accepted for publication in *ApJ*, *arXiv e-prints, arXiv:2110.03261*

Gustafsson, B. & Plez, B. 1992, in Instabilities in Evolved Super- and Hypergiants, ed. C. de Jager & H. Nieuwenhuijzen, 86

Josselin, E. & Plez, B. 2007, *A&A*, 469, 671

Kee, N. D., Sundqvist, J. O., Decin, L., de Koter, A., & Sana, H. 2 2021, *A&A*, 646, A180

Levesque, E. 2017, Astrophysics of Red Supergiants IoP ebook (IoP Publishing, Bristol)

Lucy, L. B. 1971, *ApJ*, 163, 95

Ohnaka, K., Weigelt, G., & Hofmann, K. H. 2 2017, *Nature*, 548, 310

Parker, E. N. 1958, *ApJ*, 128, 664

Paxton, B., Bildsten, L., Dotter, A., et al. 2011, *ApJS*, 192, 3

Paxton, B., Cantiello, M., Arras, P., et al. 2013, *ApJS*, 208, 4

Smartt, S. J., Eldridge, J. J., Crockett, R. M., & Maund, J. R. 2009, *MNRAS*, 395, 1409

The Origin of Outflows in Evolved Stars
Proceedings IAU Symposium No. 366, 2022
L. Decin, A. Zijlstra & C. Gielen, eds
doi:10.1017/S1743921322000114

Multiple shell ejections on a 100 yr timescale from a massive yellow hypergiant

René D. Oudmaijer[1] **and Evgenia Koumpia**[2]

[1]School of Physics & Astronomy, University of Leeds, Woodhouse Lane, LS2 9JT Leeds, UK

[2]ESO Vitacura, Alonso de Córdova 3107 Vitacura, Casilla 19001 Santiago de Chile, Chile
email: r.d.oudmaijer@leeds.ac.uk

Abstract. This contribution focuses on a rare example of the class of post-Red Supergiants, IRAS 17163-3907, the central star of the Fried Egg nebula. In particular, we discuss some of our recently published results in detail. The inner parts of the circumstellar environment of this evolved massive star are probed at milli-arcsec resolution using VLTI's GRAVITY instrument operating in the K-band (2 μm), while larger, arcsecond, scales are probed by VISIR diffraction limited images around 10 μm, supplemented by a complete Spectral Energy Distribution. The spectro-interferometric data cover important diagnostic lines (Brγ, Na I), which we are able to constrain spatially. Both the presence and size of the Na I doublet in emission has been traditionally challenging to explain towards other objects of this class. In this study we show that a two-zone model in Local Thermal Equilibrium can reproduce both the observed sizes and strengths of the emission lines observed in the K-band, without the need of a pseudo-photosphere. In addition, we find evidence for the presence of a third hot inner shell, and demonstrate that the star has undergone at least three mass-loss episodes over roughly the past century. To explain the properties of the observed non-steady mass-loss we explore pulsation-driven and line-driven mass-loss and introduce the bi-stability jump as a possible underlying mechanism to explain mass-loss towards Yellow Hypergiants.

Keywords. stars: evolution, stars: mass-loss, stars: AGB and post-AGB, stars: individual: IRAS 17163-3907, circumstellar matter

---

1. Introduction on post-Red Supergiants

Amongst the Yellow Hypergiants (YHG) which are massive evolved stars, we find objects with evidence of having gone through a previous post-Red Supergiant (post-RSG) phase. This evidence is mostly based on a very large infrared excess indicating a previous mass losing phase, as well as CO rotational line emission probing these mass loss episodes (e.g. Oudmaijer et al. 1996; Wallström et al. 2017) indicating this phase to be the Red Supergiant phase, or similar. Such stars may well be in transition from the RSG phase to the Wolf-Rayet or Luminous Blue Variable (LBV) phases and are as such excellent laboratories to study massive star evolution. Only few such objects are known, with IRAS 17163-3907, IRC +10420, and HD179821 the most famous members of the class. More background can be found in reviews on post-Red Supergiants that appeared with about a ten year frequency over the past decades (de Jager 1998; Oudmaijer et al. 2009; Gordon and Humphreys 2019).

In this contribution we focus on our work on IRAS 17163-3907 (hereafter IRAS 17163) which was published recently (see Koumpia et al. 2020 where much more detailed information than we can provide here can be found). With an IRAS 12 μm flux density in excess of 1000 Jy, it is one of the brightest infrared sources on the sky, yet dedicated

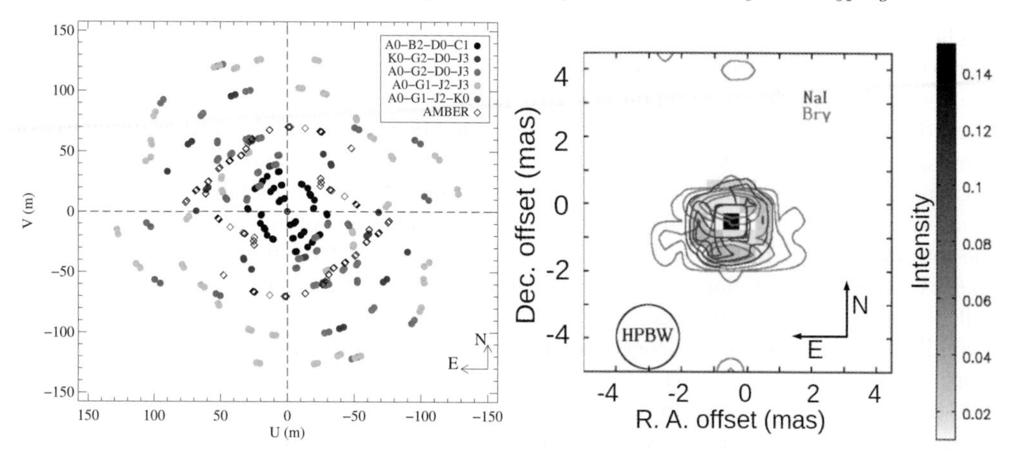

Figure 1. *Left*: Coverage of uv-plane of combined VLTI/GRAVITY and AMBER observations of IRAS 17163. *Right*: Image reconstruction of the continuum emission (greyscale) overplotted with Brγ in red contours and Na I in blue contours. Brγ arises from a larger region than the Na I emission, which in turn is more extended than the continuum.

studies of the object have been sparse. This might perhaps be due to the fact that the star is comparatively faint at optical wavelengths, and with a V band magnitude of $\sim$13 was not included in many of the mainstream optical catalogs that formed the basis for IRAS follow-up studies. Lebertre et al. (1989) did study the star and proposed it to be a low mass post-Asymptotic Giant Branch star, but many years later, Lagadec et al. (2011) suggested a more massive nature for the object based on its very high luminosity. Current estimates put the object at 1.2 kpc, and coupled with a very large optical extinction of order 11 magnitudes at V, a luminosity of $5{\times}10^5$ $L_\odot$ is derived (Koumpia et al. 2020). IRAS 17163 is surrounded by a dusty envelope, which because of its double-shell appearance was dubbed 'the Fried Egg Nebula' by Lagadec et al. (2011). In our recent work we identified an additional third hot inner shell.

2. Inner parts

To learn more about the immediate circumstellar environment, we observed IRAS 17163 with ESO's GRAVITY instrument on the VLTI (Eisenhauer et al. 2011) using the four 1.8 m Auxiliary Telescopes. A maximum angular resolution of $\lambda/2B \sim 1.7$ mas was achieved at 2 μm, covering the hydrogen Brγ line. This resolution corresponds to a spatial scale of $\sim$2 au at 1.2 kpc. The uv-plane coverage, shown in Fig. 1, is sufficiently well sampled that we could carry out a model-independent image reconstruction. The results of the image reconstruction are also presented in Fig.1. We obtain images at three different wavelengths; a line-free continuum, along the Brγ line at 2.16 μm and along the Na I doublet at 2.2 μm.

The size of the continuum at 2 μm is comparable with that of the stellar size computed from the distance, luminosity and temperature. This implies that there is no contribution of hot dust at these wavelengths, which is consistent with our dust modelling below. The Na I emission comes from a region that is slightly larger from the star, and indicates circumstellar material very close to the star. The presence of Na I 2.2 μm doublet emission had been reported towards quite a few YHGs (IRC+10420, HD 179821, HR 8752, and ρ Cas (Lambert et al. 1981; Hrivnak et al. 1994; Hanson et al. 1996; Oudmaijer and de Wit 2013), and other massive objects such as LBVs and B[e] stars (e.g. Hamann and Simon 1986; McGregor et al. 1988; Morris et al. 1996).

The origin of the Na I 2.2 μm doublet emission is not yet clear however. Various scenarios have been discussed in literature, such as disks and pseudo-photospheres (for an overview see Oudmaijer and de Wit 2013). The spatially resolved emission allows us to discard several of these based on size-scale arguments while all models either implicitly or explicitly assume the Brγ emission to be due to the recombination of hydrogen. However, perhaps the most puzzling aspect is that the emitting region of the neutral sodium is smaller than that of the hydrogen, Brγ emission. This is not what one would expect; the ionisation potential of sodium is much lower than that of hydrogen (5.1 eV versus 13.6 eV). So, if the stellar photons can ionise hydrogen, then surely the sodium atoms will be ionised. It is even more puzzling then to find neutral sodium to originate from a smaller region than the hydrogen recombination line emission, as the radiation field is more intensive closer to the star.

Although hydrogen recombination is the most prominent explanation for hydrogen emission lines in most astrophysical situations, we do note that the central star of the Fried Egg nebula is an A-star and therefore does not have as many ionising photons at its disposal as a B-type star. Indeed, strong recombination line emission in cooler stars is unusual. It is therefore prudent to reconsider whether the Brγ (and also the previously seen very strong Hα) line emission is due to recombination and investigate a route to hydrogen emission without the need for highly energetic photons capable of ionising hydrogen from the ground state. For example, in regions of high density, collisions can excite electrons to higher energy levels, while their de-excitation can give rise to line emission without the need for ionisation. Alternatively, the energies required for ionisation from these upper levels are much lower than for ionisation from the ground level. In these situations, hydrogen could be ionised using lower energy photons, yielding recombination line emission in regions where the metals would be able to stay neutral.

We investigated this notion using a simple 2-zone model in local thermodynamic equilibrium (LTE). In short, given a temperature T and density $n_{\rm H}$, the energy and ionisation levels are populated using the Saha-Boltzmann equations. Using these, the peak optical depths and flux densities of the lines of interest are then computed. A series of models were run and spectra were produced. Models with $T \approx 6750$ K, $\log n_{\rm H} = 13.2$ cm^{-3}, and a ratio of thickness to radius $\rho \approx 0.1$ yield peak flux densities which are consistent with the observed Na I and Mg II emission, for an angular diameter $\theta \approx 1.2$ mas. The peak emission in the Brγ profile can be matched with a slightly larger region, $\theta \approx 2.0$ mas, with lower temperature, $T = 5000$ to 5500 K, and density, $\log n_{\rm H} \approx 11.6$ to 12.8 cm^{-3} respectively (see Fig.2). The take-home message here is that very simple spherical shells in LTE can very well reproduce the neutral sodium and hydrogen line emission regarding both their spatial extent (with sodium originating from a smaller volume compared to Brγ) and the line fluxes, without the necessity to introduce more complex geometries or mechanisms like that of a pseudo-photosphere or a shielding dusty disk.

3. Outer parts

In this section we move on from the milli-arcsecond near-star environment to the arc-second scales. The 10 μm diffraction limited images obtained with VISIR probe dust emission at radii up to 2.5 arcsecond (corresponding to dynamical timescales up to hundreds of years, see Fig.3 - Koumpia et al. 2020). We can constrain the mass-loss history of the object by simultaneously fitting the spatial information as well as the Spectral Energy Distribution (SED). The additional information provided by the imaging helps us break the usual degeneracies associated with the fitting of the SED alone. In this particular case, only modelling the SED would not have made it possible to reveal the presence of three distinct circumstellar shells (centered at radii of ~0.37, 0.85 and 2.2

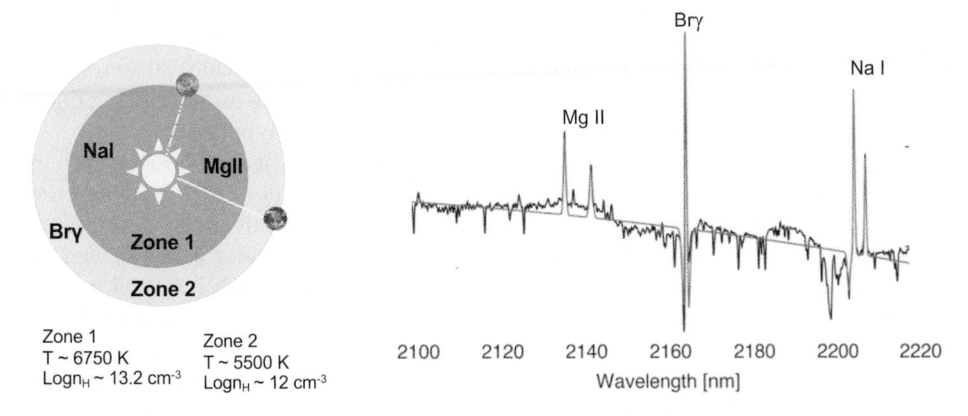

Figure 2. *Left*: a cartoon of the two-zone LTE model employed. The inner zone 1 has higher temperatures and densities compared to zone 2. *Right*: The observed X-Shooter spectrum (in black) overplotted with the model spectrum in red. The modelled intensity of both metal and hydrogen lines match the observations.

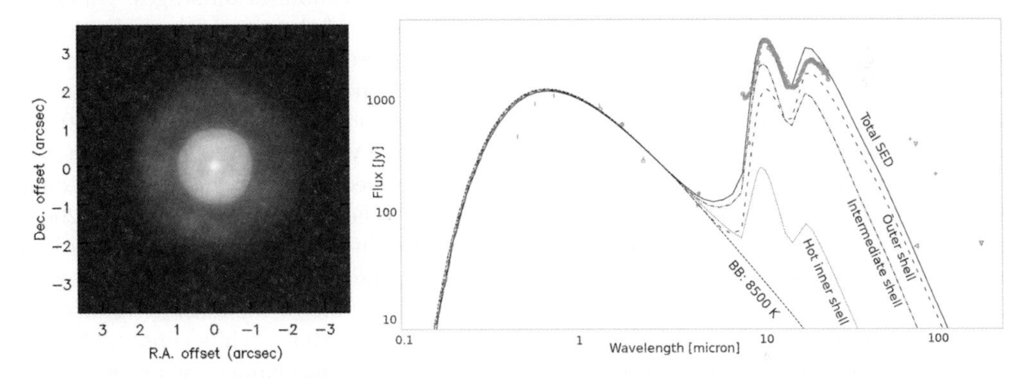

Figure 3. *Left*: Monochromatic image of the Fried Egg Nebula at 8.59 μm as taken with VISIR/VLT. *Right:* The dereddened photometry (symbols) is fitted to produce the total SED of the source. The modelled SED corresponding to the stellar and hot inner shell emission is plotted as dotted line, while the stellar and cool inner shell emission is plotted as dot-dashed line. The modelled contribution of the outer shell is plotted as a large spacing dashed line. Lastly, the combined modelled SED is shown as solid line.

arcseconds respectively), nor would we have been able to derive mass-loss rates for the respective mass-loss episodes.

The 2D radiative transfer code 2-Dust (Ueta and Meixner 2003) was used to simultaneously fit both the SED and the azimuthally averaged radial profiles of the shells at three distinct wavelengths (8.59 μm, 11.85 μm, and 12.81 μm - from the image presented in Lagadec et al. (2011)). In Figure 3 we present the SED of the three shells individually (hot inner shell, warm inner shell, and cool outer shell) and the combined SED. The total fit, also to the radial profiles (not shown here - see Koumpia et al. 2020) is of good quality as assessed by goodness-of-fit indicators. We do not fit the far-infrared points (obtained with PACS/Herschel) which trace an even older and colder shell at much larger sub-parsec scales as demonstrated by Hutsemékers et al. (2013).

One of the main results of this study is that IRAS 17163 underwent three major mass-loss episodes in a timescale of about a century. The main uncertainty in the determination of the timescales is that the expansion velocity of the dust close to the star has not yet been measured; the ALMA data by Wallström et al. (2017) probe larger scales.

Although we do not yet know the longer term history (photometrically or temperature wise) of the object, the Lebertre et al. (1989) paper seems to indicate that the star had a temperature not too dissimilar to what it is now (~8500 K), more than 30 years later. It most certainly was not a M-star in the late eighties. It is therefore an interesting question what mechanism leads to the mass-loss events in this Yellow Hypergiant.

Traditionally, pulsational instabilities are seen as the driver of mass-loss or even eruptive mass-loss events in massive evolved stars like YHGs (Lobel et al. 1994; de Jager 1998). However, among other arguments (see Koumpia et al. 2020), not much data regarding the variability of IRAS 17163 is known, which leads us to consider a line-driven mass-loss scenario that results from the bi-stability mechanism. Although it is well known for hotter objects ($>$ 20000 K), this mechanism has thus far not been considered in YHG studies. The main feature of the bi-stability jump is that stellar winds are driven by the absorption of photons at specific frequencies; in other words, they are line-driven. As the huge amount of available transitions and therefore the opacity of iron in the relevant part of the spectrum far exceeds that of the more abundant hydrogen and helium, the line-driven winds and their properties are dominated by the iron opacities. These opacities in turn depend on the temperatures and ionisation stages. Small decreases in photospheric temperature can result in the recombination of Fe to lower ionisation stages, which in turn results in an increase of the wind density, a decrease of the terminal wind velocity by about a factor of two, and ultimately a significant increase on the mass-loss rate (up to several factors; Vink et al. 1999). This process is known as the bi-stability mechanism and has been well-studied for the "first" bi-stability jump which occurs at T ~ 21000 K as a consequence of the recombination of Fe IV to III (Vink et al. 2000). However, a "second" bi-stability jump, which thus far has not received much attention, is expected to occur at T ~ 8800 K because of the recombination of Fe III to Fe II (Vink et al. 2000, 2001; Petrov et al. 2016). Similarly to the first bi-stability jump, changes in mass-loss rate and wind outflow velocities are expected. Indeed, the second bi-stability jump predicts escape and terminal wind velocities that are comparable to the outflow velocity of 30-100 km/sec observed towards IRAS 17163 (Lamers et al. 1995), while consistent with the occurrence of distinct mass-loss episodes. Exploring this mechanism, which seems to act precisely in the region where YHGs are seen to evolve to, is a potentially fruitful manner to understand the mass-loss, and therefore, the evolution, of massive stars.

4. Conclusions

We conclude by noting that the Fried Egg Nebula is a key object which is characterised by three distinct mass-loss events with varied mass-loss rates and maximum timescales from 30 yr up to 120 yr. The most recent event appears to be the least powerful in terms of mass-loss compared to the previous two known events. Here we present the key results of our study:

- The 2 μm continuum emission does not stem from dust but rather traces the star directly.
- The three distinct shells trace three mass-loss episodes with a variable mass-loss rate; the most recent mass-loss is the least powerful.
- Our two-zone LTE model can reproduce both the K-band spectrum and the sizes of the emission lines (Brγ, Na I). This approach is of great importance as it can potentially explain the enigmatic Na I doublet in emission as seen in more YHGs, without the need of introducing more complex geometries and underlying physics (e.g., no need of a pseudo-photosphere).
- We introduce the second bi-stability jump to explain the distinct mass-loss episodes observed towards a YHG and in particular towards the Fried Egg Nebula.

Acknowledgements

Credit: Koumpia et al. 2020, A&A 654, 109, Figures 1, 13, and 14 are reproduced with permission from Astronomy & Astrophysics, copyright ESO. EK was funded by the STFC (ST/P00041X/1). This project has received funding from the European Union's Framework Programme for Research and Innovation Horizon 2020 (2014-2020) under the Marie Skłodowska-Curie Grant Agreement No. 823734.

References

de Jager, C. 1998, *A&AR*, 8(3), 145–180.

Eisenhauer, F., Perrin, G., Brandner, W., Straubmeier, C., Perraut, K., & Amorim, A. et al. 2011, *The Messenger*, 143, 16–24.

Gordon, M. S. & Humphreys, R. M. 2019, *Galaxies*, 7(4), 92.

Hamann, F. & Simon, M. 1986, *ApJ*, 311, 909–920.

Hanson, M. M., Conti, P. S., & Rieke, M. J. 1996, *ApJS*, 107, 281.

Hrivnak, B. J., Kwok, S., & Geballe, T. R. 1994, *ApJ*, 420, 783–796.

Hutsemékers, D., Cox, N. L. J., & Vamvatira-Nakou, C. 2013, *A&A*, 552, L6.

Koumpia, E., Oudmaijer, R. D., Graham, V., Banyard, G., Black, J. H., Wichittanakom, C., Ababakr, K. M., de Wit, W. J., Millour, F., Lagadec, E., Muller, S., Cox, N. L. J., Zijlstra, A., van Winckel, H., Hillen, M., Szczerba, R., Vink, J. S., & Wallström, S. H. J. 2020, *A&A*, 635, A183.

Lagadec, E., Zijlstra, A. A., Oudmaijer, R. D., Verhoelst, T., Cox, N. L. J., Szczerba, R., Mékarnia, D., & van Winckel, H. 2011, *A&A*, 534, L10.

Lambert, D. L., Hinkle, K. H., & Hall, D. N. B. 1981, *ApJ*, 248, 638–650.

Lamers, H. J. G. L. M., Snow, T. P., & Lindholm, D. M. 1995, *ApJ*, 455, 269.

Lebertre, T., Epchtein, N., Gouiffes, C., Heydari-Malayeri, M., & Perrier, C. 1989, *A&A*, 225, 417–431.

Lobel, A., de Jager, C., Nieuwenhuijzen, H., Smolinski, J., & Gesicki, K. 1994, *A&A*, 291, 226–238.

McGregor, P. J., Hyland, A. R., & Hillier, D. J. 1988, *ApJ*, 334, 639–656.

Morris, P. W., Eenens, P. R. J., Hanson, M. M., Conti, P. S., & Blum, R. D. 1996, *ApJ*, 470, 597.

Oudmaijer, R. D., Davies, B., de Wit, W.-J., & Patel, M. In Luttermoser, D. G., Smith, B. J., & Stencel, R. E., editors, *The Biggest, Baddest, Coolest Stars* 2009,, volume 412 of *Astronomical Society of the Pacific Conference Series*, 17.

Oudmaijer, R. D. & de Wit, W. J. 2013, *A&A*, 551, A69.

Oudmaijer, R. D., Groenewegen, M. A. T., Matthews, H. E., Blommaert, J. A. D. L., & Sahu, K. C. 1996, *MNRAS*, 280, 1062–1070.

Petrov, B., Vink, J. S., & Gräfener, G. 2016, *MNRAS*, 458(2), 1999–2011.

Ueta, T. & Meixner, M. 2003, *ApJ*, 586, 1338–1355.

Vink, J. S., de Koter, A., & Lamers, H. J. G. L. M. 1999, *A&A*, 350, 181–196.

Vink, J. S., de Koter, A., & Lamers, H. J. G. L. M. 2000, *A&A*, 362, 295–309.

Vink, J. S., de Koter, A., & Lamers, H. J. G. L. M. 2001, *A&A*, 369, 574–588.

Wallström, S. H. J., Lagadec, E., Muller, S., Black, J. H., Cox, N. L. J., Galván-Madrid, R., Justtanont, K., Longmore, S., Olofsson, H., Oudmaijer, R. D., Quintana-Lacaci, G., Szczerba, R., Vlemmings, W., van Winckel, H., & Zijlstra, A. 2017, *A&A*, 597, A99.

The Origin of Outflows in Evolved Stars
Proceedings IAU Symposium No. 366, 2022
L. Decin, A. Zijlstra & C. Gielen, eds
doi:10.1017/S174392132200062X

BOSS-3D: A Binary Object Spectral Synthesis Code in 3D

L. Hennicker[1], N. D. Kee[2], T. Shenar[1,3], J. Bodensteiner[1,4], M. Abdul-Masih[5], I. El Mellah[6], H. Sana[1] and J. O. Sundqvist[1]

[1]Institute of Astronomy, KU Leuven, Celestijnenlaan 200D, 3001 Leuven, Belgium
email: levin.hennicker@kuleuven.be

[2]National Solar Observatory, 22 Ohi'a Ku Street, Makawao, HI 96768, USA

[3]Anton Pannekoek Institute for Astronomy, University of Amsterdam, Postbus 94249, 1090 GE Amsterdam, The Netherlands

[4]European Southern Observatory, Karl-Schwarzschild-Strasse 2, D-85748 Garching bei München, Germany

[5]European Southern Observatory, Alonso de Cordova 3107, Vitacura, Casilla 19001, Santiago de Chile, Chile

[6]Institut de Planétologie et d'Astrophysique de Grenoble, 414 Rue de la Piscine, 38400 Saint-Martin-d'Hères, France

Abstract. To decode the information stored within a spectrum, detailed modelling of the physical state is required together with accurate radiative transfer solution schemes. In the analysis of stellar spectra, the numerical model often needs to account for high velocity outflows, multi-dimensional structures, and the effects of binary companions. Focusing now on binary systems, we present the BOSS-3D spectral synthesis code, which is capable of calculating synthetic line profiles for a variety of binary systems. Assuming the state of the circumstellar material to be known, the standard *pz*-geometry is extended by defining individual coordinate systems for each object. By embedding these coordinate systems within the observer's frame, BOSS-3D automatically accounts for outflows or discs within both involved systems, and includes all Doppler shifts. Moreover, the code accounts for different length-scales, and thus could also be used to analyse transit-spectra of planetary atmospheres. As a first application of BOSS-3D, we model the phase-dependent H_α line profiles for the enigmatic binary (or multiple) system LB-1.

Keywords. radiative transfer – methods: numerical – stars: emission-line, Be – stars: black holes – binaries: spectroscopic

---

1. Introduction

Interpreting the radiation emerging from stellar or planetary atmospheres is key to understanding our Universe as a whole, where deep insight into the underlying physics of an observed object can be obtained by analysing the electromagnetic spectrum of the emitting object. Decoding the information stored within a spectrum can be considered as a three-step process. Firstly, the emitting gas needs to be modelled numerically, which generally involves the solution of the full (radiation)-hydrodynamics equations with appropriate initial and boundary conditions, and including chemical kinetics. Secondly, the emergent spectrum needs to be calculated by solving the equation of radiative transfer for many frequencies and directions to obtain the emergent flux. Finally, the numerically calculated (synthetic) spectrum needs to be compared with the observed one, yielding an estimate of the underlying physical state of the gas.

This general procedure, however, is computationally very challenging. The solution of the full radiation-hydrodynamics equations, for instance, constitutes already a 7-dimensional problem (three spatial dimensions, one time dimension, and for the radiative transfer one frequency/wavelength dimension if non-LTE effects are excluded and two angular dimensions). Moreover, the coupling of radiation and the state of the gas in full NLTE (non local-thermal-equilibrium) complicates the situation even further, requiring sophisticated iteration techniques via the so-called accelerated Λ-iteration. Thus, state-of-the-art spectral analysis tools rely on various approximations. In the OB-star regime, for instance, atmospheric modelling codes typically assume a static and spherically symmetric gas with an often prescribed parametrised velocity field (e.g. PHOENIX: Hauschildt 1992; CMFGEN: Hillier & Miller 1998, Hillier 2012; WM-BASIC: Pauldrach et al. 2001; PoWR: Hamann & Gräfener 2003, Sander et al. 2017; FASTWIND: Sundqvist & Puls 2018, Puls et al. 2020).

While these codes are very powerful in deducing the fundamental stellar parameters of OB stars (such as L_*, $T_{\rm eff}$, R_*, $\dot{M}$), they all suffer from primarily one caveat, namely when geometrical effects induced by magnetic fields, surface and wind distortions from rotation, surrounding discs, and binarity effects play a significant role. Due to the complexity of including full NLTE calculations within multi-D geometries with supersonic velocity fields, corresponding spectral synthesis codes are only gradually developed (e.g. HDUST: Carciofi & Bjorkman 2006; PHOENIX/3D: Hauschildt & Baron 2006; WIND3D: Lobel & Blomme 2008; MAGRITTE: De Ceuster et al. 2020a, De Ceuster et al. 2020b; and a code developed by Hennicker et al. 2020).

On the other hand, when the atomic level populations are known (for instance obtained by such codes), the synthetic spectra can be calculated by solving the equation of radiative transfer along the direction to the observer in a cylindrical coordinate system (Lamers et al. 1987, Busche & Hillier 2005, Sundqvist et al. 2012, see also Hennicker et al. 2018), and the emergent flux spectrum is thus determined. This last step, that is the calculation of the so-called 'formal integral' is the topic of this paper. At first, we review the basic formalism for single objects and then extend this method to a detailed radiative transfer solution framework for binary systems accounting for rays propagating through the atmospheres of both objects. Finally, we apply the newly developed code to the LB-1 system, which shows a clear anti-phase behaviour of stellar absorption lines against the line wings of a broad H_α emission (Liu et al. 2019).

2. The formal integral

As pointed out above, we are considering only the formal integral in this paper, that is, we assume the opacities and emissivities to be known, and then aim at solving the equation of radiative transfer for many frequencies to obtain the emergent fluxes. For unresolved objects at large distances, we consider only parallel rays in the direction to the observer. For a given object described in a spherical coordinate system (the object's system), $\Sigma_{\rm spc}^{\rm (obj)}$, we define a cylindrical coordinate system $\Sigma_{\rm cyc}^{\rm (ray)} = (p, \zeta, \tilde{z})$ in which the equation of radiative transfer will be discretized (see also Fig. 1). For each impact parameter p and polar angle ζ, the discretization is described through the so-called pz-geometry along the $\tilde{z}$-axis by calculating the intersection points of the ray with the object's spherical grid. With the discretization in the cylindrical coordinate system thus given, all coordinates are transformed to the object's spherical system, and the required information (i.e. opacities, emissivities, and velocity fields) are interpolated onto the ray coordinates.

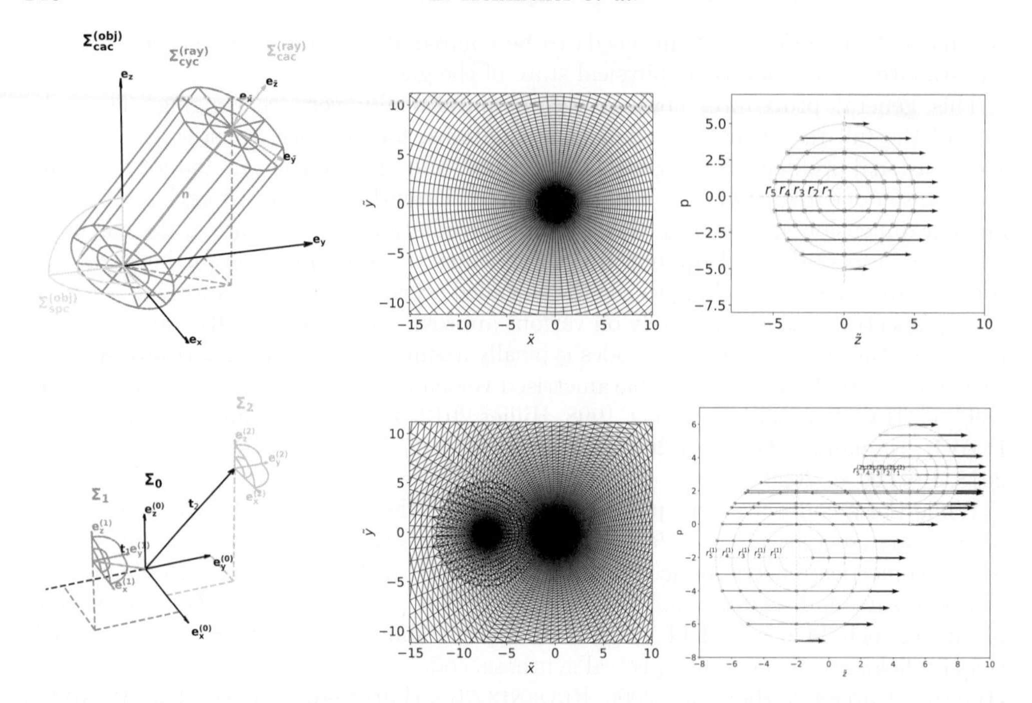

Figure 1. Geometry used for the formal integral. Top left panel: The object's spherical coordinate system $\Sigma_{\mathrm{spc}}^{(\mathrm{obj})}$ (light grey) with corresponding Cartesian reference frame $\Sigma_{\mathrm{cac}}^{(\mathrm{obj})}$ (black), and the ray's cylindrical coordinate system $\Sigma_{\mathrm{cyc}}^{(\mathrm{ray})}$ (dark grey) with corresponding Cartesian reference frame $\Sigma_{\mathrm{cac}}^{(\mathrm{ray})}$ (grey) for an observer's direction $\boldsymbol{n} \parallel \boldsymbol{e}_{\tilde{z}}$. Top middle panel: The $p-\zeta$ (or $\tilde{x}-\tilde{y}$-plane) of the cylindrical coordinate system (perpendicular to the observer's direction). Top right panel: An arbitrary $p-\tilde{z}$-slice, with $\boldsymbol{e}_{\tilde{z}}$ pointing towards the observer. The radial grid of the spherical coordinate system is indicated by the grey circles, and the individual rays for each impact parameter p is shown by the black arrows. The radiative transfer is then performed along each ray with corresponding discretized $\tilde{z}_k$ coordinates indicated by the grey dots dots. Bottom left panel: Similar to the top left panel, with two object's coordinate systems Σ_1 and Σ_2 indicated by the dark grey and light grey quarter-circles, embedded within a global coordinate system Σ_0. The vectors $\boldsymbol{t}_1$ and $\boldsymbol{t}_2$ describe the origin of the local coordinate systems in the global coordinate system. Bottom middle panel: As top middle panel, however showing the triangulation of the $\tilde{x}-\tilde{y}$-plane for an arbitrary orbital configuration in a global cylindrical coordinate system. Bottom right panel: As top right panel, but now for the binary system with the individual $\tilde{z}$-rays propagating through the system of object 1 (dark grey) and object 2 (light grey). Adapted from Hennicker et al. (2021).

To account for non-monotonic projected velocities along the ray, we apply a grid refinement of ray coordinates if required. In the observer's frame, the (time-independent) equation of radiative transfer is then easily solved for a given frequency, with appropriate boundary conditions for the specific intensity set here to the Planck function (or a photospheric line profile) if the ray hits the stellar core, and zero otherwise (thus neglecting the cosmic microwave background). We emphasize that Doppler shifts arising from (non-relativistic) velocity fields are automatically accounted for within the observer's frame formulation. Thus, by solving the equation of radiative transfer for many frequencies and by integrating the emergent intensities over the projected disc at large $\tilde{z}$, we obtain the emergent flux profile, or accordingly the corresponding line profile.

For binary systems, we can essentially apply the very same idea as above, where now each individual object in the system is described by its own spherical coordinate system, Σ_1 and Σ_2, which are embedded in a global coordinate system, Σ_0 (see Fig. 1). For each

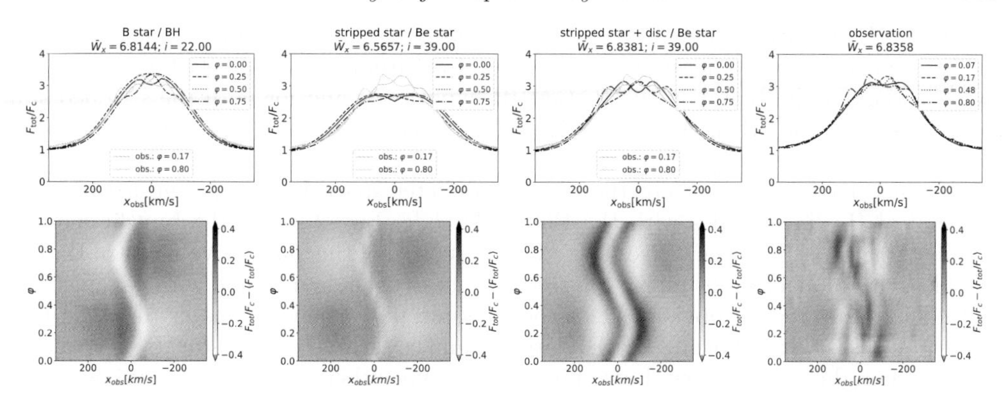

Figure 2. H_α line profiles at different phases φ (top row) with corresponding mean-subtracted dynamical spectra (bottom row), where x_{obs} describes the frequency shift from line centre in velocity space. The first three columns to the left display the solution for three different models, (primary: B star, secondary: black hole), (primary: stripped star, secondary: Be-star), (primary: stripped star + disc, secondary: Be-star), with H_α line profiles at two distinct phases from the actual observations indicated by the grey lines. The right column additionally displays the corresponding observed (see Shenar et al. 2020) H_α line profiles and mean-subtracted dynamical spectrum of LB-1. All observed line profiles have been convolved with a Gaussian filter of width $10\,\mathrm{km\,s^{-1}}$ to reduce the noise. The phase-averaged equivalent width evaluated in velocity space in units of $100\,\mathrm{km\,s^{-1}}$, $\bar{W}_x$, and the inclination, i, used within the synthetic-spectra calculations are indicated at the top. Adapted from Hennicker et al. (2021).

object, we again define individual cylindrical coordinate systems which are merged in a global cylindrical system where the equation of radiative transfer will be solved. Thus, the coordinate representation of the global cylindrical system within each individual object's system can be found by simple coordinate transformations, and all required quantities are again interpolated onto each ray. The emergent fluxes are then calculated by numerically integrating the emergent specific intensities over the triangulated area of the global cylindrical coordinate system.

Since the individual coordinate systems can have completely different length-scales (which are automatically accounted for by the transformation matrices), we are able to calculate synthetic line profiles for a variety of systems, ranging from transit spectra of planetary atmospheres, to post-asymptotic giant branch binaries including circumstellar and circumbinary discs and massive-star binaries with stellar winds and disc systems.

3. The LB-1 system

As a first application of the developed BOSS-3D code, we consider the enigmatic binary (or multiple) system LB-1. Among other hypotheses, this system has been proposed to consist of either a B-star/black-hole (BH, Liu et al. 2019) or a stripped-star/Be-star binary (Shenar et al. 2020). From an observational point of view (see also Fig. 2), there is a clear anti-phase behaviour of the H_α line wings against the line core. This manifests in the mean-subtracted dynamical profile, where the line wings are moving to the blue (left) side at phases $\varphi \in [0, 0.5]$ while the line core is moving to the red (right) side.

Based on these observations, we aim at modelling the LB-1 system within BOSS-3D focussing on the two above mentioned hypotheses. To this end, we use a parametrised form of the Shakura-Sunyaev α-disc model (see Shakura & Sunyaev 1973) based on Carciofi & Bjorkman (2006) for both the BH-accretion disc and the Be-decretion disc. Essentially, we assume Keplerian rotation and vertically hydrostatic equilibrium, with the temperature stratification either given from a blackbody reprocessing disc (for the Be-decretion disc) or assumed to be constant (for the BH-accretion disc). Further, we calculate H_α opacities and emissivities from Saha-Boltzmann statistics – assuming hence LTE – and use photospheric line profiles obtained from FASTWIND for all involved objects (i.e. for the B-star in the B-star/BH hypothesis, and for both the stripped star and Be-star in the other one) as an inner boundary condition of the specific intensity when the rays are crossing the stars.

For such models, we explore the disc's parameter space by calculating a variety of dynamical H_α line profiles with the BOSS-3D code. Fig. 2 shows the best-fit models. While the B-star/BH binary gives a good reproduction of the qualitative shape of H_α line profiles, this model cannot reproduce the anti-phase behaviour of line wings against line core. This is somewhat obvious, since the emission part solely originates from the BH-accretion disc, thus following the BH orbit. Similarly, the stripped-star/Be-star model cannot reproduce the observed anti-phase behaviour either, since the H_α emission now is following the Be-star orbit.

When introducing an (artificial) disc around the stripped star, however, both the qualitative shape of the H_α line profiles as well as the dynamical spectrum is in remarkable good agreement with the observations. Though speculative, this disc might be a remnant from previous mass-transfer phases or could have been formed from re-accretion of material from the Be-star disc. We emphasize that a similar good fit can also be found when introducing a disc-disc system within the B-star/BH hypothesis as well. Thus, while we cannot rule out the one or other hypothesis, our findings still suggest that LB-1 hosts two discs.

4. Summary

In this paper, we have presented a 3D radiative transfer code for binary systems (BOSS-3D), which calculates synthetic line profiles for given source and sink terms. By describing each object in the system within individual coordinate systems, BOSS-3D can handle completely different length-scales of the involved objects without loss of spatial resolution, and can be easily extended to triple or multiple systems. Moreover, by solving the radiative transfer equation in the observer's frame, BOSS-3D can easily handle (arbitrary, but non-relativistic) supersonic velocity fields. BOSS-3D then requires the density, temperature, velocity field, opacities, and emissivities in the rest frame of each object as input, and gives the synthetic line profile(s) as output.

As a first application of BOSS-3D, we considered the LB-1 system focussing on two hypotheses (i.e. the B-star/BH and the stripped-star/Be-star hypotheses previously proposed in the literature). The observed phase-dependent H_α line profiles, however, can only be reproduced when introducing a second disc in the system. Thus, our findings provide strong evidence that LB-1 contains a disc-disc system, with an additional disc either attached to the B star in the B-star/BH scenario or to the stripped star in the stripped-star/Be-star scenario.

References

Busche, J. R. & Hillier, D. J. 2005, *AJ*, 129, 454
Carciofi, A. C. & Bjorkman, J. E. 2006, *ApJ*, 639, 1081
De Ceuster, F., Homan, W., Yates, J., et al. 2020, *MNRAS*, 492, 1812
De Ceuster, F., Bolte, J., Homan, W., et al. 2020, *MNRAS*, 499, 5194
Hamann, W. R. & Gräfener, G. 2003, *A&A*, 410, 993
Hauschildt, P. H. 1992, *J. Quant. Spec. Radiat. Transf.*, 47, 433
Hauschildt, P. H. & Baron, E. 2006, *A&A*, 451, 273
Hennicker, L., Puls, J., Kee, N. D., & Sundqvist, J. O. 2018, *A&A*, 616, A140
Hennicker, L., Puls, J., Kee, N. D., & Sundqvist, J. O. 2020, *A&A*, 633, A16
Hennicker, L., Kee, N. D., Shenar, T., et al. 2021, *A&A*, accepted, *arXiv:2111.15345*

Hillier, D. J. & Miller, D. L. 1998, *ApJ*, 496, 407
Hillier, D. J. 2012, *in From Interacting Binaries to Exoplanets: Essential Modeling Tools, ed. M. T. Richards & I. Hubeny*, 282, 229–234
Lamers, H. J. G. L. M., Cerruti-Sola, M., & Perinotto, M. 1987, *ApJ*, 314, 726
Liu, J., Zhang, H., Howard, A. W., et al. 2019, *Nature*, 575, 618
Lobel, A. & Blomme, R. 2008, *ApJ* 678, 408
Pauldrach, A. W. A., Hoffmann, T. L., & Lennon, M. 2001, *A&A*, 375, 161
Puls, J., Najarro, F., Sundqvist, J. O., & Sen, K. 2020, *A&A*, 642, A172
Sander, A. A. C., Hamann, W. R., Todt, H., Hainich, R., & Shenar, T. 2017, *A&A*, 603, A86
Shakura, N. I. & Sunyaev, R. A. 1973, *A&A*, 500, 33
Shenar, T., Bodensteiner, J., Abdul-Masih, M., et al. 2020, *A&A*, 639, L6
Sundqvist, J. O., ud-Doula, A., Owocki, S. P., et al. 2012, *MNRAS*, 423, L21
Sundqvist, J. O. & Puls, J. 2018, *A&A*, 619, A59

The Origin of Outflows in Evolved Stars
Proceedings IAU Symposium No. 366, 2022
L. Decin, A. Zijlstra & C. Gielen, eds.
doi:10.1017/S1743921322000187

Simulations of Colliding Winds in Massive Binary Systems with Accretion

Amit Kashi[1,2] and Amir Michaelis[1]

[1]Department of Physics, Ariel University, Ariel, 4070000, Israel

[2]Astrophysics Geophysics And Space Science Research Center (AGASS), Ariel University, Ariel, 4070000, Israel
e-mails: kashi@ariel.ac.il; amirmi@ariel.ac.il

Abstract. We run numerical simulations of massive colliding wind binaries, and quantify the accretion onto the secondary under different conditions. We set 3D simulation of a LBV–WR system and vary the LBV mass loss rate to obtain different values of wind momentum ratio η. We show that the mean accretion rate for stationary systems fits a power law $\dot{M}_{\rm acc} \propto \eta^{-1.6}$ for a wide range of η, until for extremely small η saturation in the accretion is reached. We find that the stronger the primary wind, the smaller the opening angle of the colliding wind structure (CWS), and compare it with previous analytical estimates. We demonstrate the efficiency of clumpy wind in penetrating the CWS and inducing smaller scale clumps that can be accreted. We propose that simulations of colliding winds can reveal more relations as the ones we found, and can be used to constrain stellar parameters.

Keywords. stars: massive — stars: mass-loss — stars: winds, outflows — (stars:) binaries: general — stars: Wolf-Rayet — accretion, accretion disks

---

1. Introduction

Massive stars have very intense winds, especially at late evolutionary stages (Langer 2012; Smith 2014; Owocki 2015; Vink 2015). These stars can undergo giant eruptions during which the mass loss rate is as high as a few $\rm M_\odot\ yr^{-1}$. Perhaps the most extreme example is the Great Eruption of η Carinae (e.g., Davidson & Humphreys 2012, and refs. therein), an interacting colliding wind systems with stellar masses that may reach 250 $\rm M_\odot$ (Kashi & Soker 2009a). The system had two extreme outbursts in the 17$^{\rm th}$ century known as the Great Eruption and Lesser Eruption, that caused the system to brighten and expelled as much as 40 $\rm M_\odot$ to form a bipolar nebula (e.g., Davidson & Humphreys 2012). Soker (2005) suggested that the Great Eruption was the result of mass accretion onto the secondary star, and that jets from an accretion disk around the secondary formed the bipolar nebula. Kashi & Soker (2009a) showed that using accretion and mass loss it is possible to account for the peaks in the light curve of the Great Eruption and Lesser Eruption, when assuming they occur near periastron passages. This might have altered the evolution of both stars (Kashi et al. 2016), implying that accretion can play an important role in the evolution and outcome of massive star binaries. The same scenario might have caused other giant eruptions, like the 17$^{\rm th}$ century eruptions of the prototype Luminous Blue Variable (LBV) star P Cygni (Kashi 2010; Michaelis et al. 2018).

Some massive binary systems can experience accretion not only during eruptions but also during quiescence times. Kashi (2019) showed that even for the present mass loss rate of the primary of η Car, which is in the order of $\sim 1/1000$ of the mass loss rate of the Great Eruption, accretion occurs close to periastron passage. An example for accretion in younger massive binaries is the binary system HD 166734, studied in Kashi (2020). This system has an O7.5If primary and O9I(f) secondary with an orbital period of $\simeq 34.538$ days and eccentricity $\simeq 0.618$. The winds were highly unstable to the non-linear thin shell instability as they were radiative from both sides, creating long fingers instead of a smooth conical-like colliding wind structure (CWS). We obtained accretion for a long duration of the orbital period, that sums up to 1.3×10^{-8} $M_{\odot}$ each cycle. It is safe to assume that many more systems have stellar and orbital parameters that result in accretion, justifying a theoretical study of the problem.

Let us go back to the classical description of the colliding wind problem. When a massive binary system has two stars that both eject winds, the two winds collide and create a structure referred to as the CWS (Stevens et al. 1992; Usov 1992; Eichler & Usov 1993). In the most simple case, the CWS has two shocked winds separated by a contact discontinuity and curved into a conical like structure towards the star with the wind with the lower momentum. The symmetry line of the CWS in the absence of orbital motion is a line connecting the two stars. The shocked gas flows away asymptotically along the sides of the CWS towards infinity.

The main parameter that determines the shape of the contact discontinuity is the momentum ratio of the two winds

$$\eta = \frac{\dot{M}_2 v_2}{\dot{M}_1 v_1}, \tag{1.1}$$

where $\dot{M}_2$ and v_2 are the mass loss rate and the velocity of the wind of the secondary, and $\dot{M}_1$ and v_1 are the same for the primary. In the system we study here we will refer to the LBV as the primary and the WR as the secondary.

When orbital motion is added, the structure can take a conical-like shape or a spiral shape depending on the momentum ratio of the winds and the ratio between the winds velocity and the orbital velocity. At this point it is still possible to describe the system analytically, taking into account an approximation appropriate for the system and using toy-models. A well studied example is the model for η Carinae proposed in Kashi & Soker (2009b) that is based on a hyperboloid with a varying density that is rotated in a different angle at each point on an eccentric binary orbit depending on the orbital velocity and the stellar winds velocity vectors. This model was able to explain many of the observations of the system (Kashi & Soker 2007, 2008).

The next level of analysis in the attempt to understand colliding wind systems is to model them using numerical simulations. Modern simulations are performed in three dimensions and at high resolution that resolves fine details. They are composed of many ingredients and include relevant physical processes (e.g., Lamberts et al. 2011; Parkin et al. 2011; Parkin & Gosset 2011; Clementel et al. 2015; Hendrix et al. 2016; Reitberger et al. 2017; Kashi 2020; Kashi & Michaelis 2021). There is a wide space of uncertain parameters such that changing each one of them can cause observables of the CWS to change in a magnitude that is sometimes much larger than the rate of change the parameter had. An example for such a parameter is the eccentricity of the system, that is hard to constrain and can enormously change the results. To add to the complication, sometimes two different values of eccentricity can give results that are close enough indistinguishable by available observations (e.g., Davidson et al. 2017).

Though CWS have been simulated before, the first to obtain accretion were Akashi et al. (2013), who simulated η Carinae close to periastron passage. They found that dense clumps are formed by instabilities in the shocked primary wind as the winds collide. Those clumps flow towards the secondary but cannot be decelerated by the ram pressure of the secondary wind and hit the regions from where the secondary wind is launched. Our later simulations in Kashi (2017) included radiation pressure and showed that radiative breaking cannot prevent the accretion, by that confirming the theoretical prediction given by Kashi & Soker (2009c). These simulations in Kashi (2017) are also the first that solve the stars not as point sources but as approximated spheres, allowing directional analysis for accretion. In Kashi (2019) we studied four methods for treating accretion and the response of the accretor to the incoming wind, and found a numerical implementation for treating accretion and wind outflow simultaneously. We also showed that the accretion rate obtained when taking the secondary wind acceleration into account was higher than for ejecting winds at terminal speed.

While there are advanced numerical simulations studying colliding winds, most of them are focused on one specific binary system with its particular conditions (masses and radii of the stars, mass loss rates and velocities of the winds, orbital parameters, etc.). There is no set of *general* simulations that ran over a range of each of the parameters and isolated the influence of each physical effect. While obtaining such a set requires a very large number of observations, focusing on specific links between parameters through a limited set of simulations can also be very useful.

We discuss our recent set of colliding wind binaries simulations that cover part of the parameter space and focus on the shape of the colliding wind structure and accretion. Some of our results were presented in (Kashi & Michaelis 2021).

2. Accretion dependence on wind momentum ratio

We test the effect of an enhanced primary wind on the amount of accreted gas onto the primary. We neutralize the effect of orbital motion, in order to isolate the effect of the primary mass loss rate. The effect of enhancing the primary mass loss rate is equivalent to decreasing the momentum ratio.

We set up a stationary binary system with masses $M_1 = 80$ M$_\odot$ and $M_2 = 20$ M$_\odot$. For the fiducial run the primary mass loss rate is $\dot{M}_1 = 3 \times 10^{-4}$ M$_\odot$ yr^{-1} and its wind velocity has a terminal value of $v_{1,\infty} = 500$ km s^{-1} with radiative acceleration corresponding to $\beta = 1$. The secondary mass loss rate is $\dot{M}_2 = 10^{-5}$ M$_\odot$ yr^{-1}, and its wind velocity has a terminal value $v_{2,\infty} = 3000$ km s^{-1} with an acceleration parameter $\beta = 0.8$. We start the simulation with the two smooth winds (homogeneous without clumps). The momentum ratio is $\eta = \dot{M}_2 v_{2,\infty} / \dot{M}_1 v_{1,\infty} = 0.2$, but as the winds collide before reaching their terminal velocity, the effective momentum ratio is different and varies with location.

At $t = 68.5$ days the winds have filled most of the simulation volume and the CWS has reached a quasi-stable state. At this time we increase the mass loss rate of the primary to a larger value. The enhanced primary wind facing the secondary reaches the apex ≈ 14 days after its ejection. When it arrives to the colliding wind structure it disrupts its shape and induces stronger instabilities. The instabilities create dense fingers that penetrate the colliding wind structure and face the secondary.

We increase the primary mass loss rate, effectively reducing the momentum ratio η, and checking the amount of accreted mass on the companion as a result. The primary wind collides with the pre-existing CWS and changes its shape to a smaller opening angle. The side of the CWS facing the secondary shows strong instabilities and forms dense clumps and filaments. The gravity of the secondary pulls these filaments, and some of them are accreted onto the secondary. The secondary wind tries to flow against the incoming gas and forms bubbles.

We test a few values of the momentum ratio and find that for $\eta = 0.05$ accretion is first obtained. Figure 1 shows density slices with velocity vectors for six differnt times, as the enhanced primary wind propagates, interacts with the CWS and reshape it into a narrower structure. A 3D view is shown in Figure 2. We also calculate the X-ray emitted in the 2–10 KeV range (Figure 3). The emission shows a first large peak at the time the enhanced wind of the primary reaches the pre-existing colliding wind structure. After

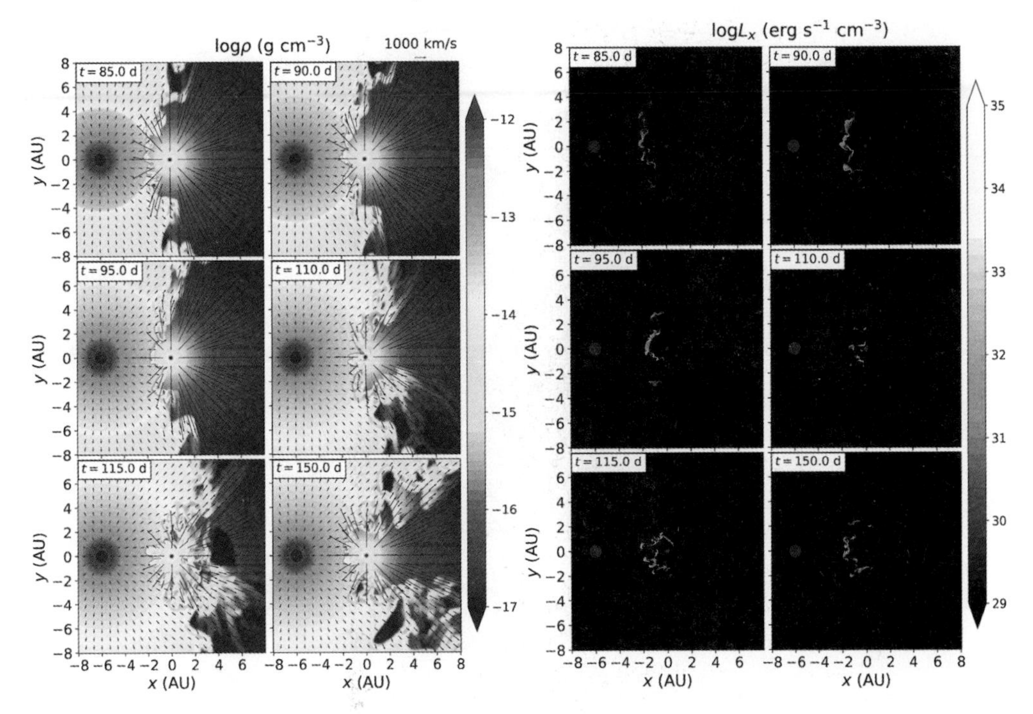

Figure 1. Left: Density maps with velocity vectors showing slices in the orbital plane ($z = 0$), for our simulation with $\eta = 0.05$. Right: X-ray emission at 2–10 KeV. Since the experiment is performed on stationary stars with no orbital motion, the slice of on the orbital plane is essentially similar to slice on any other plane around the axis joining the two stars. The secondary is at the center, marked with a small black circle while the primary, marked with a large blacked circle, is on the left side. The two winds are accelerated and collide. The primary then ejects an enhanced wind that interacts with the secondary wind at the CWS. For $\eta > 0.05$ we did not obtain accretion, however for this run clumps form the CWS penetrate into the secondary wind and accrete onto the secondary.

that, there is a quiescence level of emission, associated with the new meta-stable colliding wind structure. The actual observed X-ray flux from such a system will depend on the absorption in the line of sight, and is expected to look different from different directions. Figure 3 shows the mass accretion rate for our simulations for 400 days together with the 2–10 KeV emission. We find that the average mass accretion rate for $\eta = 0.05$ is $\dot{M}_{\rm acc,av} \simeq 1.7 \times 10^{-6}$ $M_\odot$ yr^{-1}. The accretion is intermittent, with irregular intervals and an average accretion duty cycle of $\simeq 0.055$.

We run more simulations with smaller values of η and calculate the acctetion rate onto the secondary for each of them. Figure 4 presents a simulation with mass loss rate of $\dot{M}_1 = 0.192$ $M_\odot$ yr^{-1}, and $\eta = 3.125 \times 10^{-4}$. This strong mass loss rate corresponds to a giant LBV eruption, and is an extreme case of colliding winds binary. For this simulation the secondary wind cannot blow against the strong primary wind, and is almost completely suppressed. The secondary accretes directly from the primary wind from all directions except a narrow solid angle at the side facing away from the primary. The secondary focuses the primary wind to create a narrow dense column behind the secondary, that has therefore higher density than other directions (Figure 5).

The X-ray emission curve has a strong peak when the enhanced primary wind interacts with the CWS and secondary wind, and later it is almost completely suppressed as the CWS does not exist any longer and the primary wind fills almost the entire volume, allowing only small region behind the secondary to emit.

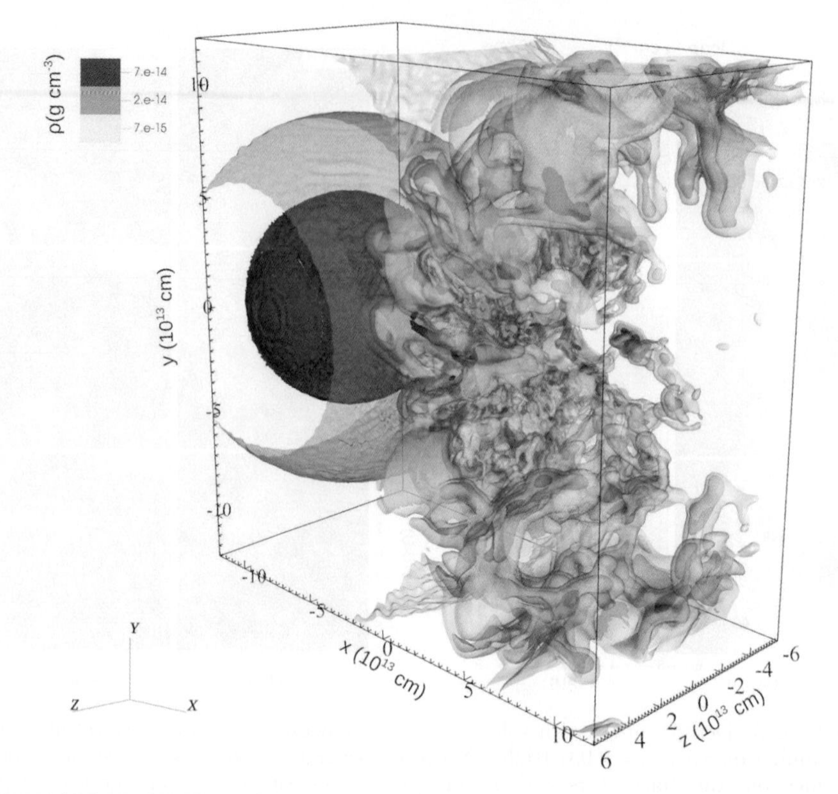

Figure 2. A three dimensional view of the colliding wind structure for $\eta = 0.05$. The stars are not shown. The view point is with the secondary closer to the observer. The primary is inside the red sphere in the far side of the figure (the sphere is a shell of the wind; not the stellar surface). The highly unstable structure of the colliding winds is revealed, with instabilities that created filaments and clumps. The flow is along the sides of the CWS, and the cavity is the secondary wind which has lower density and does not show in the image. At the point in time the figure shows there is no accretion but the CWS reaches very close to the secondary.

Figure 7 shows the average accretion rate in our simulations as a function of η. Simulations we ran with $\eta > 0.05$ did not yield any accretion and therefore do not show in the diagram as they have $\dot{M}_{\rm acc,av} = 0$. We can identify different regions in the $\dot{M}_{\rm acc,av}$–η diagram (Figure 7):

(i) No accretion: For $0.05 \lesssim \eta$ the secondary wind pushes away all the primary wind material, there is a well-defined CWS and no accretion ($\dot{M}_{\rm acc,av} = 0$).

(ii) Accretion: This region extends in the range $0.001 \lesssim \eta \lesssim 0.05$. For $0.02 \lesssim \eta \lesssim 0.05$ there is a transition region, in which accretion is very sporadic. Mass can occasionally be accreted but for most of the time the secondary wind and radiation prevent accretion. This is the region where radiative breaking is dominant. For $0.001 \lesssim \eta \lesssim 0.02$ accretion occurs most of the time. The accretion rate and the accretion duty cycle are larger as η decreases. We fit the simulations' results in region (ii) and find a power-law relation that satisfies

$$\dot{M}_{\rm acc,av} \propto \eta^{-1.6}. \tag{2.1}$$

(iii) Saturated accretion: For $\eta \lesssim 0.00125$ the accretion becomes continuous in time and the accretion rate is constant. The value of $\dot{M}_{\rm acc,av}$ approaches saturation as η continues to decrease. For $\eta \lesssim 0.0003125$ there is saturation in the accretion at $\dot{M}_{\rm acc,sat} \approx 3 \times 10^{-3}\ {\rm M}_\odot\ {\rm yr}^{-1}$.

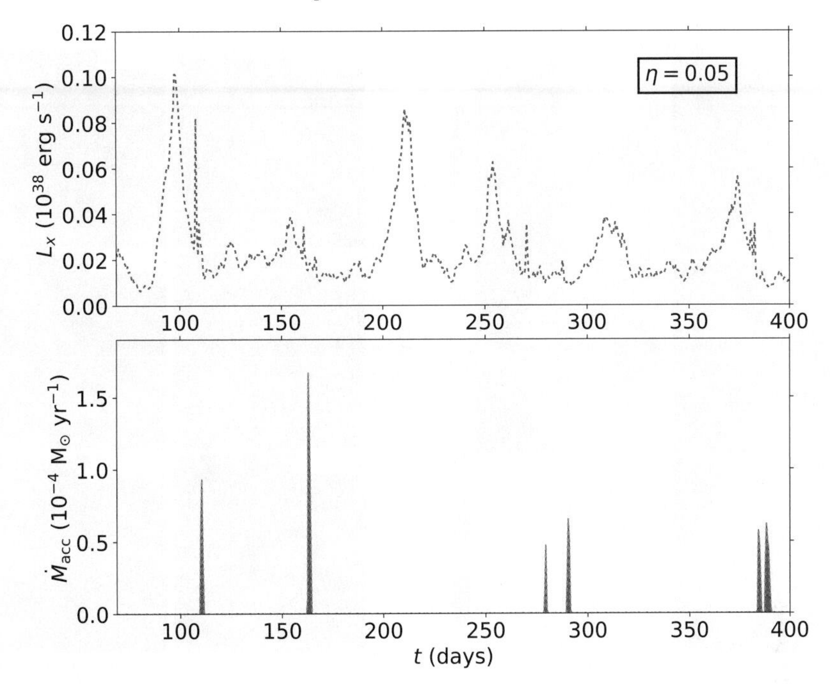

Figure 3. Lower panel: Resulted mass accretion rate onto the secondary for the simulation with $\eta = 0.05$, the highest value of η for which accretion was obtained. Accretion only happened in brief episodes that last ≈ 2 days where a clump is being accreted. In between episodes there are long duration with no accretion. ($\eta = 0.05$) Upper panel: X-ray emission at 2–10 KeV integrated over the simulations volume.

The very low values of η in region (iii) do not describe the colliding wind problem in the sense that there are no two colliding winds with post-shocked gas, and not a CWS as for larger η. Instead, there is only a small region behind the secondary where X-ray is emitted from the heated primary wind that collides with itself after being focused by the secondary, as seen in Figure 4. Our results for the very low values of η, correspond a strong mass loss rate from the LBV, as occurs during giant eruptions.

3. The opening angle for low momentum ratio

The semi-opening angle of the CWS can be obtained by solving a non-linear differential equation for the momentum ratio η. It was calculated numerically by Girard & Willson (1987), and for intermediate values of η there is an approximation derived by Eichler & Usov (1993):

$$\theta \approx 2.1 \left(1 - \frac{\eta^{2/5}}{4}\right) \eta^{1/3}, \tag{3.1}$$

which is stated to be accurate to about 1%. The expression for the shape of the CWS was later revisited a number of times (e.g., Gayley 2009; Pittard & Dawson 2018).

In fact, the actual shape of the CWS is more complicated, taking other factors into account. Among the factors we can count:

- The value of η is location dependent, due to the acceleration of the wind that results in the winds meeting at different velocities than terminal.
- Clumps in the winds (small-scale inhomogeneities common in winds of massive stars; e.g., Crowther et al. 2002; Walder & Folini 2002; Moffat 2008) flow and interact with the other wind resulting the CWS to be inhomogeneous.

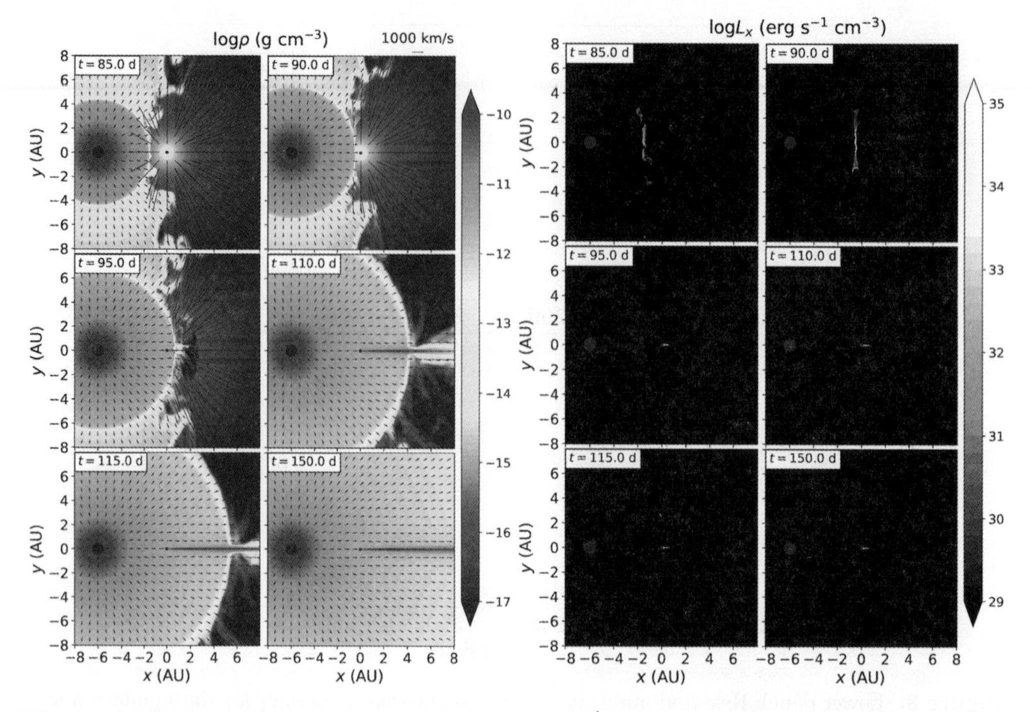

Figure 4. Same as Figure 1, but for $\eta = 3.125 \times 10^{-4}$. The secondary accretes directly from the primary wind from all directions except a narrow solid angle at the side facing away from the primary. The secondary focuses the primary wind to create a narrow dense column behind the secondary. X-ray is emitted from a very small region behind the secondary.

• Instabilities arise when the winds collide, as a result of cooling that will form clumps in the colliding wind region.

• The radiation field of the secondary (the star with the weaker wind) may decelerate the primary incoming wind (radiative inhibition; Owocki & Gayley 1995; Gayley et al. 1997).

• Mixing can result in a range of flow speeds instead of a single uniform one (e.g., Gayley 2009).

• Orbital motion completely alters the CWS shape, making it rotate and wind around itself and casing aberration. Moreover, in eccentric orbits the distance between the stars and the orbital velocity is time dependent, causing periodic changes in the CWS.

We take our simulations of static colliding winds, mostly focusing on low values of the momentum ratio (highly uneven winds that have accretion), and calculate the semi-opening angle for them. The formation of instabilities makes the shape of the winds complicated and a fitting process is needed. We use image processing methods to fit the shape line of the wind, an imaginary line dividing the turbulent flow into two domains, and the semi-opening angle. Then, based on the shape of the wind and the location of the primary and secondary, we can derive the semi-opening angle of the wind. Figure 8 shows the resulting opening angle for our for three different cases.

We plot the wind semi-opening angle as a function of η in Figure 9. For most of our simulations, the simulations obtain larger opening angles than Equation (3.1). For $\eta \lesssim 2 \times 10^{-3}$, we obtain a smaller semi-opening angle than the result from equation (3.1). The equation is actually not applicable for such a small value of η. In this run, the strong mass loss from the primary chokes the secondary wind almost completely. Therefore, the semi-opening angle does not really describe the same problem of colliding winds of two

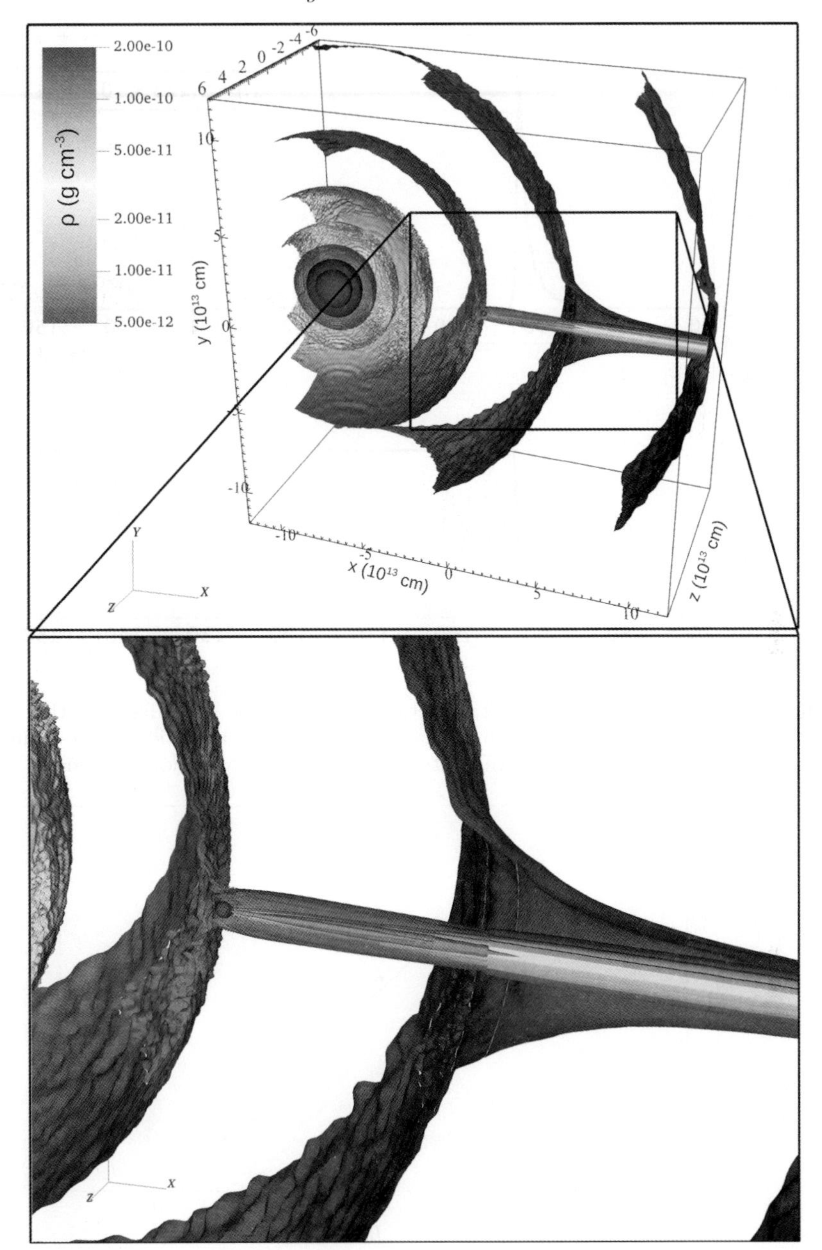

Figure 5. A three dimensional view of the colliding wind structure for $\eta = 3.125 \times 10^{-4}$. The secondary wind is almost completely suppressed. Accretion arrives from the side facing the primary. The secondary focuses the primary wind into a dense column behind the secondary.

stars, but rather the collision of the primary wind with itself after being focused by the secondary and its gravitational field.

4. Clumpy Wind and accretion

Evolved massive stars generally have isotropic winds on a large scale. Wind-clumping refers to small-scale density inhomogeneities distributed across the wind. The clumps have a much larger optical thickness than the gas among the clumps. Because clumping

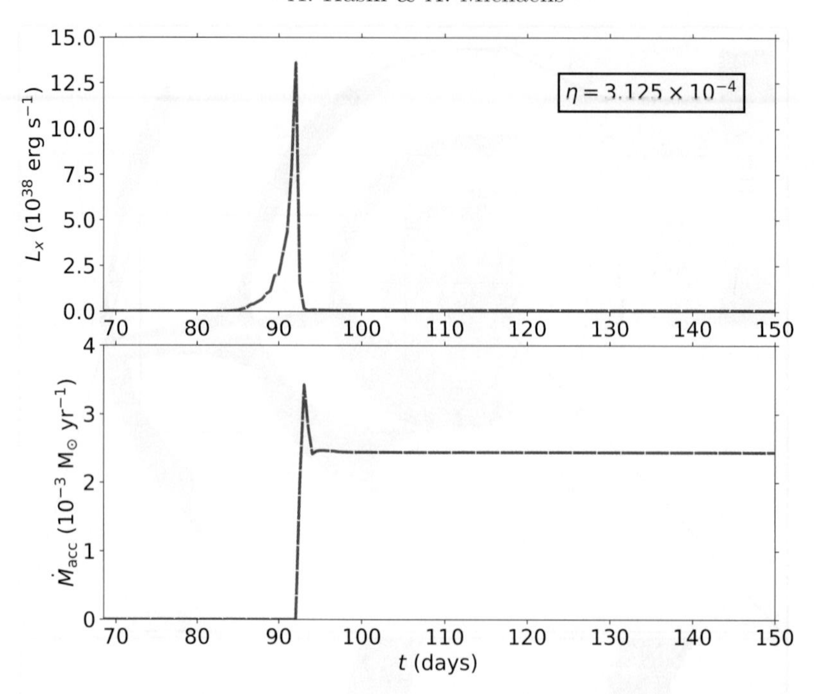

Figure 6. Lower panel: Resulted mass accretion rate onto the secondary for the simulation with $\eta = 3.125 \times 10^{-4}$. At this very high primary mass loss rate accretion occurs continuously. Upper panel: X-ray emission at 2–10 KeV integrated over the simulations volume.

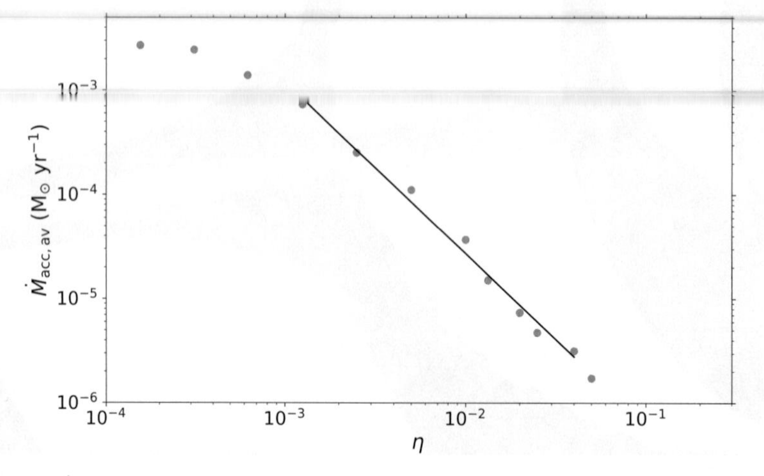

Figure 7. The $\dot{M}_{acc} - \eta$ diagram. The accretion rate obtained from our simulations. Simulations with $\eta > 0.05$ did not yield any accretion. There are different regions in the figure (see text). The straight line indicated the power law described in equation (2.1) For lower η there is transition into saturation value in the mass accretion rate.

contaminates diagnostics that are based on mass-loss, it might result in reduction of previously derived mass-loss rates for massive star (Puls et al. 2008). The effect of wind clumping on mass loss rate is non negligible. For Wolf-Rayet (WR) stars, the realization that their winds are clumped lead to a decrease in the mass-loss rate by a factor 2–4 (e.g., Moffat & Robert 1994; Hamann & Koesterke 1998; Nugis & Lamers 2000). More recent works suggest that the clumping factor is larger $\simeq 10$ (Hainich et al. 2014; Shenar et al. 2019).

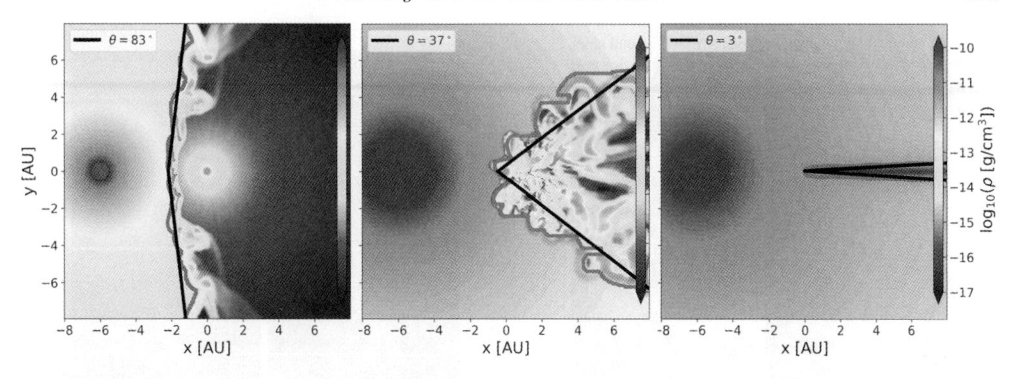

Figure 8. A fit to the colliding wind structure semi-opening angle. The blue line is the result of the wind shape detection algorithm and the black line is the optimal line of the asymptotic opening. Left: $\eta = 0.2$; Middle: $\eta = 0.01$; Right: $\eta = 1.5625 \times 10^{-4}$.

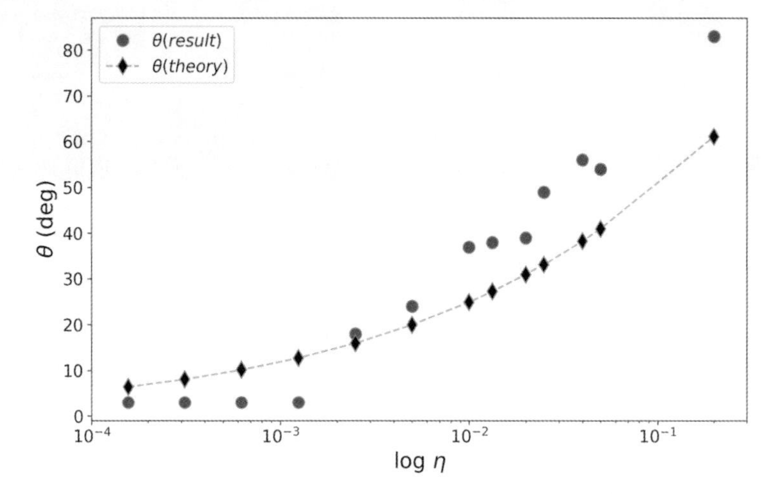

Figure 9. Wind semi-opening angle at time $t = 150$ days. We include the theoretical value according to Equation (3.1) (marked 'theory'; red diamonds) and the simulations' results (blue circles).

Recently, Zhekov (2021) studied X-ray observations of the prototype colliding wind binary WR 140, and found that a standard colliding wind model with smooth winds does not match the X-ray line profiles. They suggested that adding clumps to the WR wind can solve the discrepancy, and concluded that the clumps are efficiently dissolved in the colliding wind region when the stars are near apastron but not at periastron (the system is highly eccentric $e \approx 0.9$).

We run a simulation with a similar setting to the simulation with $\eta = 0.2$ as in section 2, but this time instead of enhancing the wind in an isotropic way we make a a sudden ejection of clumps. The degree of clumping can be measured by the clumping factor $f_{\rm cl} = \langle \rho^2 \rangle / \langle \rho \rangle^2$, where ρ is the density of the wind. The clumps are ejected symmetrically evenly spaced in volume. Obviously clumps can be ejected continuously, and at different densities and sizes. We here examine a simple setting of spherically distributed clumps. We take very large clumps that are larger than what one would expect to see in a realistic system. The symmetrical ejection of clumps in our experiment is in agreement with the findings of Gootkin et al. (2020) who found that clumps are evenly distributed around the prototype LBV star P Cygni. The clumps are ejected from a region very close to the star $\simeq 1.1 R_1$, in agreement with the results of Sundqvist & Owocki (2013).

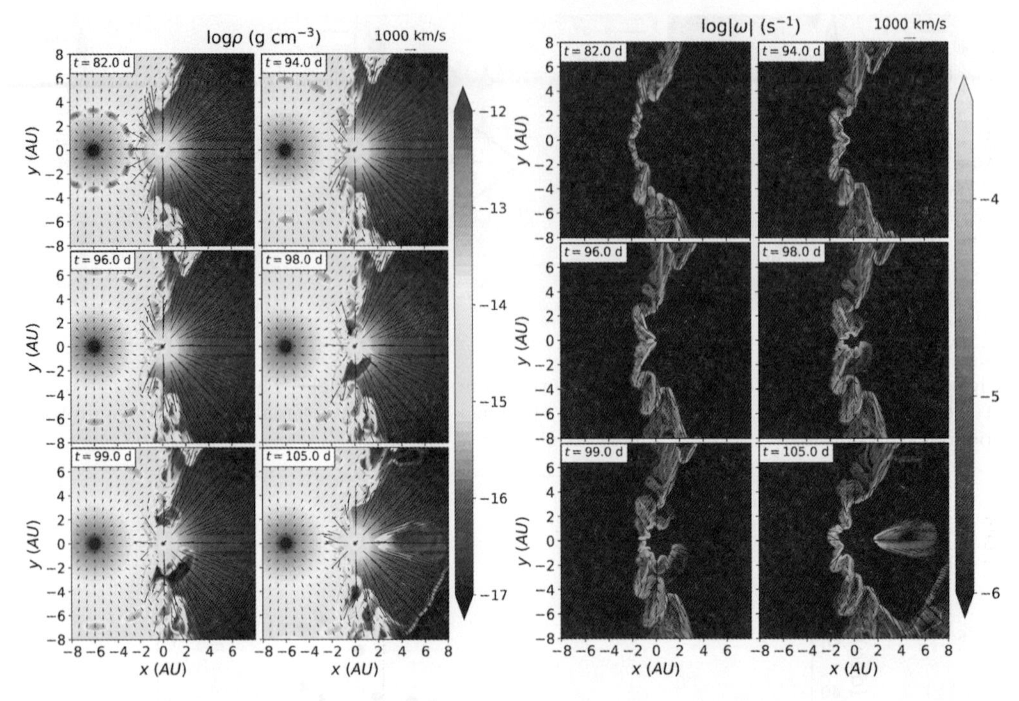

Figure 10. Clumps launching simulation with clumping factor $f_{\rm cl} = 1.66$. Left: Density maps with velocity vectors showing slices in the orbital plane ($z = 0$). The primary wind ejects clumps that interact with the secondary wind at the colliding wind region, and then penetrate into the secondary wind and accrete onto the secondary. At later times a dense clump is formed on the right side of the secondary. Right: The absolute value of the vorticity, as calculated from equation 4.1.

The clumps propagate radially, expend, and their density decreases. The clumps reach the colliding wind structure gradually, first the apex (the line connecting the primary and the secondary), and further regions at later times. The instabilities create dense fingers that penetrate the colliding wind structure and face the secondary. The secondary gravity then pulls the fingers, and the secondary wind pushes it away. Depending on the dominant force, the clump might get accreted. In our simulation with where the clumps overdensity was large enough, the clumps did penetrate the secondary wind and gas was accreted onto the secondary.

If the clumps density is not large enough, we find that the launched clumps dissipate by the time they reach the CWS and have a very small effect that can be quantified as reducing the momentum ratio η (equation 1.1). We then ran an experiment with clump densities of 20 times the density of the smooth wind, which corresponds to a larger clumping factor $f_{\rm cl} = 1.66$. The left panel of Figure 10 shows density maps for the simulation in the orbital plane ($z = 0$), at different times of the simulation.

We process our simulations' results to derive hydrodynamical quantities that can serve as indicators to the effect of the clumps on the colliding wind structure and the vicinity of the secondary, and allow us to quantify the effect of clump interaction.

For each point in our grid we calculate the vorticity, which measures the local rotation of a fluid parcel

$$\omega = \nabla \times \mathbf{v}. \tag{4.1}$$

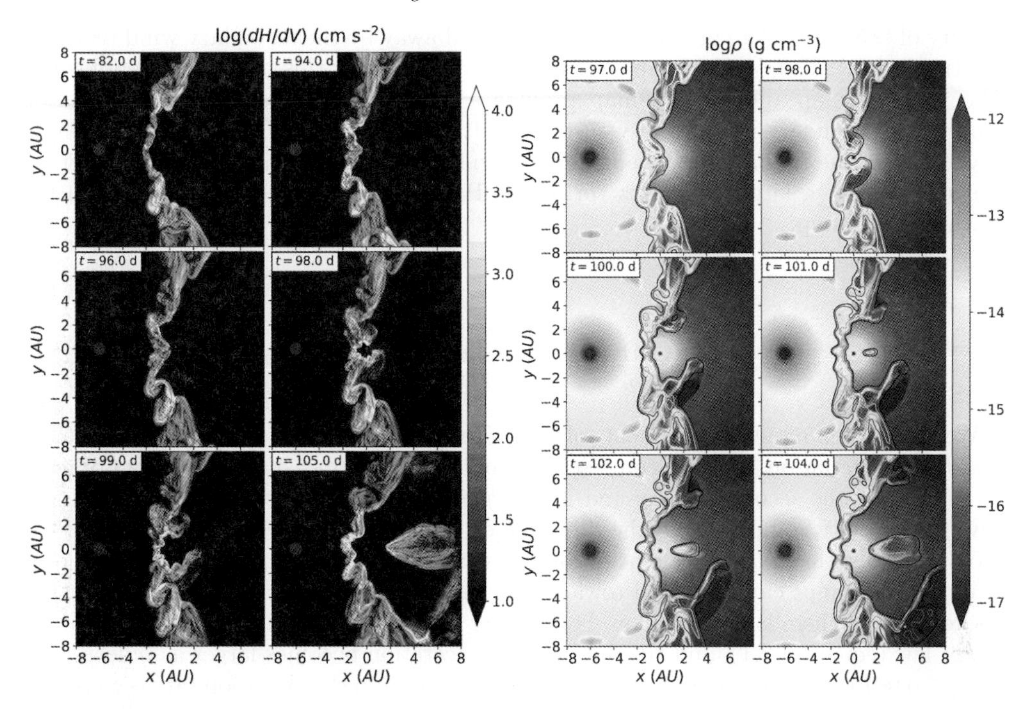

Figure 11. Left: Maps showing the specific helicity (helicity per unit volume) of the flow on the orbital plane. At later times, after the clumps had penetrated the colliding wind structure and some of the mass was accreted the downstream clumps are formed. Right: Density map together with vorticity contours showing the accretion phase ($t = 97$ days) and the following formation of the downstream blob. Vorticity contours (for the norm value) are smoothed using Gaussian smoothing and range from 10^{-6} s^{-1} (black) to 10^{-3} s^{-1} (white).

In the right pane of Figure 10 we show the absolute value of the vorticity ω. The helicity is a quantity that measures the degree of knottedness in the vortices (Moffatt 1969), defined as

$$H = \int_V \mathbf{v} \cdot (\nabla \times \mathbf{v})\, dV. \tag{4.2}$$

The integral over volume gives indication to the turbulence of the flow in that volume. The specific helicity, namely the helicity per unit volume is

$$\frac{dH}{dV} = \mathbf{v} \cdot (\nabla \times \mathbf{v}). \tag{4.3}$$

The left panel of Figure 11 shows maps of the helicity, at times similar to those in Figure 10. The vorticity and helicity maps emphasize every part of the hydrodynamical volume that does not flow in a laminar way (note that the ejected clumps are not seen until they interact with the CWS). This allows a quantitative measurement of the instabilities in the CWS, as well as detection of clumped regions.

We find that the amount of mass accreted onto the secondary is $M_{\rm acc} \simeq 3.3 \times 10^{-7}$ M$_\odot$. This amount of mass is accreted over a short period of only $\simeq 2$ days. The total energy is $\simeq 5 \times 10^{41}$ erg and the average power is $\simeq 2.9 \times 10^{36}$ erg s^{-1}. After the clumps have passed the colliding wind structure and some of the gas was accreted onto the secondary, the secondary wind has cleared the volume around the star and the secondary wind flow has been restored. However, on $t \simeq 105$ days the density map shows a massive clump that has formed on the side of the secondary opposite to the primary and the apex of the colliding wind structure. This downstream clump contains $\approx 10^{23}$ g , with a peak density of $\simeq 5 \times 10^{-13}$ g cm^{-3} and is being pushed away by the secondary wind (towards positive x axis). The density map does not easily reveal why this massive clump formed. The vorticity map and helicity map provide a hint. As seen in Figure 11 the region around the secondary became very turbulent after the accretion. This turbulence might have induced the formation of the massive downstream clump. The helicity and vorticiy maps show that a large volume is turbulent. Part of this region collapses to form the downstream clump.

5. Summary

We describe numerical experiments of a colliding wind LBV–WR system under momentum ratio η (equation 1.1). We run a set of simulations and measure the amount of mass accreted onto the secondary star and its X-ray emission. The instabilities in the CWS have the velocity component in the direction of the flow (namely, along the sides of the CWS) but also have the acceleration of the secondary. Depending on the parameters, some of the clumps can be pulled towards the secondary and get accreted. We find that above $\eta = 0.05$ there is no accretion. For smaller values the mass accretion rate follows a power law (equation 2.1), and the duty cycle of accretion becomes larger as η decreases. For $\eta \lesssim 0.001$ the accretion becomes continuous in time and the accretion rate is constant, and for $\eta \lesssim 0.0003$ there is saturation in the accretion rate, as the accretion is directly onto the secondary as there is no colliding wind structure.

The semi-opening angle for the CWS with instabilities is more complicated to define than for an adiabatic wind. We employ a numerical method to quantify it and find that it differs from the commonly used theoretical expression (equation 3.1) and for very low η it saturates (Figure 9). Our numerical method for extracting the wind angle has two major parameters: the degree of density change in the collusion region and the degree of smoothness after extracting the wind edge. We found that on most runs the parameters are very robust and changing them will result in the same angle, that gives us confidence in the method.

We showed an additional experiment in which clumps are ejected from the primary, travel towards the secondary and collide with the CWS. The interaction causes it to become unstable and form smaller clumps that are accreted onto the secondary. We demonstrated how the vorticity and helicity can be useful in quantifying the turbulent regions in the simulation, and propose they can be used for more diagnostics of the CWS and accretion.

The method demonstrated here – systematic exploration of the parameter space of massive colliding wind binaries – has shown to give quantitative relations between measurable parameters. There are many more parameters to cover, and probably more hidden relations. The observational signature of colliding wind systems spans across the spectrum. The simulations' results can be connected to observations of particular massive colliding wind binary systems, and be used to more tightly constrain stellar parameters. We therefore believe that such analysis of simulations of colliding wind binaries can be a useful tool in the study of massive stars and their evolution. This, in turn, has applications to our understanding how massive stars influence their galaxies through stellar winds, ionizing radiation, giant eruptions and supernovae explosions.

We acknowledge support from the R&D authority in Ariel University. We acknowledge the Ariel HPC Center at Ariel University for providing computing resources that have contributed to the research results reported within this paper.

References

Akashi, M. S., Kashi, A., & Soker, N. 2013, NewA, 18, 23, doi: 10.1016/j.newast.2012.05.010
Clementel, N., Madura, T. I., Kruip, C. J. H., & Paardekooper, J. P. 2015, MNRAS, 450, 1388, doi: 10.1093/mnras/stv696
Crowther, P. A., Dessart, L., Hillier, D. J., Abbott, J. B., & Fullerton, A. W. 2002, A&A, 392, 653, doi: 10.1051/0004-6361:20020941
Davidson, K., & Humphreys, R. M. 2012, 384, doi: 10.1007/978-1-4614-2275-4
Davidson, K., Ishibashi, K., & Martin, J. C. 2017, Research Notes of the American Astronomical Society, 1, 6, doi: 10.3847/2515-5172/aa96b3
Eichler, D., & Usov, V. 1993, ApJ, 402, 271, doi: 10.1086/172130
Gayley, K. G. 2009, ApJ, 703, 89, doi: 10.1088/0004-637X/703/1/89
Gayley, K. G., Owocki, S. P., & Cranmer, S. R. 1997, ApJ, 475, 786, doi: 10.1086/303573
Girard, T., & Willson, L. A. 1987, A&A, 183, 247
Gootkin, K., Dorn-Wallenstein, T., Lomax, J. R., et al. 2020, ApJ, 900, 162, doi: 10.3847/1538-4357/abad32
Hainich, R., Rühling, U., Todt, H., et al. 2014, A&A, 565, A27, doi: 10.1051/0004-6361/201322696
Hamann, W. R., & Koesterke, L. 1998, A&A, 335, 1003
Hendrix, T., Keppens, R., van Marle, A. J., et al. 2016, MNRAS, 460, 3975, doi: 10.1093/mnras/stw1289
Kashi, A. 2010, MNRAS, 405, 1924, doi: 10.1111/j.1365-2966.2010.16582.x
—. 2017, MNRAS, 464, 775, doi: 10.1093/mnras/stw2303
—. 2019, MNRAS, 486, 926, doi: 10.1093/mnras/stz837
—. 2020, MNRAS, 492, 5261, doi: 10.1093/mnras/staa203
Kashi, A., Davidson, K., & Humphreys, R. M. 2016, ApJ, 817, 66, doi: 10.3847/0004-637X/817/1/66
Kashi, A., & Michaelis, A. 2021, Galaxies, 10, 4, doi: 10.3390/galaxies10010004
Kashi, A., & Soker, N. 2007, MNRAS, 378, 1609, doi: 10.1111/j.1365-2966.2007.11908.x
—. 2008, NewA, 13, 569, doi: 10.1016/j.newast.2008.03.003
—. 2009a, NewA, 14, 11, doi: 10.1016/j.newast.2008.04.003
—. 2009b, MNRAS, 394, 923, doi: 10.1111/j.1365-2966.2008.14331.x
—. 2009c, NewA, 14, 11, doi: 10.1016/j.newast.2008.04.003
Lamberts, A., Fromang, S., & Dubus, G. 2011, MNRAS, 418, 2618, doi: 10.1111/j.1365-2966.2011.19653.x
Langer, N. 2012, ARA&A, 50, 107, doi: 10.1146/annurev-astro-081811-125534
Michaelis, A. M., Kashi, A., & Kochiashvili, N. 2018, NewA, 65, 29, doi: 10.1016/j.newast.2018.06.001
Moffat, A. F. J. 2008, in Clumping in Hot-Star Winds, ed. W.-R. Hamann, A. Feldmeier, & L. M. Oskinova, 17
Moffat, A. F. J., & Robert, C. 1994, ApJ, 421, 310, doi: 10.1086/173648
Moffatt, H. K. 1969, Journal of Fluid Mechanics, 35, 117, doi: 10.1017/S0022112069000991
Nugis, T., & Lamers, H. J. G. L. M. 2000, A&A, 360, 227
Owocki, S. P. 2015, in Astrophysics and Space Science Library, Vol. 412, Very Massive Stars in the Local Universe, ed. J. S. Vink, 113
Owocki, S. P., & Gayley, K. G. 1995, ApJL, 454, L145, doi: 10.1086/309786
Parkin, E. R., & Gosset, E. 2011, A&A, 530, A119, doi: 10.1051/0004-6361/201016125
Parkin, E. R., Pittard, J. M., Corcoran, M. F., & Hamaguchi, K. 2011, ApJ, 726, 105, doi: 10.1088/0004-637X/726/2/105
Pittard, J. M., & Dawson, B. 2018, MNRAS, 477, 5640, doi: 10.1093/mnras/sty1025
Puls, J., Vink, J. S., & Najarro, F. 2008, AAPR, 16, 209, doi: 10.1007/s00159-008-0015-8
Reitberger, K., Kissmann, R., Reimer, A., & Reimer, O. 2017, ApJ, 847, 40, doi: 10.3847/1538-4357/aa876d
Shenar, T., Sablowski, D. P., Hainich, R., et al. 2019, A&A, 627, A151, doi: 10.1051/0004-6361/201935684

Smith, N. 2014, ARA&A, 52, 487, doi: 10.1146/annurev-astro-081913-040025
Soker, N. 2005, ApJ, 635, 540, doi: 10.1086/497389
Stevens, I. R., Blondin, J. M., & Pollock, A. M. T. 1992, ApJ, 386, 265, doi: 10.1086/171013
Sundqvist, J. O., & Owocki, S. P. 2013, MNRAS, 428, 1837, doi: 10.1093/mnras/sts165
Usov, V. V. 1992, ApJ, 389, 635, doi: 10.1086/171236
Vink, J. S. 2015, in Astrophysics and Space Science Library, Vol. 412, Very Massive Stars in the Local Universe, ed. J. S. Vink, 77
Walder, R., & Folini, D. 2002, in Astronomical Society of the Pacific Conference Series, Vol. 260, Interacting Winds from Massive Stars, ed. A. F. J. Moffat & N. St-Louis, 595
Zhekov, S. A. 2021, MNRAS, 500, 4837, doi: 10.1093/mnras/staa3591

The Origin of Outflows in Evolved Stars
Proceedings IAU Symposium No. 366, 2022
L. Decin, A. Zijlstra & C. Gielen, eds.
doi:10.1017/S174392132200059X

Mass loss from binary stars approaching merger

Ondřej Pejcha

Institute of Theoretical Physics, Faculty of Mathematics and Physics, Charles University, V Holešovičkách 2, 180 00 Praha 8, Czech Republic
email: pejcha@utf.mff.cuni.cz

Abstract. Some binary stars experience common envelope evolution, which is accompanied by drastic loss of angular momentum, mass, and orbital energy and which leaves behind close binaries often involving at least one white dwarf, neutron star, or black hole. The best studied phase of common envelope is the dynamical inspiral lasting few original orbital periods. We show theoretical interpretation of observations of V1309 Sco and AT2018bwo revealing that binaries undergo substantial prolonged mass loss before the dynamical event amounting up to few solar masses. This mass loss is concentrated in the orbital plane in the form of an outflow or a circumbinary disk. Collision between this slower mass loss and the subsequent faster dynamical ejection powers a bright red transient. The resulting radiative shock helps to shape the explosion remnant and provides a site of dust and molecule formation.

Keywords. binaries: general, stars: mass loss, stars: winds, outflows

1. Introduction

Many binary stars undergo at least one episode of common envelope (CE) evolution. This short evolutionary phase causes ejection of a considerable fraction of total binary mass, significantly reduces the orbital separation of surviving bodies, or leads to a merger of the two binary components (e.g. Paczynski 1976; Iben & Livio 1993; Sana et al. 2012; Ivanova et al. 2013a). CE evolution is important for the formation of many objects of astrophysical importance, including gravitational wave sources (e.g. Dominik et al. 2012).

The binary star typically starts CE by developing a phase of unstable mass transfer. As the mass transfer rates gradually increase, the accreting star cannot accept this inflow of material and a fraction of the mass leaving the donor likely escapes the binary system altogether. Most of this material leaves the binary in the vicinity of Lagrange points L2 or L3. As the mass transfer instability runs away, the fraction of mass leaving the binary increases. Similar outcome likely occurs when the two stars begin their spiral-in due to the tidal Darwin instability. Eventually, the evolution of the two stars becomes fully dynamical, which can be viewed as an instantaneous ejection of material. The surviving binary or single merged star then relaxes to hydrodynamical and thermal equilibrium.

Traditionally, CE has been studied by comparing pre- and post-CE populations of binary and single stars. New discoveries and increasing volume data from time-domain transient surveys have opened new ways how to study CE evolution. In particular, a class of transients named Luminous red novae (LRNe) is now associated with CE events (Ivanova et al. 2013b). In this contribution, we discuss astrophysical interpretations of time-series observations before and during the merger. We study the possible outcomes in low- and high-mass stars by interpreting observations of V1309 Sco and AT2018bwo, respectively.

2. Gradual mass loss preceding dynamical phase

Recent binary evolution models suggest that the runaway binary mass transfer can last many hundreds or thousands of orbits and that mass-loss rates can exceed $\dot{M} \gtrsim 10^{-2}\,M_{\odot}\,\mathrm{yr}^{-1}$ (Blagorodnova et al. 2021). Much of the gas from the donor leaves the binary altogether. By studying trajectories of ballistic test particles leaving the L2 point, Shu et al. (1979) showed that the tidally-torqued gas either leaves to infinity or forms a circumbinary disk. Hubová & Pejcha (2019) found a wider varied range of outcomes when they considered particles with initial kicks or positional offsets from L2. Pejcha et al. (2016a,b) studied the radiative hydrodynamics of the same problem. They found that as the spiral stream expands, the spiral windings collide with themselves forming radial internal shocks. The velocity difference in the shocks Δv is closely related to the binary orbital velocity, $\Delta v \propto \sqrt{GM/a}$, where M is the binary mass. The resulting shock is radiative and powers emission with the luminosity of the order of $L \propto \dot{M}(\Delta v)^2$. For high $\dot{M}$, the outflow is optically-thick and the shock power is adiabatically degraded before it can radiate.

Ignoring viewing-angle effects, the L2 outflow is an additional source of light added on top of the central binary star. Depending on the binary and mass-loss properties, we can expect two possible outcomes. When the L2 outflow dominates, we should observe a gradual increase of L before the main outburst, although the emission from the L2 outflow might come out at mostly in the infrared. If the central binary dominates, we might observe constant pre-outburst flux or even dimming of the central binary due to dust obscuration by the outflow.

What is the dividing line between the two regimes? We can express $\dot{M} \sim M/P$, where P is the orbital period, and combine it with Kepler's laws to estimate L2 luminosity as $L \sim \dot{M}(\Delta v)^2 \sim M^2/(Pa) \sim M^{2.5}/a^{2.5}$. This approximation is very crude, because $\dot{M}$ is likely much smaller than M/P for most of the pre-CE evolution and because the actual value of $\dot{M}$ is set by the structure of the mass-losing star and binary properties. For a Roche-lobe filling primary star on the main sequence, $a \sim M^{0.8}$, which gives $L \sim M^{0.5}$. This implies that L2 luminosity grows only very slowly with the binary mass and is weaker for more evolved primaries. Luminosity of the stars on the main sequence scales as $L \propto M^{3.5}$ and the luminosity of more evolved stars of the same mass is even higher. This means that the effect of L2 mass loss will be harder to detect in high-mass binaries.

V1309 Sco was classified as a LRN by Mason et al. (2010). Tylenda et al. (2011) analyzed dense photometric dataset from the *OGLE* survey covering approximately 7 years before the explosion and found that V1309 Sco was initially a contact binary with $P \approx 1.4$ days. Orbital period experienced rapid decrease on the approach to the merger, which was accompanied by change of the orbital light curve from double-hump to single-hump profile. About a year before the peak brightness, the orbital variability disappeared, the observed flux shortly decreased and then gradually increased to the main peak.

Pejcha et al. (2017) used semi-analytic models combined with smoothed particle hydrodynamic simulations with flux-limited diffusion treatment of radiation in the vertical direction of the equatorially-concentrated outflow to explain the observed pre-explosion behavior of V1309 Sco. They explained the change of orbital light curve profile by setting the inclination angle of the binary to about 80° and viewing it through an L2 spiral outflow with gradually increasing $\dot{M}$. Mass leaving the binary also carries angular momentum, which leads to the decrease of P. Pejcha et al. (2017) showed that $\dot{M}$ inferred from changing light curve shape and $\dot{P}$ inferred from orbital period variations broadly agree with each other.

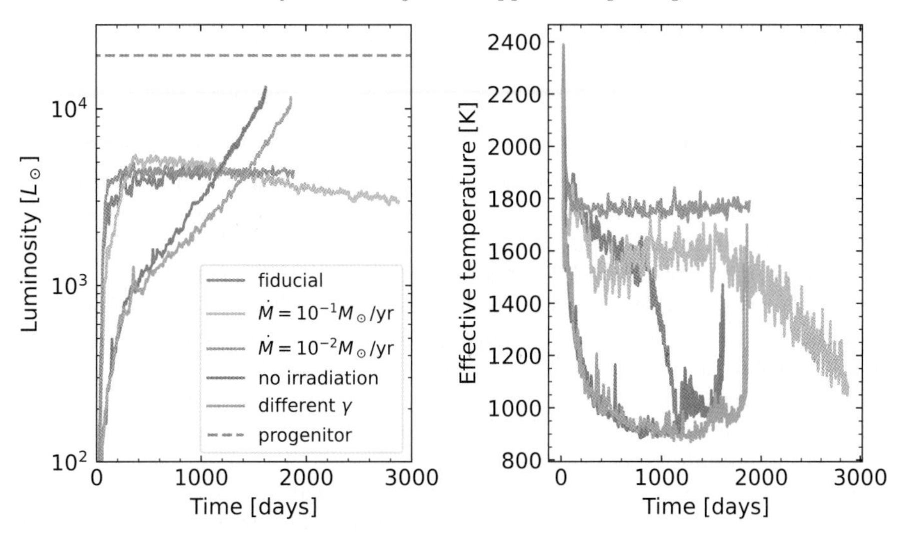

Figure 1. Radiative properties of mass ejected from L2 in a binary modeled after AT2018bwo (Blagorodnova et al. 2021). We show the evolution of luminosity (left panel) and mean effective temperature (right panel). The simulation was performed with the smoothed particle hydrodynamics code with vertical radiative diffusion presented in Pejcha et al. (2016a,b, 2017), but for a binary with $M_1 = 2.6\,M_\odot$, $M_2 = 13\,M_\odot$, $a = 1\,\mathrm{AU}$, and the binary effective temperature $T_{\rm eff} = 6000\,\mathrm{K}$, which gives binary luminosity assumed in the code $L = 4\pi a^2 \sigma T_{\rm eff}^4 \approx 5 \times 10^4\,L_\odot$. The mass loss rate of the fiducial model (blue lines) was initially set to $\dot{M} = 3 \times 10^{-2}\,M_\odot\,\mathrm{yr}^{-1}$, which increased as a power law with an index $\gamma = 3$ and with a singularity set to $t = 2000$ days. We also show modifications of the fiducial model by setting $\dot{M}$ constant (orange and green lines), no irradiation by the central binary (red lines), and $\gamma = 2$ (purple lines). For comparison, dashed vertical line in the left panel shows the progenitor luminosity of AT2018bwo, $L \approx 2 \times 10^4\,L_\odot$ observed about 14 years before the outburst by Blagorodnova et al. (2021).

For $\dot{M} \gtrsim 10^{-3}\,M_\odot\,\mathrm{yr}^{-1}$, the binary is obscured by the L2 outflow and the orbital variability was not visible anymore. After this, internal shocks in the L2 outflow provide enough luminosity to increase the observed L. Since the outflow is optically-thick, interplay of diffusion and adiabatic expansion control the amount of released radiation; Pejcha et al. (2017) found reasonably good match to the observed light curve for a prescribed evolution of $\dot{M}$ with numerical simulations. Pejcha et al. (2017) also found indications that the properties of the outflow change approximately 50 days before the merger, which can be caused either by growing temperature of the binary due to mass-loss stripping or changes to the morphology of the mass loss flow (MacLeod et al. 2018).

AT2018bwo was thoroughly analyzed by Blagorodnova et al. (2021) using a combination of pre-explosion photometry, spectroscopy, binary evolution models, and modeling of the transient. They found that the progenitor position in the Hertzprung-Russel diagram matches $M \approx 15\,M_\odot$ binary with $a \approx 1\,\mathrm{AU}$ undergoing thermal-timescale mass transfer with $\dot{M} \approx 10^{-2}\,M_\odot\,\mathrm{yr}^{-1}$.

In Figure 1, we show the results of modeling AT2018bwo using the similar assumptions as was done for V1309 Sco; more thorough discussion is in Section 4.4 of Blagorodnova et al. (2021). We see that even under optimistic assumptions the L2 luminosity does not reach the luminosity of the progenitor observed approximately 14 years before the peak of the outburst. This is different from V1309 Sco, where the L2 outflow was the dominant source of luminosity for a year before the merger, but it is also expected based on our analytic estimates.

Figure 1 also shows estimates of effective temperature $T_{\rm eff}$ of the L2 radiation. We see that the expected value is between 1000 and 2000 K, which implies that most of the luminosity will be seen in the near infrared. However, it is not clear whether the outflow is sufficiently cool for dust condensation.

3. Collision of dynamical ejecta with pre-explosion mass loss

Once the timescale of the acceleration of the binary inspiral becomes shorter than the expansion timescale of the outflow near the binary, it becomes more convenient to think about the subsequent mass loss as a nearly instantaneous mass ejection. The ejecta likely moves faster than the previous L2 outflow, because it was launched from a binary on a much tighter orbit, and it is also likely less concentrated within the equatorial plane, because of more shock heating. As the more spherical faster ejecta expands, it will radiate part of its thermal energy. Envelopes of most stars are hydrogen-rich and the resulting transient should resemble a scaled-down version of Type II-P supernova, as was first pointed out by Ivanova et al. (2013b). Multiple peaks in the light curves of LRNe are explained as individual mass ejections.

It is perhaps inevitable that the faster more spherical ejecta collides with the pre-existing equatorial outflow forming a radiative shock. The hydrodynamics of such an interaction are relatively well understood (e.g. Suzuki et al. 2019; Kurfürst & Krtička 2019; Kurfürst et al. 2020; McDowell et al. 2018), but the implications for transients are less explored. Metzger & Pejcha (2017) constructed a semi-analytic model of an equatorial radiative shock coupled to an expanding envelope. They argued that the first peak in the light curves of LRNe comes from cooling emission from the freely-expanding polar ejecta (MacLeod et al. 2017), while the second peak is caused by diffusion of light from the dense equatorial radiative shock. The reported scaling relations can explain luminosities and timescales of the observed events and suggest that some of the long infrared transients recently identified by Kasliwal et al. (2017) are CE events from evolved binaries on wide orbits.

V1309 Sco shows a single peak, which can be explained by ejecting few hundredths of $M_\odot$ of recombining hydrogen (Ivanova et al. 2013b; Nandez et al. 2014). In the shock-powered model, the second peak can be hidden behind the dust formed near the radiative shock in the equatorial plane, because we are viewing the system near the original orbital plane. Alternatively, the shock might not be energetic enough to keep the hydrogen ionized for sufficiently long (Metzger & Pejcha 2017).

AT2018bwo also showed a single peak, but the data are substantially scarcer than in V1309 Sco. Bolometric light curve of Blagorodnova et al. (2021) shows a possible brightening toward the end of the plateau, potentially resembling a second peak. The transient properties were analyzed by Blagorodnova et al. (2021) using analytic scaling relations in the Type II-P supernova and shock-powered models. They found that different Type II-P supernova scaling relations give very different inferences of the ejection radius R_0 and ejecta mass $M_{\rm ej}$, because their application to LRNe is an extrapolation from the domain where they have been validated by radiation hydrodynamics simulations. The analytic scalings of the shock powered model of Metzger & Pejcha (2017) give reasonable values for the masses of the pre-existing equatorial outflow ($M_{\rm wind}$) and of the faster ejecta ($M_{\rm ej}$). The inference shows that $M_{\rm wind} > M_{\rm ej}$, which is in agreement with binary evolution models of the same event of Blagorodnova et al. (2021). In Figure 2, we show the confidence ellipsoids of the inferred physical parameters for the two models of the transient.

4. Future outlook

Observations of LRNe can provide new insight into the open questions in the CE evolution. We have argued that the most often studied dynamical phase of CE evolution

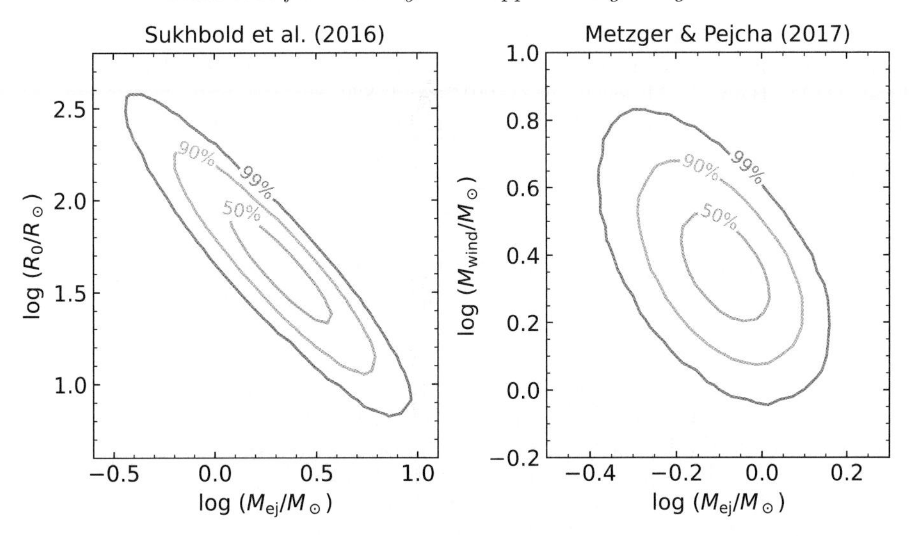

Figure 2. Ejecta properties of AT2018bwo estimated from its plateau luminosity, duration, and expansion velocity. The left panel shows the confidence ellipsoids for the initial radius R_0 and ejecta mass M_{ej} in the model of the scaled-down Type II-P supernova (Sukhbold et al. 2016). The right panel shows the confidence ellipsoids for the mass of pre-explosion outflow M_{wind} and mass of the ejecta M_{ej} in the shock collision model (Metzger & Pejcha 2017).

is preceded by a long gradual loss of mass from the binary, which can be observed as a slow rise of brightness. When the event becomes dynamical, the faster younger and more spherical ejecta should collide with the older equatorially-concentrated mass distribution. The resulting radiative shock can explain the observed luminosities and timescales as well as double peaks seen in some events.

But there remains much to be done. A predictive theory of mass-loss rate evolution before the dynamical phase remains to be found. Standard spherically-symmetric stellar evolution codes can be evolved far enough to give very high mass-loss rates, but they currently cannot reach close enough the main peak. Dust formation in the gradual equatorial outflow and its observational signatures need to be properly characterized. Modeling of the transients would benefit from calibrating the analytic scaling relation of Type II-P supernovae with radiation hydrodynamic simulations appropriate for LRNe. The shock powered model needs to be more developed to be directly comparable to observations. This is difficult, because the problem geometry deviates from spherical symmetry and it is necessary to include realistic equation of state as well as appropriate opacities and take into account dust formation.

Acknowledgements

This research was supported by Horizon 2020 ERC Starting Grant 'Cat-In-hAT' (grant agreement no. 803158).

References

Blagorodnova, N., Karambelkar, V., Adams, S. M., et al. 2020, MNRAS, 496, 5503
Blagorodnova, N., Klencki, J., Pejcha, O., et al. 2021, A&A, 653, A134
Dominik, M., Belczynski, K., Fryer, C., et al. 2012, ApJ, 759, 52
Kurfürst, P. & Krtička, J. 2019, A&A, 625, A24
Kurfürst, P., Pejcha, O., & Krtička, J. 2020, A&A, 642, A214
Hubová, D. & Pejcha, O. 2019, MNRAS, 489, 891
Iben, I. & Livio, M. 1993, PASP, 105, 1373

Ivanova, N., Justham, S., Chen, X., et al. 2013a, AAPR, 21, 59
Ivanova, N., Justham, S., Avendano Nandez, J. L., et al. 2013b, Science, 339, 433
Kasliwal, M. M., Bally, J., Masci, F., et al. 2017, ApJ, 839, 88
MacLeod, M., Macias, P., Ramirez-Ruiz, E., et al. 2017, ApJ, 835, 282
MacLeod, M., Ostriker, E. C., & Stone, J. M. 2018, ApJ, 863, 5
Mason, E., Diaz, M., Williams, R. E., et al. 2010, A&A, 516, A108
McDowell, A. T., Duffell, P. C., & Kasen, D. 2018, ApJ, 856, 29
Metzger, B. D. & Pejcha, O. 2017, MNRAS, 471, 3200
Nandez, J. L. A., Ivanova, N., & Lombardi, J. C. 2014, ApJ, 786, 39
Paczynski, B. 1976, Structure and Evolution of Close Binary Systems, 73, 75
Pejcha, O. 2014, ApJ, 788, 22
Pejcha, O., Metzger, B. D., & Tomida, K. 2016a, MNRAS, 455, 4351
Pejcha, O., Metzger, B. D., & Tomida, K. 2016b, MNRAS, 461, 2527
Pejcha, O., Metzger, B. D., Tyles, J. G., et al. 2017, ApJ, 850, 59
Sana, H., de Mink, S. E., de Koter, A., et al. 2012, Science, 337, 444
Shu, F. H., Lubow, S. H., & Anderson, L. 1979, ApJ, 229, 223
Sukhbold, T., Ertl, T., Woosley, S. E., et al. 2016, ApJ, 821, 38
Suzuki, A., Moriya, T. J., & Takiwaki, T. 2019, ApJ, 887, 249
Tylenda, R., Hajduk, M., Kamiński, T., et al. 2011, A&A, 528, A114

The Origin of Outflows in Evolved Stars
Proceedings IAU Symposium No. 366, 2022
L. Decin, A. Zijlstra & C. Gielen, eds.
doi:10.1017/S1743921322000199

Explaining the winds of AGB stars: Recent progress

Susanne Höfner and Bernd Freytag

Theoretical Astrophysics, Department of Physics & Astronomy, Uppsala University,
Box 516, SE-75120 Uppsala, Sweden
email: susanne.hoefner@physics.uu.se

Abstract. The winds observed around asymptotic giant branch (AGB) stars are generally attributed to radiation pressure on dust, which is formed in the extended dynamical atmospheres of these pulsating, strongly convective stars. Current radiation-hydrodynamical models can explain many of the observed features, and they are on the brink of delivering a predictive theory of mass loss. This review summarizes recent results and ongoing work on winds of AGB stars, discussing critical ingredients of the driving mechanism, and first results of global 3D RHD star-and-wind-in-a-box simulations. With such models it becomes possible to follow the flow of matter, in full 3D geometry, all the way from the turbulent, pulsating interior of an AGB star, through its atmosphere and dust formation zone into the region where the wind is accelerated by radiation pressure on dust. Advanced instruments, which can resolve the stellar atmospheres, where the winds originate, provide essential data for testing the models.

Keywords. stars: AGB and post-AGB, stars: atmospheres, stars: winds, outflows, stars: mass loss, circumstellar matter

1. Introduction

While going through the Asymptotic Giant Branch phase, low- and intermediate mass stars develop winds with typical mass loss rates of $10^{-7} - 10^{-5} M_{\odot}/\mathrm{yr}$ and wind velocities of about $5 - 30$ km/s, which affect their observable properties and their further evolution. AGB stars tend to show a pronounced variability of their luminosities and spectra with periods of about $100 - 1000$ days, attributed to large-amplitude pulsations. The pulsations, together with large-scale convective flows, trigger strong radiative shock waves in the stellar atmospheres, which intermittently lift gas to distances where temperatures are low enough to permit dust formation. Radiation pressure on dust grains is assumed to be the driving force behind the massive winds of AGB stars, which turn them into white dwarfs and enrich the interstellar medium with newly-produced chemical elements.

Considerable efforts have been made to understand the physics of AGB stars, and to develop quantitative models of their dynamical interiors, atmospheres and winds. A recent review by Höfner & Olofsson (2018) discusses both theoretical and observational aspects of AGB star winds in some detail. Here, we give a brief summary of recent developments, focusing on properties of wind-driving dust grains, stellar pulsation and convection, and the 3D morphology of atmospheres and winds.

2. Quantitative dynamic models: Ingredients and applications

Ideally, quantitative models of the dynamical atmospheres and winds should be based on first physical principles, describing the complex interplay of dynamical, radiative and micro-physical processes in AGB stars. By following the flow of matter in full 3D geometry, all the way from the convective, pulsating interior of an AGB star, through its atmosphere and dust formation zone into the region where the wind is accelerated by radiation pressure on dust, a predictive theory of mass loss can be developed. Taking snapshots of the resulting structures of the dynamical atmospheres and winds, synthetic observable properties may be computed and compared to observations, in order to test such models. While the most advanced 3D "star-and-wind-in-a-box" models (discussed in Sect. 5) are approaching this ideal, it has to be noted that such numerical simulations are computationally very demanding.

Current dynamic wind models for AGB stars usually assume spherically symmetric flows, which reduces computation times to a degree that makes it possible to construct large grids, covering the wide range of stellar parameters needed in stellar evolution studies. These models come in two basic categories:

- steady wind models, assuming time-independent radial outflows, and
- time-dependent atmosphere and wind models, accounting for effects of pulsation.

Historically, steady wind models have played a critical role in establishing radiation pressure on dust as the most probable driving mechanism, and in studying the composition of the dust grains by comparing synthetic spectral energy distributions to observed data. They are also widely used to derive wind properties (mass loss rates, radial velocity profiles) from observations. Their computational efficiency makes them common tools for stellar population studies, and for predicting dust yields (see, e.g., the talk by Nanni, this conference). However, it has to be kept in mind that the mass loss rate is an input parameter in these models (set via the starting conditions at the foot point of the steady outflow), not a result.

In time-dependent models, on the other hand, the physical conditions in the region, where dust forms and initiates an outflow, are a result of pulsation-induced shock waves. These models predict mass loss rates, wind velocities and dust properties for given stellar parameters, elemental abundances and pulsation properties (periods, amplitudes). The simulations describe the varying radial profiles of densities, temperatures, velocities, and dust properties (see Fig. 1 for an example). These can be used to compute a wide range of synthetic observables, e.g. spectra at various resolutions, photometric fluxes and light curves, interferometric data. Much of our current understanding of dynamical atmospheres and dust-driven winds of AGB stars is derived from such models, as discussed below.

3. Properties of wind-driving dust grains

Chemically speaking, AGB stars can be divided into two groups, defined by the value of the atmospheric C/O ratio being smaller (M-type) or larger (C-type) than 1. Due to the high binding energy of the CO molecule, the less abundant of the two elements is almost completely bound up in CO, while the more abundant element is available for forming other molecules and dust.

In C-type AGB stars (with C/O >1 in the atmosphere, due to the 3. dredge-up) amorphous carbon is the main wind-driving dust species. The mass loss rates, wind velocities, spectral energy distributions, and photometric variations resulting from time-dependent atmosphere and wind models are in good agreement with observations, see, e.g., Nowotny et al. (2011), Nowotny et al. (2013), Eriksson et al. (2014). However, due to the strong absorption by carbon dust, the stellar photospheres and inner atmospheres may be severely obscured, and detailed comparisons with spatially resolved observations can be difficult, e.g., Paladini et al. (2009); Stewart et al. (2016); Sacuto et al. (2011); Wittkowski et al. (2017).

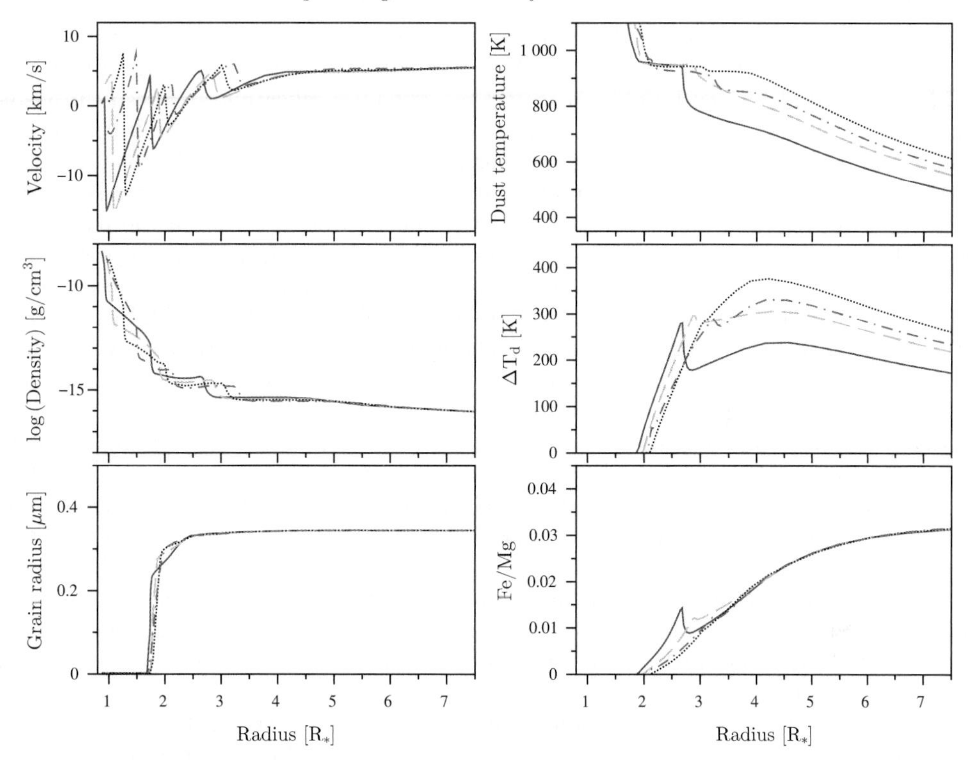

Figure 1. Time-dependent radial structure of a DARWIN model (Dynamical Atmosphere and Radiation-driven Wind model, based on Implicit Numerics; model An315u3 in (Höfner et al. 2021, Höfner et al. 2021)). *Left, top to bottom*: Flow velocity, gas density, dust grain radius. *Right, top to bottom*: Temperature of the olivine-type Mg-Fe silicate grains, difference in grain temperature with and without taking Fe-enrichment into account, and the Fe/Mg ratio in the dust grains (set by self-regulation via the grain temperature; note the plateaus in dust temperatures, top right panel). The plots show the inner parts of the model, zoomed in on the dust formation and wind acceleration region (snapshots of 4 phases during a pulsation cycle). The first snapshot (solid lines) corresponds to a near-minimum phase, the dashed lines represent the ascending part of the bolometric light curve, the dotted lines shows a phase close to the luminosity maximum, and the dash-dotted lines represent the descending part of the bolometric light curve. It should be noted that only the innermost, dust-free parts of the model structures show periodic variations that repeat every pulsation cycle, while grain growth and wind acceleration are governed by other time scales. The top right panel, showing the highest dust temperatures at maximum light, demonstrates the general effects of radiative heating. The bottom right panel shows the Fe-enrichment, and how this affects the grain temperature is displayed in the middle right panel.

The winds of M-type AGB stars (with $C/O < 1$ in the atmosphere) have long been assumed to be driven by radiation pressure on silicate dust, which consists of abundant elements and produces characteristic mid-IR features around 10 and 18 microns, observed in many such stars. Detailed models suggest a scenario where the radiative pressure, that triggers the outflows, is caused by photon scattering on Fe-free silicate grains, Höfner (2008). Such particles are highly transparent at visual and NIR wavelengths, resulting in significantly less radiative heating and smaller condensation distances than for Fe-bearing silicate grains, see Dorschner et al. (1995); Jäger et al. (2003); Woitke (2006); Zeidler et al. (2011). To create sufficient radiative pressure by scattering, the grains need to be of a size comparable to the wavelengths where the stellar flux peaks, meaning that grain radii should fall in the range of 0.1 – 1 μm. Recent observational studies have found dust grains with radii of about $0.1 - 0.5\,\mu$m, at distances below 2 – 3 stellar radii, e.g., Norris et al. (2012); Ohnaka et al. (2016, 2017), as predicted by the numerical simulations.

Models of winds driven by photon scattering on Fe-free silicate grains result in visual and near-IR spectra, light curves, variations of photometric colors with pulsation phase, and wind properties that are in good agreement with observations, see Bladh et al. (2013, 2015, 2019); Höfner et al. (2016). In order to produce the characteristic mid-IR silicate features, however, a gradual enrichment of the silicate dust with Fe in the inner wind region has to be taken into account, to keep grain temperatures from dropping too quickly with distance from the star. Figure 1 shows new DARWIN models by Höfner et al. (2021), describing the growth of olivine-type silicate grains with a variable Fe/Mg ratio, set by self-regulation via the grain temperature. The models demonstrate that the enrichment of the silicate grains with Fe is a secondary process, taking place in the stellar wind on the surface of Fe-free grains that have initiated the outflow. The self-regulating feedback between grain composition and radiative heating leads to low values of Fe/Mg, typically a few percent.

Dust formation in the atmospheres of AGB stars is a non-equilibrium process, since the time-scales of grain growth are comparable to dynamical processes (pulsation, wind acceleration) and variations in radiative flux. Due to falling densities in the outflows, grain growth rates drop strongly with distance from the star, leading to condensation of dust-forming elements that typically is far from complete. In dynamical wind models grain growth is usually treated with a kinetic description, defining the rates at which material is condensing onto the grains out of the surrounding gas. A long-standing problem in this context is a realistic description of nucleation, that is the formation of the very first condensation nuclei from the gas. The nature of these first condensates and their formation rates have been a subject of intense debate, see, e.g., Gobrecht et al. (2016); Gail et al. (2016) and references therein. In M-type AGB stars, Al_2O_3 (corundum) is one of the most promising candidates, and Gobrecht et al. (2021) report recent progress in a quantum-chemical description of this species. Condensation of silicates onto pre-existing Al_2O_3 seed particles may speed up grain growth to sizes necessary for driving a wind by photon scattering, as demonstrated by Höfner et al. (2016).

4. Stellar pulsation and convection

In the current time-dependent atmosphere and wind models discussed above, the effects of radial pulsation are introduced through variable inner boundary conditions, just below the stellar photosphere. In most cases, the periodic expansions and contractions of a star are simulated by prescribing a sinusoidal motion of the innermost layer, accompanied by a variation in luminosity (so-called piston models). The pulsation period may be set according to empirical period–luminosity relations and the amplitude can, in principle, be constrained by comparing the resulting wind properties, synthetic spectra and light curves to observations. This simple treatment of pulsation effects has the advantage of few free parameters, but it also has its limitations. Introducing a non-sinusoidal shape of the luminosity variation and a phase shift compared to the motion of the photospheric layers, as suggested by observed light curves and CO line profile variations, may lead to significant differences in mass loss rates and wind velocities, see Liljegren et al. (2017).

Ultimatively, a predictive mass loss theory requires realistic models of stellar convection and pulsation, which affect the stellar atmosphere and wind. Pulsations of AGB stars are notoriously difficult to model, but some progress has been made in recent years, regarding lower amplitude overtone pulsations using a linear non-adiabatic approach, e.g., Wood (2015); Trabucchi et al. (2017). Furthermore, Trabucchi et al. (2021) presented new results on non-linear radial pulsation, which resolve earlier problems with predicted fundamental-mode periods, compared to observations of Mira variables.

A general drawback of 1D pulsation models is that they have to rely on a simplified description of stellar convection. The classical picture of mixing-length style energy transport is probably inadequate for the extremely non-linear, non-adiabatic, large-scale interior dynamics of AGB stars. Both theory and observations indicate the existence of giant convection cells, in other words turbulent gas flows on scales comparable to the stellar radius, Schwarzschild (1975); Freytag & Höfner (2008); Freytag et al. (2017); Paladini et al. (2018). They blur the distinction between convection and pulsation, making the applicability of 1D pulsation models questionable.

A way to resolve these problems is global 3D radiation-hydrodynamical (RHD) modelling. The pioneering AGB "star-in-a-box" models created by Freytag & Höfner (2008) and Freytag et al. (2017), building on the capability of the CO5BOLD code to cover the entire outer convective envelope and atmosphere, indeed show self-excited radial pulsations, with periods that are in good agreement with observations of Mira variables. Liljegren et al. (2018) indicated how the 3D model results can be used to improve the description of pulsation effects in 1D wind models.

5. 3D morphology of atmospheres and winds

A strong incentive to develop global 3D models of AGB stars and their dust-driven winds comes from recent high-angular-resolution observations. Imaging of nearby AGB stars at visual and infrared wavelengths has revealed complex, non-spherical distributions of gas and dust in the close circumstellar environment, see, e.g., Stewart et al. (2016), Wittkowski et al. (2017). Temporal monitoring shows changes in both atmospheric morphology and grain sizes over the course of weeks or months, e.g. Khouri et al. (2016); Ohnaka et al. (2017); see also the talk by Khouri, this conference. Such phenomena cannot be investigated with the spherically symmetric atmosphere and wind models mentioned above. In the 3D RHD "star-in-a-box" models by Freytag et al. (2017), on the other hand, an inhomogeneous distribution of atmospheric gas emerges naturally, as a consequence of large-scale convective flows below the photosphere and the resulting network of atmospheric shock waves. As shown in detail by Höfner & Freytag (2019), the dynamical patterns in the gas are imprinted on the dust in the close stellar environment, due to the density- and temperature-sensitivity of the grain growth process, explaining the origin of the observed clumpy dust clouds. However, these earlier 3D simulations did not include the effects of radiation pressure on dust, and could therefore not predict the structure of the wind formation zone.

In Fig. 2, we show results of the first global 3D RHD "star-and-wind-in-a-box" models, computed with the CO5BOLD code (Freytag & Höfner, in prep.). The models explore the interplay of interior dynamics (convection, pulsation), atmospheric shocks, dust formation, and wind acceleration in full 3D geometry. They include non-grey radiative transfer, as well as a time dependent description of silicate grain growth and evaporation, and they account for the effects of radiative pressure on dust in a simple way. These new models feature a much larger computational domain than previous "star-in-a-box" models, covering the inner wind region. This allows us to follow the emerging 3D structures to a distance where the outflow is established, and to compute mass loss rates.

6. Summary and conclusions

The winds of AGB stars are driven by radiation pressure on dust grains, which form in the highly dynamical atmospheres of these pulsating, strongly convective stars. Detailed 3D RHD models of the stars and their winds are essential for understanding the physical processes that cause the outflows, for explaining the observed properties of the stars, and for developing a predictive theory of mass loss on the AGB. High angular resolution

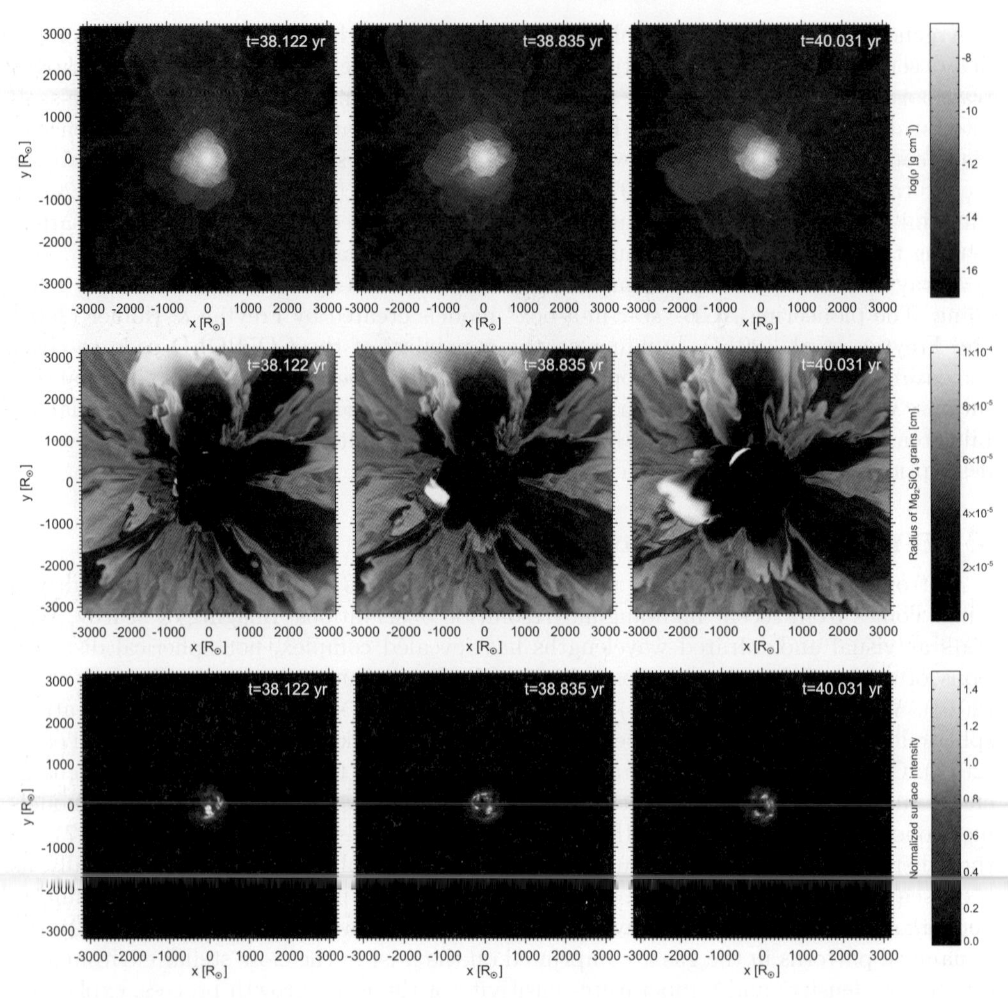

Figure 2. Time-dependent structure of a 3D radiation-hydrodynamical model of an AGB star, including interior dynamics (convection, pulsation), dust formation and wind acceleration, computed with the CO5BOLD code (Freytag & Höfner, in prep.). Time sequences (left to right) of gas density and silicate grain radius for slices through the center of the star (rows 1–2), and the variation of relative surface intensity (bottom row, indicating the size of the star). The stellar parameters of the model (st28gm06n052) are $1.0\,M_\odot$, $7000\,L_\odot$, and an effective temperature of 2700 K, and the resulting mass loss rate is about $5 \cdot 10^{-6}\,M_\odot/\mathrm{yr}$. The sequence of images shows the formation of two new distinct dust clouds (bright areas in the grain size plots, row 2, lower left and upper left quadrants), in the dense wakes of atmospheric shock waves (row 1). At the top edge of the images, a dust cloud that was formed earlier is driven outwards by radiation pressure.

imaging of scattered visual and near IR light shows clumpy dust clouds surrounding AGB stars. Such clumpy dust clouds emerge naturally in 3D "star-and-wind-in-a-box" models computed with the CO5BOLD code, as a consequence of giant convection cells and related atmospheric shock waves.

Large dust grains with radii of about 0.1 – 0.5 microns are found in scattered-light observations at distances of about 2 stellar radii around AGB stars, as required for driving winds by photon scattering on near-transparent silicate grains with low Fe/Mg ratios. New DARWIN models, describing the growth of olivine-type silicate grains with a variable Fe/Mg ratio, show that the enrichment with Fe is a secondary process, taking place in the stellar wind on the surface of Fe-free grains that have triggered the outflow. The self-regulating feedback between grain composition and radiative heating leads to low values of Fe/Mg, typically a few percent. These models show distinctive mid-IR silicate features even for low Fe/Mg, and realistic photometry from visual to mid-IR wavelengths.

While 3D "star-and-wind-in-a-box" models are an essential tool for understanding the underlying physics of AGB star winds, they are computationally very demanding. For the foreseeable future, mass-loss descriptions for stellar evolution models will depend on extensive grids of 1D wind models, covering a wide range of stellar parameters. Recent results and ongoing theoretical work will help to constrain input parameters of such models, e.g. regarding stellar pulsation properties or dust nucleation.

Finally, it should be mentioned that the wind mechanism of red supergiants is still unknown, but recent observations indicate that a significant part of the ejected mass is found in dense clouds of gas and dust. This could indicate similarities with AGB stars, regarding the mass-loss process (see talk by Chiavassa, this conference).

Acknowledgements

The authors acknowledge funding from the European Research Council (ERC) under the European Union's Horizon 2020 research and innovation programme (Grant agreement No. 883867, project EXWINGS) and the Swedish Research Council (grant number 2019-04059). The computations of 3D RHD models of AGB stars and their winds with the CO5BOLD code were enabled by resources provided by the Swedish National Infrastructure for Computing (SNIC).

References

Bladh, S., Höfner, S., Aringer, B., & Eriksson, K. 2015, A&A, 575, A105
Bladh, S., Höfner, S., Nowotny, W., Aringer, B., & Eriksson, K. 2013, A&A, 553, A20
Bladh, S., Liljegren, S., Höfner, S., Aringer, B., & Marigo, P. 2019, A&A, 626, A100
Dorschner, J., Begemann, B., Henning, T., Jaeger, C., & Mutschke, H. 1995, A&A, 300, 503
Eriksson, K., Nowotny, W., Höfner, S., Aringer, B., & Wachter, A. 2014, A&A, 566, A95
Freytag, B. & Höfner, S. 2008, A&A, 483, 571
Freytag, B., Liljegren, S., & Höfner, S. 2017, A&A, 600, A137
Gail, H.-P., Scholz, M., & Pucci, A. 2016, A&A, 591, A17
Gobrecht, D., Cherchneff, I., Sarangi, A., Plane, J. M. C., & Bromley, S. T. 2016, A&A, 585, A6
Gobrecht, D., Plane, J. M. C., Bromley, S. T., et al. 2021, arXiv e-prints, arXiv:2110.11139
Höfner, S. 2008, A&A, 491, L1
Höfner, S., Bladh, S., Aringer, B., & Ahuja, R. 2016, A&A, 594, A108
Höfner, S., Bladh, S., Aringer, B., & Eriksson, K. 2021, arXiv e-prints, arXiv:2110.15899
Höfner, S. & Freytag, B. 2019, A&A, 623, A158
Höfner, S. & Olofsson, H. 2018, AAPR, 26, 1
Jäger, C., Dorschner, J., Mutschke, H., Posch, T., & Henning, T. 2003, A&A, 408, 193
Khouri, T., Maercker, M., Waters, L. B. F. M., et al. 2016, A&A, 591, A70
Liljegren, S., Höfner, S., Eriksson, K., & Nowotny, W. 2017, A&A, 606, A6
Liljegren, S., Höfner, S., Freytag, B., & Bladh, S. 2018, A&A, 619, A47
Norris, B. R. M., Tuthill, P. G., Ireland, M. J., et al. 2012, Nature, 484, 220
Nowotny, W., Aringer, B., Höfner, S., & Eriksson, K. 2013, A&A, 552, A20
Nowotny, W., Aringer, B., Höfner, S., & Lederer, M. T. 2011, A&A, 529, A129
Ohnaka, K., Weigelt, G., & Hofmann, K.-H. 2016, A&A, 589, A91
Ohnaka, K., Weigelt, G., & Hofmann, K.-H. 2017, A&A, 597, A20
Paladini, C., Aringer, B., Hron, J., et al. 2009, A&A, 501, 1073
Paladini, C., Baron, F., Jorissen, A., et al. 2018, Nature, 553, 310

Sacuto, S., Aringer, B., Hron, J., et al. 2011, A&A, 525, A42
Schwarzschild, M. 1975, ApJ, 195, 137
Stewart, P. N., Tuthill, P. G., Monnier, J. D., et al. 2016, MNRAS, 455, 3102
Trabucchi, M., Wood, P. R., Montalbán, J., et al. 2017, ApJ, 847, 139
Trabucchi, M., Wood, P. R., Mowlavi, N., et al. 2021, MNRAS, 500, 1575
Wittkowski, M., Hofmann, K. H., Höfner, S., et al. 2017, A&A, 601, A3
Woitke, P. 2006, A&A, 460, L9
Wood, P. R. 2015, MNRAS, 448, 3829
Zeidler, S., Posch, T., Mutschke, H., Richter, H., & Wehrhan, O. 2011, A&A, 526, A68

The Origin of Outflows in Evolved Stars
Proceedings IAU Symposium No. 366, 2022
L. Decin, A. Zijlstra & C. Gielen, eds.
doi:10.1017/S1743921322000072

The onset of mass loss in evolved stars

Iain McDonald[1,2]

[1]Open University, Walton Hall, Kents Hill, Milton Keynes, MK7 6AA, UK
email: iain.mcdonald@open.ac.uk

[2]Jodrell Bank Centre for Astrophysics, Alan Turing Building, University of Manchester, Manchester, M13 9PL, UK

Abstract. To look at propagating winds from evolved stars into the interstellar medium is to look at how they are sustained. To understand their origins, we must look to the circumstances that create them in the first instance. In this article, I examine the physical conditions under which pulsation-enhanced, dust-driven winds are first generated. These initial conditions can help constrain the late evolutionary stages of these stars and provide insight into the mechanisms that cause the mass loss itself.

Keywords. stars: AGB and post-AGB, circumstellar matter, stars: fundamental parameters, stars: late-type, stars: mass loss, stars: Population II, stars: winds, outflows, stars: variables: other, globular clusters: general

---

1. Introduction

The pulsation-enhanced, dust-driven wind of thermally pulsating asymptotic giant branch (TP-AGB) stars is the last in a series of mechanisms by which low- and intermediate-mass ($0.6 \lesssim M \lesssim 8$ M$_\odot$) stars lose mass (e.g., S. Höfner, this proceedings). It replaces chromospheric mass loss, the dominant wind-driving mechanism on the red giant branch (RGB) and early AGB (e.g., Dupree, Hartmann & Avrett 1984). This transition co-incides with several points in stellar evolution: the increase in mass-loss rate it provides generally leads to the mass-loss rate exceeding the nuclear-burning rate; while strong surface pulsations begin at a similar time as the much longer (and unrelated) thermonuclear pulsations of the helium shell and the associated (third) dredge-up of nuclear-processed material to the stellar surface (e.g., Pastorelli et al. 2020). Decorrelating and disentangling cause and effect in these mechanisms and their observational tracers can therefore be difficult, and must be done over a wide range of stellar environments.

2. RGB mass loss

To understand mass loss on the TP-AGB, we need to first understand the chromospheric mass loss that precedes it. It is generally thought that magneto-acoustic waves propagate to a stellar chromosphere, where their energy disperses and drives a warm (∼5000–10 000 K), fast (∼10–100 km s^{-1}) wind from the star (Dupree, Smith & Strader 2009; Cranmer & Saar 2011; Rau et al. 2018). It is normally parameterised using some form of Reimers (1975) law, $\dot{M} = 4 \times 10^{-13}\, \eta L R / M$, for mass M, luminosity L and radius R in solar units, with a normalisation constant η.

Calibrating this law is difficult, because it is hard to measure the mass-loss rates of individual RGB stars. Chromospheric estimates are difficult and have significant uncertainty in their modelling. Instead, estimates of η are based on integrated mass loss, normally from a point on the RGB to either the horizontal branch or the early AGB. The RGB-integrated value of η has been well-constrained at $\eta \approx 0.4 - 0.5$ in globular clusters, both directly by McDonald, Johnson & Zijlstra (2011), and indirectly by McDonald & Zijlstra (2015) and, recently, by Tailo et al. (2020 and this proceedings). The latter performs the most self-consistent analysis, though defines a metallicity dependence on η (predicting $\eta \approx 0.7$ for solar metallicity). This conflicts with observations of solar-metallicity open clusters, which suggest $\eta \ll 0.3$ (Miglio et al. (2012); Handberg et al. (2017)). Consequently Tailo et al.'s may not be the most effective description of RGB mass loss in general, but remains our best description for Population II stars at this time. An alternative parameterisation,

$$\dot{M} \sim 5 \times 10^{-20}\, \mathrm{M_{\odot}\, yr^{-1}}\, \frac{L^{10/3} R^{1/2}}{M^{5/2}}, \tag{2.1}$$

fits the observed data reasonably well, including the open-cluster data. The stronger mass scaling is the only viable method of creating a low η in Galactic open clusters, while retaining high η in globular clusters, particularly the metal-rich clusters that have the lowest initial masses. The stronger luminosity scaling is required to compensate for the differences in RGB-tip luminosity at different metallicities. However, both Tailo's and the above parameterisations are empirical fits to integrated data. What are really needed are direct measurements of RGB mass-loss rates, and a re-examination of Reimers' original law. Since any RGB mass-loss law predicts mass-loss rates should be highest (therefore easiest to measure) near the RGB tip, we must concentrate our measurements of mass-loss rate near the RGB tip, and on early-AGB stars at similar luminosities. However, the prediction of the above law of a rapid rise in $\dot{M}$ with luminosity is potentially important if it can be proven: notably this parameterisation still gives the canonical $\dot{M} \sim 10^{-7}$ $\mathrm{M_{\odot}}$ $\mathrm{yr^{-1}}$ near the RGB tip.

3. The onset of AGB mass loss

The transition to a pulsation-enhanced, dust-driven wind requires several factors: pulsations to lift material from the stellar surface; a cool, dense environment in which dust can form close to the star; and sufficient radiation pressure on that dust that it can be ejected from the system.

This onset of late-AGB mass loss can be traced in two ways: by searching for mid-infrared excess, which traces the circumstellar dust, and is effective out to distances of a few Mpc; or (more accurately) by tracing molecular outflows through sub-mm rotational lines of CO, which (for low-$\dot{M}$ stars) is only effective within a few hundred pc. To sample a range of environments with confidence, both methods are needed.

One of the striking features defining the onset of AGB mass loss is its strong correlation with pulsation period, whereby stars attaining a period of ~60 days appear suddenly eligible to produce dust. This is seen in Milky Way field stars (McDonald & Zijlstra 2016), in the Magellanic Clouds (Boyer et al. 2015) and in globular clusters (e.g., Boyer et al. 2009; McDonald et al. 2009 & McDonald et al. 2011). Not all stars begin losing mass at a period of ~60 days: only low-mass stars do, such as those in globular clusters. The root significance of the 60-day period is that it traces a transition between pulsation sequences B and C', where the long secondary period (LSP) sequence is first initiated, and the strength of the shorter, primary-period pulsations grow significantly (McDonald & Trabucchi 2019). It can be concluded from this strong dependence on pulsation amplitude that the conditions required for dust formation and ejection already exist in such stars, and they wait only for the pulsations to become strong enough for them to initiate the wind. This in turn implies that the chromosphere may have been essentially disrupted (or otherwise negligible) by this point, perhaps by the pulsations and consequent increase in scale itself, in order that a cool environment exists close to the star. To examine these hypotheses, we can look to stars of different metallicity, and those below the 60-day transition.

The onset of mass loss in metal-poor stars proceeds in much the same way: stars begin to produce copious dust once they transition across the $B \rightarrow C'$ sequence boundary. Curiously, the aforementioned globular cluster studies plot these stars in the same places in the period–infrared-excess diagram as their metal-rich counterparts. That is, there is the same dust opacity around metal-poor stars as metal-rich stars, despite the lack of dust-producing elements. Strong dust production by oxygen-rich, metal-poor stars is also seen in metal-poor dwarf galaxies (e.g., Boyer et al. 2015 & Boyer et al. 2017) where its onset also appears to correlate with the onset of strong pulsation (McDonald et al. 2014). Few measurements of wind velocities and CO-based mass-loss rates exist for marginally dust-producing, metal-poor stars: those few suggest the dust opacity remains the same partly because the wind is slower (e.g., McDonald et al. 2019 & McDonald et al. 2020). Infrared spectra also show the presence of a different dust species in metal-poor stars, showing only continuum emission, which is posited to be metallic iron (McDonald et al. 2010, 2011). Few measurements exist also at higher metallicities due to lack of sources, but the little evidence we have suggests mass loss is no more effective here (van Loon, Boyer & McDonald 2008). The lack of metallicity response to the onset of strong dust production adds evidence that the ability to condense and drive dust are not limiting factors for generating AGB winds, but that it is the rate at which material is supplied to the dust-condensation zone by pulsations that largely controls the mass-loss rate.

Stars with pulsation periods near the 60-day boundary are another poorly observed class. This is partly because they tend to lie at luminosities close to or below the RGB tip, beyond the observational cutoff of most AGB-star studies, and in a region where RGB and AGB stars cannot normally be cleanly separated. Existing observations come from Groenewegen (2014) (VY Leo), McDonald et al. (2016) (EU Del) and McDonald et al. (2018) (several stars). These observations show that low mass-loss rates† ($\dot{M} \sim 10^{-9}$ to 10^{-8} $M_\odot$ yr^{-1}) can be traced from these stars via molecular CO. Interestingly, VY Leo is one of the original stars studied by Reimers (1973), where it was assigned a mass-loss rate from chromospheric indicators (Ca II H&K and Hα) of $\dot{M} = 10^{-7.15}$ $M_\odot$ yr^{-1}. Whether this rate survives modern scrutiny remains unclear, as it has not been remeasured.

Dust production by RGB stars is a controversial topic (e.g., Boyer et al. 2010, McDonald et al. 2011), partly because it is impossible to cleanly separate RGB from early-AGB stars in most cases. However, statistical indications are generally that RGB stars do not produce dust, while some AGB stars at the same luminosity do (McDonald & Zijlstra 2016). Consequently, it is an open debate about whether the dust-producing stars near the RGB tip are true RGB stars or the less-numerous early-AGB stars with similar surface properties. The strong mass scaling in the previous section gives rise to the possibility that the mass difference between the lowest-mass RGB and AGB stars could play an important role in generating a dusty wind from AGB but not RGB stars. However, the small-amplitude (SARG) pulsations from RGB-tip stars (e.g., Takayama, Saio & Ita 2013) may speculatively either lead to some enhancement of the wind, or assist in disrupting the chromospheric mass loss.

† We have recently been made aware of an error in McDonald et al. (2018), where $R_\odot$ was assumed as the unit of R_* in the scaling law of De Beck et al. 2010, where the appropriate unit was cm. This affects the mass-loss rates that were derived.

4. Discussion

We can therefore paint an evolutionary picture where the mass loss from RGB stars rises sharply towards the RGB tip, being strongest for the lowest-mass stars, and perhaps enhanced by the small-amplitude pulsations that occur there for such stars. That mass loss effectively ceases when the star transitions to the horizontal branch (or blue loop for more-massive stars), but begins to rise again as the star ascends the AGB. When the star becomes unstable to strong long-period pulsations, during the sequence $B \rightarrow C'$ crossing when the long-secondary period initiates, a pulsation-enhanced, dust-driven wind can begin in earnest. As the star exits onto the fundamental sequence (C), the pulsations become stronger, and the wind builds in strength until the entire envelope is ejected. While this paints a largely coherent picture of mass loss from low- and intermediate-mass stars, it is based on only a small amount of data.

Obtaining more data on stars undergoing this transition is crucial if we want to chart how, when and why stars initiate their final, dusty winds. This is one of the primary goals of the Nearby Evolved Stars Survey (NESS; Scicluna et al. 2022). The survey is designed as a volume-limited survey, tiered in mass-loss rate and distance. The lower tiers target low-mass-loss-rate, nearby stars, including many of those undergoing this transitionary phase. The results of this survey can be compared with higher-resolution surveys of smaller numbers of more-evolved objects (e.g., DEATHSTAR; Ramstedt et al. 2020; and ATOMIUM; Gottlieb et al. 2022), which are indicating that companions to evolved stars are also important in shaping the properties of the winds we see. It is hoped that the combination of these surveys (plus, of course, new results from the *James Webb Space Telescope*) will provide a much more rounded perspective on the reasons why, when and how AGB stars start to lose mass, with which information we can start to close one of the final unanswered problems in single-star evolution.

References

Boyer, M.L., McDonald, I., van Loon, J.Th., et al. 2009, *ApJ*, 705, 746
Boyer, M.L., van Loon, J.Th., McDonald, I., et al. 2010, *ApJ*, 711, L99
Boyer, M.L., McQuinn, K.B.W., Barmby, P., et al. 2015, *ApJ*, 800, 51
Boyer, M.L., McQuinn, K.B.W., Groenewegen, M.A.T., et al. 2017, *ApJ*, 851, 152
Cranmer, S.R., Saar, S.H. 2011, *ApJ*, 741, 54
De Beck, E., Decin, L., de Koter, A., et al. 2010, *A&A*, 523, 18
Dupree, A.K., Hartmann, L., & Avrett E.H. 1984, *ApJ*, 281, L37
Dupree, A.K., Smith, G.H., & Strader J. 2009, *AJ*, 138, 1485
Gottlieb, C.A., Decin, L., Richards, A.M.S., et al. 2022, *A&A*, in press (arXiv:/2112.04399)
Handberg, R., Brogaard, K., Miglio, A., et al. 2017, *MNRAS*, 472, 979
Groenewegen, M.A.T. 2014, *A&A*, 561, L11
McDonald, I., Boyer, M.L., Groenewegen, M.A.T., et al. 2019, *MNRAS*, 484, L85
McDonald, I., Boyer, M.L., van Loon, J.Th., et al. 2011, *ApJS*, 193, 23
McDonald, I., Boyer, M.L., van Loon, J.Th., & Sloan, G.C. 2011, *ApJ*, 730, 71
McDonald, I., De Beck, E., Lagadec, E., & Zijlstra, A.A. 2018, *MNRAS*, 481, 4984
McDonald, I., Johnson C.I., & Zijlstra, A.A. 2011, *MNRAS*, 416, L6
McDonald, I., Sloan, G.C., Zijlstra, A.A., et al. 2010, *ApJ*, 717, L92
McDonald, I., & Trabucchi, M. 2015, *MNRAS*, 484, 4678
McDonald, I., van Loon, J.Th., Decin, L., et al. 2009, *MNRAS*, 394, 831
McDonald, I., van Loon, J.Th., Sloan, G.C., et al. 2011, *MNRAS*, 417, 20
McDonald, I., Uttenthaler, S., Zijlstra, A.A., et al. 2020, *MNRAS*, 491, 1174
McDonald, I., & Zijlstra, A.A. 2015, *MNRAS*, 448, 502
McDonald, I., Zijlstra, A.A., Sloan, G.C., et al. 2014, *MNRAS*, 439, 2618
McDonald, I., Zijlstra, A.A., Sloan, G.C., et al. 2016, *MNRAS*, 456, 4542
Miglio, A., Brogaard, K., Stello, D., et al. 2012, *MNRAS*, 419, 2077

Pastorelli, G., Marigo, P., Girardi, L., et al. 2020, *MNRAS*, 498, 3283
Ramstedt, S., Vlemmings, W.H.T., Doan, L., et al. 2020, *A&A*, 640, A133
Rau, G., Nielsen, K.E., Carpenter, K.G., et al. 2020, *ApJ*, 869, 1
Reimers, D. 1973, *A&A*, 24, 79
Reimers, D. 1975, in: B. Baschek, W.H. Kegel, & G. Traving (eds.), *Problems in stellar atmospheres and envelopes* (New York: Springer–Verlag), p. 229
Scicluna, P., Kemper, F., McDonald, I., et al. 2022, *MNRAS*, 512, 1091 (arXiv:/2110.12562)
Tailo, M., Milone, A.P., Lagioia, E.P., et al. 2020, *MNRAS*, 498, 5745
Takayama, M., Saio, H., Ita, Y. 2020, *MNRAS*, 431, 3189
van Loon, J.Th., Boyer, M.L., McDonald, I. 2008, *ApJ*, 680, L49

The Origin of Outflows in Evolved Stars
Proceedings IAU Symposium No. 366, 2022
L. Decin, A. Zijlstra & C. Gielen, eds.
doi:10.1017/S1743921322000096

The distribution of carbonaceous molecules and SiN around the S-type AGB star W Aquilae

T. Danilovich[1]†, M. Van de Sande[2], A. M. S. Richards[3] and the ATOMIUM Consortium

[1]Institute of Astronomy, KU Leuven, Celestijnenlaan 200D, 3001 Leuven, Belgium
email: taissa.danilovich@kuleuven.be

[2]School of Physics and Astronomy, University of Leeds, Leeds LS2 9JT, UK

[3]JBCA, Dept Physics and Astronomy, University of Manchester, Manchester M13 9PL, UK

Abstract. S-type AGB stars, with C/O ratios close to 1, are expected to have a mixed circumstellar chemistry as they transition from being oxygen-rich stars to carbon-rich stars. Recently, several different carbonaceous molecules, thought to be more characteristic of carbon stars, have been found in the circumstellar envelope of the S-type AGB star W Aql. We have obtained new high spatial resolution ALMA images of some of these molecules, specifically HC_3N, SiC_2 and SiC, and SiN, which we present here. We report diverse behaviour for these molecules, with SiC_2 being seen with a symmetric spatial distribution around the star, SiN and SiC being asymmetrically distributed to the north-east of the star, and HC_3N being seen in a broken shell to the south-west. These differing distributions point to complex dynamics in the circumstellar envelope of W Aql.

Keywords. stars: AGB, circumstellar matter, stars: individual: W Aql

---

1. Introduction

W Aquilae (W Aql) is an S-type AGB star, meaning that it has a C/O ratio close to 1. S-type AGB stars are expected to have a mixed circumstellar chemistry as they transition from being oxygen-rich stars to carbon-rich stars. Such mixed chemistry was indeed reported by Danilovich et al. (2014), based on *Herschel*/HIFI observations. Recently, a spectral scan of W Aql with the APEX telescope by De Beck & Olofsson (2020) identified several different carbonaceous molecules, previously thought to be more characteristic of carbon stars, in the circumstellar envelope of W Aql, including C_2H, SiC_2, SiN and, tentatively, HC_3N.

Here we present high spatial resolution ALMA observations of W Aql with a focus on carbonaceous molecules and SiN. Our spatially resolved observations allow us to identify in which regions of the circumstellar envelope the various molecules are present.

2. Observational results

W Aql was observed as part of the ATOMIUM ALMA Large Programme (2018.1.00659.L) over bands across the frequency range $\sim$214–270 GHz. For medium resolution observations, the approximate synthetic beam size was $0.4 \times 0.3''$ with a maximum recoverable scale of $3.9''$, and for low resolution observations the approximate

†Senior Postdoctoral Fellow of the Fund for Scientific Research (FWO), Flanders, Belgium.

Table 1. Emission lines of carbon-bearing molecules and SiN detected towards W Aql.

Molecule	Frequency [GHz]	Transition J_{K_a,K_c}	$\rightarrow$	$J'_{K'_a,K'_c}$	E_{up} [K]	Molecule	Frequency [GHz]	Transition $J \rightarrow J'$	E_{up} [K]
SiC_2	220.774	$10_{0,10}$	$\rightarrow$	$9_{0,9}$	59.8	HC_3N	227.419	$25 \rightarrow 24$	142.0
SiC_2	222.009	$9_{2,7}$	$\rightarrow$	$8_{2,6}$	60.3	HC_3N	236.513	$26 \rightarrow 25$	153.4
SiC_2	235.713*	$10_{6,5}$	$\rightarrow$	$9_{6,4}$	132.6	HC_3N	245.606	$27 \rightarrow 26$	165.2
SiC_2	235.713*	$10_{6,4}$	$\rightarrow$	$9_{6,3}$	132.6	HC_3N	254.700	$28 \rightarrow 27$	177.4
SiC_2	237.150‡	$10_{4,7}$	$\rightarrow$	$9_{4,6}$	93.7	SiC	236.288	$6 \rightarrow 5$ ($\Omega = 2$)	22.7
SiC_2	237.331	$10_{4,6}$	$\rightarrow$	$9_{4,5}$	93.8				
SiC_2	254.982	$11_{2,10}$	$\rightarrow$	$10_{2,9}$	81.9	SiN	262.156†	$13/2 \rightarrow 11/2$ ($N = 6 \rightarrow 5$)	31.5
SiC_2	259.433*	$11_{6,5}$	$\rightarrow$	$10_{6,4}$	145.0				
SiC_2	259.433*	$11_{6,6}$	$\rightarrow$	$10_{6,5}$	145.0				

Notes:
(*) indicates blended lines; (†) indicates a line with hyperfine components. (‡) indicates a line partially truncated by the edge of the observed band. **References:** Line frequencies from Müller et al. (2012) for SiC_2, from Creswell et al. (1977); de Zafra (1971); Mallinson & de Zafra (1978); Chen et al. (1991); Yamada et al. (1995) and Thorwirth et al. (2000) for HC_3N, from Cernicharo et al. (1989) for SiC, and from Saito et al. (1983) for SiN, all via CDMS (Müller et al. 2001, 2005).

synthetic beam size was $1.0 \times 0.7''$ with a maximum recoverable scale of $8.9''$. Full details of the observations are given in Decin *et al.* (2020) and Gottlieb et al. (2021). Plots of zeroth moment maps presented here are of combined datasets, with synthetic beam sizes ranging from $0.05''$ to $0.2''$.

A total of 110 lines have been identified towards W Aql (Wallström et al, *in prep*), including lines from common molecules such as CO, HCN, SiO, SiS and CS, less-studied molecules such as the metal halides AlCl and AlF (Danilovich et al. 2021), SiN and the carbon-bearing molecules SiC, SiC_2 and HC_3N. Emission from the latter four molecules is presented and discussed below. Table 1 includes frequencies and upper energy levels of all the detected lines from these four molecules. Selected line spectra for these molecules are plotted in Fig 1.

2.1. *SiC_2*

Seven individual lines of SiC_2 were detected in the ATOMIUM observations towards W Aql. This includes two lines that are blends of two transitions, as indicated by the asterisks in Table 1, and one line that was partially truncated when falling on the edge of a band in our observations. Although there are additional SiC_2 lines that fall in the observed frequency ranges, these other lines are all predicted to be more than an order of magnitude fainter than our detected lines. Hence we did not expect to detect them at our observational sensitivity. We also checked our data for the brightest vibrationally excited SiC_2 lines in the $\nu_3 = 1$ state, but found no detections.

The line shapes across the different transitions are consistently double-peaked with the red peak generally brighter than the blue (see Fig 1). This asymmetry in velocity space is most likely owing to non-spherical dynamics in the circumstellar envelope. The spatial distribution of the SiC_2 emission in the plane of the sky is relatively symmetric, with minor asymmetries most likely attributable to noise. Three zeroth moment maps of selected lines are plotted in Fig 2 and show that the emission is approximately centred on the continuum peak.

2.2. *HC_3N*

Four lines of HC_3N were detected towards W Aql in the ATOMIUM observations. Although HC_3N is known to exhibit hyperfine splitting, this is negligible and not detectable at the spectral resolution of our observations, and hence neglected here. The

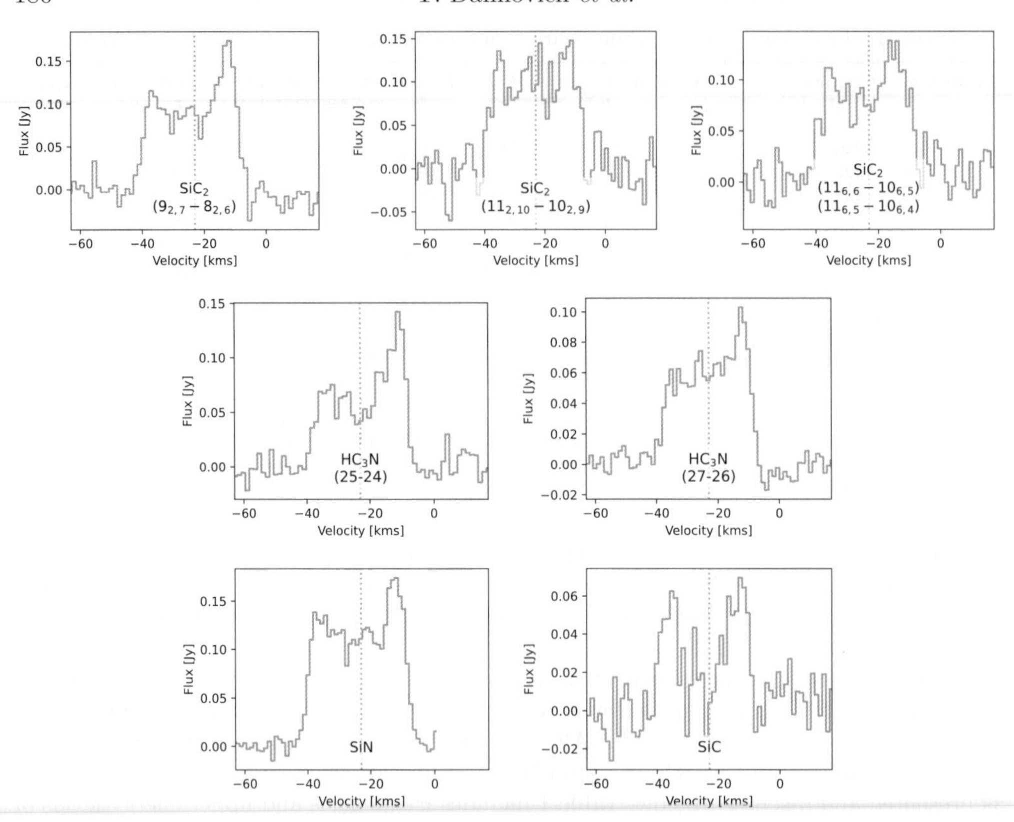

Figure 1. Spectra of selected SiC_2 (top row) and HC_3N (middle row) lines, and spectra of the only detected SiN and SiC lines (bottom row) towards W Aql. Spectra were extracted for a circular aperture centred on the continuum peak with a radius of 1.8″ from the medium spatial resolution ALMA data cubes for SiC_2 and HC_3N and for the low spatial resolution data cubes for SiC and SiN. The dotted vertical lines represent the $\upsilon_{\rm LSR} = -23$ of W Aql and the spectral resolution is ~ 1 km s^{-1}.

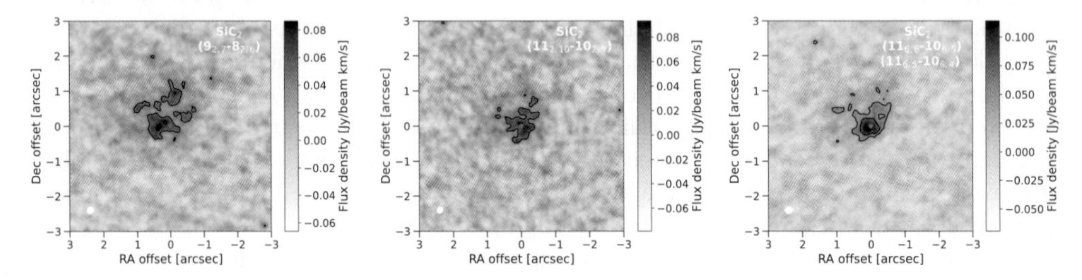

Figure 2. Selected zeroth moment maps of SiC_2 towards W Aql. Shown are maps for the $(9_{2,7} \rightarrow 8_{2,6})$ line (left), the $(11_{2,10} \rightarrow 10_{2,9})$ line (centre), and the blended $(11_{6,6} \rightarrow 10_{6,5})$ and $(11_{6,5} \rightarrow 10_{6,4})$ lines (right). North is up and east is left. The synthetic beam sizes are given by the white ellipses in the bottom left of each plot, the stellar position based on the continuum peak is indicated at (0,0) and the contours indicate levels of 3 and 5σ flux.

observed HC_3N lines are listed in Table 1 and consist of all four lines covered in our frequency settings. We also checked for lines from the singly-substituted ^{13}C isotopologues of HC_3N, but all covered lines were undetected.

The spectra of the detected HC_3N lines are similar in shape to SiC_2 (see Fig 1), with two-peaked profiles showing a brighter peak on the red side. Zeroth moment maps of two example HC_3N lines are given in Fig 3 and show a clearly asymmetric flux distribution

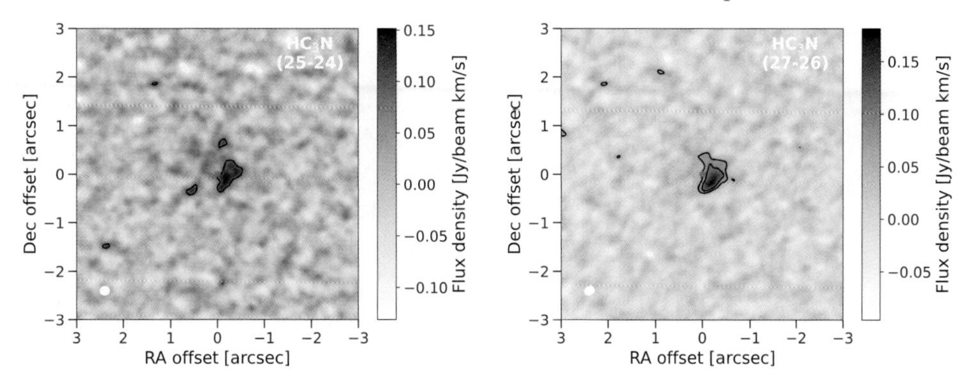

Figure 3. Selected zeroth moment maps of HC_3N towards W Aql. Shown are maps for the $(25 \rightarrow 24)$ and $(27 \rightarrow 26)$ lines in the left and right panels respectively. North is up and east is left. The synthetic beam sizes are given by the white ellipses in the bottom left of each plot, the stellar position based on the continuum peak is indicated at (0,0) and the contours indicate levels of 3 and 5σ flux.

that is not centred on the continuum peak. This asymmetric distribution resembles a partly broken (or partly undetected) shell structure surrounding the star. Thc cmission peaks around $0.3''$ from the continuum peak, when measuring for the $(27 \rightarrow 26)$ line.

Shell-like structures are expected for HC_3N based on chemical models (Van de Sande & Millar 2021). Indeed, Agúndez *et al.* (2017) found HC_3N emission present in a shell around the nearby carbon star CW Leo, as part of a detailed ALMA study. We ran some preliminary radiative transfer models assuming 1D spherical symmetry and comparing our results to azimuthally averaged observations. Based on these results, if we assume that the HC_3N emission towards W Aql is part of a shell-like structure, then the distance of this shell from the star is comparable to that of CW Leo if we also assume that the size of the shell scales linearly with mass-loss rate.

2.3. *SiN*

A single line of SiN (composed of three unresolved hyperfine components) was covered by the ATOMIUM observations and was clearly detected towards W Aql. The spectrum of SiN (see Fig 1) is again a two-peaked profile, with the red peak brighter than the blue peak. A comparison with the same SiN line as observed with APEX by De Beck & Olofsson (2020) tells us that all the flux is recovered by our ALMA observations and that no large scaled structure is resolved out. A clear asymmetry is seen in the zeroth moment map, plotted in Fig. 4, where the bulk of the emission comes from an apparent wedge in the north to north-east, and includes some emission overlapping with the continuum peak.

Prior to the current study, the only evolved star for which SiN has been detected is CW Leo (Turner 1992). The observations of Turner (1992) were performed with the NRAO 12 m telescope and hence were not spatially resolved. No interferometric observations of SiN have been published until now. This means that we are unable to compare the asymmetric distribution of SiN in the CSE of W Aql with any other stars.

2.4. *SiC*

Only one line of SiC was covered by the ATOMIUM observations and was detected with a lower signal to noise than the other lines presented here. Similar to the other lines, the spectrum is two-peaked, as can be seen in Fig 1, but the signal to noise is not adequate to judge whether the red peak is brighter than the blue peak. The zeroth moment map,

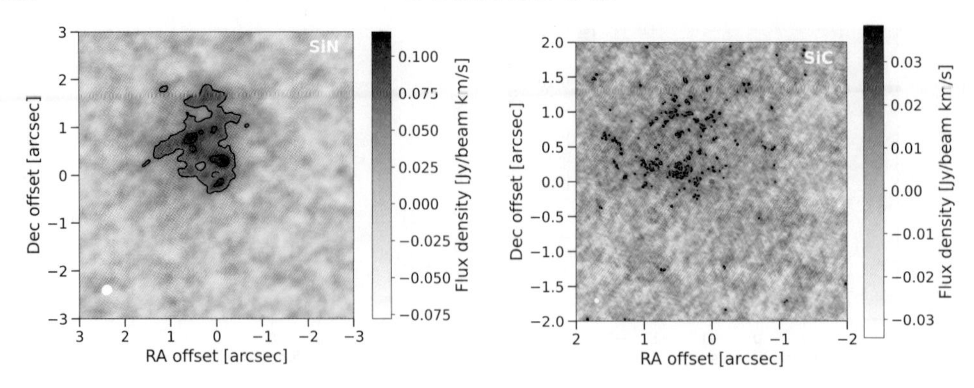

Figure 4. Zeroth moment maps of (left) SiN ($N=6\rightarrow5$, $J=13/2\rightarrow11/2$) and (right) SiC ($J=6\rightarrow5$) towards W Aql. North is up and east is left. The synthetic beam sizes are given by the white ellipses in the bottom left of each plot, the stellar position based on the continuum peak is indicated at (0,0) and the contours indicate levels of 3 and 5σ flux.

shown in Fig 4 and centred on the continuum peak, shows emission coming mainly from the north-east quadrant of the map. Overall, the distribution of SiC is similar to that of SiN, although the lower signal to noise makes a more detailed analysis difficult.

Massalkhi et al. (2018) surveyed a sample of 25 carbon stars and and detected (clearly or tentatively) a line of SiC towards 12 of their sources. In contrast, SiC_2 was more frequently detected, seen towards 22 of their sources. They also found a correlation between the intensities of their detected SiC line and a chosen SiC_2 line for the same stars, which they interpret as a correlation between the abundances of the two species. We cannot rule out a correlation between the abundances of SiC and SiC_2 for W Aql, however the fact that they are not co-located suggests a more complex relationship between the two species.

3. Discussion and conclusions

The molecules presented here all exhibit similar spectral profiles, with two peaks and a tendency for the red peak to be brighter than the blue peak. However, the spatial distribution of these molecules can be divided into three categories: SiC_2 emission is seen centrally and coinciding with the continuum peak; HC_3N emission is seen mainly to the west of the continuum peak and could represent part of a broken or not fully-detected shell structure; and SiN and SiC emission mainly originates in the north-east quadrant relative to continuum peak. The AlF emission presented in Danilovich et al. (2021) is also asymmetric with a qualitatively similar distribution to SiN and SiC. Taken together, these asymmetries suggest complex dynamics and chemistry in the circumstellar envelope of W Aql.

The CO ($2\rightarrow1$) emission of W Aql, observed as part of ATOMIUM, can be seen in Decin *et al.* (2020) and the accompanying supplementary materials, and shows a spiral-like structure with many arcs. Despite the somewhat chaotic nature of this emission, the overall circumstellar envelope of W Aql appears roughly circular on the scale of the CO emission. Note, however, that the ATOMIUM CO emission suffers from resolved out flux and hence smooth large scale structure is not recovered. Lower resolution ALMA observations of the CO ($3\rightarrow2$) emission is presented by Ramstedt *et al.* (2017). In those data, for which all the flux is recovered, individual arcs are not resolved but the larger scale structure is asymmetric on scales $\sim5-10''$, with some extended CO emission visible to the south east of the continuum peak. This represents a fourth category of asymmetry not exhibited by the molecules studied here and further enforces the need for a detailed analysis of the dynamics in the circumstellar envelope of W Aql.

There are several other molecular species detected towards W Aql, some of which may also exhibit spatial asymmetries. We plan to carefully catalogue and study these in future work, with the goal of presenting a more complete understanding of the dynamics of the circumstellar envelope of W Aql.

References

Agúndez, M., Cernicharo, J., Quintana-Lacaci, G., et al. 2017, *A&A* 601, A4

Cernicharo, J., Gottlieb, C. A., Guelin, M., Thaddeus, P., & Vrtilek, J. M. 1989, *ApJL* 341, L25

Chen, W., Bocquet, R., Wlodarczak, G., & Boucher, D. 1991, *International Journal of Infrared and Millimeter Waves* 12(9), 987

Creswell, R. A., Winnewisser, G., & Gerry, M. C. L. 1977, *Journal of Molecular Spectroscopy* 65, 420

Danilovich, T., Bergman, P., Justtanont, K., et al. 2014, *A&A* 569, A76

Danilovich, T., Van de Sande, M., Plane, J. M. C., et al. 2021, *A&A* 655, A80

De Beck, E. & Olofsson, H. 2020, *A&A* 642, A20

de Zafra, R. L. 1971, *ApJ* 170, 165

Decin, L., Montargès, M., Richards, A. M. S., et al. 2020, *Science* 369(6510), 1497

Gottlieb, C. A., Decin, L., Richards, A. M. S., et al. 2021, *A&A* Forthcoming

Mallinson, P. D. & de Zafra, R. L. 1978, *Molecular Physics* 36(3), 827

Massalkhi, S., Agúndez, M., Cernicharo, J., et al. 2018, *A&A* 611, A29

Müller, H. S. P., Cernicharo, J., Agúndez, M., et al. 2012, *Journal of Molecular Spectroscopy* 271(1), 50

Müller, H. S. P., Schlöder, F., Stutzki, J., & Winnewisser, G. 2005, *Journal of Molecular Structure* 742, 215

Müller, H. S. P., Thorwirth, S., Roth, D. A., & Winnewisser, G. 2001, *A&A* 370, L49

Ramstedt, S., Mohamed, S., Vlemmings, W. H. T., et al. 2017, *A&A* 605, A126

Saito, S., Endo, Y., & Hirota, E. 1983, *J. Chem. Phys.* 78(11), 6447

Thorwirth, S., Müller, H. S. P., & Winnewisser, G. 2000, *Journal of Molecular Spectroscopy* 204, 133

Turner, B. E. 1992, *ApJL* 388, L35

Van de Sande, M. & Millar, T. J. 2021, *MNRAS*

Yamada, K. M. T., Moravec, A., & Winnewisser, G. 1995, *Zeitschrift Naturforschung Teil A* 50(12), 1179

The Origin of Outflows in Evolved Stars
Proceedings IAU Symposium No. 366, 2022
L. Decin, A. Zijlstra & C. Gielen, eds.
doi:10.1017/S1743921322000540

Long-term light curve variations of AGB stars: episodic mass-loss or binarity?

Roberto Ortiz[1], **Alain Jorissen**[2] **and Léa Planquart**[2]

[1]Escola de Artes, Ciências & Humanidades, Universidade de São Paulo, Brazil
email: rortiz@usp.br

[2]Institut d'Astronomie et d'Astrophysique, Université Libre de Bruxelles, Belgium
emails: Alain.Jorissen@ulb.be, lea.planquart@ulb.be

Abstract. A significant fraction of the stars near the tip of the AGB phase become regular or semi-regular (Mira-type, SRs) pulsators. However, some of these light curves have shown intriguing secondary minima or sharp dips with much longer periods. Although this phenomenon shows some resemblance with the R CrB variables, the light curve is generally symmetric before and after the dip, whereas in R CrB the luminosity recovers slower after its minimum. More recently, high-resolution ALMA CO observations revealed a spiral structure around some of these stars, which suggests the presence of a stellar or sub-stellar companion. In these cases, the long-term light curve minima could be caused by periodic eclipses of the primary by a spiral circumstellar structure, and the long-period would be related to the orbital period. In this paper we discuss the pros and cons of the various proposed scenarios for the long-term minima of pulsating AGB stars.

Keywords. AGB stars, variable stars, binaries, mass-loss

1. Introduction

The asymptotic giant branch (hereafter AGB) is an evolutionary phase when an intermediate-mass star reaches a high luminosity, which favours a strong mass-loss rate (up to $10^{-4} M_{\odot}\mathrm{yr}^{-1}$). A significant fraction of the AGB stars develop radial pulsation in a scale of a few hundreds of days. Pulsation can be regular (the Mira-type variables), semi-regular (SRa, SRb types) or irregular (Lb). The high mass-loss rate often generates a optically-thick dust shell that may obscure (partial or totally) the visual spectrum of the star. The opacity of the dust shell is generally higher in carbon-rich stars (*i.e.* when [C]/[O] > 1), except in the cases where the O-rich AGB star undergoes a very high mass-loss rate (e.g. OH/IR stars, Lépine et al. 1995).

Long-term photometric observations (often limited to visual) of pulsating AGB stars obtained along the last century have revealed various kinds of inhomogeneities in their light curves, such as: variable amplitude and period, occurrence of multiple periods, changing light curve shapes, etc. However, about half of these pulsating stars show an intriguing very long secondary period (Soszyński et al. 2021), and a small fraction of them exhibit deep and sharp luminosity drops that can last one or a few pulsation cycles (Fig. 1). During these events, the visual brightness can decline 5 or 6 magnitudes, and the phenomenon can repeat in regular intervals of many years (between $18 \sim 37$ yrs in Table 1). The vast majority of these objects are C-rich.

In this paper, we discuss the most likely scenarios proposed in the literature to explain these long-term brighness decays, their pros and cons, considering multi-wavelength observations obtained at various spectral and spatial resolutions.

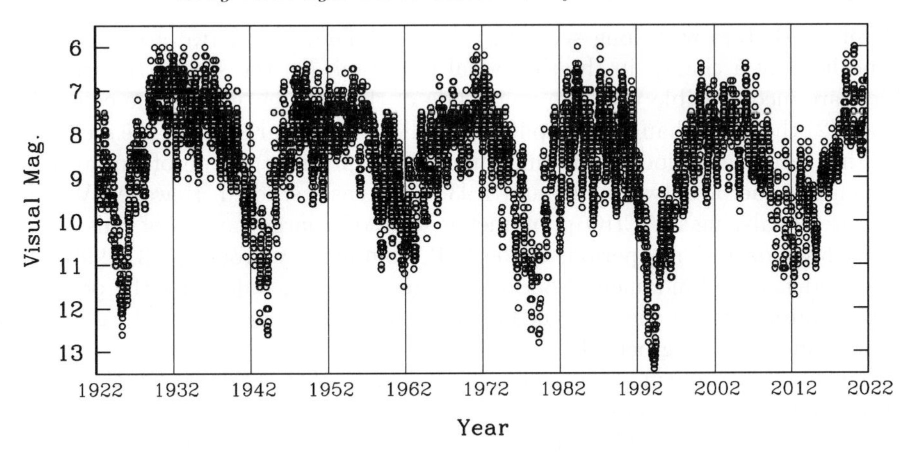

Figure 1. Visual light curve of V Hya, covering a period of 100 yrs, showing its 18-yrs eclipse period. Data collected by the AAVSO.

2. R Lep: the prototype of a new class of variable?

R Lep is a C-rich, regularly pulsating AGB star that has been closely monitored since the first half of the XIX century. According to Mayall (1963), the star "was faint before 1910, then bright until 1949, and definitely faint from 1950 through 1961". Besides R Lep, Mayall (1963) recalls that the light curve of V Hya shows a similar behaviour, and since then other pulsating variables have been added to the list (Table 1).

The case of R Lep is emblematic because some of its eclipses were closely observed, using various techniques. The first polarimetric observations were obtained starting 6 yr after its 1959–1960 eclipse by Serkowski (1966), and then by Kruszewski, Gehrels & Serkowski (1968) and Serkowski (1971). Eventually, Raveendran (2002) obtained long-term polarimetric observations, including the long photometric dim that occurred between 1993–1998. Based on these observations, Raveendran & Kameswara Rao (1989) proposed a correlation between the long-term changes in polarization and the photometric dims observed in the light curve of R Lep. Apparently, polarization follows the intensity of the eclipse, reaching its maximum percentage together with the light curve minimum. Besides, Raveendran (2002) observed that the polarization angle changed abruptly at the beginning of the 1993–1998 eclipse, from $\theta \simeq 45°$ (before) to 28° (from 1993 onwards).

Long-term *JHKL* photometry carried out by Lloyd-Evans (1997) and Whitelock et al. (2000) indicates that the photometric dims are caused by obscuration by circumstellar dust grains. The rapid change in the polarization angle and the synchronicity between the polarization degree and the light curve suggest that the circumstellar dust that causes the eclipse is condensed near the star, perhaps within a few stellar radii (Clayton, 1992).

3. Possible Mechanisms

Olivier & Wood (2003) cite two major scenarios to interpret the "eclipse" undergone by some AGB stars. In this section we briefly describe their hypotheses.

3.1. *Episodic mass-loss*

Episodic mass-loss, perhaps as a result of a change in the pulsation mode, may result into hydrodynamical or thermal instabilities that can trigger the formation of dust clouds that block the stellar radiation (Winters et al. 1994; Woitke & Niccolini 2005). Hinkle et al. (2002) and Wood et al. (2004) monitored the radial velocity of a sample of AGB stars with long secondary periods. They noted two arguments against the hypothesis of a change in the the radial pulsation: (1) the stellar $T_{\rm eff}$ values along the cycle are incompatible with the amplitude of the light curve and the colour indices observed; (2) the large change in amplitude implies a large variation in the stellar radius, and in this case the period of the fundamental (or primary) mode of pulsation would have changed a factor between 1.64 and 1.93 in the cases studied. However, Wood et al. (2004), after analysing MACHO photometric data of a large sample of pulsating AGB stars with secondary (long) periods in the LMC compiled by Wood et al. (1999), did not find any evidence for fundamental period changes. They excluded also the possibility of non-radial pulsations, as they are incompatible with the large observed radial velocity amplitudes and their long periods.

Although the mechanism of episodic mass-loss is not fully understood, there are solid evidences that it operates in some AGB stars, provided the object is observed at high spatial resolution. For example, using interferometric observations at 11 μm, Hale et al. (1997) observed a periodic dust shell structure around the AGB star IK Tau at 100, 320 and 570 mas, which correspond to intense mass-loss events occurred in intervals of 12 yrs (whilst the stellar pulsation period is $470^{\rm d}$). Unfortunately, high spatial resolution observations are available only for few AGB stars with long-period photometric dims (Sect. 5).

Whatever its cause, if episodic dust ejection operates among AGB stars with long-period dims, they must be asymmetric in order to explain the change in the polarization angle observed after the beginning of the event (Sect. 2). In this scenario, the circumstellar material expelled by the star forms an aspherical structure around the star, perhaps a clump or a disc. As the circumstellar shell moves away, its opacity decreases and the light curve returns to its pre-obscuration status.

3.2. *The binary hypothesis*

The second possibility is that the AGB star is a member of a **binary system**, and its light is obscured by a cloud of dust and gas associated with an orbiting companion (Wood et al. 1999; Soszyński & Udalski 2014). Binarity is a common phonomenon among AGB stars, but detecting visually the secondary component is generally unfeasible because the high-luminosity primary outshines its low-luminosity companion. Fortunately, in some cases this limitation can be overcome. For example, if the secondary component is relatively hot or it exhibits a chromosphere, a far-UV excess can be detected (Sahai et al. 2008; Ortiz et al. 2019). Alternatively, far-UV excess could be originated in an accretion disc around the secondary, and the UV continuum of some of these AGB stars with far-UV excess was observed to exhibit "flickering" on very short timescales (< 20 s), and the UV spectra show emission lines (SIV, CIV) with P-Cygni profiles. Altogether, these UV characteristics have been interpreted as a strong evidence of a hot accretion disc around a degenerate companion that remains overlooked in the visual spectral range (Sahai, 2018).

The far-UV features described above are sometimes accompanied by X-ray emission, even though large-scale X-ray surveys are still too shallow to allow a firm conclusion about the occurrence of X-rays among far-UV AGB stars. Two possibilities have been proposed to explain the X-ray emission associated with AGB stars: (1) coronal emission of its main-sequence or sub-giant companion; (2) a hot accretion disc around a white dwarf companion (Jorissen et al. 1996; Sahai 2015; Ortiz & Guerrero 2021). Stellar coronae are normally found over a wide variety of spectral types and luminosities, except in stars situated at the top-right corner of the HR diagram, where AGB stars belong

Table 1. Some examples of AGB stars showing regular "eclipses" or "luminosity dims".

Star's name	$P_{\rm puls}$ (days)	$P_{\rm fade}$ (yrs)	Spectral Class	References
EV Eri	226	?	C-J	Whitelock et al. (2006)
R For	387	37	C4,3e	Feast et al. (1984)
V Hya	530	18	C-N:6	Baize (1962)
R Lep	438	35	C7,6e	Lloyd-Evans (1997)
II Lup	575	19	C	Feast et al. (2003)
L_2 Pup	140	?	M5eIII	Bedding et al. (2002)

(Linsky & Haisch 1979; Ayres et al. 1981). Thus, since AGB stars do not emit X-rays, these sources must be extrinsic to the AGB star.

In the case the AGB is binary, the mass expelled by the primary component orbits the system, forming an asymmetrical envelope that is sometimes observed in the mid-infrared and radio observations (Feast et al. 2003). More recently, the binary hypothesis has been reinforced by radial observations of the star V Hya (Table 1). Monitoring of its radial velocity showed that the velocity of its primary component equals that of the centre-of-mass just in the middle of the eclipse, i.e. when the companion is in front of the AGB star (Planquart et al. 2022).

4. Phenomena associated with the eclipse

Besides the sharp luminosity decline, the main phenomena associated with the luminosity drop are summarized below:

• **Infrared Excess** is commonly observed simultaneously with the luminosity drop. Unfortunately, photometric monitoring of an eclipse has never been obtained at wavelengths beyond the L' band in the near-infrared. Near-infrared excess has been associated with the differential extinction caused either by the episodic mass-loss of the primary or debris around a secondary component.

• **Spectral changes** are observed during the eclipse. The C_2 molecular bands between 4700 $\sim$ 4730 Å and 5100 $\sim$ 5200 Å, and the SiC_2 band near 4900 Å, all of them in emission, were observed to increase their intensity during an eclipse of R Lep (Lloyd-Evans 1997).

• **The Degree of Polarization** was observed to decrease following the end of an eclipse of the star R Lep. Besides, there was a change in the polarization angle, when the pre-eclipse and post-eclipse angles are compared. This has been interpreted as the presence of asymmetric material around the star. It is not clear whether this asymmetry is caused by an asymmetric mass-loss of the AGB star or by the presence of a companion.

5. High-resolution imaging

Recent high spatial resolution observations obtained at various wavelengths have revealed the presence of asymmetric structures in the neighbourhood of the AGB star. For example, near-infrared images of II Lup (Table 1) obtained by Lykou et al. (2018) in the K_s, L and M bands showed circumstellar non-coincident structures with different shapes and sizes. However, these features are not found in all stars of this kind. For example, near-infrared, high spatial resolution images of R Lep showed no sign of asymmetry (Paladini et al. 2017).

A wider picture of II Lup was obtained by Lykou et al. (2018) at radio wavelengths, using the ALMA facility. Observations revealed a gigantic spiral structure detected at the CO (J=1−0) line, extending up to 23 arcsec from the star, amounting to 12 windings separated by $\sim$1.7 arcsec from each other. Their results have been interpreted as a strong evidence of binarity, and the spiral of gas was supposed to have been formed by the stellar wind of the II Lup, driven by the mutual movement of both members of the system around the centre of mass. Based on the ALMA data an orbital period of 128 yr was derived for that system, very different from its "eclipse" period of 19 yr (Table 1). The discrepancy suggests that the period of the deep dims may not be related to the orbital period of the system.

There are other binary AGB stars that exhibit a disc or a spiral structure, but their light curves look standard. ALMA high-resolution observations of the binary AGB star R Aqr (Ramstedt et al. 2018) revealed a ring-like structure in the orbital plane, which is an evidence for an episodic mass-loss event that occurred 100−50 yr ago. Another example, π^1 Gru, is a binary S-type AGB star with a secondary situated at ~400 AU and an orbital period of 330 yr. ALMA images of ^{12}CO (J=3−2) revealed a spiral structure as well as a bipolar outflow perpendicular to the plane of the spiral (Doan et al. 2020). The bipolar structure seems to have been formed during an episodic mass-loss event that lasted 10−15 yr. In spite of the morphological similarities between R Aqr, π^1 Gru and II Lup (a disc or spiral structure, intense mass-loss episodes), the former two stars do not show evidence of undergoing luminosity dims, either because these phenomena are not correlated, the orbital angle of the system relative to the observer does not produce eclipses, or the period between the luminosity dims is too long to be noticed.

6. Conclusions and Perspectives

Considering the evidences gathered to this date, it is not possible to draw a single scenario that definitively explains the variety of observational characteristics observed. It is possible that a fraction of the "eclipsing AGBs" consists of binary systems, whereas other stars may be single. Detecting the secondary by their UV excess (if the binary hipothesis is valid) is not an easy task, because the high opacity of the circumbinary material can absorb most of the UV flux emitted by the secondary or its accretion disc. Therefore, observations obtained at high spatial resolution, especially ALMA, may become paramount to detect evidence of binarity. Considering the small number of "eclipsing AGBs" a special effort must be done to observe every future eclipse of this kind of object.

References

Ayres, T.R., Linsky, J.L., Vaiana, G.S., Golub, L. & Rosner, R. 1981, *ApJ*, 250, 293
Baize, P. 1962, *Journal des Observateurs*, 45, 117
Bedding, T.R., Zijlstra, A.A., Jones, A., Marang, F., Matsuura, M., Retter, A., Whitelock, P.A. & Yamamura, I. 2002, *MNRAS*, 337, 79
Clayton, G.C., Whitney B. A., Stanford S. A., Drilling J. S., 1992, *ApJ*, 397, 652
Doan, L., Ramstedt, S., Vlemmings, W.H.T., Mohamed, S., Höfner, S., De Beck, E., Kerschbaum, F., Lindqvist, M., Maercker, M., Paladini, C. & Wittkowski, M. 2020, *A&A*, 633, A13
Feast, M.W., Whitelock, P.A., Catchpole, R.M., Roberts, G. & Overbeek, M.D. 1984 *MNRAS*, 211, 331
Feast, M.W., Whitelock, P.A. & Marang, F. 2003, *MNRAS*, 346, 878
Hale, D.D.S., Bester, M., Danchi, W.C., Hoss, S., Lipman, E., Monnier, J.D., Tuthill, P.G., Townes, C.H., Johnson, M., Lopez, B. & Geballe,, T.R. 1997, *ApJ*, 490, 407
Hinkle, K.H., Lebzelter, T., Joyce, R.R. & Fekel, F.C. 2002, *AJ*, 123, 1002
Jorissen, A., Schmitt, J.H.M.M., Carquillat, J.M., Ginestet, N. & Bickert, K.F. 1996, *A&A*, 306, 467
Kruszewski A., Gehrels T. & Serkowski, K. 1968, *AJ*, 73, 677
Lépine, J.R.D., Ortiz, R. & Epchtein, N. 1995, *A&A*, 299, 453
Linsky, J.L. & Haisch, B.M. 1979, *ApJ*, 229, L27

Lloyd-Evans, T. 1997, *MNRAS*, 286, 839
Lykou, F., Zijlstra, A.A., Kluska, J., Lagadec, E., Tuthill, P.G., Avison, A., Norris, B.R.M. & Parker, Q.A. 2018, *MNRAS*, 480, 1006
Mayall, M.W. 1963, *JRASC*, 57, 237
Olivier, E.A. & Wood, P.R. 2003, *ApJ*, 584, 1035
Ortiz, R., Guerrero, M.A. & Costa, R.D.D. 2019, *MNRAS*, 482, 4697
Ortiz, R. & Guerrero, M.A. 2021, *ApJ*, 912, 93
Paladini, C., Klotz, D., Sacuto, S., Lagadec, E., Wittkowski, M., Richichi, A., Hron, J., Jorissen, A., Groenewegen, M.A.T., Kerschbaum, F., Verhoelst, T., Rau, G., Olofsson, H., Zhao-Geisler, R. & Matter, A. 2017 *A&A*, 600, 136
Planquart, L., Jorissen, A., van Winckel, H. & van Eck, S. 2022 *IAU Symp.*, 366 (this conference)
Ramstedt, S., Mohamed, S., Olander, T., Vlemmings, W.H.T., Khouri, T. & Liljegren, S. 2018, *A&A*, 616, A61
Raveendran, A.V. & Kameswara Rao, N. 1989, *A&A*, 215, 63
Raveendran, A.V. 2002, *MNRAS*, 336, 992
Sahai, R., Findeisen, K., Gil de Paz, A. & Sánchez-Contreras, C. 2008, *ApJ*, 689, 1274
Sahai, R., Sanz-Forcada, J., Sánchez-Contreras, C. & Stute, M. 2015, *ApJ*, 810, 77
Sahai, R., Sánchez-Contreras, C., Mangan, A.S., Sanz-Forcada, J., Muthumariappan, C. & Claussen, M.J. 2018, *ApJ*, 860, 105
Serkowski, K. 1966, *ApJ*, 144, 857
Serkowski, K. 1971, *Kitt Peak Natl. Obs., Contrib.*, 554, 107
Soszyński, I. & Udalski, A. 2014, *ApJ*, 788, 13
Soszyński, I., Olechowska, A., Ratajczak, M., Iwanek, P., Skowron, D.M., Mróz, P., Pietrukowicz, P., Udalski, A., Szymański, M.K., Skowron, J. et al. 2021, *ApJ*, 911, L22
Whitelock, P.A., Marang, F. & Feast, M. 2000, *MNRAS*, 319, 728
Whitelock, P.A., Feast, M.W., Marang, F. & Groenewegen, M.A.T. 2006, *MNRAS*, 369, 751
Winters, J.M., Fleischer, A.J. & Gauger, A. 1994, *A&A*, 290, 623
Woitke, P. & Niccolini, G. 2005, *A&A*, 433, 1101
Wood, P.R. *et al.* (MACHO collaboration) 1999, in *IAU Symp.*, 191, 151, ed. T. Le Bertre, A. Lèbre & C. Waelkens (San Francisco: ASP)
Wood, P.R., Olivier, E.A. & Kawaler, S.D. 2004, *ApJ*, 604, 800

The Origin of Outflows in Evolved Stars
Proceedings IAU Symposium No. 366, 2022
L. Decin, A. Zijlstra & C. Gielen, eds.
doi:10.1017/S1743921322000436

MATISSE first pictures of dust and molecules around R Sculptoris

J. Drevon[1], F. Millour[1], P. Cruzalèbes[1], C. Paladini[2], J. Hron[3], A. Meilland[1], F. Allouche[1], B. Aringer[3], P. Berio[1], W.C. Danchi[4], V. Hocdé[5], K.-H. Hofmann[6], S. Lagarde[1], B. Lopez[1], A. Matter[1], R. Petrov[1], S. Robbe-Dubois[1], D. Schertl[6], P. Stee[1], F. Vakili[1] J. Varga[7,8], R. Waters[9,10], G. Weigelt[6], J. Woillez[11], M. Wittkowski[11] and G. Zins[2],

[1]Université Côte d'Azur, Observatoire de la Côte d'Azur, CNRS, Laboratoire Lagrange, France
email: julien.drevon@oca.eu

[2]European Southern Observatory, Alonso de Córdova, 3107 Vitacura, Santiago, Chile

[3]Department of Astrophysics, University of Vienna, Türkenschanzstrasse 17

[4]NASA Goddard Space Flight Center, Astrophysics Division, Greenbelt, MD 20771, USA

[5]Nicolaus Copernicus Astronomical Center, Polish Academy of Sciences, Bartycka 18, 00-716 Warszawa, Poland

[6]Max-Planck-Institut für Radioastronomie, Auf dem Hügel 69, D-53121 Bonn, Germany

[7]Konkoly Observatory, Research Centre for Astronomy and Earth Sciences, Eötvös Loránd Research Network (ELKH), Konkoly-Thege Miklós út 15-17, H-1121 Budapest, Hungary

[8]Leiden Observatory, Leiden University, Niels Bohrweg 2, NL-2333 CA Leiden, The Netherlands

[9]Department of Astrophysics/IMAPP, Radboud University, P.O. Box 9010, 6500 GL Nijmegen, The Netherlands

[10]SRON Netherlands Institute for Space Research Sorbonnelaan 2, 3584 CA Utrecht, The Netherlands

[11]European Southern Observatory, Karl-Schwarzschild-Str. 2, 85748 Garching, Germany

Abstract. Carbon-rich dust is known to form in the atmosphere of the semiregular variable star R Sculptoris. Such stardust, as well as the molecules and gas produced during the lifetime of the star, will be spread into the Galaxy via the mass-loss process. Probing this process is crucial to understand the chemical enrichment of the Galaxy. R Scl was observed using the ESO/VLTI MATISSE instrument in December 2018. Here we show the first images of the star between 3 and 10 $R_\star$. Using the complementary `MIRA` 3D image reconstruction and the `RHAPSODY` 1D intensity profile reconstruction code, we reveal the location of molecules and dust in the close environment of the star. Indeed, the C_2H_2 and HCN molecules are spatially located between 1 and 3.4 $R_\star$ which is much closer to the star than the location of the dust. The R Scl spectrum is fitted by molecules and a dust mixture of 90% of amorphous carbon and 10% of silicone carbide. The inner boundary of the dust envelope is estimated by `DUSTY` at about 4.6 $R_\star$. We derive a mass-loss rate of $1.2 \pm 0.4 \times 10^{-6} M_\odot$ yr^{-1} however no clear SiC forming region has been detected in the MATISSE data.

Keywords. instrumentation: interferometers, stars: AGB and post-AGB, stars: carbon, stars: atmospheres, stars: mass loss, stars: individual: R Scl

---

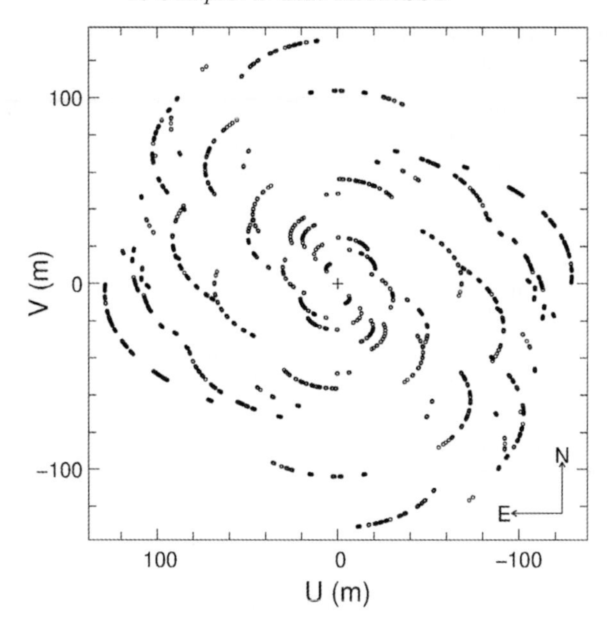

Figure 1. Final (u,v)-plane coverage on R Scl with VLTI/MATISSE. The U and V coordinates are those of the projected baseline vectors.

1. Introduction

R Scl is an Asymptotic Giant Branch (AGB) carbon-rich star known to be a bright infrared and semi-regular pulsation source with a mass-loss rate ranging from $2 \times 10^{-10}\,M_{\odot}\,\mathrm{yr}^{-1}$ (Brunner et al. 2018), up to $1.6 \times 10^{-6}\,M_{\odot}\,\mathrm{yr}^{-1}$ (De Beck et al. 2010). The conservative distance of this spectral C-N5-type star used in this proceeding is 360 ± 50 pc estimated by Maercker et al. (2018). R Scl has a pulsation period of about 370 days (Samus et al. 2009) with a C/O$\approx$1.4 (Hron et al. 1998). ALMA observations (Maercker et al. 2012) revealed spirals-like, clumpy, and non-centrosymmetric structures inside the shell (also confirmed through polarimetric observations by Yudin & Evans (2002)). These structures can be explained by the presence of an unseen companion of $0.25\,M_{\odot}$ hidden at 60 au in the dusty shell. Wittkowski et al. (2017) presented PIONIER observation of the surface of the star, reveling an extended photosphere with a dominant spot very likely of convective origin. In addition, Sacuto et al. (2011) observed the stellar environment above 3 $R_{\star}$ with the MIDI instrument but the uv-plane was not dense enough to use the observations for image reconstruction. Here we fill the gap and use MATISSE to obtain the first simultaneous $L-$ (3–4 μm) and $N-$ (8–13 μm) band observations with a complete uv-plane coverage to make image reconstruction.

2. Data analysis

The observations of R Scl were obtained using the MATISSE instrument (Lopez et al. 2022) mounted at the Very Large Telescope Interferometer (VLTI), in combination with the 4 Auxiliary Telescopes (ATs) in various configurations. The star was observed in December 2018 for about 7 nights (51 hours) optimizing the uv-coverage (see Fig. 1). The data reduction is done using the MATISSE software `drsmat` version 1.5.0 (Millour et al. 2016). MATISSE data were acquired in LOW spectral resolution $L-$ and $N-$ band (R$\sim$30).

Table 1. Parameters of the central source and the dust envelope derived respectively by COMARCS (top part) and DUSTY (bottom part).

Model parameter	Parameter value	Fixed
Distance	$360 \pm 50\,\mathrm{pc}$	Yes
Surface gravity	$-0.50 \pm 0.10\,\mathrm{cm.s^{-2}}$	No
Stellar surface temperature	$2700 \pm 100\,\mathrm{K}$	No
Stellar Luminosity	$8000 \pm 1000\,L_{\odot}$	No
Stellar mass	$2.0 \pm 0.5\,M_{\odot}$	No
C/O ratio	$2^{+2.0}_{-0.6}$	No
Microturbulent velocity	$2.5\,\mathrm{km.s^{-1}}$	Yes
Rosseland Radius	$1.91 \pm 0.20\,au$ 5.3 ± 0.6 mas	No
Shell chemical composition	AmC and SiC $88 \pm 11\%$ of AmC $12 \pm 11\%$ of SiC	No
Dust grain-size distribution	$n(a) \propto a^{-3.5}$	Yes
Optical depth at $\lambda = 1\ \mu$m	0.19 ± 0.05	No
Inner boundary temperature	$1200 \pm 100\,\mathrm{K}$	Yes
Inner radius	24.5 ± 9.0 mas $4.6 \pm 1.7\,R_{\star}$	No
Outer radius	$1000\,R_{in}$	Yes
Mass-loss rate	$1.2 \pm 0.4 \times 10^{-6} M_{\odot}\ \mathrm{yr}^{-1}$	No

3. SED fitting

We assume that the star at a given pulsation phase can be described as an hydrostatic COMARCS model (Aringer et al. 2016) and a DUSTY envelope (Ivezić & Elitzur 1996). To obtain the stellar parameters, we first fit the dereddened photometry (between V- and I-band) with 2600 synthetic observations. The latter are derived from 1D hydrostatic atmospheric models called COMARCS . Then, using Equation 2 from Cruzalèbes et al. (2013) we estimate the Rosseland radius $R_{\star} = 5.3 \pm 0.6$ mas associated to the best fitting model. This last result is statistically consistent within $1.5\,\sigma$ with the radius estimated by Cruzalèbes et al. (2013) and Wittkowski et al. (2017).
The best-fitting COMARCS model is set as the central source for DUSTY. Then, we simultaneously fit photometric data and MATISSE visibilities. The resulting parameters of the fit, as well as their respective error bars, are given in Tab 1 and are consistent with the literature.

4. RHAPSODY and MIRA reconstruction

Two independent methods are used to locate the dusty and molecular regions: the MIRA 3D image reconstruction algorithm (Thiebaut & Giovannelli 2010) and the second one is a 1D new approach based on an intensity radial profile reconstruction method using the Hankel transform. The Hankel method has the advantage of requiring less parameters than image reconstruction, leading to less problems of non-uniqueness of the solution, a better convergence, and a better dynamic range, given that the source presents no or small asymmetries. In both case we reconstruct independent images, one per each spectral channel and we revealed some structures.
In the continuum around $3.50\,\mu$m and $10.00\,\mu$m, both reconstruction techniques (see Fig. 2 and Fig. 3) show a structure between 0–5 mas which can be interpreted as the star's photosphere and spectrally dominated by HCN molecules. Out of the continuum, in the molecular band features around $3.05\,\mu$m (C_2H_2+HCN) and $3.9\,\mu$m (C_2H_2), the structure apparent angular size increases up to 10 mas. This spectral region is characterized by strong asymmetries with non-zero closure phase observations near 3.1 μm as seen in Fig. 4.

Figure 3 shows that such asymmetries are signature of clumpy environment. Then, RHAPSODY underlines a structure between 12–18 mas visible only in $L-$ band and that

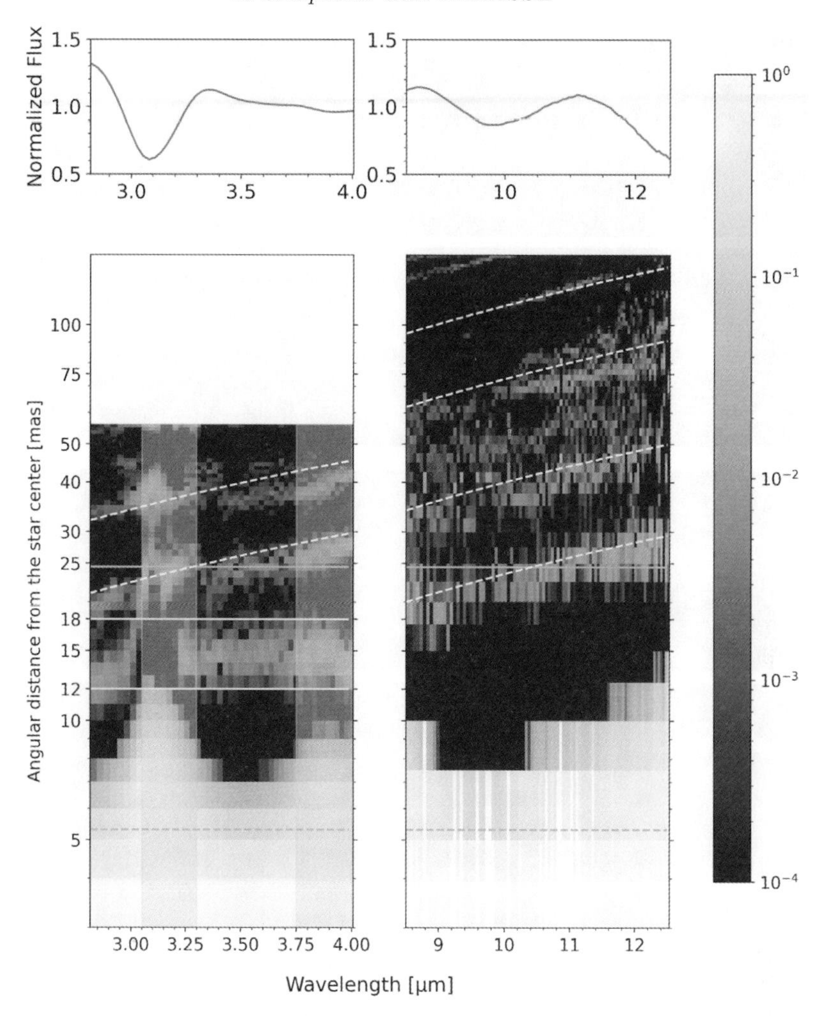

Figure 2. Upper panels correspond to the observed MATISSE spectrum normalized by the median and then divided by a black body spectra at T=2700 K to underline the observed emission and absorption features. Lower panels are the spectro-radial maps in *L*- and *N*-band obtained by plotting the best intensity Hankel profile normalized at one for each observed wavelength. The faint red structures highlighted with the inclined dashed-white lines are reconstruction artifacts. The dashed and solid blue horizontal lines show the position of the Rosseland radius and the `DUSTY` inner radius respectively. The solid green horizontal lines delimit the extension of a hot distinct molecular layer. The grey vertical bands cover the spectral ranges where the centro-symmetric Hankel profile is not able to properly reproduce the asymmetric shape of R Scl revealed by non-zero closure phases.

corresponding to the flux integration of all the clumpy structures visible on the `MIRA` image reconstruction. We speculate that such layer is composed by a mixture of molecules and first seeds of carbon dust grains. Finally, above 20 mas, `MIRA` shows asymmetric and clumpy structures as well as fuzzy emissions. Such structures could be filled by amorphous carbon (amC) or silicone carbide (SiC). However, we did not detected in the MATISSE visibilities structures related to SiC. Indeed, it is not clear at this point if the feature near 11.3 μm is induced by the superposition of continuum emission and molecular absorption or if it is truly an SiC dust feature.

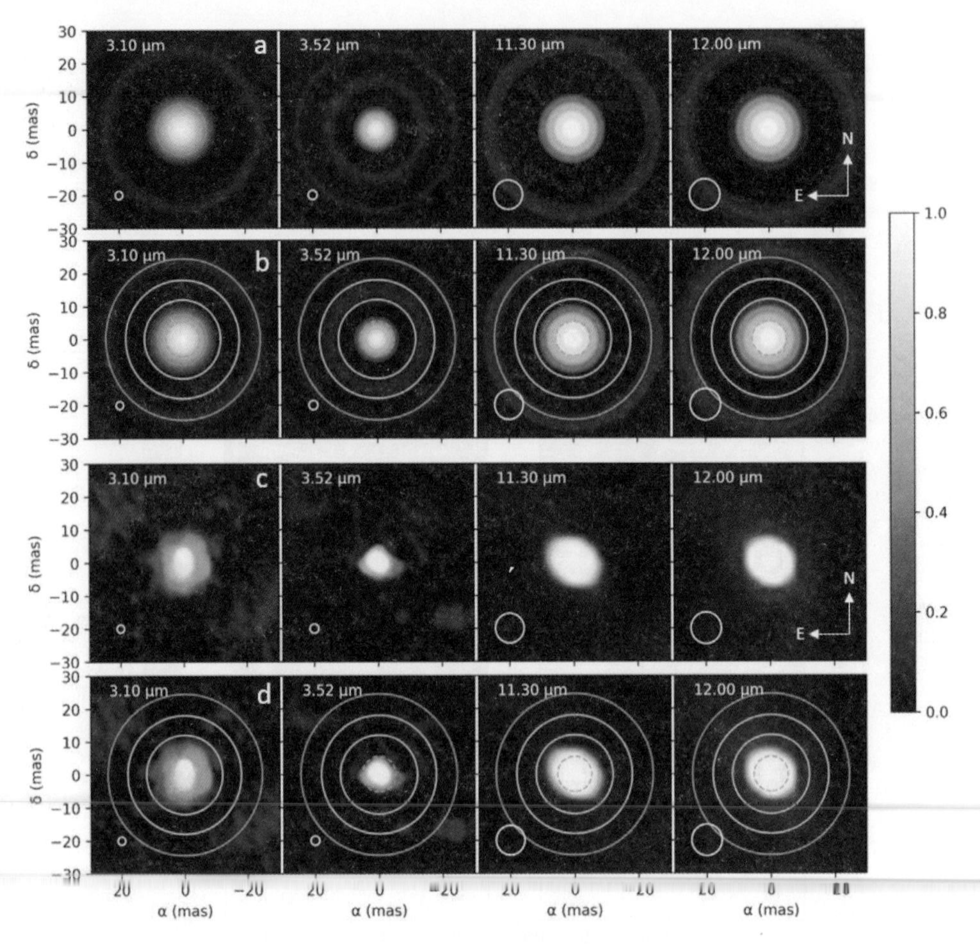

Figure 3. Reconstructed Hankel distribution from radial profiles (row a), the same with the identified features highlighted (row b), images reconstructed with `MIRA` (row c), and the same with features highlighted (row d). Each panel shows the reconstructed image at the wavelength shown at its left top corner, covering the L-band (3–4 μm) and the N-band (8–13 μm). In these panels, the blue dashed circle in the center shows the calculated Rosseland radius of the stellar photosphere, the green circles represent the inner and outer boundary of the distinct molecular shell (seen here only at 3.52 μm and only clearly identifiable in the Hankel reconstruction), the blue circle represents the inner boundary of the dust envelope predicted by `DUSTY`, while the white circle at each bottom left corner shows the theoretical angular resolution of the interferometer.

5. Results and Conclusion

The MATISSE spectra clearly show signatures possibly associated to acetylene (C_2H_2) and hydrogen cyanide (HCN) molecules at 3.1 μm as well as the solid state SiC at 11.3 μm (Yang et al. 2004). `COMARCS` and `DUSTY` allow us to estimate respectively the central source and the dust envelope properties. However this 1D composite modeling approach has strong limitations due to the strong assumptions and the complex nature of the source. Such limitations are seen for example in the quality of the L-band visibility fitting. `MIRA` and `RHAPSODY` provide good fit with MATISSE visibilities even in bandwidth where we expect strong molecular signatures. Both use different techniques: `MIRA` is a 3D image reconstruction code while `RHAPSODY` is a 1D intensity profile reconstruction code and provide consistent and complementary results between each others.

Similarities between PIONIER images and our results are found: in the continuum around

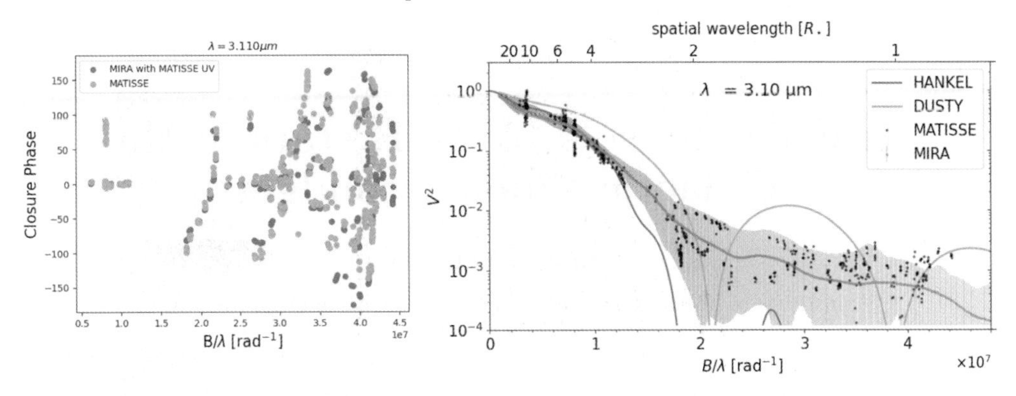

Figure 4. Left Panel: Comparison between the closure phase from the MIRA image reconstruction using the MATISSE (u,v)-plane coverage (blue dots) and the closure phase from MATISSE (orange dots). Right Panel: Plot of the MATISSE squared calibrated visibilities (grey dots) at $3.10\,\mu m$ around the C_2H_2+HCN absorption feature. We over-plotted the data with several models: Hankel profile (blue), `DUSTY` model (orange), and `MIRA` mean 1D radial profile (red) and its dispersion (solid red interval).

$3.5\,\mu$m and $10.0\,\mu$m we observe stellar disk of the same size. However, the MATISSE images are also characterize by strong asymmetries (i.e., clumpy structures) for example at $3.1\,\mu$m where C_2H_2 and HCN are located. We also observe a distinct layer between 12–18 mas formed by molecules and dust which could be the result of an episodic mass-loss event. Concerning the dust, we don't find any clear spectral and spatial signature of SiC, then according also to the small amount of SiC involved in our SED fitting, we question the presence of SiC in the stellar radii probed by our observations. A comparison of our results with extended atmosphere models taking into account the simultaneous presence of gas and dust (Paladini et al. 2009) is subject of future work.

References

Aringer, B., Girardi, L., Nowotny, W., Marigo, P., & Bressan, A. 2016, MNRAS, 457, 3611

Brunner, M., Maercker, M., Mecina, M., Khouri, T., & Kerschbaum, F. 2018, A&A, 614, A17

Cruzalèbes, P., Jorissen, A., Rabbia, Y., et al. 2013, Monthly Notices of the Royal Astronomical Society, 434, 437

De Beck, E., Decin, L., de Koter, A., et al. 2010, A&A, 523, A18

Hron, J., Loidl, R., Hoefner, S., et al. 1998, A&A, 335, L69

Ivezić, Z. & Elitzur, M. 1996, MNRAS, 279, 1011

Lopez, B., Lagarde, S., Petrov, R. G., et al. 2022, A&A, accepted

Maercker, M., Brunner, M., Mecina, M., & De Beck, E. 2018, A&A, 611, A102

Maercker, M., Mohamed, S., Vlemmings, W. H. T., et al. 2012, Nature, 490, 232

Millour, F., Berio, P., Heininger, M., et al. 2016, in Society of Photo-Optical Instrumentation Engineers (SPIE) Conference Series, Vol. 9907, Optical and Infrared Interferometry and Imaging V, ed. F. Malbet, M. J. Creech-Eakman, & P. G. Tuthill, 990723

Paladini, C., Aringer, B., Hron, J., et al. 2009, A&A, 501, 1073

Sacuto, S., Aringer, B., Hron, J., et al. 2011, A&A, 525, A42

Samus, N. N., Kazarovets, E. V., Durlevich, O. V., Kireeva, N. N., & Pastukhova, E. N. 2009, VizieR Online Data Catalog, B/gcvs

Thiebaut, E. & Giovannelli, J. F. 2010, IEEE Signal Processing Magazine, 27, 97

Wittkowski, M., Hofmann, K. H., Höfner, S., et al. 2017, A&A, 601, A3

Yang, X., Chen, P., & He, J. 2004, A&A, 414, 1049

Yudin, R. V. & Evans, A. 2002, A&A, 391, 625

The Origin of Outflows in Evolved Stars
Proceedings IAU Symposium No. 366, 2022
L. Decin, A. Zijlstra & C. Gielen, eds.
doi:10.1017/S1743921322000412

Tracing the role of AGB stars in the Galactic Fluorine budget

Maryam Saberi

Institute of Theoretical Astrophysics, University of Oslo, Norway
email: maryam.saberi@astro.uio.no

Abstract. The cosmic origin of fluorine is still under debate. Asymptotic giant branch (AGB) stars are among the few suggested candidates to efficiently synthesis F in our Galaxy, however their relative contribution is not clear. In this paper, we briefly review the theoretical studies from stellar yield models of the F synthesis and chemical equilibrium models of the F-containing molecules in the outflow around AGB stars. Previous detections of the F-bearing species towards AGB and post-AGB stars are also highlighted. We suggest that high-resolution ALMA observations of the AlF, one of the two main carriers of F in the outflow of AGB stars, can provide a reliable tracer of the F-budget in AGB stars. This will be helpful to quantify the role of AGB stars in the Galactic F budget.

Keywords. AGB stars, stellar nucleosynthesis, fluorine, circumstellar matter, abundances

---

1. Introduction

Fluorine has only one stable isotope (^{19}F) whose origin is still uncertain (e.g. Ryde 2020; Grisoni et al. 2020). The reason for this lack of understanding is the small number of measurements that have been carried out to observe the abundance evolution of F-containing molecules due to a lack in instrumental sensitivity or low spectral resolution of the previous generation of observational facilities. F can be easily destroyed by proton, neutron, and alpha particle capture reactions in stellar interiors (e.g. Ziurys et al. 1994; Abia et al. 2015). A number of scenarios have been suggested to explain the origin of Galactic F in literature. The main ones are (i) during thermal pulses and third dredge-up in Asymptotic Giant Branch (AGB) stars with initial mass 2–4 $M_{\odot}$; (ii) during neutrino process occurring in supernova explosions; (iii) during mergers between helium and carbon-oxygen white dwarfs; (iv) during He-burning phase in Wolf-Rayet (WR) stars; (v) in rapidly rotating massive stars (e.g. Woosley et al. 1995; Meynet et al. 2000; Karakas 2010; Longland et al. 2011; Abia et al. 2015; Jönsson et al. 2017; Limongi et al. 2018; Ryde 2020). Grisoni et al. (2020) has recently published an overview on the F production sites and its impact on the Galactic chemical evolution. The relative contributions of the suggested sites are still under debate (e.g. Timmes et al. 1995; Spitoni et al. 2018; Olive et al. 2019). Among these suggested sites, AGB stars are the only observationally confirmed astrophysical site to efficiently produce F (e.g. Jorissen et al. 1992; Federman et al. 2005; Werner et al. 2005; Abia et al. 2015).

2. Theoretical studies of fluorine in AGB stars

Theoretical studies suggest that the F production in AGB stars can occur by core and shell He-burning at a temperature of 1.5×10^8 K and it can be destroyed once the temperatures exceed 2.5×10^8 K. The synthesized ^{14}N in the hydrogen-burning CNO cycle can produce F by means of a chain of reactions ($^{14}N(\alpha,\gamma)^{18}F(\beta+)^{18}O(p,\alpha)^{15}N(\alpha,\gamma)^{19}F$) during the He-burning thermal pulses (e.g. Forestini et al 1992; Jorissen et al. 1992; Cristallo et al. 2014). The synthesized F can be then brought to the stellar surface by the 3rd dredge-up and be expelled into the interstellar medium by intense stellar winds or during the planetary nebula phase. In high-mass AGB stars with $M > 4M_{\odot}$, the high temperatures in stellar interiors converts F into Ne. Therefore, AGB stars that can efficiently produce F should be less massive than $4M_{\odot}$ to prevent the temperatures of hot bottom burning to destroy the freshly synthesized F (e.g. Lugaro et al. 2004; Karakas 2010).

The results from stellar yield models indicate that F nucleosynthesis is strongly dependent on the stellar mass and metallicity (e.g. Lugaro et al. 2004; Karakas 2010). They predict the highest F production to occur for stars with masses in a range of $M = 2 - 4M_{\odot}$, assuming a metallicity similar to the Solar metallicity.

Recently updated chemical equilibrium models by Agúndez et al. (2020) show that a significant amount of F will be locked into HF and AlF in the inner circumstellar envelope (CSE) of AGB stars within a radius of about $\sim 10R_{\star}$ for all chemical types. Therefore, these molecules are considered to be the best observational tracer of the gas-phase F budget in the outflow of AGB stars.

3. Detections of F-bearing species in AGB and post-AGB stars

Jorissen et al. (1992) presented spectroscopic observations of the infrared vibration-rotation lines of HF towards a sample of AGB stars, showing an over-abundance of 2-30 times larger than solar F abundance. Estimation of F based on HF lines is subject to large uncertainties due to a large contamination of telluric lines in the same wavelength region. This prevents an accurate determination of the F abundance (e.g. Abia et al. 2009, 2010 and 2015). In another study, Werner et al. (2005) reported an over-abundances of 10-250 times larger than the solar F abundance in a number of hot post-AGB stars from far-UV observations of ionized FV and FIV. They suggested the F over-abundance is most likely due to the synthesised F from the preceding AGB phase which is brought to the surface during the post-AGB phase.

Detection of rotational transitions of several AlF lines are reported towards the well-studied studied carbon-rich AGB star, IRC+10216, by Ziurys et al. (1994); Agúndez et al. (2012). They reported an average fractional abundance of $f_{AlF/H_2} \sim 10^{-8}$ which is in agreement with the solar abundance of F of $(5 \pm 2) \times 10^{-8}$ that has been recently reported by Asplund et al. (2021). The initial mass of IRC+10216 is estimated to be $1.6M_{\odot}$ by De Nutte et al. (2017). Therefore, the fractional AlF abundance is in agreement with predictions from stellar yield models for an AGB star with a solar metallically and initial mass in a range of $1 - 2M_{\odot}$.

In a recently published paper, Danilovich et al. (2021) reported detections of HF and AlF lines towards the S-type AGB star, W Aql. They have estimated fractional abundances of $f_{AlF/H_2} = 1 \times 10^{-7}$ and $f_{AlF/H2} = 4 \times 10^{-8}$ using radiative transfer analysis. Their reported value of the AlF is higher than expected AlF fractional abundance for W Aql with an initial mass within a range of $1.2 - 1.6M_{\odot}$ reported by De Nutte et al. (2017).

In a recent search from ALMA archive data, we have identified detection of AlF line emission towards a sample of five M-type AGB stars (Saberi et al. submitted to A&A). From a rotational diagram analysis of multi-line transitions, we estimated fractional abundances of $f_{AlF/H_2} \sim (2.5 \pm 1.7) \times 10^{-8}$ towards *o* Ceti and $f_{AlF/H_2} \sim (1.2 \pm 0.5) \times 10^{-8}$ towards R Leo. For the rest of sample, we only identified one line which prevents a detailed excitation analysis. We crudely approximation of the AlF fractional abundance to be in a range of $f_{AlF/H_2} \sim (0.1 - 6) \times 10^{-8}$ for W Hya, R Dor, and IK Tau. All these sources have an initial mass in a range of $1-2M_{\odot}$ estimated by Decin et al. (2010); Khouri et al. (2014); Hinkle et al. (2016); Danilovich et al. (2017); Velilla Prieto et al. (2017). Therefore, our results are consistent with the predictions from stellar yield models from (e.g. Lugaro et al. 2004; Karakas 2010) and also with the chemical models by Agúndez et al. (2020).

4. Summary

We have reviewed the theoretical and observational studies of F, the element whose origin is still under debate, in AGB stars which are the only sources with observational proof to efficiently synthesis F. However, their relative contributions to the total Galactic F budget is still unclear. The results from analysis of new ALMA observations of AlF lines towards a sample of low-mass AGB stars with initial mass in a range of $1-2M_{\odot}$ are consistent with theoretical stellar yield models and chemical models (Saberi et al. submitted to A&A). This study suggest that observations of AlF lines towards AGB stars with initial mass $2-4M_{\odot}$ can potentially provide a reliable observational proof of the F nucleosynthesis predicted by stellar yield models and quantity the role of AGB stars in the Galactic F budget. We are granted ALMA observations in Cycle 8 and will perform this analysis in an upcoming paper.

References

Abia, C., Recio-Blanco, A., de Laverny, P., et al. 2009, *ApJ*, 694, 971
Abia, C., Cunha, K., Cristallo, S., et al. 2010, *ApJ*, 715, L94
Abia, C., Cunha, K., Cristallo, S., & de Laverny, P. 2015, *A&A*, 581, A88
Agúndez, M., Fonfría, J. P., Cernicharo, J., et al. 2012, *A&A*, 543, A48
Agúndez, M., Martínez, J. I., de Andres, P. L., et al. 2020, *A&A*, 637, A59
Asplund, M., Amarsi, A. M., & Grevesse, N. . 2021, *A&A*, 653, A141
Cristallo et al., 2014, *A&A*, 570, A46
Danilovich, T., Lombaert, R., Decin, L., et al. 2017, *A&A*, 602, A14
Danilovich, T., Van de Sande, M., Plane, J. M. C., et al. 2021, *A&A*, 655, A80
Decin, L., Justtanont, K., De Beck, E., et al. 2010, *A&A*, 521, L4
De Nutte, R., Decin, L., Olofsson, H., et al. 2017, *A&A*, 600, A71
Federman, S. R., Sheffer, Y., Lambert, D. L., & Smith, V. V. 2005, *ApJ*, 619, 884
Forestini et al., 1992, *A&A*, 261, 157
Grisoni, V., Romano, D., Spitoni, E., et al. 2020, *MNRAS*, 498, 1252
Hinkle, K. H., Lebzelter, T., & Straniero, O. 2016, *ApJ*, 825, 38
Jorissen, A., Smith, V. V., & Lambert, D. L. 1992, *A&A*, 261, 164
Jönsson, H., Ryde, N., Spitoni, E., et al. 2017, *ApJ*, 835, 50
Karakas, A. I. 2010, *MNRAS*, 403, 1413
Khouri, T., de Koter, A., Decin, L., et al. 2014, *A&A*, 570, A67
Limongi, M. & Chie, A. 2018, *ApJS*, 237, 13
Longland, R., Lorén-Aguilar, P., José, J., et al. 2011, *ApJ*, 737, L34
ALugaro, M., Ugalde, C., Karakas, A. I., et al. 2004, *ApJ*,, 615, 934
Meynet, G. & Arnould, M. 2000, *A&A*, 355, 176
Olive, K. A. & Vangioni, E. 2019, *MNRAS*, 490, 4307
Ryde, N., Jönsson, H., Mace, G., et al., 2020, *ApJ*, 893, 37
Spitoni, E., Matteucci, F., Jönsson, H., Ryde, N., & Romano, D. 2018, *A&A*, 612, A16
Timmes, F. X., Woosley, S. E., & Weaver, T. A. 1995, *ApJS*, 98, 617
Velilla Prieto, L., Sánchez Contreras, C., Cernicharo, J., et al. 2017, *A&A*, 597, A25
Werner, K., Rauch, T., & Kruk, J. W. 2005, *A&A*, 433, 641
Woosley, S. E. & Weaver, 1995, *ApJS*, 101, 181
Ziurys, L. M., Apponi, A. J., & Phillips, T. G. 1994, *ApJ*, 433, 729

The Origin of Outflows in Evolved Stars
Proceedings IAU Symposium No. 366, 2022
L. Decin, A. Zijlstra & C. Gielen, eds.
doi:10.1017/S1743921322000680

Tracing the inner regions of circumstellar envelopes via high-excitation water transitions

Sandra Etoka[1], Alain Baudry[2], Anita M.S. Richards[1], Malcolm D. Gray[1], Leen Decin[3] and the ATOMIUM consortium

[1]Jodrell Bank Centre for Astrophysics, The University of Manchester, M13 9PL, Manchester, United Kingdom
email: Sandra.Etoka@googlemail.com

[2]Laboratoire d'astrophysique de Bordeaux, Univ. Bordeaux, CNRS, B18N, allée Geoffroy Saint-Hilaire, 33615 Pessac, France

[3]KU Leuven, Institute of Astronomy, 3001 Leuven, Belgium

Abstract. Water is a ubiquitous molecule in circumstellar envelopes (CSEs). Its emission has been detected at a wide range of distances from the central oxygen-rich evolved star. In particular, the water maser transition at 22 GHz, typically extending from about 5–20 stellar radii to as far as several hundred stellar radii from the star, has been commonly used to probe the structure and dynamics of the intermediate regions of the CSE where dust is condensing and the inner wind is being accelerated. The advent of ALMA has opened the door to high-angular resolution mapping of much higher excitation transitions of water, probing the inner regions of the CSEs, some of which are anticipated to exhibit maser action. The ALMA ATOMIUM large program observed many such transitions towards a sample of AGB stars & red supergiants. The preliminary results show that while some transitions depart only slightly from LTE, others clearly show signs of maser action. The Gaussian fitting of the non-diffuse/compact part of some of the (quasi) thermal & maser transitions reveal interesting velocity gradients, signatures of outflowing and infalling motions hence providing important constraints for stellar wind models.

Keywords. stars: AGB and post-AGB, stars: late-type, (stars:) supergiants, (stars:) circumstellar matter, molecular data, radiation mechanisms: thermal, radiation mechanisms: nonthermal, masers, methods: data analysis, techniques: interferometric

---

1. Introduction

Oxygen-rich stars entering the late stages of their evolution, produce a wide range of oxygen bearing molecules in their circumstellar envelopes (CSEs). Water in particular, can be found throughout the CSE and can consequently be used to investigate both the structure and the dynamics from only a few stellar radii to farther out past the dust formation region, provided that high enough angular resolution is reached.

The work presented here is based on the sample of 17 oxygen-rich Asymptotic Giant Branch (AGB) stars and red supergiants (RSGs) observed as part of the Large program entitled ATOMIUM (standing for "ALMA Tracing the Origin of Molecules in dUst-forming oxygen-rich M-type stars; Decin et al. 2020, Gottlieb et al. 2022) down to an angular resolution of 25–50 mas. The aim of the program is to unravel the molecular pathways leading to the formation of the dust precursors and to study the morphology and shaping of the wind. In order to do so, the sample was selected in such a way that it

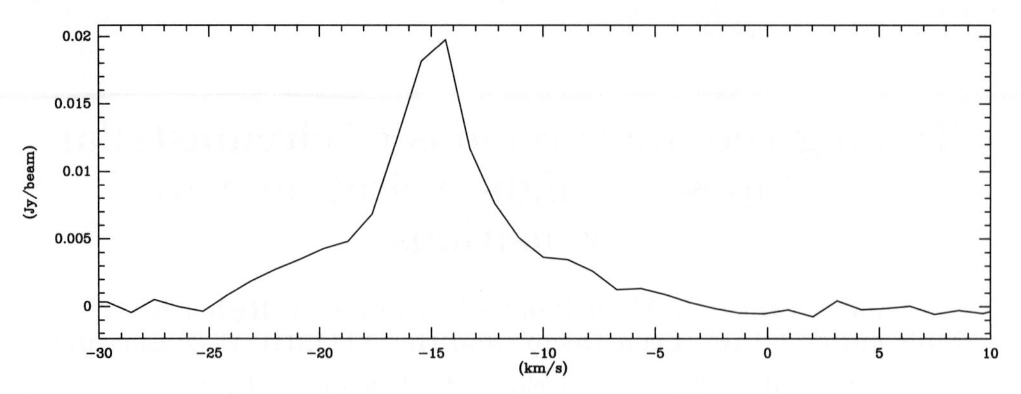

Figure 1. Spectrum of the 268.15-GHz H_2O transition of U Her.

is encompassing a wide range of physical properties (in terms of stellar masses, pulsation behaviours and evolutionary stages) and, in particular mass-loss rates which are ranging from 8×10^{-8} to 6×10^{-5} $M_\odot$ yr^{-1}.

The observations were made in tunings covering about half of the [214–270] GHz band, at a spectral resolution ranging from 1.1 to 1.3 km s^{-1}, in which 14 high-excitation H_2O transitions (i.e., with excitation temperatures 4000-9000 K) were identified. These excitation temperatures are much higher than those explored at cm wavelengths (i.e., 470-2400 K). As a consequence, these high-excitation H_2O transitions are expected to probe the inner regions of the CSE. A more general detailed analysis of the findings, including refinements of the rest frequencies of the detected transitions as well as statistics in terms of H_2O source properties is currently underway (Baudry et al. *in prep.*). Here we present preliminary results of this analysis with a focus on the non-diffuse/compact part of the emission.

2. Results

Out of the 14 transitions present in the part of [214-270] GHz band explored by our observations, 10 were detected on the basis of a quasi coincidence of an observed spectral feature with a transition in the JPL catalog† (Pickett et al. 1998) or the W2020 database (Furtenbacher et al. 2020). Fifteen out of the 17 sources show at least two transitions, with the 2 non-detected ("S-type") sources having a C/O very close to 1. The highest rate of detection is for the transition at 268.15 GHz which was detected towards all 15 sources. It is followed by the 262.90-GHz transition detected towards 12 sources and the 259.95-GHz transition detected towards 10 sources.

On top of being the most commonly detected, the 268.15-GHz transition is also always the strongest transition towards a given source when other transitions are also detected. Even in our relatively low-spectral transition datasets (i.e., $\sim$1.1 km s^{-1} at this frequency), the spectral profile of this transition shows clear signs of maser action towards some sources. An example of such a clear signature is presented in Fig. 1 showing the spectrum of U Her in this transition.

It has to be noted that the 2 other widespread transitions (i.e, at 262.90 and 259.95 GHz), though predicted to be also significantly inverted (Gray et al. 2016) do not show conclusive sign of maser action in their spectral profile, though admittedly such a signature could have been hampered by the low spectral-resolution of the datasets.

In order to investigate more precisely the structure and dynamics of the regions where the non-diffuse emission of the detected high-excitation H_2O transitions is emanating from, we performed Gaussian fittings. This analysis confirms the departure from LTE for

† https://spec.jpl.nasa.gov/ftp/pub/catalog/

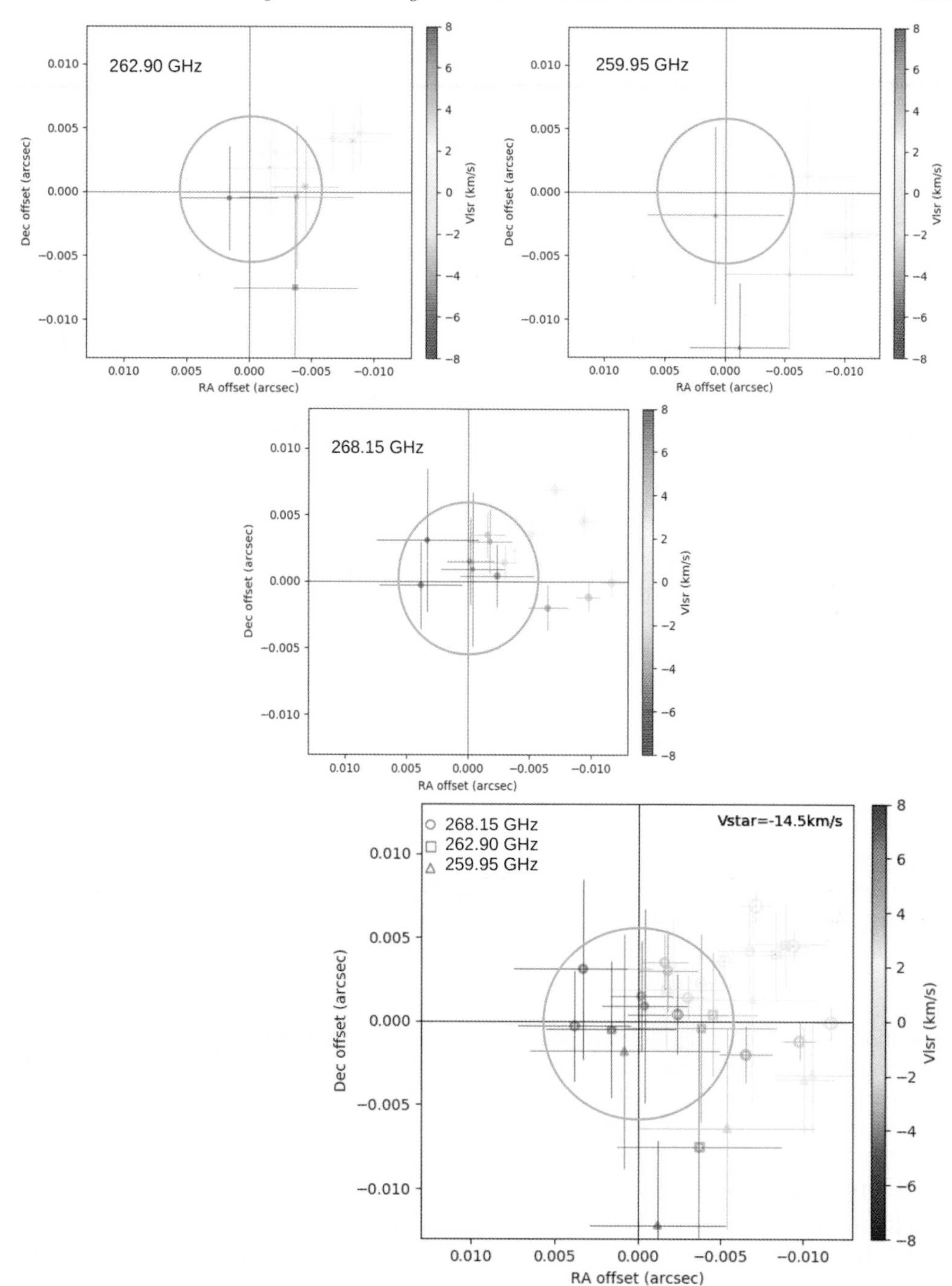

Figure 2. Top and middle: Distribution of the (Gaussian-fitted) features for the 262.90-, 259.95 and 268.15-GHz transitions of U Her. The size of the symbols is proportional to the square root of the intensity of the features while the cross give their positional uncertainty and their colour their velocity information (relative to the stellar velocity) as given by the right-hand side bar. The grey circle represents the stellar diameter as measured in the optical. bottom: The 3 transitions together.

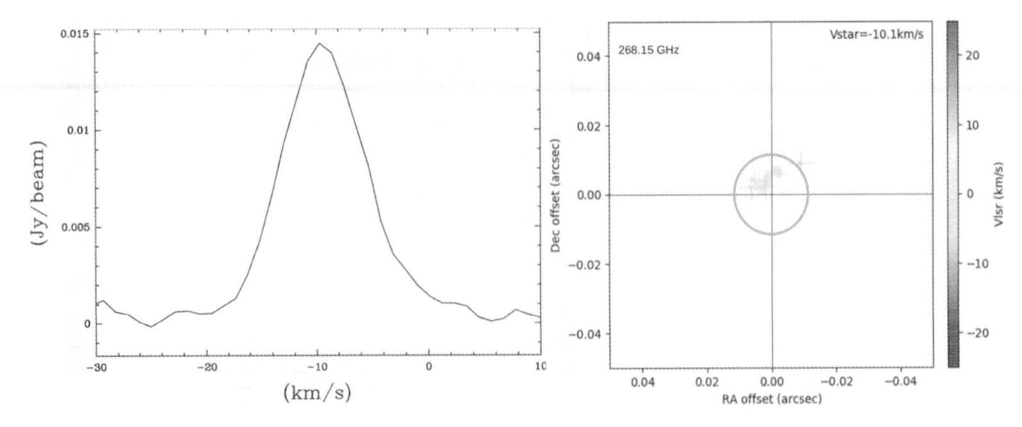

Figure 3. Left: Spectrum of the 268.15-GHz H_2O transition of R Hya. Right: Distribution of the fitted features in this transition.

the strongest and most compact fitted feature of the peak reaching a brightness temperature of $T_B \sim 10^4$ K. Figure 2 top and middle panels present the distribution of the (fitted) features for the 268.15-, 262.90- and 259.95-GHz transitions of U Her. While the three transitions probe similar radii, it can be seen from the figure that each transition probes for the most part, different regions though with a common overall velocity field signature. This later finding is evident in the bottom panel of Fig. 2 presenting the combination of the 3 transitions. This figure also highlights not only a East-West asymmetry but also reveals clear velocity gradients as well as "blue" and "red" shifted material superimposed on the stellar diameter, signatures of ouflowing and infalling motions.

R Hya is one of the sources which shows a substantial number of reasonably strong high-excitation H_2O transitions. It has to be noted that is spite of the strength of the 268.15 GHz transition towards this source, its spectral profile presented on the left panel of Fig. 3, does not show any sign of maser action. On the other hand, here again a clear velocity gradient can be observed in this transition (cf. right panel of Fig. 3). Figure 4 presents the distributions of the non-diffuse emission of the other high-excitation H_2O transitions strong enough to pass the 5σ threshold set as a minimum requirement for the Gaussian fitting. For this source too, the combination of the features reveals the presence of a quasi-linear position-velocity structure extending along a SE-NW axis and signatures of ouflowing and infalling motions particularly well traced by the 268.15-GHz transition showing all but one of the fitted features gradually changing from "blue" and "red" shifted material along the axis aforementioned superimposed on (Northern hemisphere of) the stellar diameter (cf. right panel of Fig. 3).

3. Conclusion

The ATOMIUM datasets observed towards 17 oxygen-rich evolved stars in the [214-270]-GHz band confirmed the presence of 10 out of the 14 high-excitation H_2O transitions identified as present in the part of the band in which the observations were performed. All but two sources (namely the "S" type ones, for which the C/O ratio is nearly 1), show the presence of at least two transitions.

The 268.15-GHz transition is by far the most common, since observed towards all the 15 sources. It is also always the strongest transition when more than one transition is detected towards the same source and in some of the sources, it shows signs of maser action.

Though predicted to be also significantly inverted, the low-spectral-resolution spectra of the 262.90- and 259.95-GHz transitions do not reveal any conclusive sign of

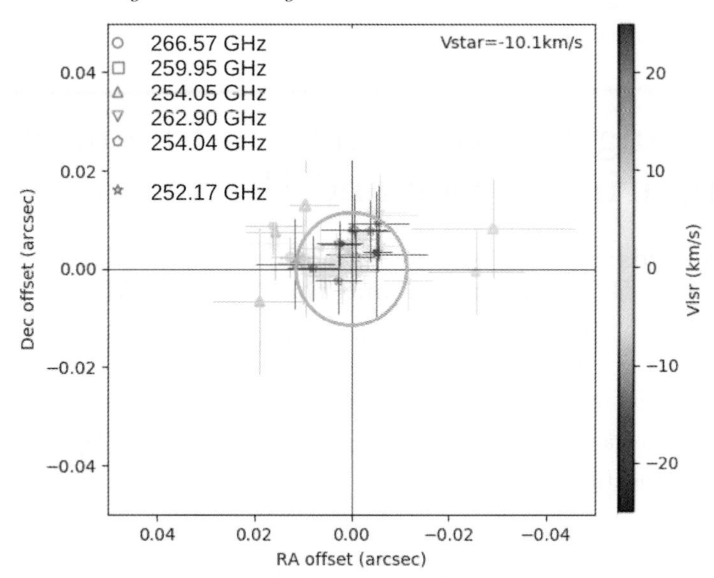

Figure 4. Distribution of the fitted features of the strongest high-excitation H_2O transitions excluding the 268.15-GHz transition of R Hya.

maser action. Higher spectral-resolution observations are needed to further analyse any potential sign of departure from LTE for these transitions.

Finally, clear velocity gradients and signatures of infalling and outflowing motions are observed towards some sources.

References

Decin, L., Montargès, M., Richards, A.M.S., *et al.* 2020, *Science*, 369, 1497

Furtenbacher, T., Tóbiás, R., Tennyson, J., Polyansky, O.L., & Császár, A.G. 2020, *J. Phys. Chem. Ref. Data*, 49, 033101

Gottlieb C.A., Decin, L., Richards, A.M.S., *et al.* 2022, *accepted for publication in A&A*, 2021arXiv211204399G

Gray, M.D., Baudry, A., Richards, A.M.S., Humphreys, E.M.L., Sobolev, A.M., Yates, J.A. 2016, *MNRAS*, 456, 374

Pickett, H.M., Poynter, R.L., Cohen, E.A., *et al.* 1998, *J. Quant. Spectrosc. Radiat. Transfer*, 60, 883

The Origin of Outflows in Evolved Stars
Proceedings IAU Symposium No. 366, 2022
L. Decin, A. Zijlstra & C. Gielen, eds.
doi:10.1017/S1743921322000837

Detailing evolved star wind complexity: comparing maser and thermal imaging

A.M.S. Richards[1], K.A. Assaf[1,2], A. Baudry[3], L. Decin[4], S. Etoka[1], M.D. Gray[1], B. Pimpanuwat[1] and the ATOMIUM collaboration†

[1]JBCA, University of Manchester, M15 9PL, UK
email: a.m.s.richards@manchester.ac.uk

[2]Dept. of Physics, College of Science, University of Wasit, Iraq

[3]Université de Bordeaux, Laboratoire d'Astrophysique de Bordeaux, 33615, Pessac, France

[4]Instituut voor Sterrenkunde, KU Leuven, Leuven, Belgium

Abstract. Maser properties can be measured with milli-arcsec precision over multiple epochs using ALMA, cm- and mm-wave VLBI and e-MERLIN. This allows: (i) Tracing SiO maser proper motions in the pulsation-dominated zone; (ii) Quantifying clumpiness, variability and asymmetry of the wind traced by masers; (iii) Contrasting behaviour from OH masers even at similar distances from the star; (iv) Measuring magnetic fields. Mass lost from the star, traced by SiO masers, is likely to take decades to reach ∼5 stellar radii. At 5–50 stellar radii, once dust is well formed, 22-GHz H_2O masers show the wind accelerating through the escape velocity; its overall direction is away from the star but the velocity field is complex. In a few cases (so far), highly-directed, localised ejecta are seen. Magnetic fields appear to be stellar-centred and strong enough to influence wind kinematics. Recent ALMA and other observations have shown that otherwise inconspicuous companions shape a majority of evolved star winds, whilst advanced models demonstrate how, for some situations, this is compatible with masers showing negligible rotation proper motions. The long-term monitoring achievable at radio frequencies complements the multi-transition maser studies and analysis of thermal lines and dust at shorter wavelengths.

Keywords. stars: AGB, stars: supergiants, masers, stars: mass loss, winds, outflows

---

1. Introduction

ATOMIUM (ALMA Tracing the Origins of Molecules In dUst forMing winds, Gottlieb et al. 2021), is a Large Programme which observed 17 O-rich evolved stars (AGB and RSG) using 3 ALMA configurations with a spectral resolution ~ 1.2 km s^{-1} and angular resolutions down to ∼20 milli-arcsec (mas), comparable to the stellar diameters. The frequency range covers about half of the 214–270 GHz band, including transitions of the main accessible molecules involved in dust formation and tracers of the wind from the stellar surface to the interstellar medium. This enables investigation of links between chemical processes and dynamics in the wind and resolves the kinematics in detail. This showed that the winds have complex structures with acceleration continuing far beyond the dust formation zone (from a few to $\sim 10\ R_\star$, stellar radii). Moreover, all stellar winds showed different degrees of axisymmetry or asymmetry, in most cases almost certainly due to interactions with companions from planetary mass upwards (Decin et al. 2020). The aim of this paper is to demonstrate the role of masers in tracing the winds on even smaller scales and exploit years of monitoring data for some objects.

†See Gottlieb et al. 2022 for list.

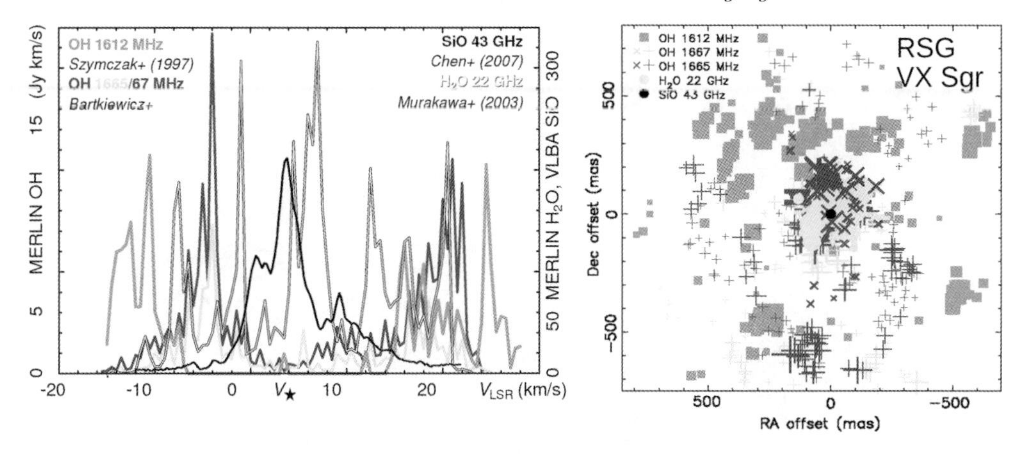

Figure 1. Velocity profiles and locations of common masers around red supergiant VX Sgr. The star is at $V_{\rm LSR}$ 5.3 km s^{-1}, position (0,0) (within the black circle of SiO masers).

The exponential nature of maser amplification means that line widths can be narrowed to a few tenths km s^{-1} and maser spot positions can be fitted with an accuracy of 1/10 of the synthesised beam or better – au scales at the distances of most of the ATOMIUM sample. Groups of spots trace clumps which share distinctive physical conditions. ATOMIUM masers are discussed in this Proceedings by Etoka et al. and others; this paper concentrates on complementary observations of masers at frequencies $<$50 GHz. Fig. 1 shows that masers located at increasing distances from the star extend to progressively higher expansion velocities, also approximately a decreasing order of excitation temperatures. The outermost (1612-MHz OH) masers have a twin-peaked profile characteristic of a fairly smoothly expanding shell at 50–100 $R_\star$. The innermost SiO masers have a central spectral peak characteristic of strong acceleration. The OH mainline and H_2O 22 GHz masers show intermediate behaviour; the 22-GHz masers are found in clumps where the pumping conditions require much higher densities and temperatures than the OH mainline masers emanating from the surrounding gas (Richards et al. 2012).

2. A slow start to the wind

Multi-epoch VLBA monitoring has been performed for 43-GHz SiO masers around a number of Northern sources; whilst R Cas is not in the ATOMIUM sample it is a typical M-type mira. Fig. 2 shows maser clumps have non-linear proper motions. Although the maximum line of sight velocity with respect to the star is $\sim$6 km s^{-1} this includes infall as well as outflow. The average net proper motion away from the star is 0.4$\pm$0.1 km s^{-1}, or up to $\sim$0.55 km s^{-1} allowing for projection effects. This suggests that the wind takes 45–70 yr to cross the SiO shell, out to 3.5 $R_\star$ ($\sim$5 au), thus providing a long timescale for dust formation and other chemistry. Once dust forms, the wind expands at least 10$\times$ faster, covering $\sim$50 au in the next 50 yr (Assaf et al. 2018). The magnetic field strength measured from Zeeman splitting of masers shows that it could contribute to shaping the wind but not play the main rôle in launching mass loss (Assaf et al. 2013).

3. Arcs and spokes

U Her is a Mira variable at about 266 pc (Vlemmings et al. 2007), stellar velocity $V_\star$ –14.9 km s^{-1}, $R_\star$ 5.5 mas (Ragland et al. 2006). Its OH mainline masers at 1.6 GHz were imaged using the EVN in 1999 (4 epochs) and using e-MERLIN in 2014. Water masers at 22 GHz were imaged with MERLIN in 1994, 2000 and 2001 (Richards et al. 2012). The W side of the shell is brighter/elongated in many species (Fig. 3). The H_2O

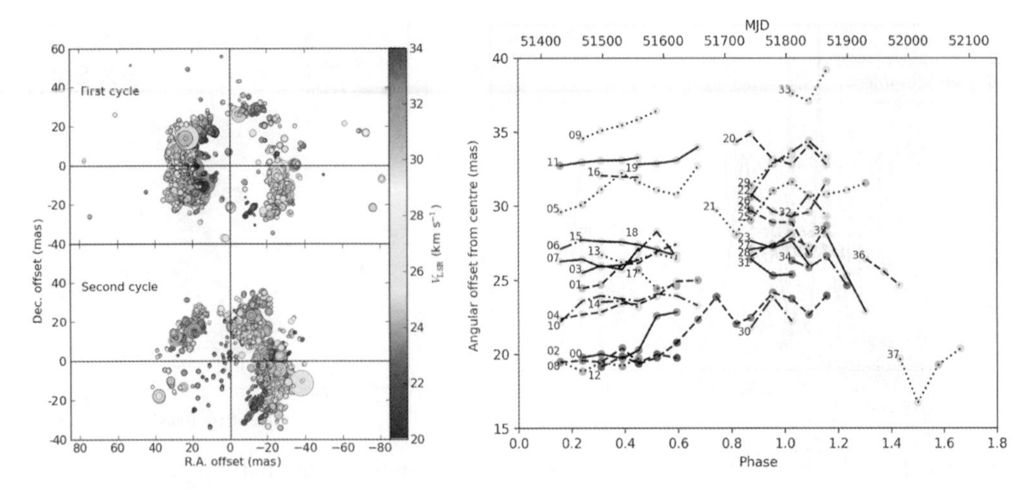

Figure 2. R Cas SiO masers observed 23 times over almost 2 stellar periods (∼2 years). Left: Maser components detected in each cycle. The stellar radius is 12.6 mas (Weigelt et al. 2000). Right: Proper motion trajectories (angular separation from the star v. date and phase) for clumps which can be matched over multiple epochs.

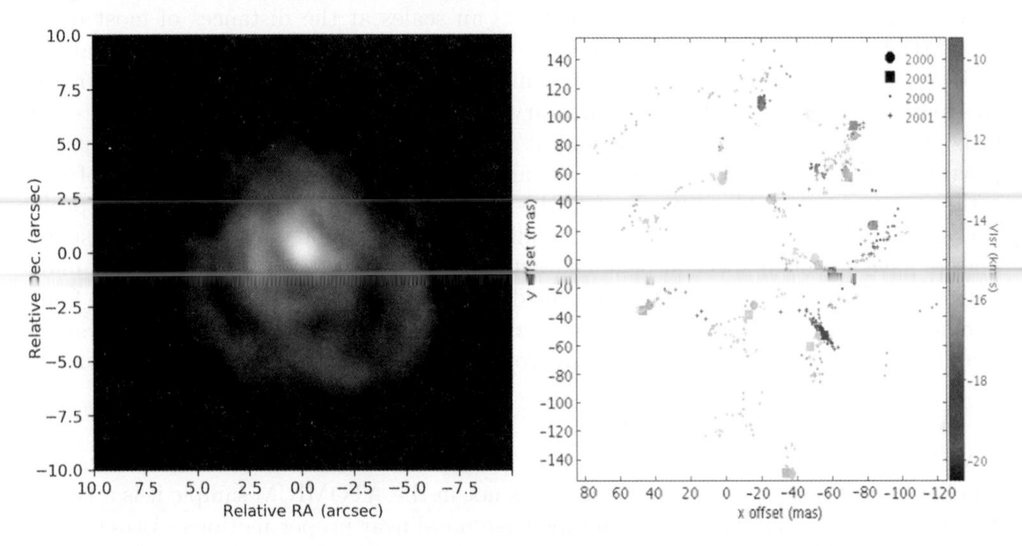

Figure 3. Left: Integrated CO emission from U Her (Decin et al. 2020). Right: The positions of 22 GHz H_2O masers observed in 2000 and/or 2001 (large/small symbols).

maser structure is similar at all epochs and proper motions show overall, accelerating expansion and exhibit radial 'spokes' at around (–50, –60) and (–60, –10) mas offset, –16 to –22 km s^{-1}. SiO spokes have been found to be aligned with the local magnetic field, e.g. studies of U Her by Cotton et al. (2010).

Fig. 4, right, shows an H_2O maser spoke in the SW (as seen in Fig 3), superimposed on thermal SiO emission at a similar velocity. The SiO spoke is seen in all the v=0 transitions in ATOMIUM: ^{28}SiO ^{29}SiO and ^{29}SiO J=5-4 and ^{28}SiO J=6-5. The angular separation between the H_2O maser and the SiO spoke is $\sim$0$''$.1, which in the 24 yr between observations corresponds to about 5 km s^{-1} proper motion. This is realistic, allowing for projection effects, as the line of sight velocity with respect to $V_\star$ is –7 to –10 km s^{-1}.

Fig. 5 shows a (pale-shaded) OH maser clump further from the star in the same direction as the SW H_2O maser spoke but at a velocity closer to $V_\star$. Left-hand circular

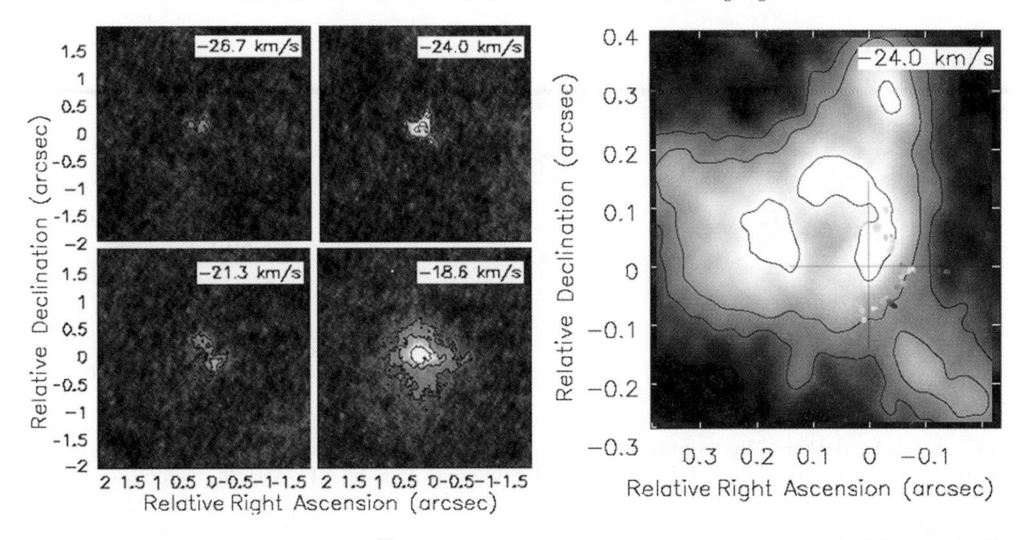

Figure 4. Left: U Her thermal ^{28}SiO J=5-4 v=0 emission, selected channels (ATOMIUM), observed in 2018. A spoke is seen at $V_{\rm LSR}$ ~–24 km s^{-1}. Right: Zoom into the SiO spoke, overlaid with the positions of 22 GHz H_2O masers (see also Fig. 3), spoke components are darkest).

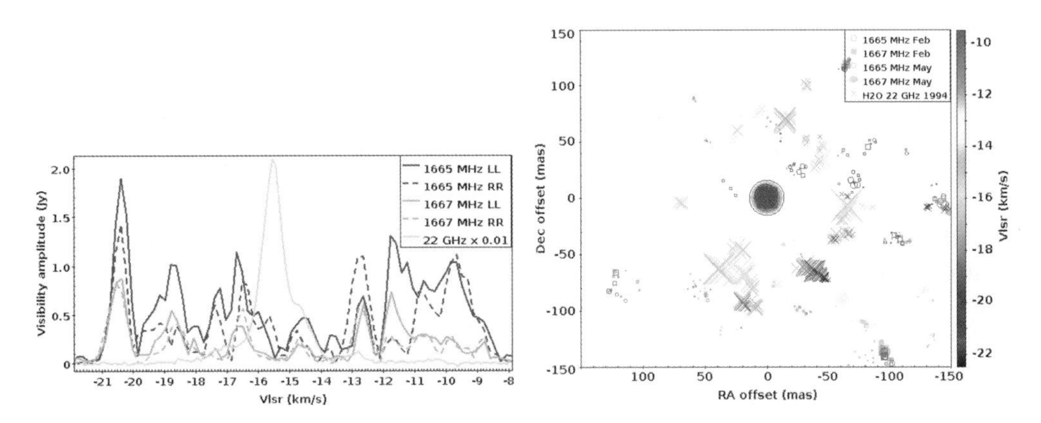

Figure 5. Left: U Her maser spectra. OH RR is only brighter than LL at ~–12.7 km s^{-1}. Right: The –12.7 km s^{-1} feature corresponds to the pale-shaded clump around (–65, –90) mas offset. The centre of expansion, assumed to be the star (behind the most blue-shifted OH), is at (0, 0). Symbols sizes are proportional to flux density.

polarization (LL) is mostly brighter than right-hand circular (RR) (Zell & Fix 1996), implying a magnetic field directed towards the observer Cook (1977) but in the direction of expansion of the H_2O maser spoke the field appears to be reversed. A ~1 G magnetic field was inferred from VLBA H_2O maser observations by Vlemmings et al. (2002) but was not detectable in later observations (Vlemmings et al. 2005). In the absence of absolute astrometry they located the magnetised feature close to the star but we suggest that it is identified with the 22 GHz clump at a similar velocity. A possible explanation is a directed, clumpy or episodic outflow, so that shock compression transiently enhances and distorts the local magnetic field, which may be frozen in to the clumps. OH 1612 MHz masers show unusual flares (Etoka et al. 1977).

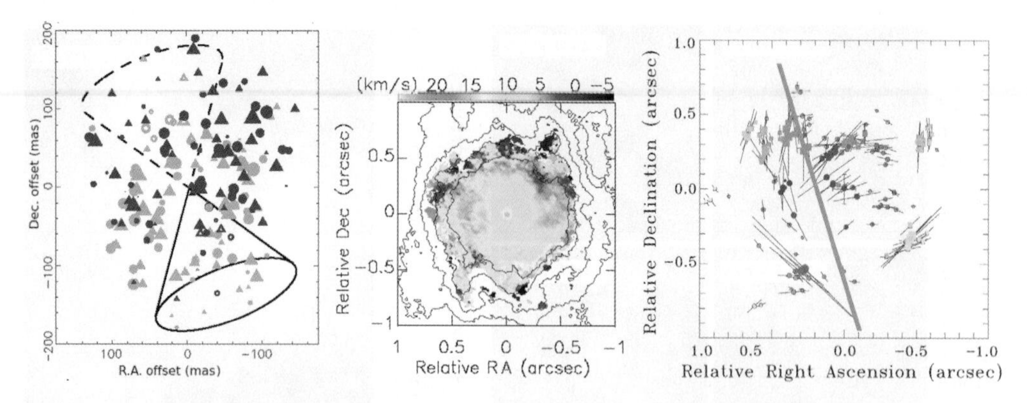

Figure 6. RSG VX Sgr. Left: 22-GHz H_2O maser positions (MERLIN), emission at velocities $<V_\star$ shown by darker (blue) symbols, $>V_\star$ lighter (pink). Masers inside the outlined bicone are weaker than those outside. Middle: Thermal SiO v=0 J=5-4 (ATOMIUM) velocity-weighted integral of emission >0.015 Jy overlaid on contours of total intensity $>3\sigma_{\rm rms}$. Right: OH 1612-MHz masers, symbol colour representing velocity as in middle panel; thin lines show polarization vectors and thick grey line is inferred magnetic dipole axis (Szymczak et al. 2001).

4. Bright 22-GHz water masers trace equatorial density enhancement

Multi-epoch 22 GHz H_2O maser studies of clump proper motions show accelerating expansion, between ~5 to $\geqslant 20\ R_\star$. The 22-GHz masers often show higher expansion velocities than interleaving OH masers which appear to be at the same distance from the star, consistent with the momentum from radiation pressure on dust being transferred more efficiently in denser clumps. There is usually no systematic rotation seen in proper motions; even where the CSE shows an offset between blue- and red-shifted emission this appears to be due to the 22-GHz regions being more extended or brighter in some directions, likely to be an equatorial density enhancement. This is well-demonstrated by the RSG VX Sgr ($V_\star$ 5.7 km s^{-1}) which has 22-GHz H_2O maser proper motions showing that the only systematic flow is accelerating, radial expansion, but Fig. 6 left shows that the masers are brightest in a thick belt and weakest in a biconical region (Murakawa et al. 2003). This is what produces the apparent blue-shifted – red-shifted NE–SW offset as the emission from receding gas (with respect to $V_\star$) is weaker in the NE, whilst the emission from approaching gas is brighter in the SW. This is aligned within uncertainties with the axis of a stellar-centred magnetic dipole, as measured by Vlemmings et al. (2005, 2011) from H_2O and SiO masers, and from OH masers (Szymczak et al. 1999, 2001). Fig. 6 (middle) shows that in a selected velocity range, thermal SiO has a similar red-/blue-shifted offset but OH (right) has the opposite offset, possibly due to the less dense regions or a smaller velocity gradient within the cones favouring its pumping conditions.

5. Conclusions

Imaging masers at <50 GHz complements the ATOMIUM studies of CSE dust formation and dynamics. The slow and irregular progress of mass loss traced by SiO masers in the inner 5 $R_\star$ provides time for complex chemistry to develop. 22 GHz H_2O masers are concentrated in dense clumps between ~5–$\geqslant 20\ R_\star$ where the wind attains escape velocity. In U Her there is an apparent correlation between H_2O maser spokes and thermal SiO clumps, which suggests that some mass is lost as localised ejecta. This has been reported in RSG e.g. VY CMa (Humphreys et al. 2021) but not well-studied previously in AGB stars. Zeeman splitting suggests that the magnetic field strength is not sufficient to drive mass loss but can shape the wind, e.g. the mild axisymmetry of VX Sgr.

Well-studied circumstellar SiO and 22-GHz H_2O masers are usually found in thick shells. These may be poorly filled but extreme-velocity emission is usually seen nearer to the direction of the star, than emission at close to $V_\star$, and long-term monitoring or comparison with other lines shows offsets in different directions, incompatible with systematic rotation. A few AGB stars have 22-GHz masers which do appear to be located in a rotating disc with a Keplerian velocity profile (Szczerba et al. 2006) but these are quite distinct. It is possible that studying the brightest 22 GHz masers selects those with more spherical shells, as strong companion interactions tend to disrupt the velocity coherence or other maser pumping conditions, as modelled for OH by Howe & Rawlings (1994). U Her and VX Sgr are the only ATOMIUM stars with multi-epoch, high-resolution 22 GHz maser imaging although the majority have single-dish detections. We are investigating the morphology and kinematics of 22-GHz masers around other ATOMIUM stars in a programme of single-dish monitoring of the objects at Declination $\geqslant -28°$ using the Medicina and Pushchino radio telescopes, to be followed up by e-MERLIN imaging.

6. Acknowledgements

We thank A. Bartkiewicz, J. Brand, F. Herpin, M. Leal-Ferreira, E.E. Lekht, M. Masheder, K. Murakawa, H.J. van Langevelde and J.A. Yates for contributing to discussions and data reduction. This paper makes use of the following ALMA data: ADS/JAO. ALMA#2018.1.00659.L. ALMA is a partnership of ESO, NSF (USA) and NINS (Japan), together with NRC (Canada), MOST and ASIAA (Taiwan), and KASI (Republic of Korea), in cooperation with the Republic of Chile. The Joint ALMA Observatory is operated by ESO, AUI/NRAO and NAOJ. We have used VLBA data, provided by the National Radio Astronomy Observatory, a facility of the National Science Foundation operated under cooperative agreement by Associated Universities, Inc. and MERLIN and e-MERLIN data. e-MERLIN is operated by the University of Manchester on behalf of STFC.

References

Assaf, K.A., Diamond, P.J., Richards, A.M.S. & Gray, M.D., 2013, *MNRAS*, 431, 1077
Assaf, K.A., 2018, *ApJ*, 869, 80
Chen, X., Shen, Z.-Q. & Xu, Y., 2007, *CJAA*, 7, 531
Cook, A.H., 1977, *Celestial Masers*, CUP
Cotton, W. D., Ragland, S.; Pluzhnik, E. A. et al., 2010, *ApJS*, 188, 506
Decin, L., Montargès, M., Richards, A.M.S. et al. , 2020, *Science*, 369, 1497
Etoka, S. & Le Squeren, A.-M., 1997, *A&A*, 321, 877
Gottlieb, C., Decin, L., Richards, A.M.S. et al., 2022, *A&A*, accepted, 2021arXiv211204399G
Howe, D.A. & Rawlings, J.M.C., 1994, *MNRAS*, 271, 1017
Humphreys, R.M., Davidson, K., Richards, A.M.S. et al., 2021, *AJ*, 161, 98
Murakawa, K., Yates, J.A., Richards, A.M.S. & Cohen, R.J., 2003, *MNRAS*, 344, 1
Richards, A.M.S., Etoka, S., Gray, M.D. et al., 2012, *A&A*, 546, A16
Ragland S., Traub, W.A. & Berger, J.-P., 2006, *ApJ*, 652, 650
Szymczak, M., Cohen, R.J. & Richards, A.M.S, 1999, *MNRAS*, 304, 877
Szymczak, M., Cohen, R.J. & Richards, A.M.S, 2001, *A&A*, 371, 101
Szczerba, R., Szymczak, M., Babkovskaia, N., et al., 2006, *A&A*, 452, 561
Vlemmings, W.H.T., Diamond, P.J. & van Langevelde, H.J., 2002, *A&A*, 394, 589
Vlemmings, W.H.T., van Langevelde, H.J. & Diamond, P.J., 2005, *A&A*, 434, 1029
Vlemmings, W.H.T. & van Langevelde, H.J., 2007, *A&A*, 472, 547
Vlemmings, W.H.T., Humphreys, E.M.L. & Franco-Hernández, R., 2011, *ApJ*, 728, 149
Weigelt G. et al. 2000, in Léna P., Quirrenbach A., eds,*Proc. SPIE*, Vol. 4006, Bellingham, 617
Zell, P.J. & Fix, J.D., 1996, *AJ*, 112, 252

The Origin of Outflows in Evolved Stars
Proceedings IAU Symposium No. 366, 2022
L. Decin, A. Zijlstra & C. Gielen, eds.
doi:10.1017/S1743921322001326

Mass–loss rates of cool evolved stars in M 33 galaxy

Atefeh Javadi[1] and Jacco Th. van Loon[2]

[1]School of Astronomy, Institute for Research in Fundamental Sciences (IPM), P. O. Box 19395-5531, Tehran, Iran
email: atefeh@ipm.ir

[2]Lennard-Jones Laboratories, Keele University, ST5 5BG, UK
email: j.t.van.loon@keele.ac.uk

Abstract. We have conducted a near-infrared monitoring campaign at the UK InfraRed Telescope (UKIRT), of the Local Group spiral galaxy M 33 (Triangulum). In this paper, we present the dust and gas mass-loss rates by the pulsating Asymptotic Giant Branch (AGB) stars and red supergiants (RSGs) across the stellar disc of M 33.

Keywords. stars: evolution–stars: mass–loss–galaxies: individual: M 33–galaxies: star formation

---

1. Introduction

On the AGB, more than half of the mass is lost to the interstellar medium (ISM) in the form of a dusty wind (van Loon et al. 2005). Mass loss is of great importance for stellar evolution and the end products including supernovae, but also for the chemical enrichment of a galaxy. AGB stars are the principal contributors of molecules and dust, and a major source of carbon and nitrogen. The *Spitzer* mid–IR data allow us to derive accurate mass–loss rates. The luminosities and amplitudes will then provide a relation between the mass–loss rate and the mechanical energy involved in the pulsation (van Loon et al. 2006). Mass loss affects the pulsation period, which also depends on the mantle mass. The amount of mass that has already been lost can thus be estimated from the period and luminosity (Wood 2000). A statistical inventory of the mass loss along the AGB in different metallicity range will yield the duration and strength of the mass loss, and thus provide feedback intensities and timescales for chemical evolution models. The low–mass stars lose most of their mass through dusty stellar winds, but even super–AGB stars and red supergiants lose $\sim 40\%$ of their mass via a stellar wind (Javadi et al. 2013). Furthermore, while more massive stars (with birth masses $\gtrsim 8$ M$_\odot$) are incapable of avoiding core collapse, mass loss during the red supergiant (RSG) phase can severely deplete the mantle of the star and even force a return to the blue (Georgy 2012; Georgy et al. 2012). Fortunately, AGB stars and RSGs are relatively easy to detect, as they become not only very luminous ($\sim 10^{3.5-5.5}$ L$_\odot$) but also very red, and thus stand out at infrared (IR) wavelengths above other types of stars within galaxies (Davidge 2000, 2018).

In this project, we aim to understand how galaxies such as our own have evolved to look the way they do today. Our position within its dusty disc precludes such study in the Milky Way, hence we turn to nearby spiral galaxy M 33. We exploit the cool variable stars that trace the endpoints of stellar evolution and are major sources of dust. We monitored M 33 with the UK InfraRed Telescope. Following our work on the nucleus

(Javadi et al. 2011), we will now [1] perform a census of cool variable stars across the disc of M 33; [2] reconstruct the star formation history across M 33 (and other nearby galaxies) and [3] quantify the return of matter throughout M 33.

2. Why M 33 galaxy?

M 33 is the nearest spiral galaxy besides the Andromeda galaxy, and seen under a more favourable angle. This makes M 33 ideal to study the structure and evolution of a spiral galaxy. We will thus learn how our own galaxy the Milky Way formed and evolved, which is difficult to do directly due to our position within its dusty disc.

The methodology consists of three different phases: [1] firstly, we identify long period variables stars (LPVs) (Javadi et al. 2015); [2] secondly, we uniquely relate their brightness to their birth mass, and use the birth mass distribution to reconstruct the star formation history (SFH) (Javadi et al. 2017); [3] thirdly, we measure the excess infrared emission from dust produced by these stars, to estimate the amount of matter they return to the interstellar medium in M 33 (Javadi et al. 2013).

3. The data we use

To derive the mass–loss rates of evolved stars we make use of two data sets; our own near–IR data in the J, H and K_s bands (Javadi et al. 2015) and archival mid–IR *Spitzer* data at 3.6, 4.5 and 8 μ m (McQuinn et al. 2007).

3.1. *Near–IR data*

The project exploits our large observational campaign between 2003-2007, over 100 hr on the UK InfraRed Telescope. The observations were done in the K_s–band (λ= 2.2 μm) with occasionally observations in J– and H–bands (λ= 1.28 and 1.68 μm, respectively) for the purpose of obtaining colour information. The photometric catalogue comprises 403 734 stars, among which 4643 stars were identified as LPVs – AGB stars, super–AGB stars and RSGs.

3.2. *Mid–IR data*

Using five epochs of *Spitzer* Space Telescope imagery in the 3.6–, 4.5– and 8 μm bands, variables have been identified by McQuinn et al. (2007), using a similar method to that we used ourselves.

Of the stars in common, 985 stars were identified as variables in both surveys, but two were saturated and therefore excluded from further analysis. This means that 3658 of the WFCAM variable stars were not identified as variables in the *Spitzer* survey, which is mainly because of the limitation of *Spitzer* in detecting the fainter, less dusty variable red giants. On the other hand, the *Spitzer* survey identified 2923 variables, suggesting a one–third completeness level of the WFCAM variable star survey – this agrees with our internal assessment from a comparison between the WFCAM and UIST data on the central square kpc (Javadi et al. 2015). Generally, both surveys do well in detecting dusty variable AGB stars (and RSGs); this is crucial to estimate mass–loss rates based on IR photometric data.

4. From LPVs luminosities to the star formation history

In the final stage of stellar evolution, low– and intermediate mass (0.8–8 $M_\odot$) stars enter the AGB phase (Marigo et al. 2017) and high mass ($M \gtrsim 8$ $M_\odot$) stars enter the RSG phase (Levesque 2010). These two phases of stellar evolution are characterized by strong radial pulsation of cool atmosphere layers, making them identifiable as LPVs in the photometric light curves (Ita et al. 2004; Yuan et al. 2018; Goldman et al. 2019). The LPVs (AGB–stars, super–AGB stars and RSGs) are at the end–points of their evolution, and their luminosities directly reflect their birth mass (Javadi et al. 2011). Stellar evolution models provide this relation. The distribution of LPVs over luminosity can thus be translated into the star formation history, assuming a standard initial mass function. Because LPVs were formed as recently as < 10 Myr ago and as long ago as $>$ 10 Gyr, they probe almost all of cosmic star formation. We have successfully used this new technique in M 33 (Javadi et al. 2011, 2017) using the Padova models which also provide the lifetimes of the LPV phase (Marigo et al. 2017).

5. Modelling the spectral energy distribution

Spectral energy distributions (SEDs) contain information about the stellar luminosity, temperature, metal content, surface gravity and extinction. If sampled over a sufficient range in wavelength, employing accurate stellar spectral templates allows to retrieve some or all of these parameters. To model SEDs of WFCAM variables we used the publicly available dust radiative transfer code DUSTY (Ivezić & Elitzur 1997). All variables with at least two measurements in near–IR bands (K_s and J and/ or H) and two mid–IR bands (3.6, 4.5 and/or 8 μm) were modelled ($\sim$ 2000 stars). DUSTY calculates the radiation transport in a dusty envelope. We fixed the input temperatures of the star and of the dust at the inner edge of the circumstellar envelope, at 3000 and 900 K, respectively. The density structure is assumed to follow from the analytical approximation for radiatively driven winds (Ivezić, Nenkova & Elitzur 1999). This obviates the need to assume or measure the outflow velocity, as it is implicit in the relation between luminosity, optical depth, gas–to–dust mass ratio and mass–loss rate. We used amorphous carbon dust (Hanner 1988) and a small amount (15 per cent) of silicon carbide (Pégourié 1988) for carbon stars, and astronomical silicates (Draine & Lee 1984) for M–type stars (Fig. 1). Because a sub–set of AGB stars, carbon stars have a different type of circumstellar dust, we must try to identify which stars are likely to be carbon stars. In the absence of spectroscopic confirmation for most of these, and the limited constraints we have from photometry, we resort to making use of theoretical expectations. Correcting the observed colours for the effect of circumstellar dust, we obtain an intrinsic K–band brightness. Using stellar evolution models (Marigo et al. 2017) we convert this into a birth mass, given that these are highly evolved stars that will not evolve much in luminosity. The mass range for AGB stars to become carbon stars spans $\sim$ 1.5–4 $M_\odot$.

5.1. *Mass–loss rate*

For our complete sample (Javadi et al. 2015), some dependence of mass–loss rate on luminosity is seen (Fig. 2); the maximum mass–loss rate increases with luminosity and the highest mass–loss rates are generally achieved by the most luminous, most massive large–amplitude variable stars. This confirms earlier studies in the central region of M 33 (Javadi et al. 2013) and in the Magellanic Clouds (Srinivasan et al. 2009). The mass–loss rates for M–type AGB stars and RSGs are similar to those found in the Solar Neighbourhood (a few $\times 10^{-5}$ and 10^{-7}–10^{-4} $M_\odot$ yr^{-1}, respectively; Jura & Kleinmann 1989). The mass–loss rates for presumed carbon stars are also in good agreement with those found in the Milky Way (a few $\times 10^{-5}$ $M_\odot$ yr^{-1}; Whitelock et al. 2006) and in the Magellanic Clouds ($\sim 10^{-5}$ $M_\odot$ yr^{-1}; Gullieuszik et al. 2012).

It is reassuring to see that the RSGs (certainly stars well above the AGB limit of $\log L/L_\odot = 4.73$ – Wood, Bessell & Fox (1983)) are generally oxygenous; that the least luminous stars are too, and that the maximum mass–loss rate increases with luminosity (in fact rather steeply). Oxygenous stars around – or slightly fainter than – the AGB

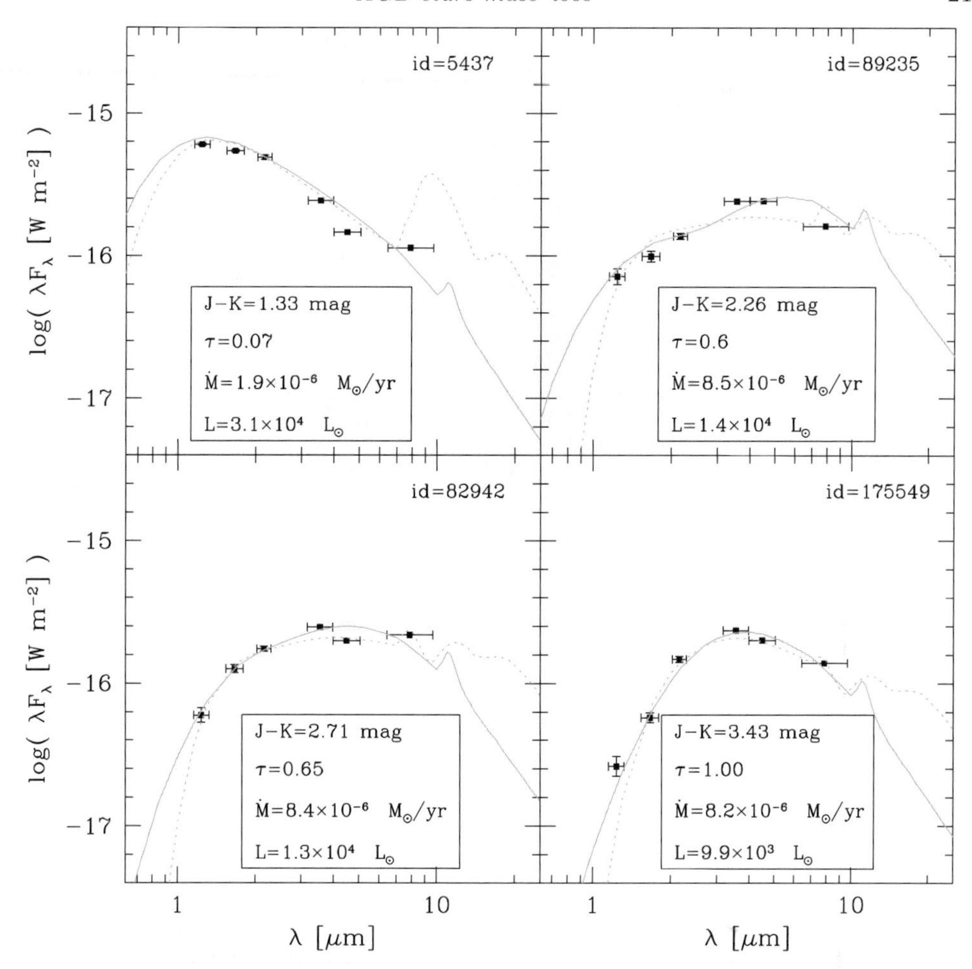

Figure 1. Example SEDs of presumed carbon stars. The horizontal "errorbars" on the data represent the width of the photometric passbands. The best matching SEDs modelled with DUSTY are shown with solid lines. For comparison, the best matching fits using silicates are shown with dotted lines.

limit with very high mass–loss rates are probably massive AGB stars, the equivalent of (most of) the OH/IR stars that are found in the LMC (Marshall et al. 2004).

6. On going works and conclusion remarks

Comparison of the total mass return rate from dusty evolved stars across the galactic disc of M 33 ($\approx$0.1 $M_{\odot}yr^{-1}$; Fig. 2) and recent star formation rate ($\xi = 0.45 \pm 0.10$ $M_{\odot}yr^{-1}$; Javadi et al. 2017), suggests that for star formation to continue beyond the next Gyr or so, gas must flow into the disc of M 33, via cooling flows from the circum–galactic medium and/or by inward migration from gas reservoirs in the outskirts of the disc (Javadi et al. in prep).

In order to gain a comprehensive understanding of galaxy formation and evolution in the Local Group, recently we have conducted an optical monitoring survey of the majority of nearby dwarf galaxies with Isaac Newton Telescope (INT) to identify LPVs (Saremi et al. 2019, 2020). This research is very important from both theoretical and observational perspectives: First, it will give an unprecedented map of the temperature and radius variations as a function of luminosity and metallicity for mass-losing

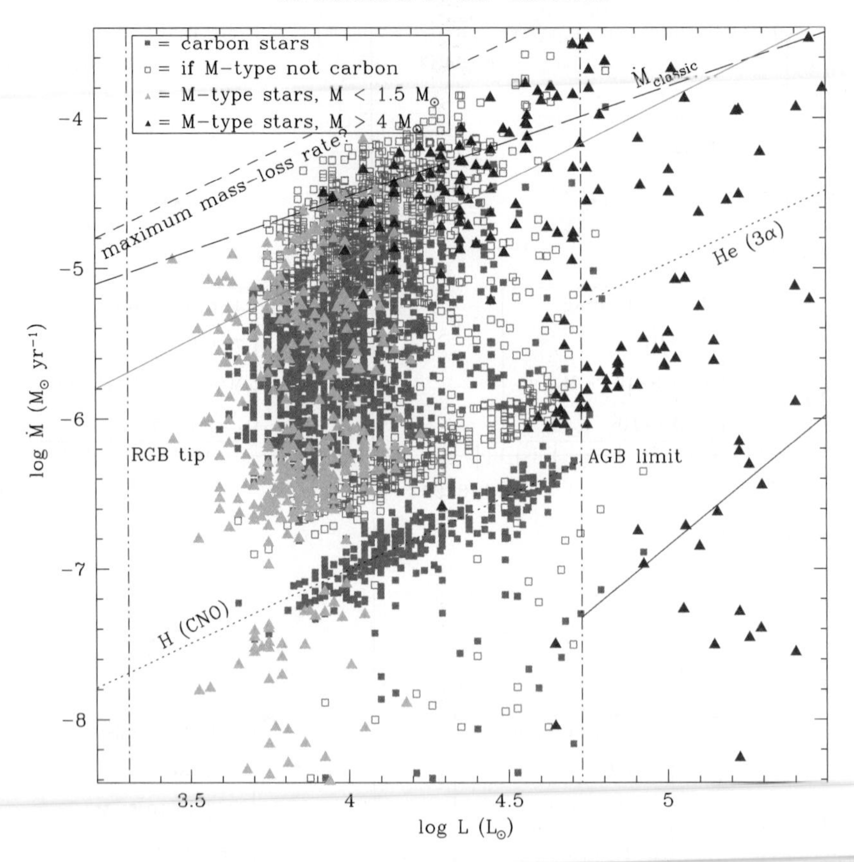

Figure 2. Mass–loss rate vs. luminosity, modelled with DUSTY for low–mass AGB stars (green triangles), intermediate–mass carbon stars (red squares) and massive AGB stars and RSGs (blue triangles). The open red squares show the results if the presumed carbon stars are presumed to be oxygen–rich instead. The vertical dash–dotted lines mark the tip luminosity of the first ascent red giant branch (RGB) and the classical limit of the most massive AGB stars (excluding the effects of Hot Bottom Burning). The dotted lines trace the mass–consumption rates by shell hydrogen burning (CNO cycle) on the AGB and core helium burning (triple–α reaction) in RSGs. The dashed lines trace the limits to the mass–loss rate in dust-driven winds due to single scattering (classic) and multiple scattering (maximum?; see van Loon et al. 1999). The magenta line traces the fit to the mass–loss rate vs. luminosity presented in Goldman et al. (2017), whilst the blue line traces the relation found by Verhoelst et al. (2009) for Galactic RSGs.

stars at the end of their evolution, which places important constraints on stellar evolution models and which is a vital ingredient in the much sought-after description of the mass-loss process. Second, from observational prospective, this research will gather independent diagnostics of the SFHs of different types of dwarf galaxies found in different environments, which help build a detailed picture of galaxy evolution in the nearby galaxies.

References

Davidge T., 2000, AJ, 119, 748
Davidge T., 2018, ApJ, 856, 129
Draine B. T., Lee H. M., 1984, ApJ, 285, 89
Georgy C., 2012, A&A, 538, L8
Georgy C., Ekström S., Meynet G., Massey P., Levesque E. M., Hirschi R., Eggenberger P., Maeder A., 2012, A&A, 542, A29

Goldman S. R. et al., 2017, MNRAS, 465, 403
Goldman S. R., 2019, ApJ, 877, 49
Gullieuszik M. et al., 2012, A&A, 537A, 105
Hanner M. S., 1988, NASA Conf. Pub. 3004, 22
Ita Y., et al. 2004, MNRAS, 353, 705
Ivezić Ž, Elitzur M., 1997, MNRAS, 287, 799
Ivezić Ž, Nenkova M., Elitzur M., 1999, DUSTY User Manual (University of Kentucky)
Javadi A., van Loon J. Th., Mirtorabi M. T., 2011, MNRAS, 414, 3394
Javadi A., van Loon J. Th., Khosroshahi H. G., Mirtorabi M. T., 2013, MNRAS, 432, 2824
Javadi A., Saberi M., van Loon J. Th., Khosroshahi H. G., Golabatooni N., Mirtorabi M. T., 2015, MNRAS, 447, 3973
Javadi A., van Loon J. Th., Khosroshahi H. G., Tabatabaei F., Hamedani Golshan R., Rashidi M., 2017, MNRAS, 464, 2103
Jura M., Kleinmann S. G., 1989, ApJ, 341, 359
Levesque E. M., Massey P., Olsen K. A. G., Plez B., Josselin E., Maeder A., Meynet G., 2005, ApJ, 628, 973
Marigo P. et al., 2017, ApJ, 835, 19
Marshall J. R., van Loon J. Th., Matsuura M., Wood P. R., Zijlstra A. A., Whitelock P. A., 2004, MNRAS, 355, 1348
McQuinn K. B. W. et al., 2007, ApJ, 664, 850
Pégourié B., 1988, A&A, 194, 335
Saremi E., Javadi A., van Loon J. Th., Khosroshahi H. G., Rezaeikh S., Hamedani Golshan R., Hashemi S. A., 2019, Proceedings of IAU Symposium, 344, 125
Saremi E. et al., 2020, ApJ, 894, 135
Srinivasan S. et al., 2009, AJ, 137, 4810
van Loon J. Th., Groenewegen M. A. T., de Koter A., Trams N. R., Waters L. B. F. M., Zijlstra A. A., Whitelock P. A., Loup C., 1999, A&A, 351, 559
van Loon J. Th., Cioni M.-R. L., Zijlstra A. A., Loup C., 2005, A&A, 438, 273
van Loon J. Th., Marshall J. R., Cohen M., Matsuura M., Wood P. R., Yamamura I., Zijlstra A. A., 2006, A&A, 447, 971
Verhoelst T., Van der Zypen N., Hony S., Decin L., Cami J., & Eriksson K., 2009, A&A, 498, 127
Whitelock P. A., Feast M. W., Marang F., Groenewegen M. A. T., 2006, MNRAS, 369, 751
Wood P. R., Bessell M. S., Fox M. W., 1983, ApJ, 272, 99
Wood P. R., 2000, PASA, 17, 18
Yuan W., Macri L. M., Javadi A., Lin Z., Huang J. Z., 2018, AJ, 156, 112

The Origin of Outflows in Evolved Stars
Proceedings IAU Symposium No. 366, 2022
L. Decin, A. Zijlstra & C. Gielen, eds.
doi:10.1017/S1743921322000497

Shaping the initial-final mass relation of white dwarfs with AGB outflows

Paola Marigo

Department of Physics and Astronomy G. Galilei, University of Padova,
Vicolo dell'Osservatorio 3,i IT-35136, Padova, Italy
email: paola.marigo@unipd.it

Abstract. A recent analysis of a few carbon-oxygen white dwarfs in old open clusters of the Milky Way (MW) identified a kink in the initial-final mass relation (IFMR), located over a range of initial masses, $1.65 \lesssim M_{\rm i}/M_{\odot} \lesssim 2.10$, which unexpectedly interrupts the commonly assumed monotonic trend. The proposed interpretation links this observational fact to the formation of carbon stars and the modest outflows (with mass loss rate $< 10^{-7}\,M_{\odot}$/yr) that are expected as long as the carbon excess remains too low to produce dust grains in sufficient amount. Under these conditions the mass of the carbon-oxygen core can grow more than is generally predicted by stellar models. We discuss these new findings also in light of a new systematic follow-up investigation, based on Gaia EDR3, of evolved giants (13 carbon stars, 3 S stars and 4 M stars) belonging to intermediate-age open clusters.

Keywords. stars: evolution, stars: AGB and post-AGB, stars: mass loss, stars: winds, outflows, stars: white dwarfs, stars: abundances, stars: atmospheres

---

1. Introduction

The IFMR connects the mass of a star on the main sequence, $M_{\rm i}$, with the mass, $M_{\rm f}$, of the WD left at the end of its evolution. This fate (Herwig 2005) is common to low- and intermediate-mass stars ($0.9 \lesssim M_{\rm i}/M_{\odot} \lesssim 6-7$) that, after the exhaustion of helium in the core, experience the AGB phase and produce carbon-oxygen WDs. Quasi-massive stars ($8 \lesssim M_{\rm i}/M_{\odot} \lesssim 10$) that, after the carbon burning phase evolve through the Super-AGB phase, may also produce oxygen-neon-magnesium WDs. In both evolutionary scenarios stellar winds play a key role in determining the final mass of the compact remnant.

To derive the semi-empirical IFMR, singly-evolved WDs that are members of star clusters are ideally used (Cummings et al. 2018). Spectroscopic analysis provides their atmospheric parameters, that is, surface gravity, effective temperature, and chemical composition. Coupling this information to appropriate WD cooling models provides the WD mass, its cooling age, and additional parameters for testing cluster membership and single-star status. Finally, subtracting a WD's cooling age from its cluster's age gives the evolutionary lifetime of its progenitor, and hence its $M_{\rm i}$.

2. The detection of IFMR kink

A recent work (Marigo et al. 2020) identified a kink in the IFMR at $M_{\rm i} \simeq 2\,M_{\odot}$, following updated analyses of a few WDs members of intermediate-age open clusters (NGC 7789, NGC 752, Ruprecht 147, with ages $\simeq 1.5-2.5$ Gyr). The new results were obtained with a novel analysis technique that combines photometric and spectroscopic data to better constrain the WD parameters. The data are shown in Fig. 1 (diamonds with error bars). While the steep increase in the IFMR near $M_{\rm i} \simeq 1.65\,M_{\odot}$ is at present well

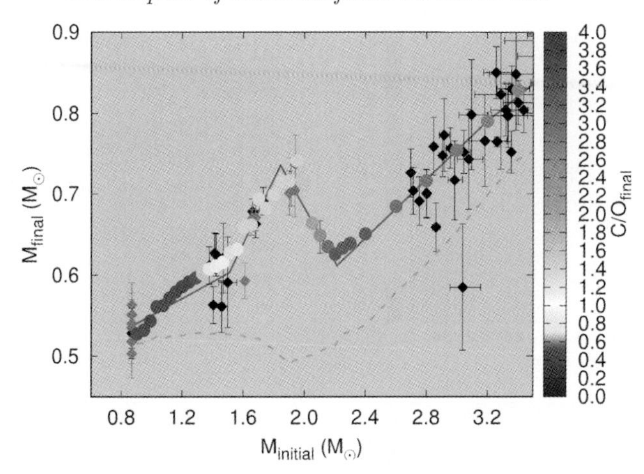

Figure 1. The semi-empirical IFMR (diamonds with errors bars) from Cummings et al. (2018) and Marigo et al. (2020), with the 7 newly discovered and 12 newly analysed WDs shown in green. Error bars cover a range of $\pm 1\,\sigma$. Superimposed is the theoretical IFMR obtained with state-of-the art mass-loss prescriptions for mass loss in carbon stars, in combination with calibrated efficiency λ of the third dredge-up, color-coded as a function of the final C/O at the end of the TP-AGB phase (right color bar). The theoretical core mass at the first thermal pulse is also shown (magenta dashed line).

constrained, further data are needed to better probe the decreasing IFMR for $M_{\rm i} \gtrsim 2\,M_{\odot}$. However, we underline that after this rapid rise such a temporary decrease is necessary to keep consistency with the observed field-WD mass distribution. For example, if this steep rise was instead followed by a plateau at $\rm M_f \sim 0.7\,M_{\odot}$, then every progenitor with $1.9 \lesssim M_{\rm i}/M_{\odot} \lesssim 2.8$ would create a $\sim 0.7\,M_{\odot}$ WD, substantially overproducing field WDs at this mass compared with observations.

3. The physical interpretation of the IFMR: the role of AGB winds

We interpret the kink in the IFMR as the signature of the lowest-mass stars in the MW that became carbon stars during the TP-AGB phase. The proposed explanation is as follows. At solar-like metallicity low-mass carbon stars ($1.65 \lesssim M_{\rm i}/M_{\odot} \lesssim 1.90$) attain low C/O ratios ($\lesssim 1.3$) and low values of the excess of carbon compared to oxygen, C − O, in the atmosphere. The quantity C − O is particularly relevant as it measures the budget of free carbon, not locked in the CO molecule, available to condense into dust grains. In fact, state-of-the-art dynamical models for carbon stars (Bladh et al. 2019; Eriksson et al. 2014; Mattsson et al. 2010) predict that a minimum carbon excess, $({\rm C-O})_{\rm min}$, is necessary to generate dust-driven winds, with mass-loss rates exceeding a few $10^{-7}\,M_{\odot}\,{\rm yr}^{-1}$. We recall that according to a standard notation, ${\rm C-O} = \log(n_{\rm C} - n_{\rm O}) - \log(n_{\rm H}) + 12$, where $n_{\rm C}$, $n_{\rm O}$, and $n_{\rm H}$ denote the number densities of carbon, oxygen, and hydrogen, respectively.

The existence of a threshold in carbon excess impacts on the TP-AGB evolution and hence on the IFMR. In TP-AGB stars the surface enrichment of carbon is controlled by the 3DU, a series of mixing episodes that happen each time the base of the convective envelope is able to penetrate into the inter-shell region left at the quenching of a thermal pulse. The efficiency of a 3DU event is commonly described by the dimensionless parameter $\lambda = \Delta M_{\rm 3DU}/\Delta M_{\rm c}$, defined as the amount of dredged-up material, $\Delta M_{\rm 3DU}$, relative to the growth of the core mass, $\Delta M_{\rm c}$, during the previous inter-pulse period.

Figure 2 illustrates how the efficiency of 3DU regulates the increase of the surface C/O and hence the carbon excess; how the latter, in turn, affects the mass-loss rate, hence the lifetime of a carbon star and eventually the final mass of the WD. The models refer

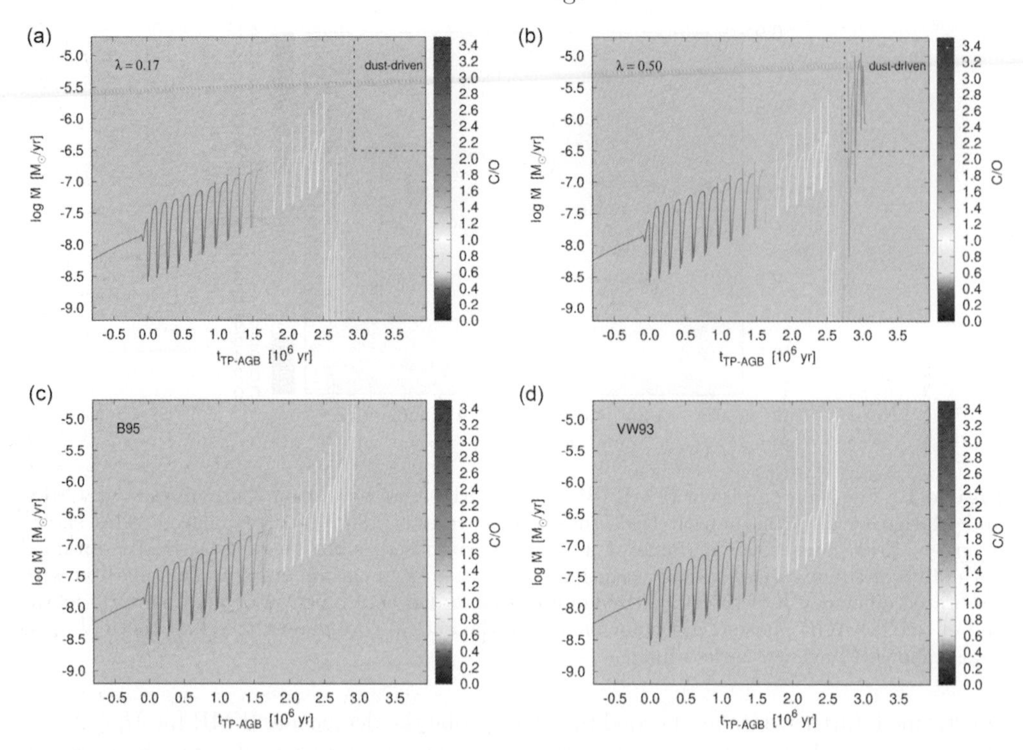

Figure 2. Evolution of the mass-loss rate during the whole TP-AGB evolution of a star with $M_{\rm i} = 1.8\,M_\odot$ and solar metallicity. Time is set to zero at the first TP. The tracks are colour-coded according to the current photospheric C/O ratio. Calculations differ in the treatment of mass loss and/or in the efficiency λ of the 3DU. **a** - **b**, Models in which mass loss during the carbon star phase does depend on the carbon excess (Mattsson et al. 2010; Bladh et al. 2019). The 3DU is less efficient in model **a** compared to model **b**. Inside the region delimited by the dotted lines stellar winds are driven by carbonaceous dust grains. **c** - **d**, Results obtained with widely-used mass-loss formalisms (Bloecker 1995) that, unlike in **a** and **b**, do not contain an explicit dependence on the carbon abundance.

to a star with $M_{mi} = 1.8\,M_\odot$ and $Z = 0.014$, near the kink's peak of the semi-empirical IFMR. The two cases in Fig. 2 share the same set of input prescriptions, except that the 3DU is shallow in the model **a** ($\lambda \simeq 0.17$), and much more efficient in the model **b** ($\lambda = 0.5$ as the star becomes C-rich). In both cases, as soon as the star reaches $\mathrm{C/O} > 1$, a sudden drop in the mass-loss rate is expected to occur. This prediction deserves to be explained in detail. The transition from $\mathrm{C/O} < 1$ to $\mathrm{C/O} > 1$ marks a radical change both in the molecular abundance pattern of the atmosphere (shifting from O-bearing to C-bearing species; see Marigo & Aringer 2009), and in the mineralogy of the dust that could in principle condense in the coolest layers.

When a carbon star is born, the silicate-type dust that characterises the circumstellar envelopes of M-type stars is no longer produced and the composition of the grains that may actually form suddenly changes, switching mainly to silicon carbide and amorphous carbon (Ferrarotti & Gail 2006). The key point is that the growth of carbonaceous dust requires suitable physical conditions (e.g., temperature, density and chemical composition of the gas, stellar radiation field), and these may not always be fulfilled as soon as $\mathrm{C/O} \gtrsim 1$. In addition, carbonaceous grains are expected to drive a wind only if they form in sufficient amount (Mattsson et al. 2010; Bladh et al. 2019), a condition expressed by the threshold $(\mathrm{C-O})_{\rm min}$. It follows that, as long as the atmospheric abundance of free carbon is small, typically during the early carbon-star stages, dust grains cannot be abundantly produced.

The natural conclusion is that if carbonaceous dust grains are not abundant enough to drive a powerful wind, when C/O just exceeds unity, then only a modest outflow may be generated, possibly sustained by small-amplitude pulsations, like those of Semi-regular variable stars (McDonald & Trabucchi 2019). These conditions apply to the model **a** of Fig. 2, which experiences a shallow 3DU. The C/O ratio grows slowly (maximum of $\simeq 1.33$), the star stays in a phase of very low mass loss until the threshold in carbon excess is slightly overcome and a moderate dust-driven wind is eventually activated, with mass-loss rates not exceeding few $10^{-6}\,M_{\odot}\,\mathrm{yr}^{-1}$. Also the model **b** enters a phase of low mass loss soon after the transition to carbon star, but then its evolution proceeds differently. As the 3DU is more efficient, C/O increases more rapidly (maximum of $\simeq 1.91$), so that the threshold in carbon excess is largely overcome, a powerful dust-driven wind is generated, and the mass-loss rate rises up to $\simeq 10^{-5}\,M_{\odot}\,\mathrm{yr}^{-1}$.

These model differences affect the carbon star lifetimes and, in turn, the final masses left after the TP-AGB phase. In the model **a** the carbon star phase lasts $\simeq 1.15$ Myr and produces a WD with a final mass of $\simeq 0.732\,M_{\odot}$. In the model **b** the duration of the carbon star phase is halved, $\simeq 0.52$ Myr, and terminates with a WD mass of $\simeq 0.635\,M_{\odot}$. Likewise, shorter lifetimes and lower final masses are obtained if we adopt mass-loss formulations that do not depend on the carbon abundance. In Fig. 2 we also show two examples (panels **c, d**) in which we use the Bloecker (1995) and Vassiliadis & Wood (1993) relations. Both predict a systematic increase of the average mass-loss rate as the star evolves on the TP-AGB. The resulting WD masses are $0.646\,M_{\odot}$ and $0.615\,M_{\odot}$, respectively.

4. Calibration of the 3DU efficiency

The analysis presented above indicates that the final WD masses of the progenitor carbon stars are the result of the interplay between mass loss and third dredge-up. In light of this, we computed a large grid of TP-AGB models with `COLIBRI` (Marigo et al. 2013) that cover a relevant region of the parameter space $(M_{\rm i}, \lambda)$, assuming solar initial metallicity. $M_{\rm f}$ of carbon stars anticorrelates with λ. At this point, the natural step is to pick up the $(M_{\rm i}, \lambda)$ combinations that best approximate the semi-empirical IFMR. The detected IFMR kink over the range $1.65 \lesssim M_{\rm i}/M_{\odot} \lesssim 2.0$ is well recovered by assuming that these stars experience a shallow 3DU during the TP-AGB phase, typically with $0.1 \lesssim \lambda \lesssim 0.2$. In this regard, we note that observations of TP-AGB stars in Galactic open clusters (see Sect. 5) indicate that carbon stars slightly less massive than $\simeq 1.65\,M_{\odot}$ can be formed at solar metallicity, with an initial mass $M_{\rm i} \simeq 1.5\,M_{\odot}$. At this $M_{\rm i}$ the semi-empirical IFMR show white dwarfs with $M_{\rm f} \lesssim 0.62\,M_{\odot}$, therefore without clear evidence of an excess of mass, unlike at higher $M_{\rm i}$. From a theoretical point of view this is easily explained by assuming that the transition to C-rich phase occurred quite late during the TP-AGB evolution, when a substantial fraction of the envelope had already been expelled by the O-rich winds. Under these conditions, the growth of the core mass could not proceed significantly. Overall, stars with $1.5 \lesssim M_{\rm i}/M_{\odot} < 1.8$ are those just massive enough to become carbon stars. They are little enriched in carbon, with low final ratios $(1.1 \lesssim \mathrm{C/O} \lesssim 1.4)$ and low carbon excesses. At larger masses the 3DU becomes more efficient and powerful dust-driven winds are activated. The results of our calibration are shown in Fig. 1.

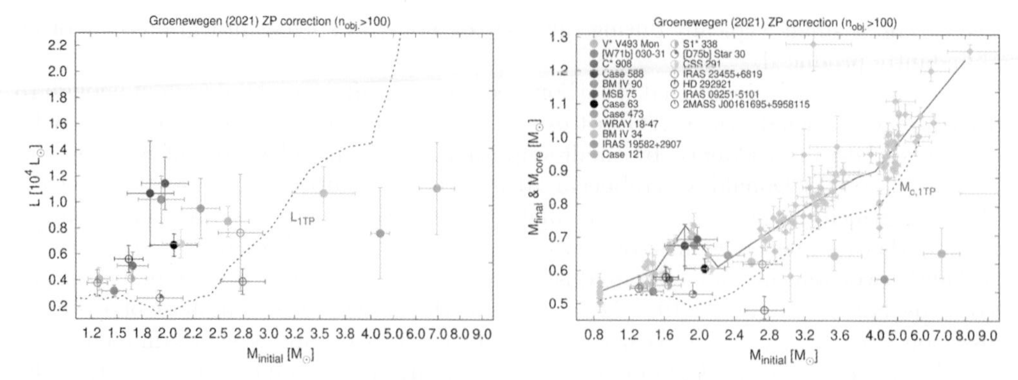

Figure 3. Luminosities and core masses of TP-AGB stars in open clusters as a function of the initial stellar mass. The M, S, and C stars (reported in the legend) are marked with colored symbols and error bars. For comparison we over-plot the luminosity and core mass at the first thermal pulse, L_{1TP} and $M_{c,1TP}$, predicted by the PARSEC stellar models (Bressan et al. 2012) at solar metallicity (dashed line). The cluster ages and visual extinctions are taken from the work of Cantat-Gaudin et al. (2020). Gaia EDR3 parallaxes are corrected for the zero-point offset following Groenewegen (2021). Note that the X-axis is stretched over the range $0.8 \leq M_i/M_\odot \leq 4.0$. *Left panels*: Bolometric luminosties derived from the fitting of the spectral energy distributions. *Right panels*: The initial-final mass relation of white dwarfs in the Milky Way is compared to the current core masses of TP-AGB stars in open clusters. The semi-empirical IFMR (gray diamonds with error bars) is taken from Marigo et al. (2020) and Cummings et al. (2018). The solid line is a fit to the IFMR data.

5. AGB stars in Galactic open clusters

Benefiting from the *Gaia* DR2 and EDR3 releases of photometric and astrometric data we examine the population of asymptotic giant branch stars that appear in the fields of intermediate-age and young open star clusters (Marigo et al. (2022)). We identify 49 AGB star candidates, brighter than the tip of the red giant branch, with a good-to-high cluster membership probability. Among them we find 19 TP-AGB stars with known spectral type: 4 M stars, 3 MS/S stars and 12 C stars. By combining observations, stellar models, and radiative transfer calculations that include the effect of circumstellar dust, we characterize each star in terms of initial mass, luminosity, mass-loss rate, core mass, period and mode of pulsation (see Fig. 3). The information collected helps us shed light on the TP-AGB evolution at solar-like metallicity, placing constraints on the third dredge-up process, the initial masses of carbon stars, stellar winds, and the initial-final mass relation (IFMR). In particular, we find that two bright carbon stars, MSB 75 and BM IV 90, members of the clusters NGC 7789 and NGC 2660 (with similar ages of $\simeq 1.2-1.6$ Gyr and initial masses $2.1 \gtrsim M_i/M_\odot \gtrsim 1.9$), have unusually high core masses, $M_c \approx 0.67-0.7\,M_\odot$. These results support the detection of the IFMR kink and its proposed interpretation in terms of carbon-star formation and AGB outflows.

6. Conclusions

A new thorough analysis of a few WDs in old open clusters with turn-off masses over the range from $1.6\,M_\odot \lesssim M_i \lesssim 2.1\,M_\odot$ has revealed that the IFMR exhibits a non-monotonic component, with a peak of $M_f \approx 0.70-0.75\,M_\odot$ at $M_i \simeq 1.8-2.0\,M_\odot$. It happens just in proximity of the transition mass, $M_i \simeq M_{HeF}$, between low-mass stars that experience the He-flash in their degenerate He-cores at tip of the red giant bracng phase, and intermediate-mass stars that avoid electron degeneracy.

The proposed physical interpretation is that the IFMR kink marks the formation of solar-metallicity low-mass carbon stars. These latter experienced a shallow 3DU

($\lambda \simeq 0.1 - 0.3$) during the TP-AGB phase, so that the amount of carbon dust available to trigger a radiation-driven wind was small and $\dot{M}$ remained mostly below a few $10^{-6}\, M_{\odot}\, \mathrm{yr}^{-1}$. These circumstances led to a prolongation of the TP-AGB phase with the consequence that fairly massive WDs ($M_{\mathrm{f}} > 0.65\, M_{\odot}$), larger than commonly expected, were left at the end of the evolution. The progenitors that populate the IFMR kink are expected to be potentially important contributors to the galaxy emitted light and modest sources of carbon (in the form of gas and dust) at the same time.

In a systematic follow-up study we fully characterized the AGB star population in open clusters, exploiting the new *Gaia* data. Several findings emerge from our study. The minimum initial mass for carbon star formation at solar-like metallicity should not be higher than $\simeq 1.5\, M_{\odot}$, while the maximum mass should not be lower than $3.0 - 4.0\, M_{\odot}$. The 3 stars of type MS and S provide information about the onset of the 3DU and the transition to the C-star domain. The 12 carbon stars are all optically visible, and none appear truly dust-enshrouded. The mass-loss rate for most of them is very low ($\dot{M} \approx 10^{-8}\, M_{\odot}/\mathrm{yr}$), below the typical values that characterize a dust-driven wind, except for two carbon stars of low initial mass (V* V493 Mon and [W71b]).

The photometric variability data we retrieved suggest the stars in the sample are LPVs. The observed periods, in combination with derived absolute magnitudes, are consistent with Mira-like or semi-regular variability. Most of the C-stars appear to be fundamental mode pulsators, while M-, MS- and S-type stars pulsate predominantly in the first overtone mode (consistent with the fact that they are less evolved), except for the S-star S1* 338 whose primary period is attributed to pulsation in the second overtone mode.

The comparison of the estimated M_{c} with the IFMR of the white dwarfs has highlighted a striking fact: the presence of almost dust-free bright carbon stars with $0.65 \lesssim M_{\mathrm{c}}/M_{\odot} \lesssim 0.70$, and initial masses of $\approx 1.9 - 2.0\, M_{\odot}$.

Therefore, our study on AGB stars in open clusters (Marigo et al. (2022)) not only support the existence of the IFMR kink, but also the underlying interpretative hypotheses: the progenitors are 1) carbon stars that 2) experienced modest outflows for a significant fraction of their C-rich phase, 3) with inefficient dust production. In fact, the carbon stars MSB 75 and BM IV 90 ($L \approx 10\,000 - 13\,000\, L_{\odot}$), have an estimated mass-loss rate of $\approx$ few $10^{-8}\, M_{\odot}/\mathrm{yr}$, while their variability is characterized by low-amplitude pulsation. Overall, these new findings are of great significance to constrain the contribution from AGB stars to the chemical enrichment and integrated ligth of galaxies.

References

Bladh, S., Eriksson, K., Marigo, P., et al. 2019, *A&A*, 623, A119
Bloecker, T. 1995, *A&A*, 297, 727
Bressan, A., Marigo, P., Girardi, L., et al. 2012, *MNRAS*, 427, 127
Cantat-Gaudin, T., Anders, F., Castro-Ginard, A., et al. 2020, *A&A*, 640, A1
Cummings, J. D., Kalirai, J. S., Tremblay, P.-E., et al. 2018, *ApJ*, 866, 21
Eriksson, K., Nowotny, W., Höfner, S., et al. 2014, *A&A*, 566, A95
Ferrarotti, A. S. & Gail, H.-P. 2006, *A&A*, 447, 553
Groenewegen, M. A. T. 2021, *A&A*, 654, A20
Herwig, F. 2005, *ARAA*, 43, 435
Lambert, D. L., Gustafsson, B., Eriksson, K., et al. 1986, *ApJS*, 62, 373
McDonald, I. & Trabucchi, M. 2019, *MNRAS*, 484, 4678
Marigo, P., Bossini, D., Trabucchi, M., et al. 2022, *ApJS*, 258, 43
Marigo, P., Cummings, J. D., Curtis, J. L., et al. 2020, *Nature Astron.*, 4, 1102
Marigo, P., Bressan, A., Nanni, A., et al. 2013, *MNRAS*, 434, 488
Marigo, P. & Aringer, B. 2009, *A&A*, 508, 1539
Mattsson, L., Wahlin, R., & Höfner, S. 2010, *A&A*, 509, A14
Schöier, F. L. & Olofsson, H. 2001, *A&A*, 368, 969
Vassiliadis, E. & Wood, P. R. 1993, *ApJ*, 413, 641

The Origin of Outflows in Evolved Stars
Proceedings IAU Symposium No. 366, 2022
L. Decin, A. Zijlstra & C. Gielen, eds.
doi:10.1017/S1743921322001168

Mass transfer in AGB binaries - uncovering a new evolution channel by 3D radiation-hydrodynamic simulations

Zhuo Chen[1], **Natalia Ivanova**[2] **and Jonathan Carroll-Nellenback**[3]

[1]Department of Astronomy, Tsinghua University Beijing 100084, China
email: zc10@ualberta.ca

[2]Department of Physics, University of Alberta
Edmonton, AB T6G 2E1, Canada

[3]Department of Physics and Astronomy, University of Rochester Rochester, NY 14627, USA

Abstract. The origin of chemically peculiar stars and nonzero eccentricity in evolved close binaries have been long-standing problems in binary stellar evolution. Answers to these questions may trace back to an intense mass transfer during the asymptotic-giant-branch (AGB) binary phase. We use AstroBEAR to solve the 3D radiation hydrodynamic equations and calculate the mass transfer rate in AGB binaries that undergo the wind-Roche-lobe overflow or Bondi-Hoyle-Lyttleton (BHL) accretion. One of the goals of this work is to illustrate the transition from the wind- Roche-lobe overflow to BHL accretion. Both circumbinary disks and spiral structure outflows can appear in the simulations. As a result of enhanced mass transfer and angular momentum transfer, some AGB binaries may undergo orbit shrinkage, and some will expand. The high mass transfer efficiency is closely related to the presence of the circumbinary disks.

Keywords. Asymptotic giant branch; Symbiotic binary stars; Hydrodynamics

---

1. Introduction

Non-conservative mass transfer is an important but poorly understood picture in low-mass binary evolution. A low-mass binary may enter one or two asymptotic-giant-branch (AGB) binary phases as one of the stars ages. During this phase, the interaction between the stars becomes stronger because of the large size of the AGB star and the relatively slow (~ 10km/s) and dense AGB wind.

Previously, two fundamental non-conservative mass transfer modes during the AGB binary phase have been identified: Bondi-Hoyle-Littleton (BHL) accretion (Bondi & Hoyle 1944) and wind-Roche-lobe-overflow (WRLOF) (Podsiadlowski & Mohamed 2007). Without surprise, Mohamed & Podsiadlowski (2012) showed that WRLOF can enhance the mass transfer between the binary. Inspired by their work, we set up a 3D radiation hydrodynamic model and show that the WRLOF mass transfer mode may naturally lead to a new evolution morphology - circumbinary disks (CBDs). The inner radius of this kind of CBD is comparable to the binary separation. Possible candidates and examples of such CBDs include L2 Pup (Kervella et al. 2015) and AR Pup (Ertel et al. 2019).

2. Overview

The deciding factor in distinguishing the BHL accretion and WRLOF is the ratio ζ of the radius of the dust formation zone $r_{\rm dust}$ and the radius of the inner Roche-lobe $r_{\rm inner}$

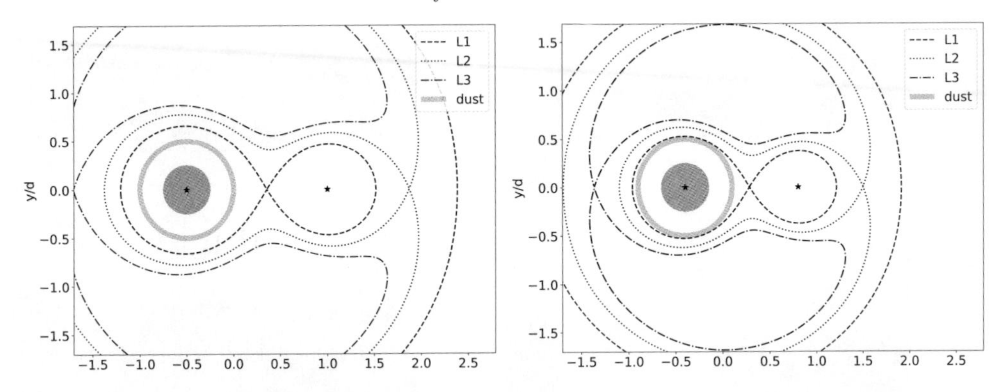

Figure 1. Illustrative plots of AGB binaries with L1, L2, and L3 potentials. Coordinate units are scaled to binary separations d. Left: AGB binary may undergo BHL accretion. Right: AGB binary will likely experience WRLOF. The grey rings represent the dust formation region near the AGB stars. The AGB star is represented by a big circle with a star at its center.

of the AGB star. When $\zeta < 1$, the BHL accretion occurs; when $\zeta \sim 1$, WRLOF may take place. Figure 1 illustrates the two scenarios.

The physical process that happen when $\zeta \sim 1$ is described as follows (see also Mohamed & Podsiadlowski (2012)):

(1) Pulsation near the surface of the AGB star drives a shock into its upper atmosphere, creating a dense and *shell* like environment.
(2) Dust condensates near $r_{\rm dust}$, the opacity of the gas and dust fluid increase by roughly 2 orders of magnitudes.
(3) The radiation pressure from the AGB star nearly balances its gravitational pull. The gas can become subsonic during this balanced period.
(4) The additional gravitational pull from the companion drags a significant amount of the gas of the *shell* toward it.
(5) The matter forms an accretion disk around the companion and is accreted onto the companion over time.

The oversimplified description of the process still points out several critical characteristics of WRLOF in AGB binaries. They are the pulsating AGB star, the dust formation zone, and the accretion disk around the companion. Together they should determine the mass transfer efficiency in large. However, the existence of a CBD cannot be straightforwardly inferred from this physical picture, and its existence will change the global evolution. To show that such a CBD may form, we dive into some physics.

One important physical aspect that is frequently missing in many AGB binary simulations is non-local radiation transfer. It is most convenient not to model the non-local radiation transfer if the outflow from the AGB binary is optically thin. On the contrary, as dust blocks light from the AGB star, the optical depth increases gradually and the balance between the radiation pressure and the gravitational force breaks down. An overall physical picture of the formation of CBD is described as follows.

(1) The AGB binary has a small binary separation so that the companion can pull the wind from higher latitudes to the equator, leading to a focused wind.
(2) The focused wind forms a dense tail behind the companion as the binary orbits.
(3) The tail blocks radiation from the AGB star and the dust in the shadow of the tail experiences smaller radiation pressure.

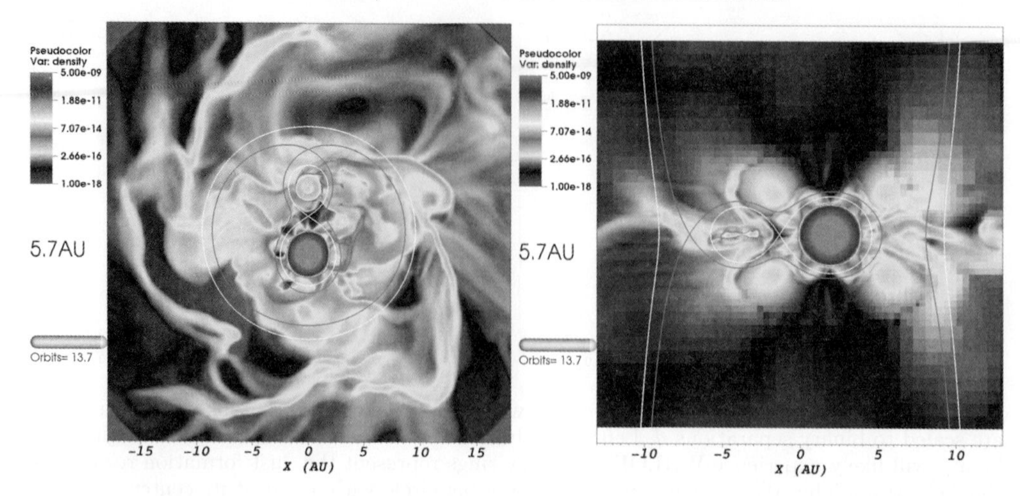

Figure 2. Left: equator density plot. Right: side view density plot. The binary separation is 5.7AU. The AGB star with a mass of $1.02\mathrm{M}_{\odot}$ is the red circle, which is pulsating. The companion, located above the AGB star, has a mass of $0.51\mathrm{M}_{\odot}$. An accretion disk forms around the companion and some gas material become gravitationally bound at a larger radius in the equator. The white and red contours are the L1 and L2 potential contours, respectively. The snapshot is taken at 13.7 binary orbits.

(4) Eventually, the radiation pressure cannot push the gas away from the AGB star and the companion. The gas starts to orbit the binary and forms a CBD.

We call this scenario "focused wind-induced CBD". We show in Figure 2 a snapshot of the simulations with a CBD formation, using a simulation presented in Chen, Ivanova, & Carroll-Nellenback (2020). A circumbinary structure as shown in the left panel forms; we think it represents the onset of the formation of a CBD.

A CBD may have many impacts on the evolution of the AGB binary. A direct result is the significantly enhanced mass transfer. Such an enhancement may be stronger than the normal WRLOF because the gas can stay in a CBD and be accreted by the companion after several orbits.

Another physical process that can become important after the formation of a CBD is angular momentum transfer between the binary and CBD (Muñoz, Miranda & Lai 2019).

Intense mass transfer not only changes the chemical composition of the companion, it may also have an impact on the long-term orbital evolution. In the mass transferring binary, the limiting cases of long-term evolution are the BHL accretion and conservative mass transfer. The BHL accretion incurs the least binary interaction and usually leads to a widening of the binary separation. On the other hand, the conservative mass transfer from the massive donor usually leads to a shrink of the binary separation. We plot $\dot{P}$ (P is the orbital period) of the two limiting cases in Figure 3 under the assumption that the mass-loss rate of the donor star is $\dot{M}_{\mathrm{AGB}} = 2.31 \times 10^{-7}\mathrm{M}_{\odot}\cdot\mathrm{year}^{-1}$, the same as in Chen et al. (2018). Simulated results are plotted as red stars and blue triangles. All simulations have a mass ratio $q = m_s/m_{\mathrm{AGB}} = 0.5$, where m_s and m_{AGB} are the mass of the companion and the AGB star, respectively.

Although the mass-loss rate and terminal wind speed of the AGB star of every simulation may be somewhat different, both results show a clear trend: $\dot{P}$ approaches the BHL accretion scenario when the binary separation is large and approaches the conservative mass transfer scenario when the binary separation is small. The transition of $\dot{P}$ from

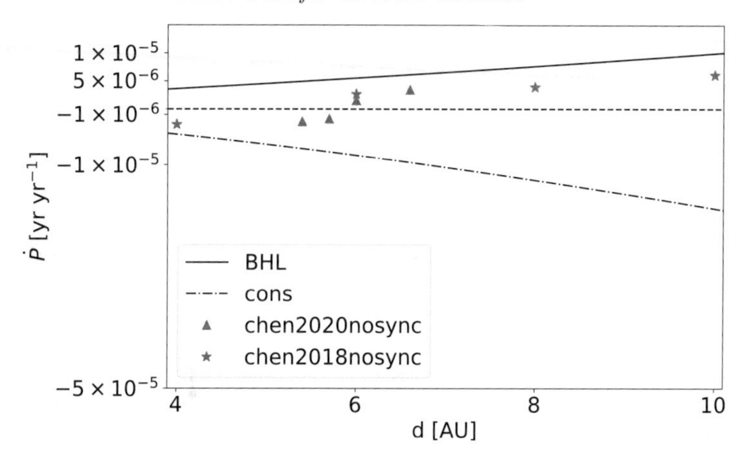

Figure 3. The solid line and dash-doted line show the $\dot{P}$ of BHL and conservative transfer case. The dashed line marks the $\dot{P} = 0$. The red stars and the blue triangles are the simulation results from Chen et al. (2018) and Chen, Ivanova, & Carroll-Nellenback (2020), respectively. "Nosync" means the AGB star does not rotate in the lab-frame by assumption.

positive to negative happens around 6AU in our models. Generally, this number should change if one adopts another AGB star model, which is worth further studying.

3. Challenges in AGB binary evolution

In this section, we discuss some of the major challenges in AGB binary simulations.

Viscous accretion disk. If there is no viscosity, the accretion rate will be extremely low. Viscosity in an accretion disk may come from magneto-rotational instability (Balbus & Hawley 1991), but we could also use an α viscous disk model (Shakura & Sunyaev 1973) for simplicity. Currently, the grid based code cannot explicitly specify the viscosity due to unknown numerical viscosity. Smooth-particle hydrodynamic (SPH) simulations can specify viscosity and we hope to see more realistic SPH simulations in the future.

Dust formation. Dust formation is strongly coupled to radiation transfer and shock physics (Höfner & Freytag 2019). First principle simulations that start from a chemical network would be computationally expensive. However, due to its importance in WRLOF, we expect this challenge to be the next to solve.

Coevolution of the CBD and binary. In our simulations, we cannot resolve the long-term dynamics between the CBD and the binary as the angular momentum conservation is not precise. However, as an initial attempt to uncover the existence of CBDs, we hope our work can provoke more new ideas and attention to CBDs around AGB binaries. The thermodynamics and angular momentum transfer between the binaries and CBDs are potentially interesting topics.

References

Balbus, S. A. and Hawley J. F. 1991, *ApJ*, 376, 214

Bondi, H. and Hoyle, F. 1944, *MNRAS*, 104, 273

Podsiadlowski, Ph. and Mohamed, S. 2007, *Baltic Astronomy*, 16, 26

Chen, Z., Blackman, E. G., Nordhaus, J., Frank, A., and Carroll-Nellenback, J. 2018, *MNRAS*, 473, 747

Chen, Z., Ivanova, N., and Carroll-Nellenback, J. 2020, *ApJ*, 892, 110

Ertel, S., Kamath, D., Hillen, M. et al 2019, *AJ*, 157, 110

Höfner, S. and Freytag, B. 2019, *A&A*, 623, A158
Kervella, P., Montargès, M., Lagadec, E., et al. 2015, *A&A*, 578, A77
Mohamed, S. and Podsiadlowski, Ph. 2012, *Baltic Astronomy*, 21, 88
Muñoz, D., Miranda, R. and Lai, D. 2019, *ApJ*, 871, 84
Shakura, N. I. and Sunyaev, R. A. 1973, *A&A*, 500, 33

Discussion

ORSOLA: You have been mostly working on relatively wide binaries and binary interactions that leads to CBDs, can you venture to guess what differences the CBDs could have if they were form via closer binaries?

ZHUO: Thank you very much for the question. You just mentioned a very good research project. If the binary is at the very edge of Roch-lobe overflow, we may find an accretion disk and a CBD at a larger radius. The qualitative morphology may not change much but the thermodynamics and radiation hydrodynamics should be more violent in those binaries.

The Origin of Outflows in Evolved Stars
Proceedings IAU Symposium No. 366, 2022
L. Decin, A. Zijlstra & C. Gielen, eds.
doi:10.1017/S1743921322000217

Route towards complete 3D hydro-chemical simulations of companion-perturbed AGB outflows

Silke Maes[1], Lionel Siess[2], Ward Homan[2], Jolien Malfait[1], Frederik De Ceuster[1,3], Thomas Ceulemans[1], Dion Donné[1], Mats Esseldeurs[1] and Leen Decin[1,4]

[1]Instituut voor Sterrenkunde, KU Leuven, Celestijnenlaan 200D, 3001 Leuven, Belgium
email: silke.maes@kuleuven.be

[2]Institut d'Astronomie et d'Astrophysique, Université Libre de Bruxelles (ULB), CP 226, 1050 Brussels, Belgium

[3]Department of Physics and Astronomy, University College London, Gower Place, London, WC1E 6BT, UK

[4]School of Chemistry, University of Leeds, Leeds LS2 9JT, UK

Abstract. Low- and intermediate mass stars experience a significant mass loss during the last phases of their evolution, which obscures them in a vast, dusty envelope. Although it has long been thought this envelope is generally spherically symmetric in shape, recent high-resolution observations find that most of these stars exhibit complex and asymmetrical morphologies, most likely resulting from binary interaction. In order to improve our understanding about these systems, theoretical studies are needed in the form of numerical simulations. Currently, a handful of simulations exist, albeit they mainly focus on the hydrodynamics of the outflow. Hence, we here present the pathway to more detailed and accurate modelling of companion-perturbed outflows with PHANTOM, by discussing the missing but crucial physical and chemical processes. With these state-of-the-art simulations we aim to make a direct comparison with observations to unveil the true identity on the embedded systems.

Keywords. Stars: AGB – Stars: winds, outflows – Methods: numerical

---

1. Introduction

Asymptotic giant branch (AGB) stars are cool, evolved stars with an initial mass between about 0.8 and 8 solar masses, experiencing a strong mass loss due to a stellar wind (e.g. Ramstedt *et al.* 2009). The launching mechanism of the wind is believed to be caused by surface pulsations in combination with dust formation. Accordingly, microscopic processes initiate a macro-scale movement of the surface material (Höfner & Olofsson 2018). In short, pulsational instabilities inside the AGB star create shocks that propagate through the stellar atmosphere, lifting dense material to cooler regions. Hence, favourable conditions arise for dust grains to condense from molecules. The grains are able to absorb the infrared radiation coming from the AGB star, contrary to the molecules, thereby gaining radial momentum and accelerating outwards. Along their way, they collide with the gas molecules, transfer momentum and drag them along. Hence, a macroscopic outward flow emerges (Liljegren *et al.* 2016; Freytag *et al.* 2017). This type of wind is commonly called 'dust-driven'. Consequently, AGB stars are embedded in a vast and dense, dusty envelope and exhibit a rich chemistry (e.g. Gail & Sedlmayr 2013).

These dusty envelopes have long thought to be spherically symmetric in shape. However, recent high-resolution observations reveal the presence of a variety of asymmetric patterns, such as spirals, arcs, disks, and bipolarity (e.g. Mauron & Huggins 2006; Decin *et al.* 2012, 2020). The observed morphologies show a large resemblance with structures found in post-AGB stars (e.g. Cohen *et al.* 2004) and planetary nebulae (e.g. Guerrero *et al.* 2003), suggesting a similar formation mechanism along this evolutionary sequence. It is believed that the morphologies of the AGB environment stem from the interaction with an undetected binary companion, orbiting in the vast stellar outflow and hence shaping it (Decin *et al.* 2020).

Nevertheless, a lot remains to be known about the interaction between the AGB outflow and the binary companion, such as its effects on the chemical composition, mass-loss rate, orbital evolution, etc. A complementary approach, including observational surveys and theoretical/numerical studies, is needed to uncover the true identity of the AGB star and its close surroundings. Over the past few decades companion-perturbed AGB outflows have been modelled (see e.g. Theuns & Jorissen 1993; Mastrodemos & Morris 1999; Chen *et al.* 2017), but a complete, systematic survey is still lacking. Hence, here we take a step forwards on the numerical side and present a route towards complete 3D hydro-chemical simulations.

2. Current numerical models

For our simulations, we use the 3D smoothed particle hydrodynamics (SPH) code PHANTOM, developed by Price *et al.* (2018) and adapted by Siess *et al.* (2022) to include stellar wind physics. Currently, the outflow is described as a polytropic gas and evolves purely according to the laws of hydrodynamics, while being influenced by the orbital motion and the gravitational potential of the two components. The wind is launched in a radiation-free way, meaning that the gravity of the AGB star is artificially reduced for the gas particles, by including the parameter α (Eq. 2.1). By setting α to 1, a macroscopic gravity-free wind is launched from the surface of the AGB star.

$$F_r = -\frac{GM_{\rm AGB}}{r_1^2}(1-\alpha) - \frac{GM_{\rm comp}}{r_2^2} \tag{2.1}$$

In Eq. (2.1) the radial force term of the general equation of motion is given, where r_1 and r_2 are the distances to the AGB star and the companion, respectively, and $M_{\rm AGB}$ and $M_{\rm comp}$ their masses, G is the gravitational constant.

For this basic setup, already interesting results are obtained. Malfait *et al.* (2021) described in detail the cause of the complexity of a variety of morphological structures in systems of different outflow velocity and orbital setup, and identified the distinction between an 'equatorial density enhancement' and a 'global flattening' of the circumstellar envelope. The former is a focussing of the wind towards the orbital plane due to the gravitational pull of the companion, the latter is due to the centrifugal force resulting from the orbital motion of the AGB star. Maes *et al.* (2021) showed the growth in morphological complexity with increasing companion mass, decreasing orbital separation, and decreasing outflow velocity (Fig. 1) agreeing with earlier studies (e.g. Mastrodemos & Morris 1999), and found that the velocity evolution in the outflow is governed by the gravitational slingshot when the companion is massive enough. They also introduced the classification parameter ε, which helps to predict the the system's morphology.

Although these results are able to account for most of the observed morphologies, it is now the time to go beyond hydrodynamics-only models. We aim to implement the missing physical and chemical processes of companion-perturbed AGB outflows, following a step-by-step approach.

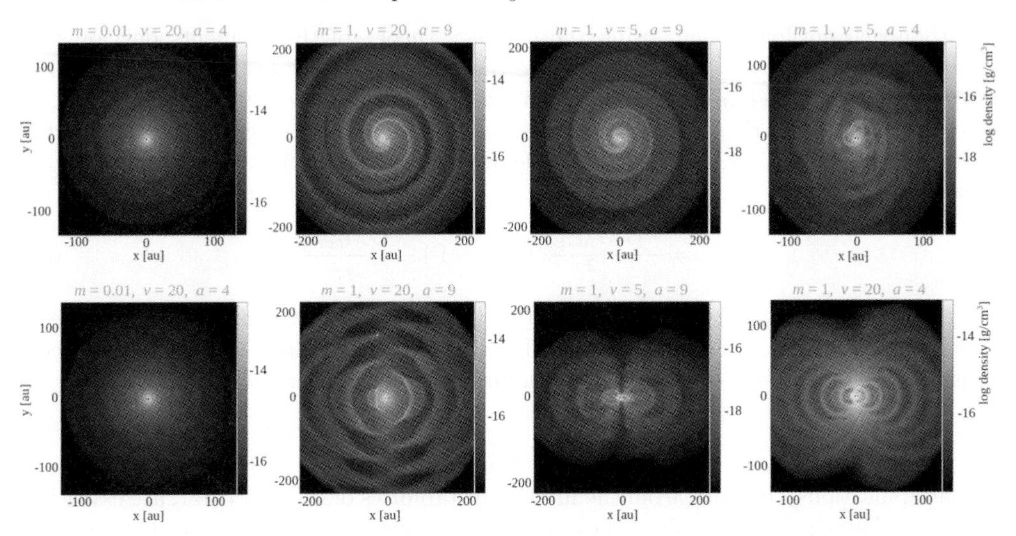

Figure 1. Gallery of companion-perturbed AGB outflow simulations, adapted from Maes *et al.* (2021). The upper row presents the density distribution in the orbital plane (x, y), the bottom row in the meridional plane (x, z). The left black dot indicates the position of an AGB star of mass $1.5\,\mathrm{M}_{\odot}$, the right dot locates the companion with mass m in $\mathrm{M}_{\odot}$ as specified in the label. v gives the initial outflow velocity of the AGB star in $\mathrm{km\,s^{-1}}$ and a the orbital separation in astronomical units. All orbits are circular.

3. Implementing missing ingredients

In general, AGB outflows emerge and evolve as a result of an interplay of dynamical, radiative, and chemical processes. Although these three pillars are each essential in order to correctly model outflows, implementing them all together can be very computationally demanding, especially in 3D. However, with advancing numerical techniques and computing power, we are now able to unite these three pillars.

There are two main mechanisms upon which we need to improve, namely the wind-launching and the evolution of the outflow itself. For the former, the crucial elements are the formation of dust grains and surface pulsations, since these initiate the outflow. For the latter, the change of radiative acceleration throughout the outflow, accurate cooling, and the chemical diversity need to be included, in order to be able to make a one-on-one comparison with observations. These different 'ingredients' are shortly discussed in the next sections.

3.1. *Wind acceleration by dust*

As explained in Sect. 2, the wind is launched by setting the parameter α to 1, to cancel out the gravitational potential of the AGB star. Siess *et al.* (2022) are now improving the wind-launching mechanism by including a treatment of dust nucleation and dust growth. Contrary to the previous approach, the wind is accelerated via the factor Γ, which represents the ratio of radiative and gravitational accelerations assuming an optical thin wind. The expression for Γ is given by

$$\Gamma = \frac{(\kappa_{\mathrm{dust}} + \kappa_{\mathrm{gas}})L_{\mathrm{AGB}}}{4\pi c G M_{\mathrm{AGB}}}, \tag{3.1}$$

where κ_{dust} and κ_{gas} are the dust and gas opacities, respectively, and L_{AGB} is the AGB star's luminosity.

The computation of Γ requires to solve a small chemical network in order to get the abundances of the carbon-bearing molecules that constitute the dust building block in a carbon-rich outflow. Then, using the complex theory of the moments developed by Gail & Sedlmayr (2013), one can estimate the nucleation and dust growth rates which are used to evolve the moments of the grain size distribution. According to this theory, the dust opacity $\kappa_{\rm dust}$ is then proportional to the 3rd moment of the grain size distribution (details in Siess *et al.* 2022). Γ is then implemented in the equation of motion in a similar way as the radiation-free wind:

$$F_r = -\frac{GM_{\rm AGB}}{r_1^2}(1-\alpha-\Gamma) - \frac{GM_{\rm comp}}{r_2^2}. \tag{3.2}$$

Dust accelaration is effective only beyond the dust condensation radius, which is typically a few times the stellar radius. Hence, within the first few stellar radii in the absence of pulsations, the wind is accelerated in a radiation-free way with $\alpha = 1$.

3.2. *Surface pulsations*

Surface pulsations, resulting from the interior instabilities of the AGB star, are essentially needed to launch the wind, since they cause the favourable conditions for dust grains to form. However in this study, the interior of the AGB star is not included in the modelling. Hence, we aim to use an approximate prescription for the dynamics of the surface of the AGB star, build upon the results obtained by Freytag *et al.* (2017)'s 'star in a box' simulations. As we expect this adaptation to the PHANTOM code will be relatively simple and will not drastically change the outcome, this ingredient will be implemented in one of the last stages.

3.3. *Radiative transfer*

Radiative transfer throughout the outflow is not taken into account in the current PHANTOM models. Approximately, this can be stated as assuming an optically thin regime throughout the outflow. However, if the wind material is accumulating due to the interaction with the companion, the opacity can locally change drastically, which in turn affects the structure of the outflow. When radiative transfer is carefully accounted for, this can give rise to the formation of rotating circumbinary structures and possibly disks (Chen *et al.* 2017, 2020). Such circumbinary disks are common around post-AGB systems (e.g. Van Winckel 2003), and recently have also been observed around AGB binary systems (e.g. Homan *et al.* 2017). Moreover, in current simulations, the so-called 'equatorial density enhancements' found by El Mellah et al. (2020), Maes *et al.* (2021) and Malfait *et al.* (2021) could be indications of the precursor of circumbinary disks. Therefore, it is important to include radiative transfer in PHANTOM.

In order to do so, we will use the 3D radiative transfer solver MAGRITTE, developed by De Ceuster *et al.* (2020a,b). We aim to extract MAGRITTE's ray tracer and plug it in PHANTOM, so that the degree by which the stellar light is attenuated can be calculated and we can improve on the optically thin assumption.

3.4. *Chemistry*

Accurate chemistry needs to be coupled to the hydrodynamics simulations, in order to investigate the effect of the companion on the 3D chemical structure of the outflow. Up until now, AGB chemistry has only been studied in a 1D-context, thus assuming a spherically symmetric outflow. Recently, Van de Sande *et al.* (2018) considered the effects of a porous outflow on the chemical abundances and Van de Sande & Millar (2022) of different companion types. Although restricted to a 1D-framework, these studies showed crucial changes compared to spherical symmetry, e.g. an overall increase in abundances of unexpected species (i.e. not predicted by thermodynamic equilibrium) due to the larger UV radiation field (because of the permeability of the porous medium or the radiating companion, respectively) in the inner wind. Thus, this demonstrates the need to step away from the 1D-context and to investigate AGB chemistry properly in 3D.

In AGB stars, surface pulsations take the chemistry out of equilibrium (e.g. Cherchneff 2006) so that local thermodynamic equilibrium (LTE) does not hold and chemical equilibrium models cannot be used to determine the abundances. Therefore, chemical kinetics methods have to be used instead. However, they require to evolve a chemical reaction network which contributes significantly to the increase in storage space and computation time. Hence, only for a small, approximate reaction network this approach is feasible in 3D, but here we aim to include an extended chemical network in order to carry out a thorough investigation. Luckily with the current advances in computational techniques, this issue can be overcome.

Indeed, machine learning can be used to construct a 'chemistry emulator' that we will be trained to produce the same results as a chemical kinetics code, only in a much faster way. It has been shown by de Mijolla *et al.* (2019) that emulation can be used to speed up chemistry calculations up to a factor of 10^5. Moreover, this technique is already been used to calculate chemistry in dark clouds (Holdship *et al.* 2021). We plan to construct a large grid of 1D chemical kinetics simulations, covering the parameter space found in AGB outflows with a chemical network containing over 400 species and 6000 reactions, based on the UMIST database RATE12 (McElroy *et al.* 2013). The grid will serve as the input to train an artificial neural network, so that the emulator learns to predict the chemistry without having to solve explicitly the kinetics equations.

Once the chemistry emulator has been tested and its performance validated, it will be coupled to PHANTOM. At every timestep in the hydrodynamics calculations, the emulator will be called and will receive a certain density, temperature, and chemical abundances, so that it is able to update the abundances together with the temperature and the polytropic index γ (more in Sect. 3.5). The latter two are needed to account for chemical heat exchange. These updates will be fed back to PHANTOM, so that PHANTOM can evolve the AGB wind with the updated values.

3.5. *Cooling*

Currently, the heat exchange in the outflow is accounted for via the polytropic index of the gas, γ:

$$T \propto \rho^{\gamma-1}, \tag{3.3}$$

where T is the gas temperature and ρ the density. However, in reality it is known that additional processes can contribute significantly, for example radiative cooling by line transitions, chemical, and continuum cooling, which alter the internal energy of the outflow. This again can have a strong effect on the morphological structure. For example, Theuns & Jorissen (1993) showed that some degree of cooling is needed in order to form an accretion disks around the companion.

An approximate cooling formalism has been implemented in PHANTOM based on the cooling tables provided by Omukai *et al.* (2010) that consider cooling via atomic and molecular line photons, cooling by continuum radiation, and the chemical cooling/heating (Donné 2021). PHANTOM also includes a basic HI-cooling prescription (Spitzer & Jura 1978) and when this option is activated, an accretion disk naturally forms around the companion, as shown in Fig. 2.

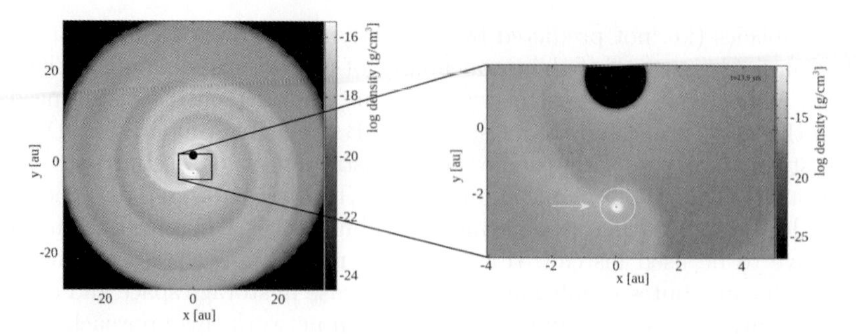

Figure 2. Density distribution in the orbital plane of a PHANTOM simulation of the outflow of an AGB star with mass $1.5\,M_\odot$ and a $1.0\,M_\odot$-companion, including HI-cooling. The circular orbit has a separation of 4 au and the initial outflow velocity is $7.0\,\mathrm{km\,s^{-1}}$. The right panel shows a zoom-in on the companion, where an accretion disk is visible in the circle indicated by the arrow. Simulation made by D. Donné.

In order to implement cooling in a more accurate and systematic way, our project envisions to use MAGRITTE to calculate the line cooling for some of our 3D hydrodynamics simulations and use this to train another emulator to simulate the line cooling, in a similar way as will be done for the chemistry. Lastly, the chemical cooling will be included in the chemistry emulator as mentioned in the previous section, by accounting for the reaction enthalpies in an additional chemistry equation.

4. Conclusions

To summarise, we are currently working on improving our numerical models of companion-perturbed AGB outflows to better describe the physical reality, using the SPH code PHANTOM. Different missing, but crucial, ingredients (including a dust and pulsations wind-launching prescription, and treatments for radiative transfer, cooling and chemistry throughout the outflow) are being developed and implemented in PHANTOM.

The ultimate goal is to construct a large grid of complete 3D hydro-chemical-radiation simulations of companion-perturbed AGB outflows, that can serve as a theoretical context in which the existing observational data can be interpreted. This grid will provide a basis on which our understanding of the shaping and evolution of these objects can be further developed. Moreover, these simulations will yield a framework for new estimates of the mass-loss rate of AGB stars and mass-accretion rates on the companion, as well as rates of angular momentum transfer that are needed to explain for example the orbital evolution properties (as in particular the eccentricity) of chemical peculiar systems, such as barium stars or carbon-enhanced metal-poor (CEMP) stars.

References

Chen, Z., Frank, A., Blackman, E. G. *et al.* 2017, *MNRAS*, 468, 4465

Chen, Z., Ivanova, N., & Carroll-Nellenback, J. 2020, *ApJ*, 892, 110

Cherchneff, I. 2006, *A&A*, 456, 1001

Cohen, M., Van Winckel, H., Bond, H. E. *et al.* 2004, AJ, 127, 2362

De Ceuster, F., Homan, W., Yates, J. *et al.* 2020, *MNRAS*, 492, 1812

De Ceuster, F., Bolte, J., Homan, W. *et al.* 2020, *MNRAS*, 499, 5194

Decin, L., Cox, N. L. J., Royer, P. *et al.* 2012, *A&A*, 548, A113

Decin, L., Montargès M., Richards, A. M. S. *et al.* 2020, *Science*, 369, 1497

de Mijolla, D., Viti, S., Holdship, J. *et al.* 2019, *A&A*, 630, A117

Donné, D. 2021, *Master's thesis*, KU Leuven, Belgium

El Mellah, I., Bolte, J., Decin, L. *et al.* 2020, *A&A*, 637, A91

Freytag, B., Liljegren, S., & Höfner, S. 2017, *A&A*, 600, A137

Gail, H. P. & Sedlmayr, E. 1988, *A&A*, 206, 153
Gail, H. P. & Sedlmayr, E. 2013, *Physics and Chemistry of Circumstellar Dust Shells*
Guerrero, M. A., Chu, Y. H., Manchado, A. *et al.* 2003, *AJ*, 125, 3213
Höfner, S. & Olofsson, H. 2018, *A&AR*, 26, 1
Holdship, J., Viti, S., Haworth, T. J. *et al.* 2021, *A&A*, 653, A76
Homan, W., Richards, A., Decin, L. *et al.* 2017, *A&A*, 601, A5
Liljegren, S., Höfner, S., Nowotny, W. *et al.* 2016, *A&A*, 589, A130
Maes, S., Homan, W., Malfait, J. *et al.* 2021, *A&A*, 653, A25
Malfait, J., Homan, W., Maes, S. *et al.* 2021, *A&A*, 652, A51
Mastrodemos, N. & Morris, M. 1999, *ApJ*, 523, 357
Mauron, N. & Huggins, P. J. 2006, *A&A*, 452, 257
McElroy, D., Walsh, C., Markwick, A. J. *et al.* 2013, *A&A*, 550, A3
Omukai, K., Hosokawa, T., & Yoshida, N. 2010, *ApJ*, 722(2), 1793.
Price, D. J., Wurster, J., Tricco, T. S. *et al.* 2018, *PASA*, 35, e031
Ramstedt, S., Schöier, F. L., & Olofsson 2009, *A&A*, 499, 515
Siess, L., Homan, W., Toupin, S. *et al.* 2022, *submitted to A&A*
Spitzer, L. & Jura, M. 1978, *Physics Today*, 31(7), 48.
Theuns, T. & Jorissen, A. 1993, *MNRAS*, 265, 946
Van de Sande, M., Sundqvist, J. O., Millar, T. J. *et al.* 2018, *A&A*, 616, A106
Van de Sande, M. & Millar, T. J. 2022, *MNRAS*, 510, 1204
Van Winckel, H. 2003, *ARA&A*, 41, 391

The Origin of Outflows in Evolved Stars
Proceedings IAU Symposium No. 366, 2022
L. Decin, A. Zijlstra & C. Gielen, eds
doi:10.1017/S1743921322000229

3D hydrodynamical survey of the impact of a companion on the morphology and dynamics of AGB outflows

Jolien Malfait[1], Silke Maes[1], Ward Homan[2], Jan Bolte[1], Lionel Siess[2], Frederik De Ceuster[1,3] and Leen Decin[1,4]

[1]Institute of Astronomy, KU Leuven,
Celestijnenlaan 200D, 3001 Leuven, Belgium

[2]Institut d'Astronomie et d'Astrophysique, Université Libre de Bruxelles (ULB),
CP 226, 1050 Brussels, Belgium

[3]Department of Physics and Astronomy, University College London,
Gower Place, London, WC1E 6BT, United Kingdom

[4]School of Chemistry, University of Leeds,
Leeds LS2 9JT, United Kingdom
email: jolien.malfait@kuleuven.be

Abstract. With the use of high-resolution ALMA observations, complex structures that resemble those observed in post-AGB stars and planetary nebulae are detected in the circumstellar envelopes of low-mass evolved stars. These deviations from spherical symmetry are believed to be caused primarily by the interaction with a companion star or planet. With the use of three-dimensional hydrodynamic simulations, we study the impact of a binary companion on the wind morphology and dynamics of an AGB outflow. We classify the wind structures and morphology that form in these simulations with the use of a classification parameter, constructed with characteristic parameters of the binary configuration. Finally we conclude that the companion alters the wind expansion velocity through the slingshot mechanism, if it is massive enough.

Keywords. Stars: AGB – Stars: winds, outflows – Hydrodynamics – Methods: numerical

1. Introduction

Low- to intermediate-mass stars shed their outer layers during the asymptotic giant branch (AGB) evolutionary phase through a dust-driven pulsation-enhanced wind (Lamers & Cassinelli 1999; Höfner & Olofsson 2018). High-resolution observations reveal that these outflows contain a large diversity of complex structures, such as spirals, arcs, bipolarity, disks, etc. (Ramstedt *et al.* 2014; Kervella *et al.* 2016; Decin *et al.* 2020; Homan *et al.* 2020a,b). These observed AGB circumstellar envelopes (CSE) resemble the morphologies of planetary nebulae (PNe), and thereby help us fill the current knowledge gap about how the complex-structured planetary nebulae are shaped (Sahai *et al.* 2011). The observed structures, that make the AGB wind deviate from spherical symmetry, are believed to be formed primarily by the interaction of the wind with a companion star or planet, that often remains undetected (Decin *et al.* 2020). A better understanding of how a companion can shape the winds of evolved stars is needed, since not accounting for the three-dimensional structures and the impact of a companion may lead to systematic errors in the estimate of critical stellar parameters such as the mass-loss rate.

Table 1. Characteristic model input parameters.

$v_{\rm ini}$ [km s^{-1}]	$M_{\rm comp}$ [M$_\odot$]	a [au]	e
5	1	2.5	0.00
10	0.01	4.0	0.25
20		6.0	0.50
		9.0	

Three-dimensional hydrodynamic simulations confirm that complex structures such as spirals and arcs form in stellar outflows when the impact of a companion is taken into account (Theuns & Jorissen 1993; Theuns *et al.* 1996; Mastrodemos & Morris 1998). Depending on the wind characteristics and properties of the binary system, flattened or bipolar morphologies, and density enhancements around the orbital plane are predicted to form (Mastrodemos & Morris 1999; Kim & Taam 2012; El Mellah et al. 2020). To improve our understanding on which binary and wind configurations create which type of wind structures and global morphologies, additional studies of high-resolution 3D hydrodynamic simulations are required. Here we discuss the main findings of such a study by Malfait *et al.* (2021) and Maes *et al.* (2021), in which the wind structure formation of a set of simulations is studied in detail.

2. Model grid

The simulations are constructed with the three-dimensional smoothed-particle hydrodynamic (SPH) code PHANTOM (Price *et al.* 2018), which solves the fluid dynamic equations in a mesh-free way. The models are purely hydrodynamic, without the inclusion of dust, chemistry, radiation, and pulsations, and the cooling is regulated by the polytropic equation of state for an ideal gas, given by

$$P = (\gamma - 1)\rho u, \tag{2.4}$$

with polytropic index $\gamma = 1.2$, and in which P is the pressure, ρ the gas density, and u the specific internal energy. To improve these models and study the impact of dust, chemistry, radiation, pulsations and cooling, these missing ingredients are currently being implemented into PHANTOM by L. Siess, W. Homan and collaborators. The simulations contain an AGB star with mass $M_{\rm AGB} = 1.5\,{\rm M}_\odot$ that launches a wind of SPH gas particles. The grid of models consist of simulations characterised by a specific initial velocity $v_{\rm ini}$, companion mass $M_{\rm comp}$, orbital separation a, and orbital eccentricity e, as indicated in Table 1. The detailed setup of these simulations, together with an analyses of their wind morphology and dynamics, is described by Malfait *et al.* (2021) and Maes *et al.* (2021).

3. Morphology classification

The companion shapes the AGB wind morphology by two primary effects, namely the induced orbital motion of both components around their center-of-mass (CoM), and the gravitational attraction of wind particles. The physical properties of the binary system and the AGB wind will determine the relative strength of these effects and the resulting global shape of the outflow. In general, the perturbation by the companion is stronger when its mass $M_{\rm comp}$ is large, the orbital separation a small, the AGB wind velocity $v_{\rm ini}$ low, and the eccentricity e high. To estimate the degree of complexity induced by the companion, these parameters are combined in a classification parameter ε, which is defined as

$$\varepsilon = \frac{e_{\rm grav}}{e_{\rm kin}} = \frac{\frac{GM_{\rm comp}\rho}{R_{\rm Hill}}}{\frac{1}{2}\rho v_{\rm w}^2} = \frac{(24G^3 M_{\rm comp}^2 M_{\rm AGB})^{1/3}}{v_{\rm w}^2 a(1-e)}, \tag{3.1}$$

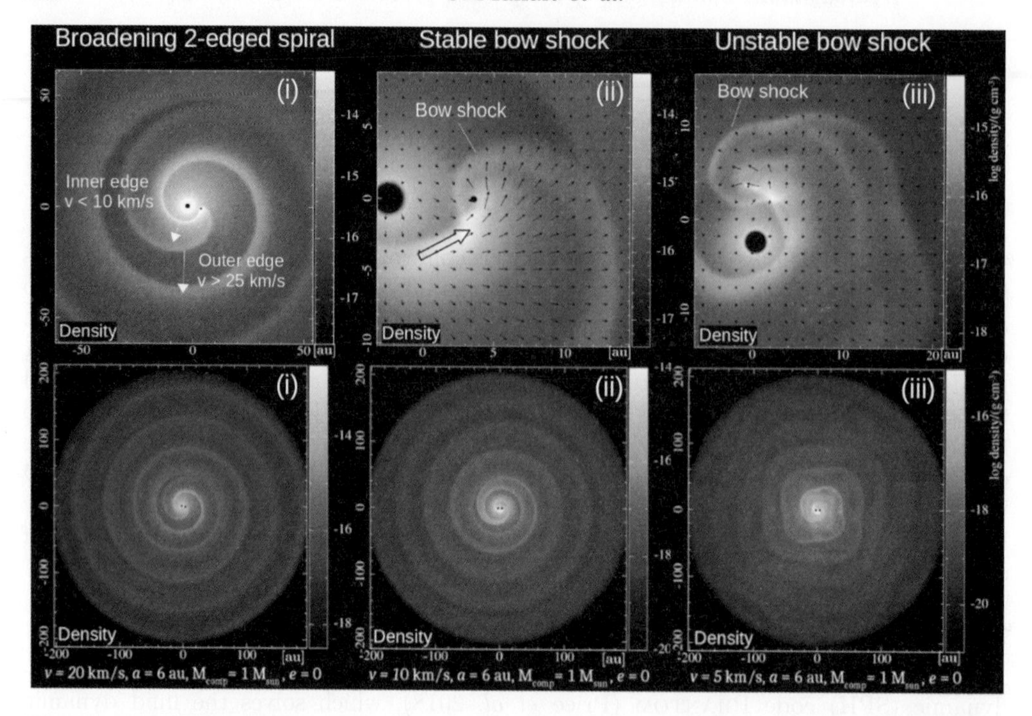

Figure 1. Density distribution in a slice through the orbital plane of three simulations with $a = 6\,\mathrm{au}$, $M_{\mathrm{comp}} = 1\mathrm{M}_{\odot}$, $e = 0$ and from left to right $v_{\mathrm{ini}} = 20, 10, 5\,\mathrm{km\,s^{-1}}$. The upper layer shows the inner density structures formed around the companion, the lower layer shows the global orbital plane morphology. Figure adapted from Malfait *et al.* (2021).

so the ratio of the gravitational energy density of the companion to the kinetic energy of the wind (Maes *et al.* 2021). A low ε value corresponds to a limited impact of the companion on the wind dynamics and morphology, whereas high ε values indicate that the wind will be strongly perturbed.

3.1. *Wind structure around companion*

The simulations are categorised according to the inner wind structure, that forms around the companion star and shapes the global wind morphology, which becomes more complex for increasing ε value. Note that the exact ε values delimiting these three categories are uncertain, and based on the available simulations of the studies by Malfait *et al.* (2021) and Maes *et al.* (2021).

By studying the wind structures in a slice through the orbital plane of the 3D morphology, we find the following classification: (i) For configurations with a classification parameter $\varepsilon \lesssim 1$ (illustrated in the left column of Fig. 1), a broadening spiral structure forms attached to the companion, that is delimited by a slow, dense inner edge and a higher-velocity, less dense outer edge. This inner wind structure shapes the outflow into an approximate Archimedes global spiral structure. (ii) In case of a stronger wind-companion interaction intensity, so higher ε-value (illustrated in the second column of Fig. 1), there is one dense spiral flow behind the companion and a second spiral emerging from a bow shock in front of the moving companion. This stable bow shock again shapes the global morphology into an approximate Archimedes spiral structure. (iii) In the simulations in which $\varepsilon \gtrsim 3$ (illustrated in the right column of Fig. 1), an unstable bow shock forms in front of the companion, which results in a global morphology with irregular spiral structures.

Table 2. Morphology classification.

Model	Global density distribution		Meridional plane structure	Orbital Plane structure	ε
v20e00	Flattened	no EDE	Concentric arcs	Spiral - Archimedes	1.0
v20e25	Flattened, asymmetric	no EDE	Arcs	Spiral - Perturbed	1.3
v20e50	Flattened, asymmetric	no EDE	Ring-arcs	Spiral - Perturbed	2.0
v10e00	Flattened	with EDE	Bicentric rings - Peanut-shape	Spiral - Archimedes	2.6
v10e25	No flattening, irregular	with EDE	irregular	irregular	3.4
v10e50	No flattening, irregular	with EDE	irregular	irregular	5.1
v05e00	No flattening, irregular	with EDE	Rose	Spiral - Squared	4.1
v05e25	No flattening, irregular	with EDE	Bipolar outflow	irregular	5.5
v05e50	No flattening, irregular	with EDE	Bipolar outflow	irregular	8.3

Notes:
Morphology classification of the global density distribution, meridional plane and orbital plane structures, and value of classification parameter ε (Eq. 3.1) of simulations with orbital separation $a = 6$ au and $M_{\rm comp} = 1 M_{\odot}$. The model names give the values of the input wind velocity and eccentrictiy, with 'vXX' denoting the input wind velocity in km s^{-1} and 'eXX' the value of the eccentricity of the system multiplied by a factor 100. Table adapted from Malfait *et al.* (2021).

In general more complex wind morphologies result in case the orbit is eccentric, as the phase-dependency makes the inner wind structure vary between the three types of inner wind structures described above throughout one orbital period. The details of how these different inner wind structures are formed and how they result in different global morphologies for both circular and eccentric configurations is described in detail by Malfait *et al.* (2021).

3.2. *Vertical wind extent & distribution*

Next, the simulations can be categorised according to their three-dimensional density distribution. In the successors of AGB stars, being Post-AGB stars and PNe, circumbinary disks and bipolar outflows are observed, of which the formation mechanism is still uncertain (Van Winckel 2003; Bujarrabel *et al.* 2013; Oomen *et al.* 2020; Manick *et al.* 2021). Studying the vertical extent of AGB winds may provide important information about the origin of these circumbinary disks and bipolar outflows.

There are two effects that can make the global wind distribution deviate from spherical symmetry. Firstly, the orbital movement of the stars around the CoM induces a centrifugal force on the wind particles. This force gives the wind particles an additional acceleration in the orbital plane direction, which causes an elongation of the entire morphology, which we will refer to as flattening. Secondly, while the companion moves on its orbit, it gravitationally attracts matter. If this effect is strong, this can result in a density enhancement around the orbital plane, referred to as an equatorial density enhancement (EDE). We define an EDE to be present if the density around the orbital plane is strongly enhanced with the respect to a non-perturbed isotropic single star simulation, and if the density around the poles is decreased with respect to a single star simulation. A more detailed explanation on how to determine if an EDE or flattening is present in the 3D simulations can be found in Malfait *et al.* (2021) and Maes *et al.* (2021).

Table 2, adapted from Malfait *et al.* (2021), illustrates that there is a flattening present in simulations where the gravitational pull of the companion is limited ($\varepsilon \lesssim 1$), and an EDE without flattening is found in more complex-structured simulations in which there is a strong gravitational impact of the companion ($\varepsilon \gtrsim 3$). Hence, this indicates that the impact of the orbital motion of the AGB star dominates when $\varepsilon \lesssim 1$, and the impact of the gravitational attraction of matter by the companion dominates in the simulations with high ε. Furthermore, it is important to note that an EDE and flattening occur for different simulation setups, and thereby a distinction should always be made.

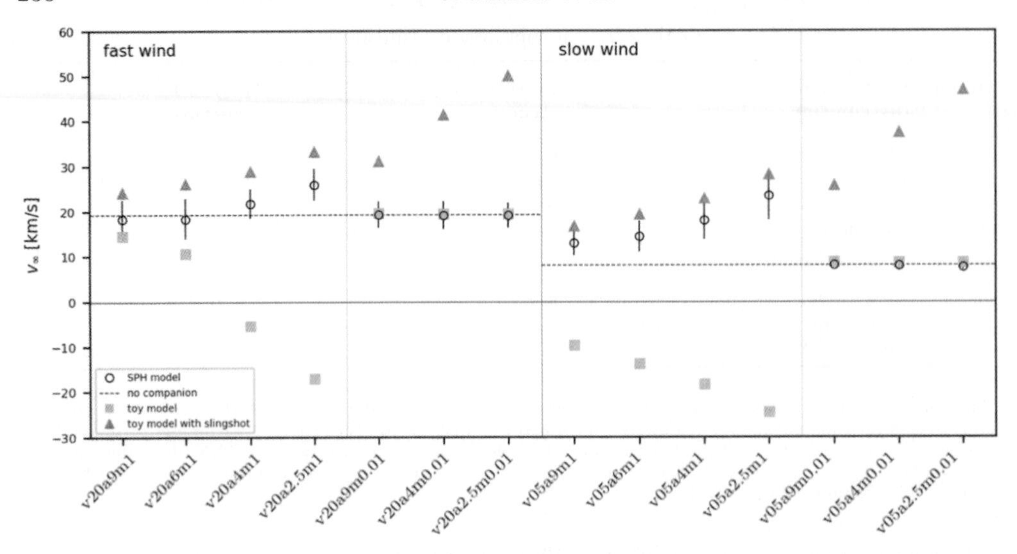

Figure 2. Terminal velocity for simulations with eccentricity $e = 0$, as calculated from the toy model with and without a slingshot in triangles and squares, respectively, and from the simulation in empty circles. The terminal velocity of the corresponding single star model is given by the black dashed lines. The model names give the values of the input parameters, with 'vXX' denoting the input wind velocity in km s^{-1}, 'aXX' the orbital separation in au, and 'mXX' the companion mass in M$_\odot$. Figure adapted from Maes *et al.* (2021).

4. Impact of companion on terminal wind velocity

Maes *et al.* (2021) investigated the effect of a companion on the terminal expansion velocity of the wind, for simulations with a stellar or planetary companion, and with different initial wind velocity and orbital separation. Fig. 2 presents the results of their analysis. By comparing the terminal velocities of the simulations (indicated by empty dots) to the terminal velocities of a single star model (dashed line), it is clear that, whereas the impact of a planetary companion appears neglectable, a stellar companion does affect the expansion velocity of the wind. To investigate the cause of this deviation, a toy model was constructed in which the terminal velocity is calculated analytically, by only taking into account the effect of the gravitational potential of the companion. From the resulting toy model terminal velocities (squares in Fig. 2) it can be concluded that for the case of a stellar companion, an important acceleration mechanism is missing, since too low and even negative terminal velocities result. Therefore, the toy model is extended by including the gravitational slingshot mechanism, which states that by conservation of momentum and energy, a small object, moving past a larger body in motion, is accelerated or decelerated. With inclusion of the slingshot mechanism (triangles in Fig. 2), the terminal velocity of the stellar models is a relatively good approximanion of the measured expansion velocity of the simulation. For more details see Maes *et al.* (2021).

5. Conclusion

The impact of a binary companion on the outflow of an AGB star is studied using a grid of 3D hydrodynamic simulations constructed with the SPH code PHANTOM. From these simulations it is concluded that depending on the binary configuration, different inner wind structures and global morphologies result, varying from a regular Archimedes spiral with a spherically symmetric global density distribution, to highly perturbed spiral structures with equatorial density enhancements and flattened global morphologies. The classification parameter ε is used to classify the morphology based on the characteristics of the binary system. Finally, we found that when the companion is massive enough, the terminal expansion velocity of the AGB wind is altered by the gravitational slingshot mechanism that acts on the wind particles.

References

Bujarrabal, V., Alcolea, J., Van Winckel, H., *et al.* 2013, *A&A*, 557, A104
Decin, L., Montargès, M., Richards, A. M. S. *et al.* 2020, *Science*, 369, 1497
El Mellah, I., Bolte, J., Decin, L. *et al.* 2020, *A&A*, 637, A91
Höfner, S., & Olofsson, H. 2018, *A&ARv*, 26, 1
Homan, W., Cannon, E., Montargès, M., *et al.* 2020a, *A&A*, 642, A93
Homan, W., Montargès, M., Pimpanuwat, B., *et al.* 2020b, *A&A*, 644, A61
Kervella, P., Homan, W., Richards, A. M. S., *et al.* 2016, *A&A*, 596, A92
Kim, H., & Taam, R. E. 2012, *ApJ*, 759, 59
Lamers, H. J. G. L. M., & Cassinelli, J. P. 1999, *Introduction to Stellar Winds*
Maes, S., Homan, W., Malfait, J. *et al.* 2021, *A&A*, 653, A25
Malfait, J., Homan, W., Maes, S. *et al.* 2021, *A&A*, 652, A51
Manick, R., Miszalski, B., Kamath, D., *et al.* 2021, *MNRAS*, 508, 2226
Mastrodemos, N., & Morris, M. 1998, *ApJ*, 497, 303
Mastrodemos, N., & Morris, M. 1999, *ApJ*, 523, 357
Oomen, G.-M., Pols, O.,Van Winckel, H., Nelemans, G. 2020, *A&A*, 642, A234
Price, D. J., Wurster, J., Tricco, T. S., *et al.* 2018, *PASA*, 35, e031
Ramstedt, S., Mohamed, S., Vlemmings, W. H. T., *et al.* 2014, *A&A*, 570, L14
Sahai, R., Morris, M. R., & Villar, G. G. 2011, *AJ*, 141, 134
Theuns, T., & Jorissen, A. 1993, *MNRAS*, 265, 946
Theuns, T., Boffin, H. M. J., & Jorissen, A. 1996, *MNRAS*, 280, 1264
Van Winckel, H. 2003, *ARA&A*, 41, 391

The Origin of Outflows in Evolved Stars
Proceedings IAU Symposium No. 366, 2022
L. Decin, A. Zijlstra & C. Gielen, eds
doi:10.1017/S1743921322000898

Morpho-kinematics around cool evolved stars Unveiling the underlying companion

I. El Mellah[1], J. Bolte[2], L. Decin[2], W. Homan[3] and R. Keppens[4]

[1]Institut de Planétologie et d'Astrophysique de Grenoble, UGA-CNRS, rue de la Piscine, 38400
email: ileyk.elmellah@univ-grenoble-alpes.fr

[2]Institute of Astronomy, KU Leuven, Celestijnenlaan 200D, 3001 Leuven, Belgium

[3]Institut d'Astronomie et d'Astrophysique, campus Plaine, Boulevard du Triomphe, Brussels, Belgium

[4]Centre for mathematical Plasma-Astrophysics, Celestijnenlaan 200B, 3001 Leuven, Belgium

Abstract. Because they lose tremendous amounts of mass, cool evolved stars are major sources of dust and molecules for the interstellar medium. Spectro-imaging of the dust-driven winds around these stars has enabled us to identify recurring nonspherical patterns (e.g. spirals, arcs, compressed wind). We use radiative-hydrodynamic simulations of dust-driven winds to study the imprints left in the wind by an orbiting stellar or sub-stellar companion. We designed 3D numerical setup to solve the wind dynamics beyond the dust condensation radius and follow the flow up to several hundreds of stellar radii. Non-uniform grids enable us to capture small scale features such as shocks or disks forming around the orbiting object. Depending on its mass and orbital parameters, we reproduced typical non-spherical features such as arcs, spirals, petals and orbital density enhancements, and identified patterns associated to eccentric orbits.

Keywords. stars:evolution - stars:winds,outflows - stars:AGB and post-AGB - methods:numerical - binaries:general

---

1. Introduction

In the stages following main sequence, mass loss plays a major role in the evolution of the star itself. As the star expands, it cools down but the luminosity remains approximately the same. As a consequence, the effective gravity drops and stellar material can more easily detach from the star. This is the case for red giant and supergiant stars (RSGs) and asymptotic giant branch stars (AGBs). These stars loose mass at rates which peak at 10^{-7}-10^{-5}M$_{\odot}$/year via a wind whose speed typically ranges from 5 km/s to 20km/s.

The mechanism which brings stellar material in the outermost layers at escape speed is still poorly understood. It hampers our attempts to determine the final fate of stars and thus leading to inaccurate occurrence rates for compact objects. However, wind launching is believed to be made possible by the combined action of convection, pulsations, molecular line absorption and continuum dust opacity (Höfner & Olofsson 2018). In the piston model, which proves to be accurate for AGB stars, the material is lifted by radial pulsations controlled by the κ-mechanism. Internal shocks form and as the temperature decreases, molecules form and condensate into dust grains. The latter absorb the continuum stellar spectrum which produces a net outward force due to radiative pressure (Freytag, Liljegren & Höfner 2017). Depending on the chemical content of the dust (carbon-rich or oxygen-rich), the opacity can be more or less high, with important consequences on the terminal wind speed and on the velocity profile. Alternatively, for RSGs, Kee et al. (2021) developed a model where turbulence pressure in the atmosphere provides enough outward momentum without the need for dust opacity.

On the other hand, stellar multiplicity has been recognized as a ubiquitous feature: stars seldom live an effectively single life. Due to their low speed, dust-driven winds can lead to significant mass transfer to the orbiting companion in binary systems where the donor star (the primary) has evolved beyond the main sequence. This mechanism is thought to be responsible for the peculiar chemical composition of certain stars (e.g. CEMP stars, Barium stars, blue stragglers). During this IAU Symposium, the evolutionary importance of chemical contamination was highlighted by Morgan Deal (for Lithium depletion) and Anke Andersen (for very metal-poor stars in the inner regions of the Milky Way). Through its gravitational influence on the primary and on the wind itself, the companion (the secondary) would also leave imprints in the fraction of the wind which escapes the binary. The nature of these marks depends on the companion mass, on its orbit and on the chemical composition of the circumbinary envelope.

The advent of a new generation of high spatial and spectral resolution instruments has ushered in a gold rush to understand mass loss of cool evolved stars. In the optical, the Hubble Space Telescope monitors dust emission while the Herschel Space Observatory and the (sub)millimeter interferometer ALMA capture the emission associated to molecular rotational and vibrational transitions. Through spectro-imaging, they grant us access to the 3D morpho-kinematics of the wind with multi-channel molecular line emission maps. They have not only revealed the complexity of the astrochemistry at work but also identified nonspherical features in these cool winds. The ALMA large program ATOMIUM has identified recurrent axisymmetric patterns like spirals, arcs or petal-like patterns (Decin *et al.* 2020). Although alternative scenarios exist, these features have been ascribed to the presence of an underlying stellar or sub-stellar companion, too dim to be directly detected.

2. Model

2.1. *Wind launching*

We developed a simplified prescription for the launching of the wind from an isolated cool star which enables us to differentiate between carbon-rich and oxygen-rich outflows. The wind from carbon-rich AGB stars reaches its terminal speed within a few dust condensation radii at most. On the contrary, for oxygen-rich outflows, the acceleration is much more progressive due to the lower opacity of the dust grains formed. We solve the 1D steady spherical equation of motion, neglecting the thermal pressure:

$$v\frac{dv}{dr} = -\frac{1}{r^2} + \frac{\Gamma}{r^2}\kappa\left(r\right) \tag{2.1}$$

where length and speed were normalized to the dust condensation radius and the escape speed at the dust condensation radius respectively. Γ is a constant ratio of luminosity similar to the Eddington parameter, but with a reference opacity κ_0 defined as:

$$\Gamma = \frac{\kappa_0 L_*}{4\pi G M_* c} \tag{2.2}$$

where M_* is the mass of the donor star, L_* is its luminosity, c is the speed of light and G is the gravitational constant. The effective opacity κ_0 stands for the normalization of the dust opacity but it also represents the coupling between the dust and the gas. It is typically much higher than the dust-free molecular gas opacity.

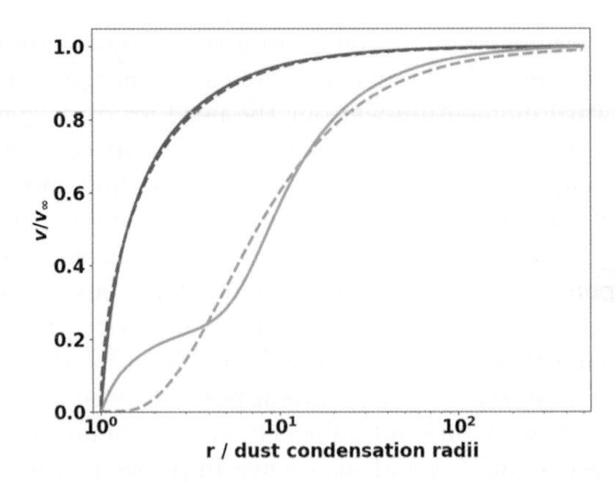

Figure 1. Velocity profiles normalized to escape speed v_∞ numerically integrated (solid lines) and fitted by *beta*-velocity profiles (dashed lines) for carbon and oxygen-rich outflows (upper and lower curves respectively).

To determine the dust opacity profile as a function of distance to the star, we connect it to the local equilibrium temperature T of the star given by:

$$T(r) = \frac{T_*}{2}\left[1 - \sqrt{1 - (R_*/r)^2}\right]^{1/4} \tag{2.3}$$

where T_* is the stellar effective temperature and R_* is the stellar radius. Bowen (1988) suggested the following dimensionless opacity profile to mimic the dust growth in carbon-rich outflows:

$$\kappa = \frac{1}{1 + \exp\left(T - T_c\right) / \Delta T}, \tag{2.4}$$

with T_c the dust condensation temperature and ΔT a range of temperature where the grains grow. For realistic values for stellar parameters and opacity of carbon-rich winds ($\kappa_0 \sim 5-10$ g·cm^{-2}), we obtain the solid upper line shown in Figure 1, very well fitted by a β-law with $\beta \sim 0.6$ (dashed upper line).

For oxygen-rich outflows, we assume that a two-stages acceleration mostly driven by two dust species, 1 and 2. Dust species 1 condensates at a higher temperature ($T_{c,1} > T_{c,2}$) but its opacity is not sufficient to bring the material to escape speed. Instead, it lifts the material up to distances where dust species 2 condensates and drives a proper wind. We obtained the solid lower velocity profile in Figure 1 which can be fitted by a β-law with $\beta \sim 5$ (dashed lower line).

2.2. *Numerical setup*

We inject the 1D acceleration profiles derived in the previous section into a 3D numerical setup centered on the donor star. We used the `MPI-AMRVAC` code (Xia et al. 2017) in order to solve the equations of hydrodynamics on a spherical grid radially stretched (El Mellah et al. 2015). For circular orbits, we work in the co-rotating frame such as the source terms are those induced by the non-inertial forces, the gravitational pull from the two bodies and the radiative pressure due to continuum opacity on dust grains. Instead of solving the equation of energy, we assume that the expansion of the wind is adiabatic with a polytropic index of 1.3. With this value, we match the observed temperature profiles $T \propto r^{-0.6}$. With 4 levels of adaptive mesh refinement, the effective resolution of these simulations is 1,280×512×1,024. The 3 dimensionless parameters of our model are:

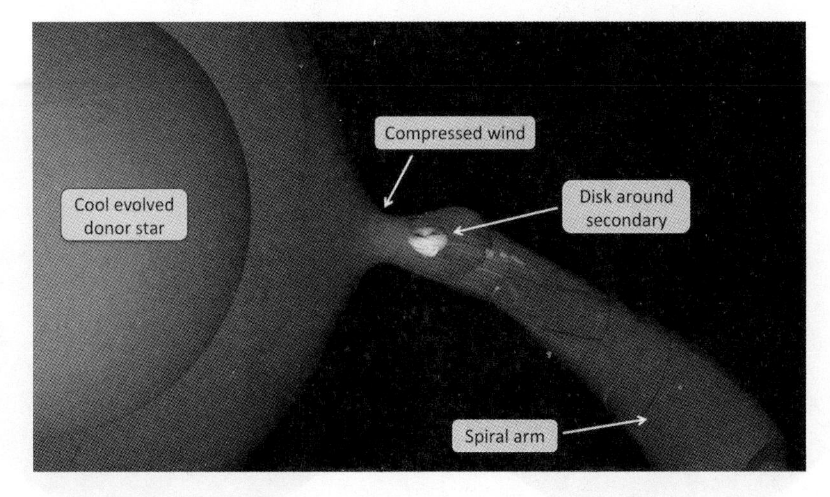

Figure 2. Structure of the flow at the orbital scale. Iso-density surfaces are represented.

(1) the ratio η of the terminal wind speed v_∞ to the orbital speed $a\Omega$, with a the orbital separation and Ω the angular orbital speed.
(2) the mass ratio q (primary to secondary).
(3) the filling factor f (ratio of dust condensation radius to Roche lobe radius).

and the donor star can be either carbon or oxygen-rich. We explored 70 different configurations whose parameters can be found in El Mellah et al. (2020).

3. Results

3.1. *Wind-captured disks*

Although the dust condensation radius is much smaller than the Roche lobe radius of the primary ($f \ll 1$), we do observe features reminiscent of Roche lobe overflow (RLOF) mass transfer (see Figure 2). The wind is beamed in the orbital plane and a tidal arm develops towards the secondary. A nozzle appears near the inner Lagrangian point and a tidal arm forms in the wake of the secondary. In some cases, the flow even gains enough angular momentum to form a wind-captured disk around the secondary, similar to what is observed in high-mass X-ray binaries (El Mellah et al. 2018). These properties are characteristic of the wind-RLOF mechanism highlighted by Mohamed et al. (2007) where the outflow is significantly affected by the presence of the secondary. In our simulations, it typically corresponds to low values of η, and, to a lesser extent, to high values of q and f. Because acceleration is more progressive for oxygen-rich outflows, they are more prone to undergo this mass transfer mechanism. A contrario, winds which quickly reach a terminal speed much higher than the orbital speed are more radial and less affected by the presence of the secondary.

This wind-RLOF mechanism is also associated to an increasing fraction of the wind being accreted by the secondary, which leads to a slower widening of the orbit compared to a pure mass loss configuration. For sufficiently low mass companion (e.g. a massive planet) and/or for sufficiently high fraction of the wind captured, the orbit can even shrink at a quick pace, which leads to an enhanced fraction of the wind being captured and a runaway spiral in.

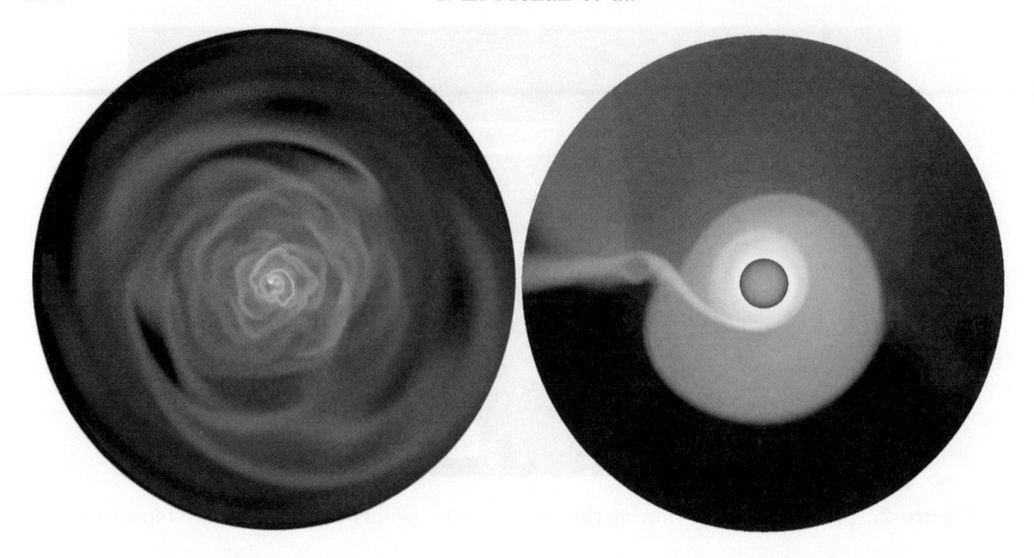

Figure 3. (Left panel) Global flow morphology for a wind computed for a companion on a circular orbit around an O-rich star. In the center, the two bodies are visible. Petal-like patterns are imprinted in the circumbinary envelope up to distances much larger than the orbital separation. (Right panel) Slice in the orbital plane of the wind density with the donor star in the center and the companion at apoastron.

3.2. *Morpho-kinematics of the circumbinary envelope*

At a broader scale, we reproduce the nonspherical features observed like spirals, arcs, petal-like patterns (Figure 3, left panel) and equatorial density enhancement. The latter has been shown by Decin et al. (2019) to be responsible for the overestimated mass loss rates derived when overlooking this compression of the wind. The spiral shocks develop due to the wobbling of the primary induced by the secondary. In these 3D simulations, we can also study the impact of the line-of-sight. Spiral shocks appear as concentric arcs when seen edge-on while the petal-like patterns emerge when the terminal wind speed is of the order of the orbital speed, leading to a marginally unbound outflow.

In order to confront these simulations to observations, we are currently extracting from these simulations synthetic molecular line emission maps. The density and temperature maps can be used to compute 1D chemical abundance profiles (Van de Sande et al. 2019). Combined with the velocity maps, we can use a radiative transfer code to obtain from each cell the Doppler-shifted molecular line emission for each frequency bin.

3.3. *Eccentric orbits: companion-driven winds*

For eccentric orbits, we reach within a few orbital periods a regime where the morphology of the outflow is set by the orbital phase. We ran a simulation with an eccentricity of 0.5 and a $f = 1$ at periastron (Figure 3, right panel). In this case, we found that the gravitational slingshot induced by the passage of the secondary at periastron drives an increase in the instantaneous mass loss rate. If confirmed, a correlation between the mass loss rate and the eccentricity of the companion's orbit would point in favor of a boosting mechanism for wind launching and new types of companion-stimulated winds.

4. Conclusions

We showed that an orbiting companion could reproduce most nonspherical features observed in the circumbinary envelope of cool stars. The main parameter driving the wind dynamics is the ratio η of the terminal wind speed to the orbital speed. As η decreases toward 1, the pitch angle of the spiral shock decreases. For $\eta \lesssim 1$, the spiral fragments into arcs which are intermittently ejected by incoming material. It produces a petal-like pattern which is reminiscent, when observed face-on, of the morphology of the C-rich AGB star CW Leo for instance.

Acknowledgments

The author wishes to thank the ATOMIUM ALMA Large Programme collaboration (2018.1.00659, PI. L. Decin) for the observational consequences of the present analysis. IEM has received funding from the Research Foundation Flanders (FWO) and the European Union's Horizon 2020 research and innovation program under the Marie Skłodowska-Curie grant agreement No 665501. The simulations were conducted on the Tier-1 VSC (Flemish Supercomputer Center funded by Hercules foundation and Flemish government).

References

Bowen, G. 1988, *ApJ*, 329, 9, 299

Decin, L., Montargès, M., Richards, A. M.S., Gottlieb, C. A., Homan, W., McDonald, I., El Mellah, I., Danilovich, T., Wallström, S. H.J., Zijlstra, A., Baudry, A., Bolte, J., Cannon, E., De Beck, E., De Ceuster, F., de Koter, A., De Ridder, J., Etoka, S., Gobrecht, D., Gray, M., Herpin, F., Jeste, M., Lagadec, E., Kervella, P., Khouri, T., Menten, K., Millar, T. J., Müller, H. S.P., Plane, J. M.C., Sahai, R., Sana, H., Van de Sande, M., Waters, L. B.F.M., Wong, K. T., Yates, J. 2020, *Science*, 369 , 6509, 1497

Decin, L., Homan, W., Danilovich, T., de Koter, A., Engels, D., Waters, L.B.F.M., Muller, S., Gielen, C., García-Hernández, D. A., Stancliffe, R. J., Van de Sande, M., Molenberghs, G., Kerschbaum, F., Zijlstra, A. A., El Mellah, I. 2019, *Nature Astronomy*, 3, 5, 408

El Mellah, I. and Casse, F. 2015, *MNRAS*, 454, 3, 2657

El Mellah, I., Sander, A. A. C., Sundqvist, J. O., Keppens, R. 2019, *A&A*, 622, A189

El Mellah, I. et al. 2020, *A&A*, 637, A91

Freytag, B., Liljegren, S., Höfner, S. 2017, *A&A*, 600, A137

Höfner S., Olofsson H. 2018, *AAR*, 26, 1

Kee, N. D., Sundqvist, J. O., Decin, L., De Koter, A., Sana, H. 2021, *A&A*, 646, A180

Mohamed, S. and Podsiadlowski, P. 2007, PhD Manuscript

Van de Sande, M., Walsh, C., Mangan, T. P., Decin, L. 2019, *MNRAS*, 490, 2, 2023

Xia, C., Teunissen, J., El Mellah, I., Chané, E., Keppens, R. 2017, *ApJS*, 234, 2, 30

The Origin of Outflows in Evolved Stars
Proceedings IAU Symposium No. 366, 2022
L. Decin, A. Zijlstra & C. Gielen, eds.
doi:10.1017/S1743921322001156

Accretion-powered Outflows in AGB Stars

Raghvendra Sahai[1], Jorge Sanz-Forcada[2] and Carmen Sanchez-Contreras[2]

[1]Jet Propulsion Laboratory, Pasadena, CA, USA
email: raghvendra.sahai@jpl.nasa.gov

[2]Centro de Astrobiología (CSIC-INTA), ESAC, Villanueva de la Cañada, Madrid, Spain

Abstract. One of the big challenges for 21st century stellar astrophysics is the impact of binary interactions on stellar evolution. Such interactions are believed to play a key role in the death throes of 1-8 $M_{\odot}$ stars, as they evolve from the AGB stars into Planetary Nebulae. X-ray surveys of UV-emitting AGB stars show that ~40% of objects with FUV emission and GALEX FUV/NUV flux ratios $\gtrsim$0.2 have variable X-ray emission characterized by very high temperatures (Tx~35-160 MK) and luminosities (Lx~0.002-0.2 $L_{\odot}$). We hypothesize that such AGB stars have accretion and (accretion-powered) outflows associated with a close binary companion. UV spectroscopy with HST/STIS of our brightest object (Y Gem) shows the presence of infalling and outflowing gas, providing direct kinematic confirmation of this hypothesis. However, the UV-emitting AGB star population is dominated by objects with little or no FUV emission, and we do not know whether the UV emission from these is intrinsic to the AGB star or extrinsic (i.e., due to binarity). Here we present the first results from a large grid of simple chromospheric models to help discriminate between the intrinsic and extrinsic mechanisms of UV emission for AGB stars.

Keywords. (stars:) circumstellar matter, (stars:) binaries (including multiple): close, stars: evolution, stars: AGB and post-AGB

---

1. Introduction

Most stars in the Universe that leave the main sequence in less than a Hubble time (i.e., stars in the 1–8 $M_{\odot}$ range) undergo extraordinary deaths, expelling half or more of their masses at rates up to $\sim 10^{-4} M_{\odot}$ yr^{-1}, as they evolve from the Asymptotic Giant Branch (AGB), through the pre-Planetary Nebula (PPN) to the Planetary Nebula (PN) evolutionary phase. Almost all of our current understanding of this late evolutionary stage of these stars is based on single-star models. However, imaging surveys of (i) the mass-ejecta in PPNe and PNe that show dramatic departures from spherical symmetry (elliptical, bipolar and multipolar morphologies) (e.g., Sahai & Trauger 1998; Sahai et al. 2006, 2007; Sahai, Morris & Villar 2011), and of (ii) AGB circumstellar envelopes that reveal spiral patterns (Decin et al. 2020) have made it increasingly clear that strong binary interactions play a major role in the deaths of low and intermediate-mass stars.

Unfortunately, observational evidence for close binary companions in AGB stars has been generally scarce because AGB stars are very luminous and variable, invalidating standard techniques for binary detection (e.g., radial-velocity and photometric variations due to a companion star, and direct imaging.) However, UV photometric observations can be used to search for binarity and associated accretion activity in AGB stars because most of these are relatively cool ($T_{eff} \lesssim 3000$K) objects (spectral types ~M6 or later), whereas any stellar companions and/or accretion disks around them are likely significantly hotter

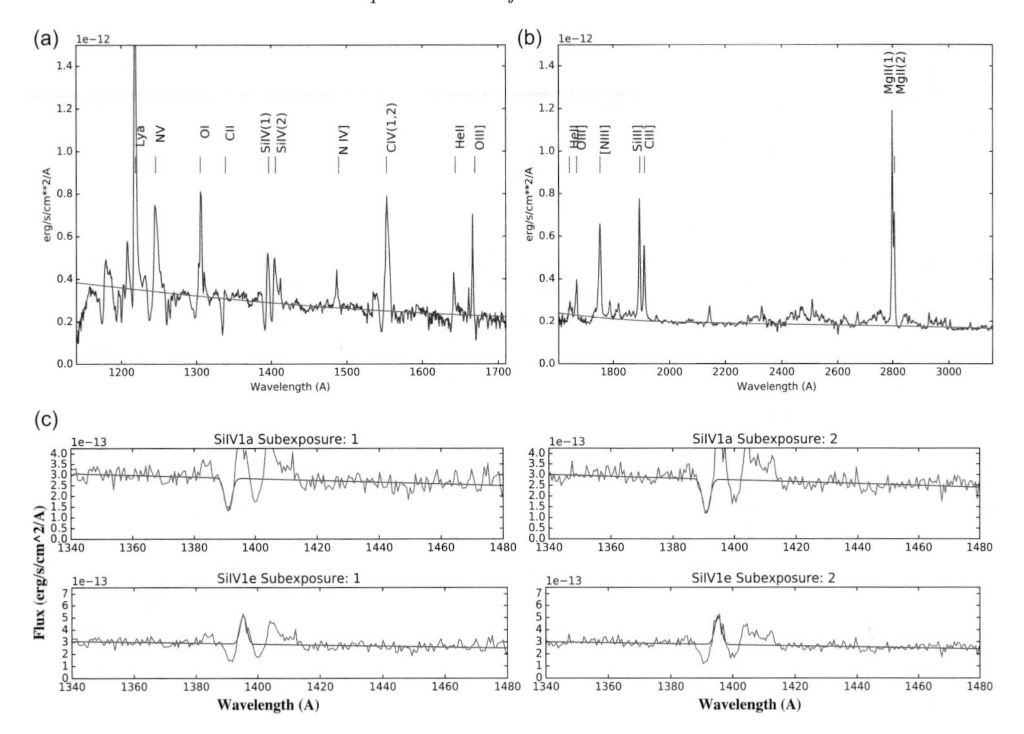

Figure 1. (a,b) Representative HST/STIS spectra of Y Gem, together with a model fit (green curve) consisting of two blackbody components, characterised by $T_{\rm eff}$= 35, 500 K, $L = 6.3\,L_{\odot}$, and $T_{\rm eff}$= 9, 400 K, $L = 6.7\,L_{\odot}$. (c) STIS/UV spectra of Y Gem in the vicinity of the Si IV(1) line for the first two 20 s subexposures. The blue curves show gaussian line-profile fits (together with a linear baseline) to the absorption (top) and emission (bottom) features (*adapted from Sahai et al. (2018)*).

($T_{eff} \gtrsim 6000$K). Hence, favorable secondary-to-primary flux contrast ratios ($\gtrsim$10) are reached in the GALEX FUV (1344 – 1786 Å) and NUV (1771 – 2831 Å) bands for a source (companion or disk) with $T_{eff} \gtrsim$6000 K and luminosity $L \gtrsim 1L_{\odot}$. The feasibility of this technique has been demonstrated in a number of recent studies (Sahai et al. 2008; Sahai, Morris & Villar 2011). From an application of this technique to a volume-limited sample (<0.5 kpc) of 58 AGB stars, Ortiz & Guerrero (2016) conclude that the detection of FUV emission, or an observed-to-predicted ratio for NUV emission of $Q_{NUV} > 20$, are criteria for binarity.

UV spectroscopy can provide an unambiguous probe of accretion-related activity because the latter is expected to produce UV lines with large widths and large Doppler shifts. An HST/STIS spectroscopic study of the prototype fuvAGB star, Y Gem, shows the presence of UV emission lines (Fig. 1) with red (blue)–shifted emssion and absorption features implying high-velocity infall and outflows, as well as short-time-scale variations (flickering). These data directly support the binary/accretion hypothesis (Sahai et al. 2018). The UV spectrum of Y Gem shows lines from species such as Si IV and C IV, with broad emission and absorption features (FWHM$\sim$ 300 – 700 km s^{-1}) that are respectively, red- and blue- shifted by velocities of $\sim$500 km s^{-1} from the systemic velocity. The UV continuum reveals strong flickering on time-scales of $\lesssim$ 20 s, characteristic of an active accretion disk.

X-ray searches in selected samples of UV-emitting AGB stars, intentionally biased to have relatively high values of the GALEX FUV/NUV ratio ($R_{fuv/nuv}$), as well as serendipitous detections in an archival survey (Ortiz & Guerrero 2021) have also provided strong supporting evidence for accretion activity in these objects. The X-ray emission is variable, both on long (months–year) and short ($\lesssim few \times 100$ s)) time-scales. APEC model fits to sources with the highest S/N X-ray spectra show that the observed X-ray luminosity (Lx) and temperature (Tx) lie in the range Lx$\sim (0.002 - 0.2)\, L_\odot$ and Tx$\sim (3.5 - 16) \times 10^7$ K. Dividing the sample of UV-emitting AGB stars into two broad sets, one with high values of $R_{fuv/nuv}$($\gtrsim 0.15$, hereafter fuvAGB stars), and one with low values of $R_{fuv/nuv}$ ($\lesssim 0.06$, hereafter nuvAGB stars), we find that X-ray emission is generally detected in fuvAGB stars with $R \gtrsim 0.2$ (including 2 ROSAT detections reported by Ramstedt et al. (2012)). The fraction of X-ray emitting objects for the $R_{fuv/nuv} > 0.17$ sample is 0.4. For stars not detected in X-rays, we find $R_{fuv/nuv} < 0.12$ (with 2 exceptions).

The primary AGB star is very unlikely to be the source of the X-ray emission, since it would require rather strong magnetic fields to confine the hot plasma, and sensitive searches for X-rays in two AGB stars with strong magnetic fields have been unsuccessful (Kastner and Soker 2004) – but the presence of strong local fields confining clumpy hot plasma that can produce the X-ray emission cannot be ruled out. Furthermore, in a recent study, Montez et al. (2017) argue that the origin of the GALEX-detected UV emission is most likely due to a combination of photospheric and chromospheric emission from the AGB star. Their argument is based on finding that for a sample of 179 AGB stars, the NUV emission is correlated with the optical to near-infrared emission. In this study, we investigate whether simple models of chromosphere around AGB stars can reproduce the UV-emission properties of UV-emitting AGB stars.

2. Results

We searched for UV detections by GALEX in a statistical sample of $\sim$3500 AGB stars with spectral types M4. Our sample was compiled from the Simbad database. A total of 316 objects were detected; $> 20\%$ in one or both of the GALEX FUV and NUV bands, and about $> 9\%$ in both. These percentages are lower limits because most of the GALEX data comes from the All-Sky Imaging Survey (AIS), which had relatively short exposure times, $\sim (1 - few) \times 100$ s.

2.1. *Red Leak in the GALEX bands*

The filters used to obtain UV photometry are often affected by a "red leak" issue – the leakage of red light through these filters. Generally these leaks are relatively low and do not significantly contaminate the UV flux determinations. According to GALEX documentation, there is no measurable red leak in either the FUV or NUV bands. However for very red stars, such as AGB stars, for which the bulk of the energy is emitted in the red-infrared wavelength region, even a small leak can contaminate the UV flux measurements. We can constrain the "red leak" very sensitively using our sample of AGB stars. We determine the ratios of the FUV and NUV fluxes to the V-band, R-band fluxes, and I-band fluxes, for our sample (and red supergiant stars). The lowest of these ratios can then be considered an upper limit to the red-leak.

The maximum possible red leak for the GALEX FUV and NUV as a fraction of the V-band flux, R-band flux, and I-band flux for AGB O-rich sources and M-type red supergiants stars. We find that the FUV-to-optical band flux ratios are 1.3×10^{-7}, 1.2×10^{-8}, and 1.2×10^{-8} (1.9×10^{-6}, 1.1×10^{-6}, and 3.7×10^{-7}) for AGB stars (red supergiant stars) in the V, R, and I-bands. The NUV-to-optical band flux ratios are 4.0×10^{-7}, 5.8×10^{-8}, and 8.5×10^{-8} (1.5×10^{-5}, 2.7×10^{-7}, and 9.1×10^{-7}) for AGB stars (red supergiant stars) in the V, R, and I-bands. AGB stars thus provide the lowest values of

Table 1. FUV and NUV Variability.

Class	FUV Var			NUV Var			FUV/NUV		
	Ave	Min	Max	Ave	Min	Max	Ave	Min	Max
AGB O-Rich	44%	0.96%	383%	50%	0.05%	306%	0.19	0.003	2.36
Symbiotic	61%	0.8%	214%	61%	0.2%	535%	2.02	0.37	6.93
M supergiants	23%	5.6%	44%	22%	1.4%	120%	0.051	0.021	0.088

the red-leak for both the FUV and NUV bands. The red leak is very low, and does not affect the results of our study.

We compare the UV properties of these 316 AGB stars with two related classes of objects, one in which accretion activity in a binary is known to be important (symbiotic stars) and M-type supergiants, that are believed to possess chromospheres. The sample of supergiants was extracted from the catalog by Hohle et al. (2010). The sample of the symbiotic stars was gathered from the catalog of symbiotic stars by Belczyński et al. (2000). GALEX FUV and NUV data were then extracted for these samples of the supergiants and symbiotic stars as for our AGB stars sample.

2.2. *Comparing the UV Properties of AGB Stars with Other Stellar Classes*

The typical, minimum and maximum values of the variability in the FUV and NUV fluxes, and FUV/NUV flux ratios of these three classes are given in Table 1. The symbiotic sources have the highest average FUV and NUV flux variability (61%), as well as the largest average FUV/NUV ratios. The AGB O-rich sources have FUV and NUV flux variability similar to the symbiotic sources, but their average FUV/NUV flux ratio is about an order of magnitude smaller: ~ 0.19 (O-rich AGB stars) compared to ~ 2 (symbiotic stars). The similarity of the FUV and NUV variability between the symbiotic sources and the AGB O-rich sources reinforcse the idea that the FUV and NUV variability is related to accretion related activity from a binary companion. For our sample of M-type red supergiant stars, which, as a class, are known to possess chromospheres, the average FUV/NUV ratio is signficantly lower, 0.051.

2.3. *Chromospheric Emission*

The mean FUV (f_{FUV}) and the mean NUV (f_{NUV}) for all UV-emitting AGB stars in our sample which were detected in both these bands shows a linear relationship. We find $f_{FUV} = R_{FUV/NUV} * f_{NUV}$ provides a good fit to the bulk of the data, with $R_{FUV/NUV} = 0.061 \pm 0.0015$. Outliers, defined as data where the observed f_{FUV} value was $\geqslant 5\sigma$ away from the model fit, were removed iteratively until the slope and number of outliers converged (in 6 iterations), resulting in the removal of 17 outliers. Restricting our dataset to objects where the SNR for both the mean FUV and NUV was $\geqslant 5$ (a total of 184 objects: Fig. 2(*left*)) did not affect the results significantly – we found $R_{FUV/NUV} = 0.060 \pm 0.0022$ after 5 iterations, with 14 outliers (preliminary results were presented in Sahai et al. 2020).

Previous observations of NUV emission from AGB stars with the IUE, including the presence of strong MgII emission lines, suggested the presence of chromospheres in these stars. Detailed modeling of the NUV IUE spectra, for an M6 AGB star by Luttermoser et al. (1994), shows that a hot chromosphere surrounding the cool photosphere, with temperatures that rise to about 10,000 K at the outer edge of the former.

We have made a large grid of simple chromospheric models ($\sim 50,000$) using the CLOUDY code (Ferland et al. 2017) and computed the UV emission from a collisionally

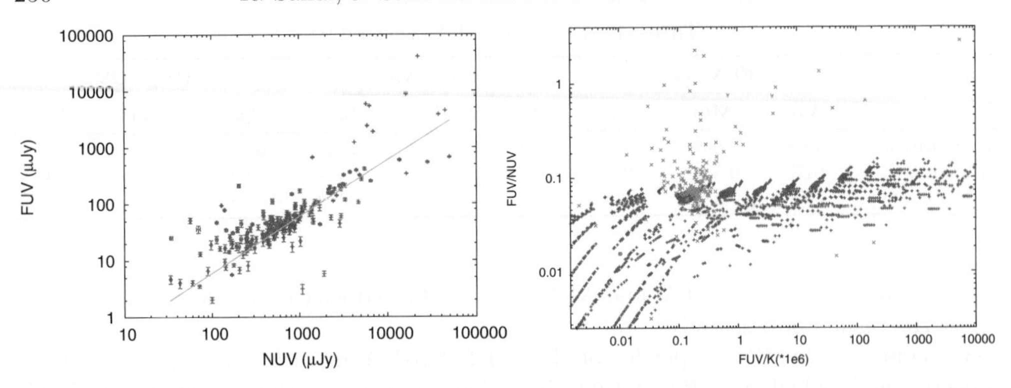

Figure 2. (*left*) The mean FUV– and NUV– band fluxes of AGB stars observed with GALEX, for sources that were detected in both UV bands with a signal-to-noise ratio, SNR$\geqslant$5. The green line shows a linear least-squares fit with outlier rejection, $f_{FUV} = R_{FUV/NUV} * f_{NUV}$, with $R_{FUV/NUV} = 0.06 \pm 0.002$. (*right*) The observed (green symbols) and model FUV/NUV flux ratio (purple symbols), $R_{FUV/NUV}$, versus the FUV/K-band (2MASS) flux ratio for AGB stars that were detected in both UV bands with SNR of $\geqslant 5$.

ionized plasma with temperature T_{chrom} and density nH contained within a layer of gas of thickness $\Delta\,R$ surrounding the AGB star. These models are characterized by two parameters that describe the AGB star (effective temperature T_{AGB}, and luminosity L_{AGB}), and three parameters that describe the chromosphere (the gas temperature T_{chrom}, the gas density nH_{chrom}, and the thickness of the layer of hot gas, $\Delta\,R$). The range covered by our model grid is as follows: $2500 < T_{AGB}\,(K) < 3300$, $3500 < L_{AGB}\,(L_{\odot}) < 15000$, $3000 < T_{chrom}\,(K) < 16000$, $10.0 < log(nH\,(cm^{-3})) < 16.0$, and $0.0 < log(\Delta\,R\,(cm)) < 7.0$. We adopt a nominal distance of 250 pc and a stellar radius of 10^{13} cm – these parameters do not affect our analysis.

We compute the model FUV and NUV-band fluxes by convolving the model spectra with the GALEX FUV and NUV passbands. We show the observed $R_{fuv/nuv}$ versus the FUV-to-K band flux ratio for our sample of UV-emitting AGB stars, $R_{FUV/K}$, together with model values, in Fig. 2 (*right*) – the latter is a good proxy for the fractional FUV luminosity, since the K-band flux is expected to result almost entirely due to the emission from the AGB photosphere, excluding the effect of the presence of dust that may be present in a stellar wind.

We find that our chromosphere models can produce $R_{fuv/nuv} \sim 0.06$ for temperatures in the range $9,000 \leqslant T_{chrom}\,(K) \leqslant 11,000$. Higher values of $R_{fuv/nuv}$ require higher temperatures, e.g., for $R_{fuv/nuv} \sim 0.1$, we require $11,000 \leqslant T_{chrom}\,(K) \leqslant 15,000$. The allowed values of the parameters $nH\,(cm^{-3})$ and $\Delta\,R$ (cm) are not independent; we find $29.5 \geqslant 2 * log(nH) + log(\Delta\,R) \leqslant 31$. This is because the total emission from a collisionally-excited plasma (as assumed for the CLOUDY models) is expected to be proportional to the square of the density, and the total volume of emitting material – the latter is proportional to the thickness since the chromospheric layer is geometrically thin.

We compare our model results to the detailed chromospheric models by Luttermoser et al. (1994), based on fitting near-UV lines from IUE data of the M6 star, g Her, with the greatest weight given to Mg II h and k doublet (~ 2800 Å), followed by Mg I ($\lambda\,2852$ Å) and the CII] UV0.01 multiplet (~2325 Å). The temperature distribution of their best-fit model, T10, ranges from 11,270 K to 3,200 K (Fig. 3). The model chromospheric layer extends from the photosphere to a height of ~ 1.7 AU, with temperatures in the range $\sim 9,000 - 11,270$ K at heights $\gtrsim 0.45$ AU. Our simple models thus appear to

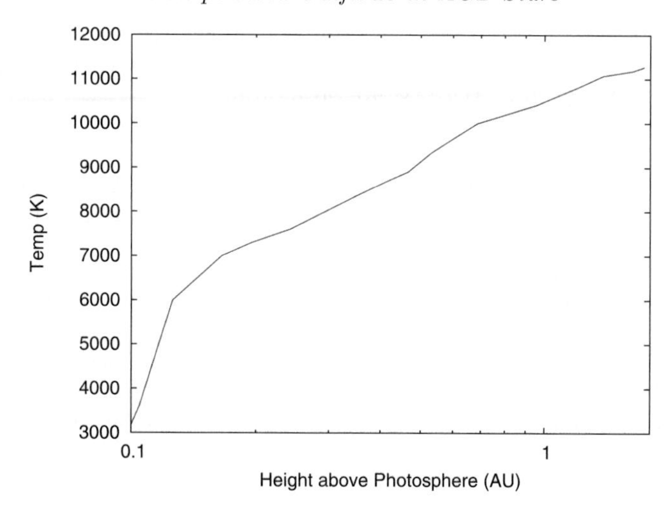

Figure 3. Temperature (T_{chrom}) vs. height above the photosphere for g Her (M6 III) from the best-fit (to NUV lines) chromosphere model of this object (*adapted from Luttermoser et al. (1994)*).

be a reasonable approximation to a more detailed model of the chromosphere for the purpose of investigating the dependence of $R_{fuv/nuv}$ on T_{chrom}.

3. Summary

High-energy observations, at UV and X-ray wavelengths, have provided a new and unique probe into binarity and binary interactions in AGB stars in recent years. The development of models that can fit these data is sorely needed, especially to distinguish between different mechanisms such as accretion-related activity and chromospheric emission. Our simple modeling of hot plasma in AGB stars detected with GALEX in the FUV and NUV bands shows that relatively low values of the FUV-to-NUV flux ratio ($\lesssim 0.06$) may be explained by gas at typical chromospheric temperatures ($\sim 10,000$ K). However, stars with higher FUV/NUV flux ratios require significantly hotter gas, presumably resulting from infall onto an accretion disk around a companion, as confirmed by UV spectroscopic observations of the most prominent UV-emitting star.

Acknowledgements

R.S.'s contribution to the research described here was carried out at the Jet Propulsion Laboratory, California Institute of Technology, under a contract with NASA, and funded in part by NASA via ADAP awards, and multiple HST GO awards from the Space Telescope Science Institute. CSC's work is supported through I+D+i project PID2019-105203GB-C22 funded by the Spanish MCIN/ AEI/10.13039/501100011033. JS-F's work is supported through I+D+i project PID2019-109522GB-C51 funded by the Spanish MCIN/ AEI/10.13039/501100011033.

References

Belczyński, K., Mikołajewska, J., Munari, U., Ivison, R. J., Friedjung, M. 2000. *A&AS* 146, 407–435.
Decin, L. and 34 colleagues 2020. *Science* 369, 1497–1500.
Ferland, G. J. and 10 colleagues 2017. *Rev. Mexicana AyA* 53, 385–438.
Hohle, M. M., Neuhäuser, R., Schutz, B. F. 2010. *AN* 331, 349.
Kastner, J. H., Soker, N. 2004. *ApJ* 608, 978-982.

Luttermoser, D. G., Johnson, H. R., Eaton, J. 1994. *ApJ* 422, 351.
Montez, R., Ramstedt, S., Kastner, J. H., Vlemmings, W., Sanchez, E. 2017. *ApJ* 841, 33.
Ortiz, R., Guerrero, M. A. 2016. *MNRAS* 461, 3036 3046.
Ortiz, R., Guerrero, M. A. 2021. *ApJ* 912, 93.
Sahai, R. & Trauger, J. T. 1998. *AJ* 116, 1357–1366.
Sahai, R., Morris, M., Sánchez Contreras, C., Claussen, M. 2006. in: M.J. Barlow & R.H. Mendéz (eds.), *Planetary Nebulae in our Galaxy and Beyond*, Proc. IAU Symposium No. 234, 499–500.
Sahai, R., Morris, M., Sánchez Contreras, C., Claussen, M. 2007. *AJ* 134, 2200–2225.
Sahai, R., Findeisen, K., Gil de Paz, A., Sánchez Contreras, C. 2008. *ApJ* 689, 1274–1278.
Sahai, R., Morris, M. R. & Villar, G. G. 2011. *AJ* 141, 134–164.
Sahai, R., Sánchez Contreras, C., Mangan, A. S., Sanz-Forcada, J., Muthumariappan, C., Claussen, M. J. 2018. *ApJ* 860, 105.
Sahai, R., Young, O., Sanchez Contreras, C., Sanz-Forcada, J. 2020. *AAS Meeting Abstracts* #235.

The Origin of Outflows in Evolved Stars
Proceedings IAU Symposium No. 366, 2022
L. Decin, A. Zijlstra & C. Gielen, eds
doi:10.1017/S1743921322000448

Learning about AGB stars by studying the stars polluted by their outflows

Ana Escorza and **Robert J. De Rosa**

European Southern Observatory, Alonso de Córdova 3107, Vitacura, Santiago, Chile
email: ana.escorza@eso.org

Abstract. A rich zoo of peculiar objects forms when Asymptotic Giant Branch (AGB) stars, undergo interactions in a binary system. For example, Barium (Ba) stars are main-sequence and red-giant stars that accreted mass from the outflows of a former AGB companion, which is now a dim white dwarf (WD). Their orbital properties can help us constrain AGB binary interaction mechanisms and their chemical abundances are a tracer of the nucleosynthesis processes that took place inside the former AGB star. The observational constraints concerning the orbital and stellar properties of Ba stars have increased in the past years, but important uncertainties remained concerning their WD companions. In this contribution, we used HD 76225 to demonstrate that by combining radial-velocity data with Hipparcos and Gaia astrometry, one can accurately constrain the orbital inclinations of these systems and obtain the absolute masses of these WDs, getting direct information about their AGB progenitors via initial-final mass relationships.

Keywords. white dwarfs - stars: late-type - stars: chemically peculiar - binaries: spectroscopic - astrometry - stars: evolution

---

1. Introduction

Asymptotic Giant Branch (AGB) stars are the main producers of about half of the chemical elements heavier than iron via the slow neutron capture (s-) process of nucleosynthesis (e.g. Lugaro et al. 2003; Karakas 2010; Käppeler et al. 2011). However, the determination of individual atomic abundances on the surface of AGB stars is complicated by their complex convective envelopes and circumstellar environments, their high mass-loss rates, and the presence of broad molecular bands on their spectra (e.g. Busso et al. 2001; Van Eck et al. 2017; Shetye et al. 2018). Additionally, a significant fraction of AGB stars might have stellar or sub-stellar companions, which are difficult to detect and characterise also due to the previously mentioned factors (e.g. Mayer et al. 2014; Decin et al. 2020). With this contribution, we want to convince the reader that we can use AGB binary interaction products to learn about the AGB phase, specifically about binary interaction and nucleosynthesis processes.

Barium (Ba) stars are a prototypical example of AGB binary interaction products. These chemically peculiar stars formed when an AGB star transferred mass to its unevolved companion in a binary system (e.g. McClure 1984; Udry et al. 1998; Jorissen et al. 1998). The former AGB star evolved long ago and is now a dim white dwarf (WD) and its s-process enriched companion, the Ba star, is now the most luminous star in the system and can be observed at different evolutionary phases (e.g. Escorza et al. 2017, 2019; Jorissen et al. 2019; Shetye et al. 2020).

The fact that Ba stars are known products of binary interaction means that their orbital properties can provide constraints to binary interaction and evolution models (e.g.

Table 1. Overview of the main stellar properties of the HD 76225 (Escorza et al. 2019).

T_{eff} [K]	logg [dex]	[Fe/H]	[s/Fe][(1)]	$L/L_{\odot}$	$M_1/M_{\odot}$
6340 ± 50	3.9 ± 0.2	-0.37 ± 0.08	1.25 ± 0.08	5.9 ± 0.7	1.21 ± 0.06

(1) [s/Fe] from Allen & Barbuy (2006b).

Table 2. Overview of the main orbital properties of HD 76225 (Escorza et al. 2019).

P [days]	ecc	T_0 [HJD]	ω [°]	K_1 [km s^{-1}]	γ [km s^{-1}]	f(m) [$M_{\odot}$]
2410 ± 2	0.098 ± 0.005	2451159 ± 400	267 ± 3	6.11 ± 0.04	30.34 ± 0.02	0.0561 ± 0.0010

Bonačić Marinović et al. 2008; Dermine et al. 2013; Escorza et al. 2020). Additionally, the surface s-process abundances of Ba stars are a tracer of the chemical production inside the AGB star that polluted them (e.g. De Castro et al. 2016; Karinkuzhi et al. 2018; Roriz et al. 2021). Keeping mixing and dilution processes (e.g. Charbonnel et al. 2007; Stancliffe et al. 2007; Aoki et al. 2008) in mind, one can combine the properties of the two stellar components and the Ba star abundances to learn about nucleosynthesis models as well (e.g. Cseh et al. 2022). When one tries to use Ba star observations to constrain models, the largest observational uncertainty comes from the WD companions since they are cool, dim, and directly undetectable in most systems. Their masses are a key input parameter to both binary and nucleosynthesis models, but very few absolute masses have been determined since these are single-lined spectroscopic systems, and there is normally no information about the orbital inclinations (Pourbaix & Jorissen 2000, for example, published a few exceptions based on Hipparcos astrometry). Escorza & De Rosa (in prep) combined radial-velocity (RV) data with astrometric measurements from the Hipparcos and Gaia missions to determine the astrometric orbital parameters and the companion masses of the Ba stars studied by Jorissen et al. (2019) and Escorza et al. (2019). This short contribution presents a proof-of-concept of this methodology using HD 76225.

2. Stellar and binary properties of HD 76225

HD 76225 (HIP 43703) is a main-sequence Ba star, first proposed as such by North et al. (1994). Allen & Barbuy (2006a,b) determined its surface chemical abundances, including the s-process elements, and Escorza et al. (2019) determined the spectroscopic stellar parameters, the luminosity and the mass of the Ba star primary, as well as the spectroscopic orbital elements of the system. These characteristics have been summarised in Tables 1 and 2. Additionally, Fig. 1 highlights the location of HD 76225 on the Hertzsprung-Russel diagram (left) and on the eccentricity-period diagram (right) together with other well-known members of the Ba star family.

As far as the authors are aware, there is no direct evidence of the presence of a WD companion in the system, i.e. no UV excess has been reported as it has been the case for other systems (e.g. Böhm-Vitense et al. 2000; Gray et al. 2011). However, its high s-process enhancement suggests that HD 76225 underwent mass transfer from an AGB star and the spectroscopic mass-function determined by Escorza et al. (2019) is compatible with the presence of a WD in the system (see Table 2).

These observational constraints were obtained from spectra from the HERMES high-resolution spectrograph (Raskin et al. 2011), the CORAVEL spectrometer (Baranne et al. 1979), and the FEROS spectrograph (Kaufer et al. 2000). However, in order to get the mass of the invisible companion in a single-lined spectroscopic binary, one needs the orbital inclination of the system and spectra alone is not enough to determine this.

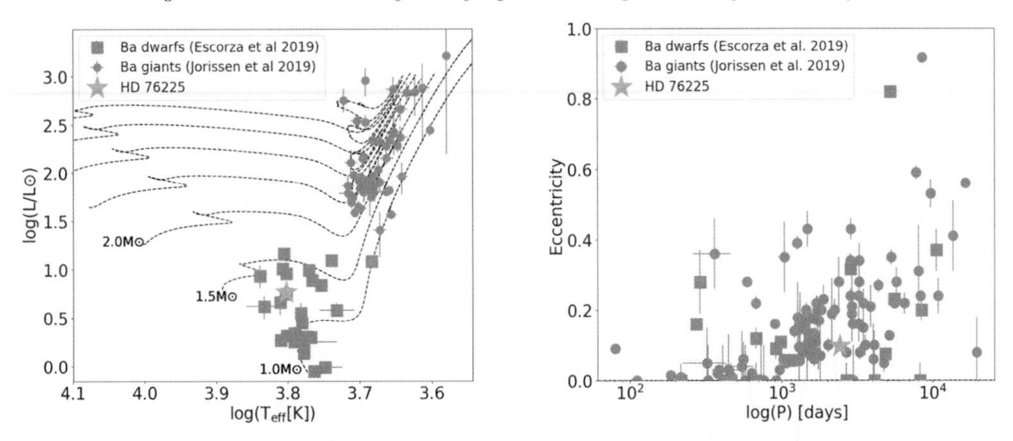

Figure 1. Hertzsprung-Russel diagram (left) and eccentricity-period diagram (right) of Ba dwarf and giants highlighting the location of the Ba dwarf HD 76225.

3. The determination of the WD companion's mass

For this contribution, we combined the individual radial-velocity points from HERMES and CORAVEL published by Escorza et al. (2019) with astrometric data from Hipparcos and Gaia. We used the catalogue positions from the Hipparcos mission (Perryman et al. 1997) and the re-reduction of the intermediate astrometric data (IAD; van Leeuwen 2007). Additionally, since the latest data release of the Gaia mission, Early DR3 or EDR3 (Lindegren et al. 2021), does not contain individual astrometric measurements yet, we used the catalogue positions and proper motions and the Gaia predicted scan epochs and angles†. Following a similar methodology to that used by De Rosa et al. (2020); Kervella et al. (2020) and Venner et al. (2021), among others, to determine the masses of exoplanets and using the code ORVARA, developed by Brandt et al. (2021b), we fit a single Keplerian model to all the different data sets at the same time employing a parallel-tempering Markov chain Monte Carlo (PTMCMC, Foreman-Mackey et al. 2013).

ORVARA also uses the Hipparcos-Gaia Catalog of Accelerations (HGCA, Brandt 2018) to ensure the proper cross-calibration of the two astrometric data sets when comparing their proper motions. HD 76225 shows a significant astrometric acceleration (proper motion difference between Hipparcos and Gaia) which is a key constraint to obtain an accurate and precise measurement of the two stellar masses. The code first fits the RV data, allowing RV points from each instrument to have a different RV zero point. Then the absolute astrometry is included and fit for the five astrometric parameters (positions, α and δ, proper motions, μ_α and μ_δ, and parallax, ϖ) using HTOF Brandt et al. 2021a) at each MCMC step. On top of the five astrometric parameters, we fit 10 parameters: the six Keplerian orbital elements (semimajor axis, a, eccentricity, e, time of periastron passage, T_0, argument of periastron, ω, orbital inclination, i, and longitude of the ascending node, Ω), the masses of the two components (M_{Ba} and M_{WD}) and a RV jitter per instrument (one for CORAVEL and one for HERMES) to be added to the RV uncertainties.

We assumed uninformative priors for all the orbital elements, but we adopted a Gaussian prior for the primary mass. We used the value given in Table 1 but used 3 times the error bar as sigma to account for systematic errors not accounted for in the statistical uncertainty. We used 15 temperatures and for each temperature we use 100 walkers with 100,000 steps per walker. The MCMC chains converged quite quickly, but we discarded the first 300 recorded steps (the first 15000 overall, as we saved every 50) as the burn-in phase to produce the results presented in Sect 3.3.

† Queried using the Gaia Observation Forecast Tool: https://gaia.esac.esa.int/gost/

Table 3. Overview of the MCMC results.

Parameter	Median $\pm$ 1σ	Parameter	Median $\pm$ 1σ
Period, P [days]	2405 $\pm$ 2	**Parallax, ϖ [mas]**	5.91 $\pm$ 0.03
Eccentricity, e	0.094 $\pm$ 0.003	**Ascending node, Ω [°]**	49 $\pm$ 2
Semimajor axis, a [AU]	4.3 $\pm$ 0.2	**Inclination [°]**	102 $\pm$ 3
Argument of periastron, ω [°]	266 $\pm$ 2	**Primary mass [$M_\odot$]**	1.2 $\pm$ 0.2
Time of periastron, T_0 [HJD]	2455974 $\pm$ 11	**Secondary mass [$M_\odot$]**	0.58 $\pm$ 0.06

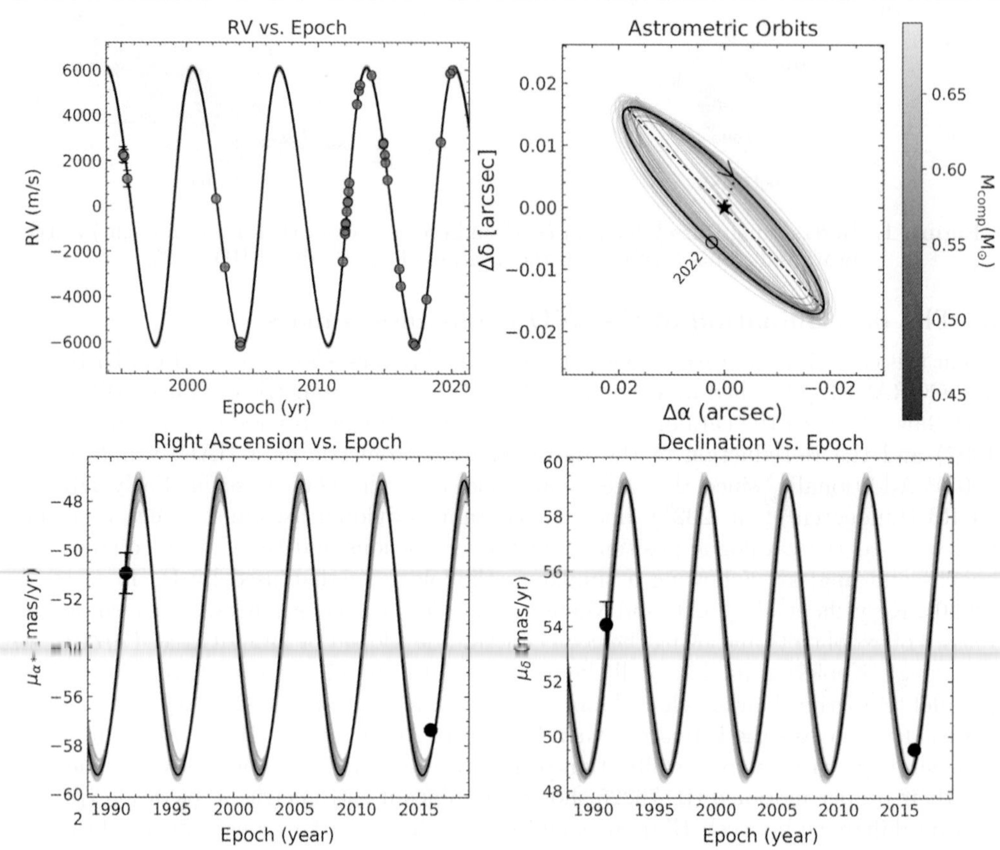

Figure 2. Fit to the RV data (top left, with CORAVEL data prior to 2005 and HERMES data newer than 2010) and to the Hipparcos (at the 1991.25 epoch) and Gaia (at the 2015.5 epoch) proper motions (bottom plots). The top right plot shows the projection of the orbit on the sky. The orbit with the highest likelihood is plotted as a thicker black line and 40 additional orbits are included, colour-coded as a function of the WD mass.

For more details about the computational implementation in ORVARA and HTOF and for case studies showing the performance of the code we refer to the mentioned publications.

4. Results and conclusions

Table 3 shows the derived posterior probabilities obtained from the MCMC orbital fit. Figure 2 shows the fit to the RV data (top left) and to the absolute Hipparcos-Gaia astrometry (bottom plots) as well as the projection of the orbit on the sky (top right). In each plot, the orbit with the maximum-likelihood is plotted with a thicker black line, but we included 40 additional solutions for different WD masses. Finally, Fig. 3 is a corner plot that shows the correlations among some of the derived parameters, especially the masses and the semimajor axis.

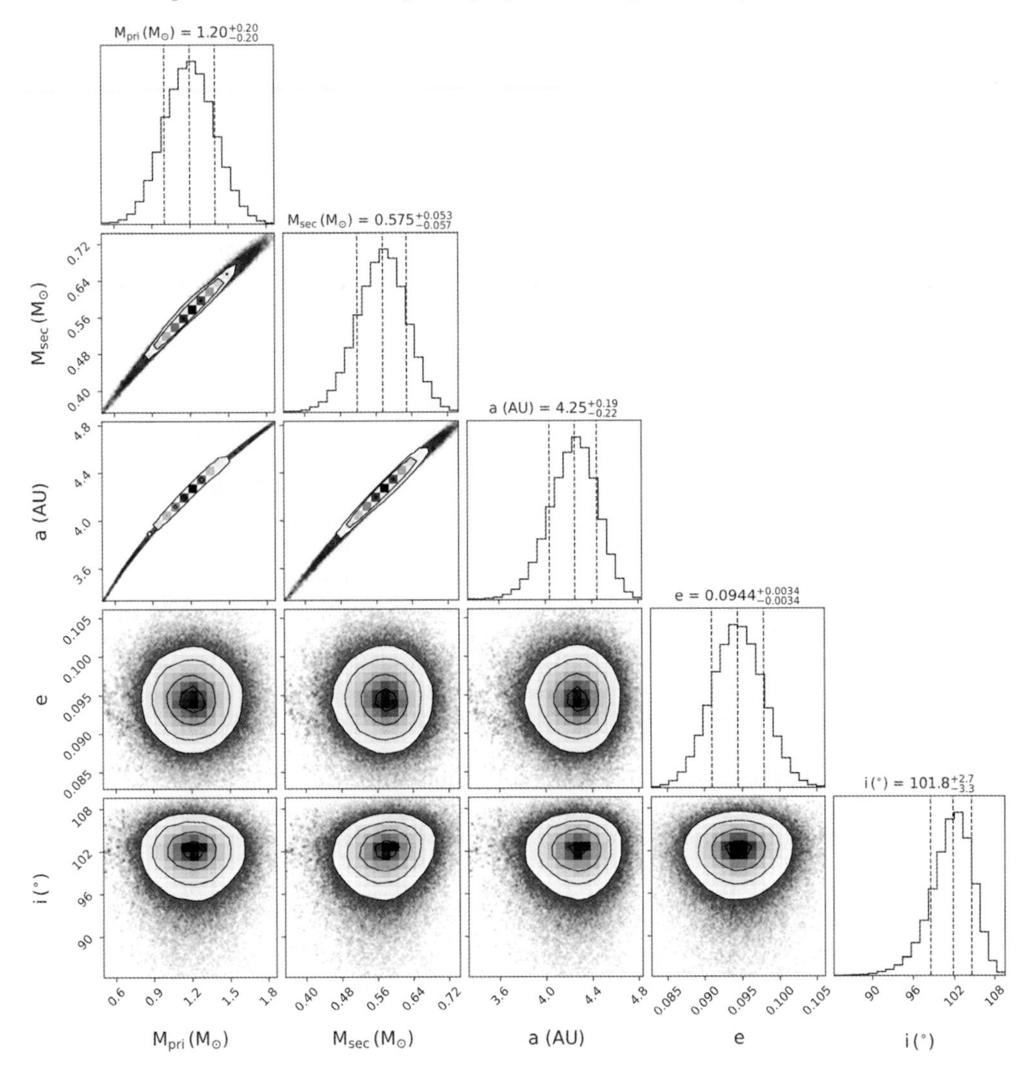

Figure 3. Corner plot of some of the derived parameters.

Our joint orbital fit of the RV and astrometric data yields a period of 2405 $\pm$ 2 days and an eccentricity of 0.094 ± 0.003, which, in spite of the small uncertainties, are almost compatible with the results published by Escorza et al. (2019) using only RV data (see Table 2). Additionally, we get a well constrained inclination for the system of $102^\circ \pm 3^\circ$. This information combined with the primary mass, for which we had prior independent information, led to a secondary mass of 0.58 ± 0.06 $M_\odot$. Now, using an initial-final mass relationship (IFMR), we can estimate the initial mass of the AGB companion that polluted HD 76225. Using the IFMR published by El-Badry et al. (2018) from Gaia DR2, we obtained 1.8 ± 0.7 $M_\odot$. This mass is compatible with the idea that low-mass AGB stars ($< 3\,\mathrm{M}_\odot$; Lugaro et al. e.g. 2003; Karinkuzhi et al. e.g. 2018) are responsible for the pollution of Ba stars. One could now use this information together with the abundance pattern on HD 76225 and put constraints on mass transfer and dilution mechanisms (as done for example by Stancliffe 2021 and Cseh et al. 2022). Finally, when the masses of the full sample are available, we will be able to look for correlations between the orbital parameters of Ba star systems and the WD masses and hopefully put constraints on mass-transfer mechanisms too.

References

Allen, D. M. & Barbuy, B. 2006a, *A&A*, 454, 895
Allen, D. M. & Barbuy, B. 2006b, *A&A*, 454, 917
Aoki, W., Beers, T. C., Sivarani, T., Marsteller, B. et al. 2008, *ApJ*, 678, 1351
Baranne, A., Mayor, M., & Poncet, J. L. 1979, *Vistas in Astronomy*, 23, 279
Böhm-Vitense, E., Carpenter, K., Robinson, R., Ake, T., & Brown, J. 2000, *ApJ*, 533, 969
Bonačić Marinović, A. A., Glebbeek, E., & Pols, O. R. 2008, *A&A*, 480, 797
Brandt, T. D. 2018, *ApJS*, 239, 31
Brandt, G. M., Michalik, D., Brandt, T. D., et al. 2021a, *AJ*, 162, 230
Brandt, T. D., Dupuy, T. J., Li, Y., Brandt, G. M., et al. 2021b, *arXiv e-prints*, arXiv: 2105.11671
Busso, M., Gallino, R., Lambert, D. L., Travaglio, C., & Smith, V. V. 2001, *ApJ*, 557, 802
Charbonnel, C. & Zahn, J. P. 2007, *A&A*, 467, L15
Cseh, B., Világos, B., Roriz, M. P., et al. 2022, *arXiv e-prints*, arXiv: 2201.13379
de Castro, D. B., Pereira, C. B., Roig, F., Jilinski, E. et al. 2016, *MNRAS*, 459, 4299
De Rosa, R. J., Dawson, R., & Nielsen, E. L. 2020, *A&A*, 640, A73
Decin, L., Montargès, M., Richards, A. M. S., Gottlieb, C. A. et al. 2020, *Science*, 369, 1497
Dermine, T., Izzard, R. G., Jorissen, A., & Van Winckel, H. 2013, *A&A*, 551, A50
El-Badry, K., Rix, H.-W., & Weisz, D. R. 2018, *ApJ*, 860, L17
Escorza, A., Boffin, H. M. J., Jorissen, A., Van Eck, S., et al. 2017, *A&A*, 608, A100
Escorza, A., Karinkuzhi, D., Jorissen, A., Siess, L., et al. 2019, *A&A*, 626, A128
Escorza, A., Siess, L., Van Winckel, H., & Jorissen, A. 2020, *A&A*, 639, A24
Escorza, A. & De Rosa, R. J. 2022, *in prep*
Foreman-Mackey, D., Hogg, D. W., Lang, D., & Goodman, J. 2013, *PASP*, 125,306
Gray, R. O., McGahee, C. E., Griffin, R. E. M., & Corbally, C. J. 2011, *AJ*, 141,160
Jorissen, A., Van Eck, S., Mayor, M., & Udry, S. 1998, *A&A*, 332, 877
Jorissen, A., Boffin, H. M. J., Karinkuzhi, D., Van Eck, S., et al. 2019, *A&A*, 626, A127
Käppeler, F., Gallino, R., Bisterzo, S., & Aoki, W. 2011, *Reviews of Modern Physics*, 83, 157
Karakas, A. I. 2010, *MNRAS*, 403, 1413
Karinkuzhi, D., Van Eck, S., Jorissen, A., Goriely, S. et al. 2018, *A&A*, 618, A32
Kervella, P., Arenou, F., & Schneider, J. 2020, *A&A*, 635, L14
Kaufer, A., Stahl, O., Tubbesing, S., et al. 2000, *SPIE*, Vol. 4008, 459–466
Lindegren, L., Klioner, S. A., Hernández, J., Bombrun, A., et al. 2021, *A&A*, 649, A2
Lugaro, M., Herwig, F., Lattanzio, J. C., Gallino, R., & Straniero, O. 2003, *ApJ*, 586, 1305
Mayer, A., Jorissen, A., Paladini, C., Kerschbaum, F. et al. 2014, *A&A*, 570, A113
McClure, R. D. 1984, *PASP*, 96, 117
North, P., Berthet, S., & Lanz, T. 1994, *A&A*, 281, 775
Perryman, M. A. C., Lindegren, L., Kovalevsky, J., Hog, E., et al. 1997, *A&A*, 500, 501
Pourbaix, D. & Jorissen, A. 2000, *A&AS*, 145, 161
Raskin, G., van Winckel, H., Hensberge, H., Jorissen, A., et al. 2011, *A&A*, 526, A69
Roriz, M. P., Lugaro, M., Pereira, C. B., Sneden, C., et al. 2021, *MNRAS*, 507, 1956
Shetye, S., Van Eck, S., Jorissen, A., Van Winckel, H., et al. 2018, *A&A*, 620, A148
Shetye, S., Van Eck, S., Goriely, S., Jorissen, A., Escorza, A., et al. 2020, *A&A*, 635, L6
Stancliffe, R. J., Glebbeek, E., Izzard, R. G., & Pols, O. R. 2007, *A&A*, 464, L57
Stancliffe, R. J. 2021, *MNRAS*, 505, 5554
Udry, S., Jorissen, A., Mayor, M., & Van Eck, S. 1998, *A&AS*, 131, 25
Van Eck, S., Neyskens, P., Jorissen, A., Plez, B., Edvardsson, B. et al. 2017, *A&A*, 601, A10
van Leeuwen, F. 2007, *A&A*, 474, 653
Venner, A., Vanderburg, A., & Pearce, L. A. 2021, *AJ*, 162, 12

The Origin of Outflows in Evolved Stars
Proceedings IAU Symposium No. 366, 2022
L. Decin, A. Zijlstra & C. Gielen, eds
doi:10.1017/S1743921322000369

Tc-rich M stars: platypuses of low-mass star evolution

Shreeya Shetye[1,2], **Sophie Van Eck**[2], **Alain Jorissen**[2], **Lionel Siess**[2] **and Stephane Goriely**[2]

[1]Institute of Physics, Laboratory of Astrophysics, École Polytechnique Fédérale de Lausanne (EPFL), Observatoire de Sauverny, 1290 Versoix, Switzerland
email: shreeya.shetye@epfl.ch

[2]Institute of Astronomy and Astrophysics (IAA), Université Libre de Bruxelles (ULB), CP 226, Boulevard du Triomphe, B-1050 Bruxelles, Belgium

Abstract. The technetium-rich (Tc-rich) M stars reported in the literature (Little-Marenin & Little 1979; Uttenthaler *et al.* 2013) are puzzling objects since no isotope of technetium has a half-life longer than a few million years, and ^{99}Tc, the longest-lived isotope along the s-process path, is expected to be detected only in thermally-pulsing stars enriched with other s-process elements (like zirconium). Carbon should also be enriched, since it is dredged up at the same time, after each thermal pulse on the asymptotic giant branch (AGB). However, these Tc-enriched objects are classified as M stars, meaning that they neither have any significant zirconium enhancement (otherwise they would be tagged as S-type stars) nor any large carbon overabundance (in which case they would be carbon stars).

Here we present the first detailed chemical analysis of a Tc-rich M-type star, namely S Her. We first confirm the detection of the Tc lines, and then analyze its carbon and s-process abundances, and draw conclusions on its evolutionary status. Understanding these Tc-rich M stars is an important step to constrain the threshold luminosity for the first occurrence of the third dredge-up and the composition of s-process ejecta during the very first thermal pulses on the AGB.

Keywords. Stars: abundances - Stars: AGB and post-AGB - Hertzsprung-Russell and C-M diagrams - Nuclear reactions, nucleosynthesis, abundances - Stars: interiors

1. Introduction

The resonance lines of the radioactive element technetium (Tc) (i.e., having no stable isotopes) were first identified in several S-type stars by Merrill (1952). The isotope ^{99}Tc is the only one to be located along the path of the slow neutron-capture nucleosynthesis (s-process) occurring in the interior of asymptotic giant branch (AGB) stars. ^{99}Tc has a half-life of about 2 x 10^5 yr. Along with carbon and other s-process elements, ^{99}Tc is brought to the surface of AGB stars by a mixing process called the *third dredge-up* (TDU; Iben & Renzini 1983). The detection of Tc in some – but not all – S stars led to the discovery that these evolved stars come in two flavors: the Tc-rich stars (also known as 'intrinsic' S stars) are genuine thermally-pulsing AGB stars, while the Tc-poor stars† (also known as 'extrinsic' S stars) owe their s-process enhancement to a binary interaction with a former AGB companion (Iben & Renzini 1983; Jorissen *et al.* 1993). Hence, the presence of Tc is a very sensitive probe of the TDU and AGB nucleosynthesis.

† More information on Tc-poor S stars can be found in Jorissen *et al.* (1988, 1993); Van Eck *et al.* (1999, 2000b,a) and Shetye *et al.* (2018).

Stars experiencing the TDU are expected to show at their surface an amount of carbon and of s-process elements increasing with the number of thermal pulses and should evolve according to the spectral sequence M $\rightarrow$ MS $\rightarrow$ S $\rightarrow$ SC $\rightarrow$ C stars. The spectral classification from M to S is based on the identification in M stars of TiO bands only and of TiO and ZrO bands in S-type stars (where Zr originates from the s-process enhancement). The Tc-rich nature of some Mira variables of spectral type M was discovered by Little-Marenin & Little (1979) and Little *et al.* (1987). The Tc-rich M stars constitute an interesting group of stars because they show clear signatures of Tc but no presence of ZrO bands. Hence, the Tc-rich M stars could be the first objects on the AGB to undergo a TDU, where the Tc can be detected in the spectrum before enhancements of other s-process elements become measurable (Goriely & Mowlavi (2000)).

Despite their importance, a detailed spectroscopic investigation of the Tc-rich M stars is still lacking. The spectroscopic investigation of a large-sample of M stars is currently on-going, and will be published in a forthcoming paper (Shetye *et al.*. in prep.). In the current work, we discuss the pilot study of the Tc-rich M star S Her. S Her was first identified as a Tc-rich MS star by Little *et al.* (1987). Keenan *et al.* (1974) classified it as M4-7.5Se. In the current work, we confirm the presence of Tc from a high-resolution spectrum of S Her. We then examine its s-process element abundances and report our findings in Section 3.

2. Spectral Analysis

We collected high-resolution spectra of S Her with the HERMES spectrograph (Raskin *et al.* 2011) mounted on the 1.2m Mercator Telescope at the Roque de Los Muchachos Observatory, La Palma (Canary Islands). HERMES spectra have a spectral resolution of R = 85 000 and a wavelength coverage 380-900 nm. The HERMES spectra of S Her used in the current analysis have a signal-to-noise ratio of 60 in the V band to ensure the accurate determination of the stellar parameters and of the chemical abundances. It is important to keep in mind that the Tc-rich Mira M stars experience severe stellar variability, making their spectral analysis difficult with 1D, static model atmospheres. According to the AAVSO† International Database, S Her has a peak-to-peak variability in the V band of roughly 5 magnitudes.

The initial step in the spectral analysis of S Her was to confirm the presence of Tc in its spectrum. We used the three Tc I resonance lines located at 4238.19, 4262.27 and 4297.06 Å. In Figure 1, we compare the three Tc lines of S Her with an extrinsic (top panel) as well as an intrinsic (middle panel) S star from Shetye *et al.* (2021). From Figure 1, we report an unambiguous detection of Tc in the spectrum of S Her, hence making it a good candidate for our case study of Tc-rich M stars.

The stellar parameters of S Her are determined using the method described in Shetye *et al.* (2018). This method was developed for the stellar-parameter derivation of S stars and was applied on a large sample of Tc-rich S stars (Shetye *et al.* 2019, 2020, 2021). We refer the reader to Shetye *et al.* (2018) for details. In summary, this method compares the high-resolution HERMES spectra to a grid of synthetic S-star MARCS model spectra (Van Eck *et al.* 2017). This grid spans a large range in effective temperatures, in surface gravities, and most importantly in the carbon/oxygen ratios and the s-process enhancement levels, the latter two being relevant parameters since they impact significantly thermally-pulsing AGB star spectra. Furthermore, we located S Her in a color-color diagram using the photometric indices $J-K$ and $V-K$ as described in Section 3 and Figure 5 of Van Eck *et al.* (2017). This photometric estimate of the stellar

† https://www.aavso.org/

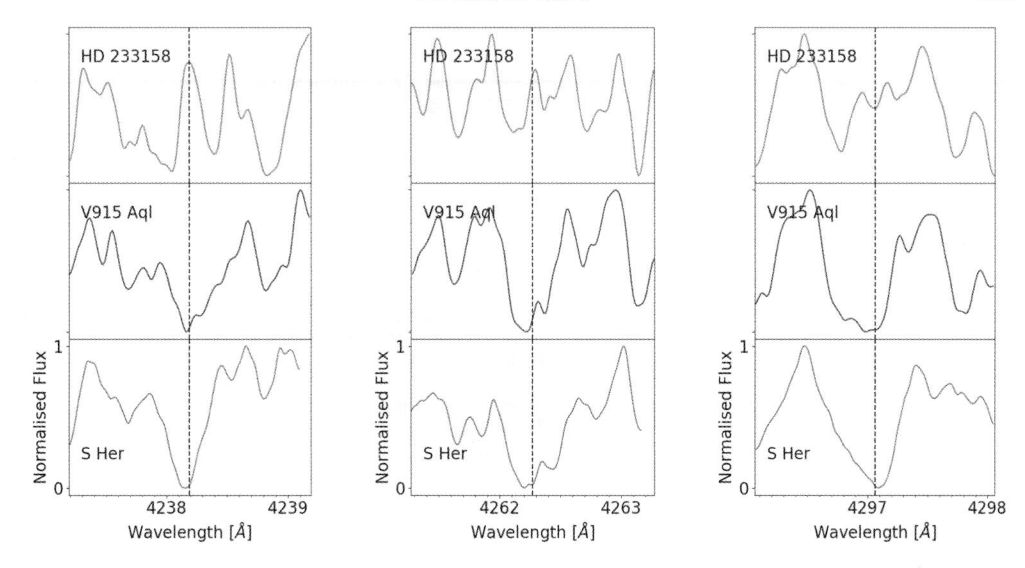

Figure 1. Spectral region around the three Tc I lines (4238.19, 4262.27 and 4297.06 Å) for the Tc-rich M star S Her. For comparison purposes, the spectra of a Tc-poor S star (HD 233158, in the top panels) and of a Tc-rich star (V915 Aql, in the middle panels) from **?** are also plotted. The spectra have been arbitrarily normalized and binned by a factor of 1.5 to increase the S/N ratio.

Table 1. The stellar parameters of S Her used in the current analysis.

Teff (K)	**log g (dex)**	**[Fe/H] (dex)**	**C/O**	**[s/Fe] (dex)**	L **($L_\odot$)**
2700	0.0	0.0	0.5	1.0	4200

Notes:
[1] The method to derive the atmospheric parameters is described in Section 2.
[2] The luminosity was derived using the Gaia eDR3 parallax and the same method as described in Shetye *et al.* (2018, 2021).

parameters is consistent with the spectroscopic one. The adopted parameters are listed in Table 1.

We used the atomic line lists provided by Shetye *et al.* (2018, 2021) for the investigation of s-process abundances in S Her. The results of the abundance analysis of a selection of heavy elements are described in the next section.

3. Implications

3.1. *The chemical status of S Her*

According to Figure 1, S Her is undoubtedly a Tc-rich M star. The detection of Tc is an unambiguous evidence for the occurrence of the third dredge-up and s-process nucleosynthesis. The presence of Tc in AGB stars undergoing TDU episodes is usually accompanied by an overabundance in other s-process elements. Even though S Her shows clear Tc lines, our analysis could not detect any enrichment in other s-process elements. The match between the observed and synthetic spectra in many spectral windows is however far too bad to derive abundances. Conversely, in the wavelength regions where the match is acceptable, there are either very few s-process lines present or they are not good abundance probes given the low temperature of S Her. In conclusion, the low temperature of S Her and its large stellar variability make the derivation of s-process abundances extremely challenging (even almost impossible!).

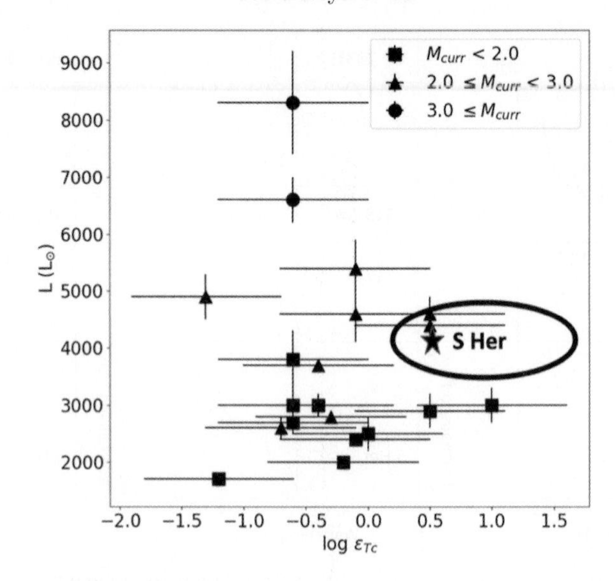

Figure 2. Figure adapted from Figure 9 of Shetye *et al.* (2021) to compare the stellar luminosity of S Her and its Tc abundance with that of Tc-rich S stars of various masses.

The spectral window including the Tc line regions is well reproduced by the synthetic spectra. We used the 4262.27 Å Tc I line to derive the Tc abundance. This line is the least blended line amongst the three available Tc lines (Little-Marenin & Little 1979; Shetye *et al.* 2021). We derived an upper limit of log $\epsilon_{Tc} \sim +0.5$ dex on the Tc abundance of S Her. In Figure 2, we compare the location of S Her in the luminosity vs Tc abundance diagram from Shetye *et al.* (2021). Surprisingly, this upper limit for the Tc abundance in the M star S Her is comparable with the highest Tc abundances encountered in S-type stars from the sample of Shetye *et al.* (2021).

3.2. *S Her and R Dor*

Considering the above-mentioned difficulties faced by spectral synthesis to derive abundances we decided to consider a differential analysis instead. We looked for an M-type giant star with properties similar to those of S Her (mainly in terms of its effective temperature and variability). We found that the stellar parameters of R Dor, classified as M8III:e by Keenan & McNeil (1989), are very similar to those of S Her. The UVES† spectrum of R Dor shows exceptional resemblance with that of S Her, according to Figure 3 which compares lines from Tc and other s-process elements in R Dor and S Her. From the first three panels of Figure 3, it is obvious that the spectra of R Dor and S Her are identical except for the Tc lines. As shown above, S Her is clearly a Tc-rich M star while R Dor is not Tc-rich. For Ba and Zr lines, the difference is not as clear (panels 4 and 5 of Figure 3). We also inspected some r-process element lines and found no clear difference either between R Dor and S Her.

4. Prospects

The Tc-rich M stars constitute an intriguing class of AGB stars. It remains difficult to decide whether the non-detection of overabundances for other s-process elements (apart from Tc) in S Her is spurious and results from the complexity of optical spectra of Mira variables or if Tc-rich M stars are truly AGB stars caught just after their first thermal

† https://www.eso.org/sci/observing/tools/uvespop/interface.html

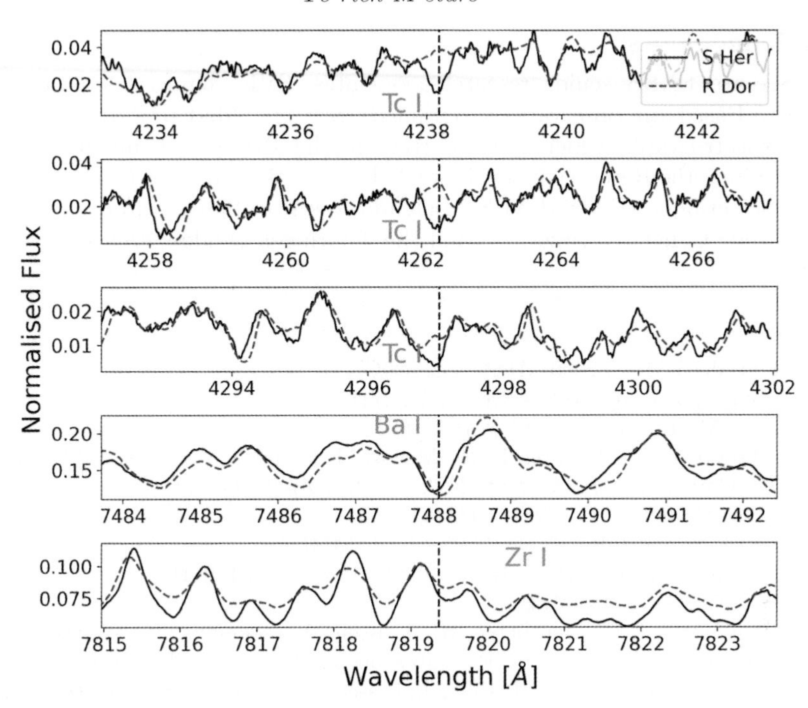

Figure 3. Comparison of s-process lines in S Her (black, continuous line) and R Dor (dashed line). From top to bottom: Tc I line at 4238.191 Å, Tc I line at 4262.27 Å, Tc I line at 4397.06 Å, Ba I line at 7488.077 Å, and Zr I line at 7819.37 Å.

pulse and with no s-process signature other than the presence of Tc. Nevertheless, Tc-rich M stars might carry clues about the luminosity of the first occurrence of the TDU and its dependence on the mass and metallicity of the stars. But such clues remain to be unfolded.

In the future, investigating the Tc-rich M stars should proceed along the following avenues:

- *Going to infrared spectroscopy*

The optical spectra of oxygen-rich AGB stars are dominated by molecules making the atomic diagnostics quite difficult. The near-infrared regions have been successfully used for the chemical analysis of AGB stars (Smith & Lambert 1985, 1990). It will be interesting to study the near-IR spectral windows, with the goal of finding some s-process lines as well as getting constraints on the carbon isotopic ratio. A combined study of optical (for Tc) and infrared spectra thus appears essential.

- *Large-scale chemical investigation of Tc-rich M stars*

The chemical pattern of S Her is quite puzzling. The presence of Tc without any other s-process overabundances might either be a special characteristic of S Her or be common among Tc-rich M stars. Hence, a large-scale study of M stars (Shetye *et al.*, in prep.) will be important to check for a possible diversity among the Tc-rich M stars.

Furthermore, the derived chemical abundances of such a sample of Tc-rich M stars can be used for a quantitative comparison between observations and stellar evolution models.

- *A spectroscopic investigation using non-static AGB model atmospheres*

The current poor agreement between the synthetic and observed spectra of Tc-rich M stars could be resolved by making use of non-static AGB models. In the future, abundance analysis using 3D hydrodynamic models should be attempted.

Acknowledgements

LS and SG are FNRS senior research associates. SVE thanks Fondation ULB for its support. Based on observations obtained with the Mercator Telescope and the HERMES spectrograph, which is supported by the Research Foundation - Flanders (FWO), Belgium, the Research Council of KU Leuven, Belgium, the Fonds National de la Recherche Scientifique (F.R.S.-FNRS), Belgium, the Royal Observatory of Belgium, the Observatoire de Genève, Switzerland and the Thüringer Landessternwarte Tautenburg, Germany.

References

Goriely, S. and Mowlavi, N. 2000, *A&A*, 362, 599G

Iben, I., Jr. and Renzini, A. 1983, *ARA&A*, 21, 271I

Jorissen, A. and Mayor, M. 1988, *A&A*, 198, 187J

Jorissen, A., Frayer, D. T., Johnson, H. R., Mayor, M. & Smith, V. V. 1993, *A&A*, 271, 463J

Keenan, P. C., Garrison, R. F., & Deutsch, A. J. 1974, *ApJS*, 28, 271

Keenan, P. C. & McNeil, R. C. 1989, *ApJS*, 71, 245

Little-Marenin, I. R. and Little, S. J. 1979, *AJ*, 84, 1374L

Little, S. J., Little-Marenin, I. R., & Bauer, W. H. 1987, *AJ*, 94, 981L

Merrill, P. W. 1952, *ApJ*, 116, 21M

Raskin, G., van Winckel, H., Hensberge, H., Jorissen, A., Lehmann, H., Waelkens, C., Avila, G., de Cuyper, J. -P., Degroote, P., Dubosson, R., Dumortier, L., Frémat, Y., Laux, U., Michaud, B., Morren, J., Perez Padilla, J., Pessemier, W., Prins, S., Smolders, K., van Eck, S., & Winkler, J. 2011, *A&A*, 526A, 69R

Shetye, S., Van Eck, S., Jorissen, A., Van Winckel, H., Siess, L., Goriely, S., Escorza, A., Karinkuzhi, D., & Plez, B. 2018, *A&A*, 620A, 148S

Shetye, S., Goriely, S., Siess, L., Van Eck, S., Jorissen, A. & Van Winckel, H. 2019, *A&A*, 625L, 1S

Shetye, S., Van Eck, S., Goriely, S., Siess, L., Jorissen, A., Escorza, A., & Van Winckel, H. 2020, *A&A*, 635L, 6S

Shetye, S., and Van Eck, S., Jorissen, A., Goriely, S., Siess, L., Van Winckel, H., Plez, B., Godefroid, M., & Wallerstein, G. 2021, *A&A*, 650A, 118S

Smith, V. V. and Lambert, D. L. 1985, *ApJ*, 294, 326S

Smith, V. V. and Lambert, D. L. 1990, *ApJ*, 294, 326S

Uttenthaler, S. 2013, *A&A*, 556A, 38U

Van Eck, S. and Jorissen, A. 1999, *A&A*, 345, 127V

Van Eck, S. and Jorissen, A. and Udry, S. and Mayor, M. and Burki, G. and Burnet, M. & Catchpole, R. 2000a, *A&AS*, 145, 51V

Van Eck, S. and Jorissen, A. 2000b, *A&A*, 360, 196V

Van Eck, S., Neyskens, P., Jorissen, A., Plez, B., Edvardsson, B., Eriksson, K., Gustafsson, B., Jørgensen, & Nordlund, A. 2018, *A&A*, 601A, 10V

The Origin of Outflows in Evolved Stars
Proceedings IAU Symposium No. 366, 2022
L. Decin, A. Zijlstra & C. Gielen, eds.
doi:10.1017/S1743921322000138

The impact of UV photons from a stellar companion on the chemistry of AGB outflows

M. Van de Sande[1,2] and T. J. Millar[3]

[1]School of Physics and Astronomy, University of Leeds, Leeds LS2 9JT, UK
email: m.vandesande@leeds.ac.uk

[2]Institute of Astronomy, KU Leuven, Celestijnenlaan 200D, 3001 Leuven, Belgium

[3]Astrophysics Research Centre, School of Mathematics and Physics, Queen's University Belfast, University Road, Belfast BT7 1NN, UK
email: tom.millar@qub.ac.uk

Abstract. Binary interaction with a stellar or planetary companion has been proposed to be the driving mechanism behind large-scale asymmetries, such as spirals and disks, observed within AGB outflows. We developed the first chemical kinetics model that takes the effect of a stellar companions's UV radiation into account. The presence of a stellar companion can initiate a rich photochemistry in the inner wind. Its impact is determined by the intensity of the UV radiation and the extinction the radiation experiences. The outcome of the inner wind photochemistry depends on the balance between two-body reactions and photoreactions. If photoreactions dominate, the outflow can appear molecule-poor. If two-body reactions dominate, chemical complexity within the outflow can increase, yielding daughter species with a large inner wind abundance. A comprehensive view on the molecular content of the outflow, especially combined with abundance profiles, can point towards the presence of a stellar companion.

Keywords. Stars: AGB and post-AGB, circumstellar matter, astrochemistry, molecular processes

1. Introduction

Spherical asymmetry is prevalent within AGB outflows. Both small-scale asymmetries, such as density-enhanced clumps (e.g., Leão *et al.* (2006); Khouri *et al.* (2016); Agúndez *et al.* (2017)), and large-scale structures, such as spirals (e.g., Mauron & Huggins (2006); Maercker *et al.* (2016)) and disks (e.g., Kervella *et al.* (2014); Homan *et al.* (2018)), have been widely observed. These asymmetries are thought to be induced by binary interaction with a stellar or planetary companion (e.g., Decin *et al.* (2015); Ramstedt *et al.* (2017); Moe & Di Stefano (2017); Decin *et al.* (2020)). A stellar companion will radiate part of its energy in the ultraviolet (UV). These photons may disrupt the chemistry throughout the outflow, potentially allowing molecular abundances to be a tool to distinguish between a stellar and planetary companion.

We developed the first chemical kinetics model that takes the effect of UV radiation from a nearby stellar companion into account. Our one-dimensional model is a first approximation of the effects on the chemistry within an AGB outflow, paving the way for future model development. In Sect. 2, we summarise the main characteristics and limitations of the model. Highlights of our results are given in Sect. 3, followed by our

Table 1. Physical parameters of the grid of chemical models.

Density structures	Smooth outflow Two-component clumpy outflow f_{ic}= 0.3, f_{vol}= 0.3, $l_* = 4 \times 10^{12}$ cm One-component clumpy outflow f_{ic}= 0.0, f_{vol}= 0.3, $l_* = 4 \times 10^{12}$ cm
Outflow density, $\dot{M}$ - v_∞	10^{-5} M$_\odot$ yr^{-1} - 15 km s^{-1} 10^{-6} M$_\odot$ yr^{-1} - 5 km s^{-1} 10^{-7} M$_\odot$ yr^{-1} - 5 km s^{-1}
Stellar radius, R_*	2 $\times 10^{13}$ cm
Stellar temperature, T_*	2330 K
Exponent temperature power-law $T(r)$, ϵ	0.7
Onset of dust extinction, R_{dust}	2, 5 R_*
Companion temperature, T_{comp}, and radius, R_{comp},	4000 K - 1.53 $\times$ 10^{10} cm 6000 K - 8.14 $\times$ 10^{10} cm 10 000 K - 6.96 $\times$ 10^{8} cm
Inner radius of the model	1.025 $\times$ R_{dust}

conclusions in Sect. 4. A full description of the model and all our results can be found in Van de Sande & Millar (2022).

2. Chemical model

2.1. *Physics*

The physical parameters of the model are listed in Table 1. The chemical model is an extension of Van de Sande & Millar (2019), where we included the effect of stellar UV photons. Internal photons, i.e. stellar and companion photons, are extinguished by dust as well as geometrically diluted. The visual extinction experienced by internal photons depends on the onset of dust extinction, R_{dust}. Since the gas-phase model does not treat dust formation, dust is assumed to be present throughout the model. To simulate different dust condensation radii, we choose R_{dust} = 2 and 5 R_*.

To include the effects of an inhomogeneous outflow in our one-dimensional model, we use the porosity formalism (Van de Sande *et al.* (2018)). The outflow is divided into a stochastic two-component medium, composed of overdense clumps and a tenuous interclump region. The clumpiness of the outflow is set by three parameters: the interclump density contrast, f_{ic}, the clump volume filling factor, f_{vol}, and the clump's size at the stellar surface, l_*. Besides a smooth outflow, we consider a two-component outflow and a one-component outflow (void interclump). These three density structures allow us to probe different behaviours of extinction experienced by interstellar and internal photons.

2.2. *Chemistry*

The parent species and their initial abundances for the O-rich and C-rich outflows are based on observations. They are taken from Agúndez *et al.* (2020), who compiled (ranges of) observed abundances in the inner regions of AGB outflows.

The stellar companion's UV flux is approximated by blackbody radiation. We vary over three types of companion, characterised by their blackbody temperature, T_{comp}, and radius, R_{comp}: a red dwarf (T_{comp} = 4000 K, R_{dust} =1.53 $\times$ 10^{10} cm), a solar-like star (T_{comp} = 6000 K, R_{dust} = 8.14 $\times$ 10^{10} cm), and a white dwarf (T_{comp} = 10 000 K and R_{dust} = 6.96 $\times$ 10^{8} cm). Cross sections are used to calculate the unshielded photodissociation and photoionisation rates. These are mainly taken from the Leiden Observatory

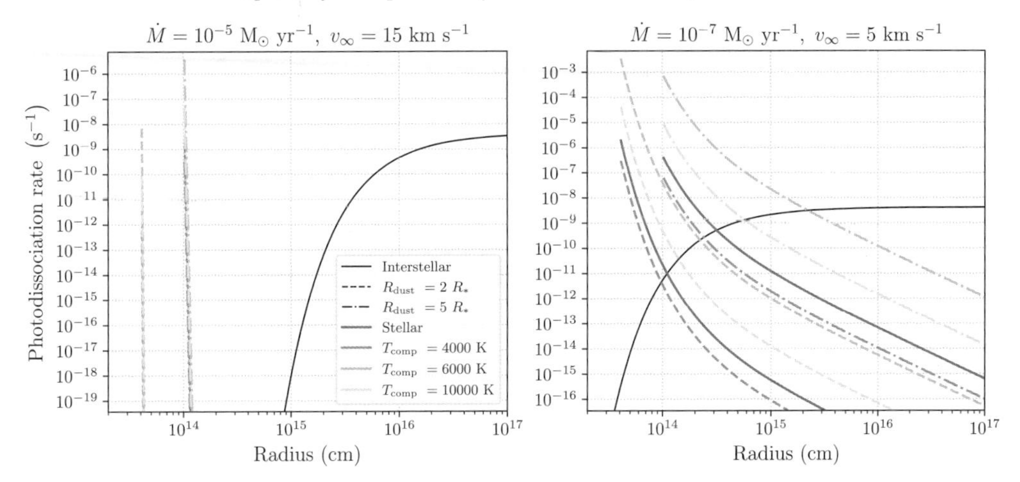

Figure 1. Photodissociation rates of SO in different outflow densities (left: highest outflow density, right: lowest outflow density). Black solid line: rate caused by interstellar UV photons. Different shades of grey show the rate caused by stellar or companion UV photons. From darkest to lightest: stellar UV photons ($T_* = 2330$ K), red dwarf companion UV photons, solar-like companion UV photons, white dwarf companion UV photons. Different line styles show the location of the onset of dust extinction. Dashed: $R_{\rm dust} = 2$ R$_*$, dotted: $R_{\rm dust} = 5$ R$_*$.

Database (Heays *et al.* (2017))†. If cross section are not available, the rate coefficients are estimated by scaling the unshielded interstellar rate by the ratio of the integrated fluxes of companion photons to interstellar photons over the 912 – 2150 Å range. This scaling, which is often used in astrochemistry, can lead to large errors, particularly for photoionisation rates at low black body temperatures. The specific scalings adopted for $T_{\rm comp} = 4000$ and 6000 K can be found in Van de Sande & Millar (2022).

Figure 1 shows the photodissociation rate of SO in a high density and low density smooth outflow, caused by interstellar, stellar and companion UV outflows and varying over $R_{\rm dust}$. The onset of photodissociation is set by the outflow density and $R_{\rm dust}$, as they determine the extinction experienced. The rate is determined by the intensity of the companion's radiation, set by $T_{\rm comp}$ and $R_{\rm comp}$. A solar-like companion yields the largest photodissociation rates; a red dwarf yields photodissociation rates comparable to those of stellar photons.

2.3. *Limitations*

Several assumptions go into our model. Since we build on our previous work, the companion is assumed to be located at the centre of the AGB star. While unrealistic, the effects of misplacing the companion within 5 R_* (the largest value of $R_{\rm dust}$) is negligible compared to the scale of the outflow. Additionally, because the model is one-dimensional, orbital effects cannot be taken into account. The underlying assumption of a continuously radiating companion is relatively reasonable for close-by companions, considering the limited fraction of the entire outflow that experiences occultation of the companion by the AGB star: gas at 10 R_* will see the UV field of a companion orbiting at 2 R_* for 80% of its orbital period. For more details, we refer to Van de Sande & Millar (2022).

3. Results

The presence of a stellar companion can initiate a rich photochemistry in the dense inner wind, photodissociating and photoionising parents. This type of chemistry is typical

† `https://home.strw.leidenuniv.nl/~ewine/photo/`

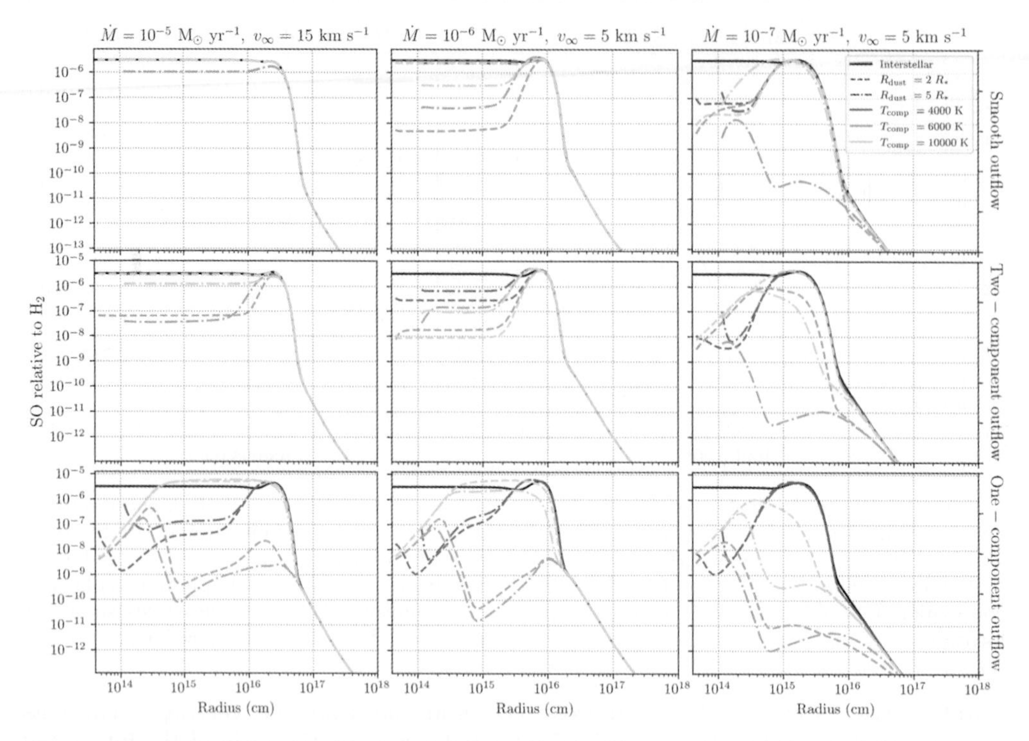

Figure 2. Fractional abundance of SO in O-rich outflows with different outflow densities (columns) and density structures (rows). Black, solid lines: results including interstellar photons only. Different shades of grey show the results included stellar and companion photons. From darkest to lightest: red dwarf companion UV photons, solar-like companion UV photons, white dwarf companion UV photons. Different line styles show the location of the onset of dust extinction. Dashed: $R_{\rm dust} = 2$ R$_*$, dotted: $R_{\rm dust} = 5$ R$_*$.

for the outer, tenuous regions of the outflow. The outcome of the inner wind photochemistry and its effects on the chemistry throughout the entire outflow depend mainly on the extinction experienced by internal photons (set by the outflow density and structure), as well as on the intensity of the companion's UV radiation (set by $T_{\rm comp}$ and $R_{\rm comp}$).

As the outflow density decreases and/or the outflow becomes more porous, the amount of UV radiation throughout the outflow increases. The extinction in the smooth outflow with $\dot{M} = 10^{-5}$ M$_\odot$ yr^{-1} is too large for internal photons to have any impact. We can distinguish between high UV outflows, where internal photons experience little extinction, and low UV outflows, where they experience moderate amount of extinction. High UV outflows have a low outflow density and/or are highly porous. In our grid, these are the one-component outflow with $\dot{M} = 10^{-6}$ M$_\odot$ yr^{-1} and all outflows with $\dot{M} = 10^{-7}$ M$_\odot$ yr^{-1}. In these outflows, two-body reactions are slower than photoreactions. This inhibits reformation of newly photodissociated parent species and any subsequent reactions between daughter species, reducing the outflow to an apparently molecule-poor state. Consequently, low UV outflows have a higher outflow density and/or a lower porosity. These are all porous outflows with $\dot{M} = 10^{-5}$ M$_\odot$ yr^{-1} and the smooth and two-component outflow with $\dot{M} = 10^{-6}$ M$_\odot$ yr^{-1}. In these outflows, two-body reactions are faster than photoreactions. Chemical complexity increases, as newly formed daughter species can readily react before further photoreactions. Species that are otherwise produced in the outer regions can now be abundantly present in the inner wind. This can result in abundance profiles similar to those of parent species, with a large inner wind abundance followed by a gaussian decline. A red dwarf companion does not significantly

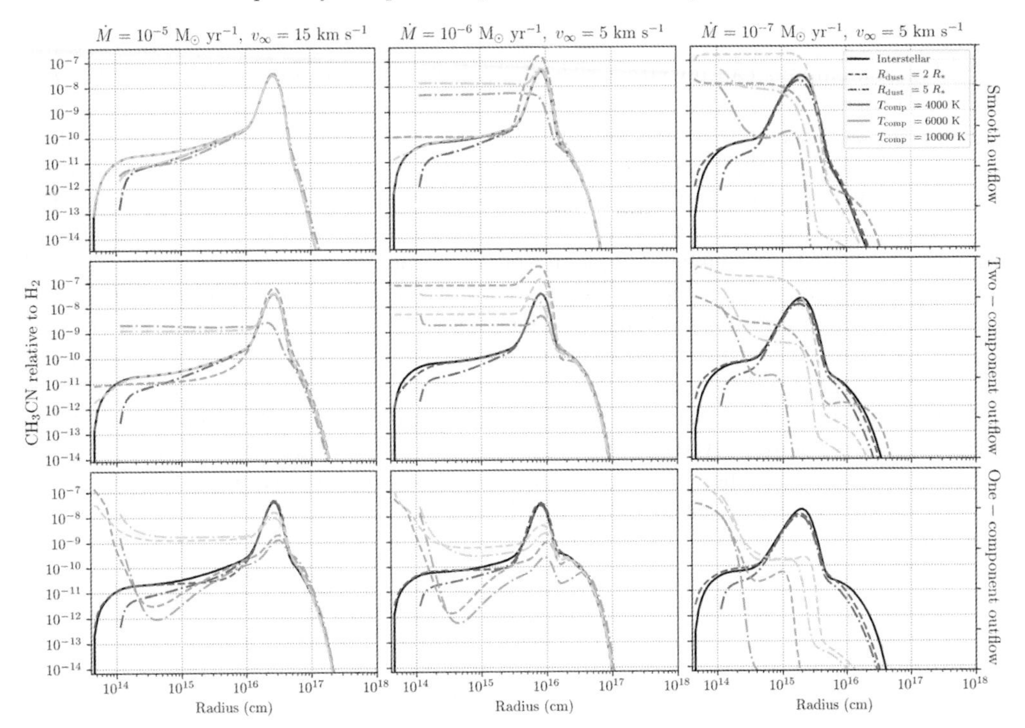

Figure 3. Fractional abundance of CH_3CN in C-rich outflows with different outflow densities (columns) and density structures (rows). Black, solid lines: results including interstellar photons only. Different shades of grey show the results included stellar and companion photons. From darkest to lightest: red dwarf companion UV photons, solar-like companion UV photons, white dwarf companion UV photons. Different line styles show the location of the onset of dust extinction. Dashed: $R_{\text{dust}} = 2$ R$_*$, dotted: $R_{\text{dust}} = 5$ R$_*$.

influence the chemistry, except in outflows with low outflow density and a porous structure. A solar-like companion shows the largest impact. Despite the larger temperature of the white dwarf companion, its compact size reduces the impact on the photochemistry.

Figure 2 shows the abundance of the parent SO in O-rich outflows. In most low UV outflows, the initial abundance of SO appears to be reduced by up to three orders of magnitude, followed by an increase in its abundance forming a bump. The bump is caused by reformation of SO via S + OH, two photodissociation products. This type of profile has been observed around higher mass-loss rate O-rich AGB stars (Danilovich *et al.* (2016, 2020)). In high UV outflows with a bright solar-like or white dwarf companion, as well as the low UV one-component outflow with $\dot{M} = 10^{-5}$ M$_\odot$ yr^{-1} and a solar-like companion, OH itself is photodissociated, reducing or eliminating the bump and rendering the outflow apparently poor in SO.

Figure 3 shows the abundance of the daughter CH_3CN in C-rich outflows. In low UV outflows, CH_3CN is formed in the innermost regions via the reaction HCN + CH_3^+, followed by dissociative recombination with an electron. This reaction channel highlights the delicate balance between two-body reactions and photoreactions: enough UV radiation is necessary to produce CH_3^+, a second-generation daughter of the parent CH_4, while the parent HCN should still be abundantly present. The resulting parent-like abundance profile (large initial abundance followed by a gaussian decline) has been observed around IRC+10216 (Agúndez *et al.* (2015)). In high UV outflows, any newly formed CH_3CN is rapidly photodestroyed.

4. Conclusions

Stellar companions can initiate a rich photochemistry in the inner regions of the outflow. The effects on the chemistry throughout the outflow depend mainly on the outflow density and structure, as well as on the companion's UV radiation field. The resulting chemistry is determined by the balance between two-body reactions and photoreactions. In outflows where internal photons experience an intermediate amount of dust extinction and where mass-loss rates are higher, two-body reactions are faster than photoreactions. This can increase the chemical complexity of the outflow, with daughter species showing a parent-like abundance profile of a large inner abundance followed by a gaussian decline. In outflows where internal photons experience little dust extinction, photoreactions are faster. Parents and newly formed daughters can be rapidly photodissociated, rendering the outflow apparently molecule-poor.

Our chemical model is the first to take any effects of a stellar companion on the chemistry throughout an AGB outflow into account. We find that the resulting chemistry is highly dependent on the outflow's density and structure, the type of stellar companion, as well as the assumed parent species and their initial abundances. Nonetheless, our simplified model shows that the chemical composition of the outflow can be used as a tool to distinguish between a stellar and substellar companion, and paves the way for further (three-dimensional) model development.

Acknowledgements

MVdS acknowledges support from the Research Foundation Flanders (FWO) through grant 12X6419N and the European Union's Horizon 2020 research and innovation programme under the Marie Skłodowska-Curie grant agreement No 882991. TJM gratefully acknowledges the receipt of a Leverhulme Emeritus Fellowship and the STFC for support under grant reference ST/P000312/1 and ST/T000198/1.

References

Agúndez, M., Cernicharo, J., Quintana-Lacaci, G. *et al.* 2015, *ApJ*, 814, 143
Agúndez, M., Cernicharo, J., Quintana-Lacaci, G. *et al.* 2017, *A&A*, 601, A4
Agúndez, M., Martínez, J. I., de Andres, P. L. *et al.* 2020, *A&A*, 637, A59
Danilovich T., De Beck E., Black J. H. *et al.* 2016, *A&A*, 588, A119
Danilovich T., Richards A. M. S., Decin L. *et al.* 2020, *MNRAS*, 494, 1323
Decin L., Richards A. M. S., Neufeld D. *et al.* 2015, *A&A*, 574, A5
Decin, L., Montargès M., Richards, A. M. S. *et al.* 2020, *Science*, 369, 1497
Heays, A. N., Bosman, A. D. and van Dishoeck, E. F. 2017, *A&A*, 602, A105
Homan W., Danilovich T., Decin L. *et al.* 2018, *A&A*, 614, A113
Kervella, P., Montargès, M., Ridgway, S. T. *et al.* 2014, *A&A*, 564, A88
Khouri, T., Maercker, M., Waters, L. B. F. M. *et al.* 2016, *A&A*, 591, A70
Leão, I. C., de Laverny, P., Mékarnia, D. *et al.* 2006, *A&A*, 455, 187
Maercker M., Danilovich T., Olofsson H. *et al.* 2016, *A&A*, 591, A44
Mauron N. & Huggins P. J. 2006, *A&A*, 452, 257
Moe M. & Di Stefano R. 2017, *ApJS*, 230, 15
Ramstedt, S., Mohamed, S., Vlemmings, W. H. T. *et al.* 2017, *A&A*, 605, A126
Van de Sande, M., Sundqvist, J. O., Millar, T. J. *et al.* 2018, *A&A*, 616, A106
Van de Sande, M. & Millar, T. J. 2019, *ApJ*, 873, 36
Van de Sande, M. & Millar, T. J. 2022, *MNRAS*, 510, 1204

The Origin of Outflows in Evolved Stars
Proceedings IAU Symposium No. 366, 2022
L. Decin, A. Zijlstra & C. Gielen, eds.
doi:10.1017/S1743921322001119

Chemistry on hot astrochemical dust surfaces: Sulfur in AGB outflows

Amy Wolstenholme-Hogg[1], **Alexander D. James**[1], **John M. C. Plane**[1] **and Marie Van de Sande**[2]

[1]School of Chemistry, University of Leeds, Leeds, LS2 9JT, United Kingdom
email: A.James1@leeds.ac.uk

[2]School of Physics and Astronomy, University of Leeds, Leeds, LS2 9JT, United Kingdom

Abstract. Astrochemical models treat dust surfaces as ice covered. We investigate the effects of implementing increased bare dust binding energies of CO and S-bearing species on the chemistry in the outflows of asymptotic giant branch (AGB) stars. We demonstrate the potential for improving agreement with observations in the outflow of IK Tau.

Increasing the binding energies to measured and computationally derived values in high mass-loss AGB outflows increased the production of daughter species. Switching from a high binding energy on bare dust to weaker binding to ice, the gas phase abundance increased at a radius in agreement with observations of IK Tau, suggesting that displacement of bound species could contribute to this observational puzzle. Using a strong binding to bare dust, a gas phase increase was not observed, however parent species concentrations had to be increased by around a factor of four to explain observed concentrations.

Keywords. Surface Binding, Kinetics, AGB Outflows, Sulfur

---

1. Introduction

Chemistry on hot, bare dust surfaces affects the chemical constituents of a variety of astrochemical environments, from the outflows of stars on the Asymptotic Giant Branch (AGB), through the InterStellar Medium (ISM) to ProtoPlanetary Disks (PPDs). Some binding and reaction processes on ice surfaces have been studied Cuppen *et al.* (2017), but chemistry on bare dust is poorly constrained by experiment Penteado *et al.* (2017).

Most stellar dust is formed in the outflows of Asymptotic Giant Branch (AGB) stars Zhukovska & Henning (2013), where temperatures from 10s to 1000s K, interstellar radiation, and gas concentrations from 10 to 10^{13} cm^{-3} produce bare dust surfaces Van de Sande *et al.* (2019). There is widespread interest in new oxidation pathways to CO_2 and SO_2 Gobrecht *et al.* (2016), chemical routes to Complex Organic Molecules (COMs) from e.g. acetylene Millar (2016), and modelling concentrations of observed species in the outflows of AGB stars Danilovich *et al.* (2020).

Chemical kinetic models designed to investigate processes in these environments typically use chemical networks of many reactions to consume a relatively small number of "parent" species, which have initial concentrations constrained by observations, and produce "daughter" species McElroy *et al.* (2013).

Here we report on development of a technique to treat the binding of gas phase CO and sulfur bearing species to bare dust in a chemical model of AGB outflows. We examine the impact of bare dust binding energies, E_{bind}, on gas phase (observable) composition in outflows with generic initial parent species abundance (Carbon or Oxygen rich outflows)

Table 1. Physical parameters used in model outflows.

Physical Parameters	Carbon-Rich	Oxygen-Rich	IK Tau
Mass-loss rate, $M_\odot$ / yr^{-1}	1×10^{-5}	1×10^{-5}	5×10^{-6}
Stellar Temperature, T* / K	2000	2000	2100
Outflow velocity, ν_∞ / km s^{-1}	10	10	17.5
Stellar Radius, R* / cm	5×10^{13}	5×10^{13}	2×10^{13}
Drift velocity, ν_{drift} / km s^{-1}	15	15	10
Dust temperature, $T_{dust,*}$ / K	1950	1110	1110
Exponent of T, ϵ [a]	0.7	0.7	0.7
Exponent of T_{dust}(r)/dust temperature profile, s [b]	0.9	1.3	1.3

[a] See equation 2.1, [b] See equation 2.2.

Table 2. Grid of chemical models calculated.

Outflow	Species	E_{bind} treatment
C-rich	CO	Range from 855 to 10,300 K
C-rich	SO	Switch from 1,800 to 15,000 K
O-rich	SO	Switch from 1,800 to 15,000 K
IK Tau	SO, CO, SiO, SiS, CS, H_2S, SO_2, HS	Switch from ice to dust[a]
IK Tau	SO, CO, SiO, SiS, CS, H_2S, SO_2, HS	Separate dust bound species

[a] See Section 3, Table 3 for values.

and for oxygen rich AGB star IK Tau, where we attempt to explain the observed radial concentration profiles of SO and SO_2.

2. Methods

To identify targets for laboratory investigation of binding energies, electronic structure calculations with hybrid density functional/Hartree-Fock theory with the B3LYP functional and the 6-311+G(2d,p) basis set were used to estimate the binding energies of a range of species to a model Fe_2SiO_4 molecule and a cluster of 12 H_2O molecules, as proxies for astrochemical dust and ice surfaces, respectively. This is a relatively flexible basis set with both diffuse and polarisation functions. A CO or sulfur bearing molecule was allowed to relax into a cluster with the model surface and the reduction in energy of the system taken to be the binding energy.

The estimated binding energies were then implemented in an existing chemical kinetics model AGB outflows (Van de Sande *et al.* 2019). Tables 1 to 2 list the physical parameters and the grid of models calculated. All model runs used initial and final radii of 10^{15} and 10^{18} cm, respectively and a gas temperature, T, defined by equation 2.1:

$$T(r) = T_* \left(\frac{r}{R_*}\right)^{-\epsilon} \tag{2.1}$$

where, T_* and R_* are the stellar temperature and radius, respectively, and ϵ is a dimensionless power law exponent. The dust temperature is then determined by the energy balance between gas and dust, given by equation 2.2:

$$T_{dust}(r) = T_{dust*} \left(\frac{2r}{R_*}\right)^{-\frac{2}{(4+s)}} \tag{2.2}$$

where T_{dust*} is the stellar temperature determined from fits to radiative transfer modelling, and s describes the wavelength dependency of the dust opacity. Parent species concentrations were set as in Van de Sande & Millar (2022). The "C-rich" and "O-rich" model conditions used here represent generic initial parent species concentrations (Van de Sande, Walsh, & Millar 2021), but with high mass loss rates to maximise dust-gas interactions.

Table 3. Calculated binding energies of CO and S bearing species on Fe-silicate, as a model of bare dust, and ice surfaces.

Species	Binding energy on Fe-silicate kJ mol^{-1}	on H_2O Ice kJ mol^{-1}
SO	124	13
CO	86	7
SiO	245	41
SiS	116	43
CS	157	17
H_2S	74	20
SO_2	83	39
HS	218	21

As a first approximation, the binding energies for CO and SO in the chemical network were adjusted to those estimated by electronic structure calculations. This assumes that all dust in the outflow is bare, the opposite extreme case to previous work where all dust is implicitly assumed to be covered by ice (Van de Sande *et al.* 2019). To improve upon this, a switch in binding energy was implemented when more than two monolayers of ice were present on the dust. This approximates displacement of material strongly bound to the dust surface by the bulk ice species (mostly H_2O). Finally, material bound to bare dust was treated as a separate, non-interacting species, which could only form from gas uptake and be destroyed by thermal or photo-desorption. This represents the assumption that material bound to the bare dust surface can become trapped underneath layers of ice. If a significant effect is observed, this study can indicate the likely effect of each of these scenarios, without the involved process of implementing a full treatment such as a layered ice model.

3. Results and Discussion

Table 3 shows the results of the electronic structure calculations of binding energies. Binding to dust was always found to be significantly stronger than binding to H_2O ice, in the case of SO, CO, CS, and HS with E_{bind} larger by a factor of as much as 10. This suggests that implementation of binding to bare dust in chemical kinetic networks could have a significant effect on observable species.

3.1. *Bare dust in generic outflows*

In this case the binding energy used in previous work (Van de Sande *et al.* 2019) is increased to account for a bare dust surface. Increased binding energy for the parent CO leads to a longer lifetime and higher concentration on the dust surface, which allows reactions on the surface to compete with other routes to chemical synthesis. This results in competitive chemical pathways to surface bound and gas phase product daughter species. Figure 1 shows the effect on surface and gas phase parent CO and daughter CO_2 for a high mass-loss rate C-rich outflow. As the parent binding energy increases, the surface and gas phase concentrations of the CO_2 daughter increase, with a relatively small (< 1 %) impact on the parent concentration. Whilst in this case the gas phase CO_2 daughter concentration remains low, this does demonstrate that a treatment of binding to bare dust can alter the balance of competitive chemical pathways to more daughter species, e.g. including oxidation of CO under C-rich conditions.

A similar effect was found for the oxidation of SO to SO_2, both in C-rich and O-rich outflows, with gas phase SO_2 increasing by more than two orders of magnitude. This again demonstrates an altered balance of competing chemical pathways, and suggests that dust surfaces might be able to act as reservoirs for chemical species of increasing chemical complexity.

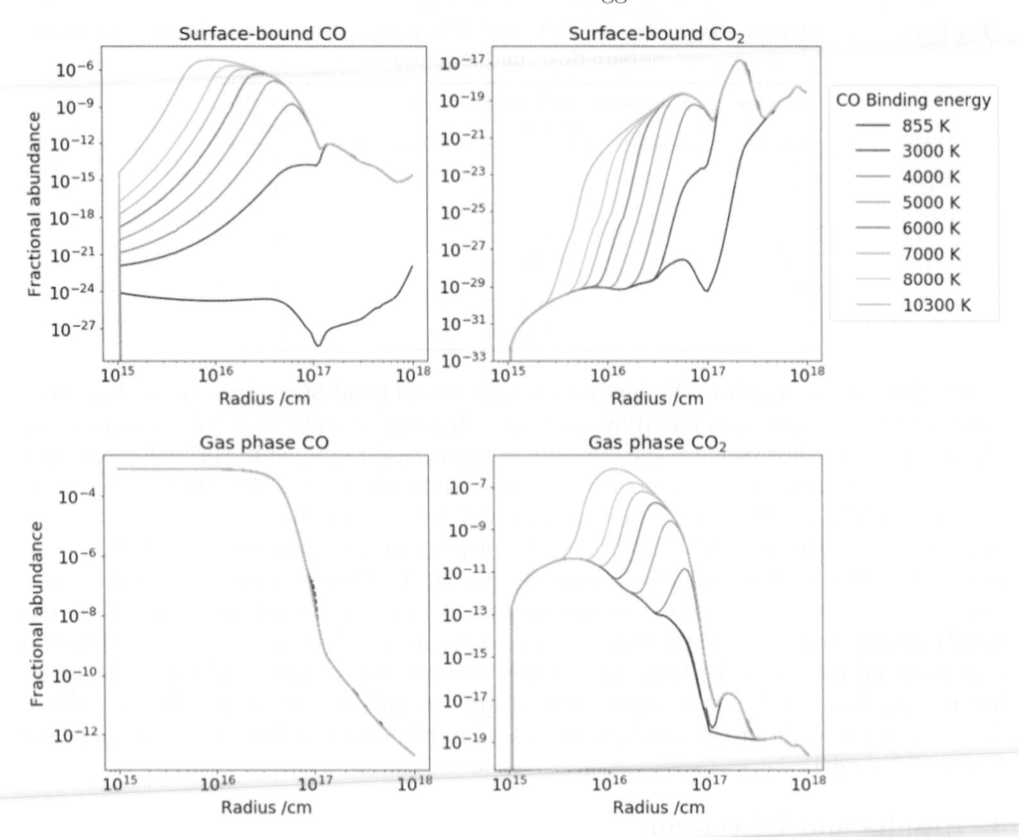

Figure 1. The fractional abundance of surface bound CO, surface-bound CO_2, gas phase CO and gas-phase CO_2 against radius with various binding energies of CO for a generic C-rich stellar outflow.

3.2. *Binding energy switch in IK Tau*

Switching the binding energy from a value representing binding to bare dust to one which represents binding to ice once two ice monolayers have formed on the surface approximates displacement of material strongly bound to the dust surface by the bulk ice species. This could lead to an increased gas phase concentration such as observed at around 10^{16} cm radius in the outflow of IK Tau (Danilovich *et al.* 2020). Figure 2 shows the effect on SO and SO_2 of implementing such a switch for IK Tau. In the relatively low radiation field less than 10^{16} cm from the star, SO and other species are able to accumulate on the dust surface, leading to a depletion of the initial parent abundance. When the surface becomes coated and the binding energy of the sulfur species switches to the lower ice-surface value, the temperature is still sufficiently high ($>$100 K) for thermal desorption to release the SO into the gas phase. This could be a mechanistic explanation for some puzzling observations, although other mechanisms which allow increased UV radiation field have also been proposed (Danilovich *et al.* 2020).

As shown in Figure 3, treating the bare dust-bound material as a separate species leads to a rather similar gas-phase abundance profile to the assumption of ice only binding, though the concentrations of all species in Table 3 were found to decrease by a factor of approximately four. This suggests that dust could act as a significant chemical reservoir. In the case of these species in the outflow of IK Tau, photo-desorption removes the ice layer in the outer reaches of the outflow. However for species which chemisorb strongly to the surface (essentially becoming part of the dust) material may remain bound to the

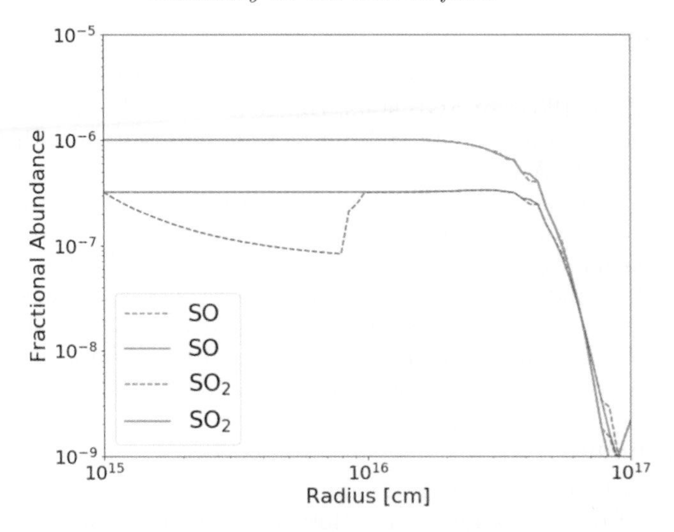

Figure 2. The fractional abundance of SO (green, initial concentration 3×10^{-7}) and SO_2 (red, initial concentration 1×10^{-6}) when using a binding energy switch at two monolayers ice coverage (dashed lines) compared to assuming an ice surface binding energy for IK Tau.

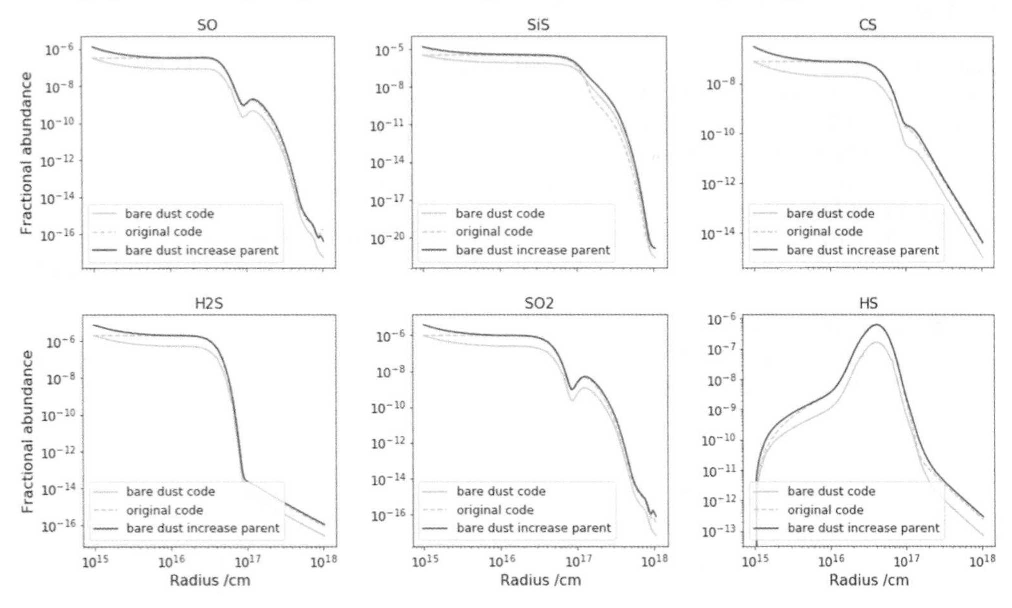

Figure 3. The gas-phase fractional abundance of sulfur species in the simulated outflow of IK Tau.

surface until its supply to the ISM. If reaction occurs in the warmer part of the outflow, complex products could remain bound to the dust, protected by layers of ice, and be carried to other astrochemical environments.

Measurement of binding to bare dust, rather than ice covered surfaces, using relevant laboratory techniques would be required to constrain the potential impact of bare dust in astrochemical environments. Stable species such as CO, CO_2 and SO_2 seem good candidates for initial investigations, since there are open questions in the literature where these species are relevant. For more reactive and hazardous compounds which are difficult to handle in the laboratory, computational methods to constrain binding energies may also be appropriate.

4. Conclusions

We have presented an initial assessment of the impact of treating binding to bare dust surfaces in the outflows of AGB stars. Different chemical routes were found to be competitive in the oxidation of CO and SO to CO_2 and SO_2 in both C-rich and O-rich high mass-loss rate AGB outflows. The implications of binding to bare dust were also investigated under several different assumptions for IK Tau. An SO abundance profile similar to observations, with an increase towards the end of the outflow, could be produced by assuming that ice which forms on the surface is able to displace other material, which then binds less strongly to the ice. If bare dust-bound species were treated as strongly bound to the surface, remaining protected by the ice, this allowed initial parent concentrations of sulfur species to be increased by a factor of four, suggesting that bare dust surfaces can act as significant chemical reservoirs.

Acknowledgements

The authors wish to acknowledge funding and excellent collaboration which was provided by the AEROSOL project (ERC Horizons 2020, 646758).

References

Cuppen H. M., Walsh C., Lamberts T., Semenov D., Garrod R. T., Penteado E. M., Ioppolo S., 2017, SSRv, 212, 1. doi:10.1007/s11214-016-0319-3

Penteado E. M., Walsh C., Cuppen H. M., 2017, ApJ, 844, 71. doi:10.3847/1538-4357/aa78f9

Zhukovska S., Henning T., 2013, A&A, 555, A99. doi:10.1051/0004-6361/201321368

Gobrecht D., Cherchneff I., Sarangi A., Plane J. M. C., Bromley S. T., 2016, A&A, 585, A6. doi:10.1051/0004-6361/201425363

Van de Sande M., Walsh C., Mangan T. P., Decin L., 2019, MNRAS, 490, 2023. doi:10.1093/mnras/stz2702

Millar T. J., 2016, JPhCS, 728, 052001. doi:10.1088/1742-6596/728/5/052001

Danilovich T., Richards A. M. S., Decin L., Van de Sande M., Gottlieb C. A., 2020, MNRAS, 494, 1323. doi:10.1093/mnras/staa693

Van de Sande M., Millar T. J., 2022, MNRAS, 510, 1204. doi:10.1093/mnras/stab3282

Van de Sande M., Walsh C., Millar T. J., 2021, MNRAS, 501, 491. doi:10.1093/mnras/staa3689

McElroy D., Walsh C., Markwick A. J., Cordiner M. A., Smith K., Millar T. J., 2013, A&A, 550, A36. doi:10.1051/0004-6361/201220465

The Origin of Outflows in Evolved Stars
Proceedings IAU Symposium No. 366, 2022
L. Decin, A. Zijlstra & C. Gielen, eds.
doi:10.1017/S1743921322000175

Circumbinary discs around post-AGB binaries as a result of binary interactions: an infrared interferometric view

Akke Corporaal, Jacques Kluska and Hans Van Winckel

Institute of Astronomy, KU Leuven, Celestijnenlaan 200D, 3001 Leuven, Belgium
email: akke.corporaal@kuleuven.be

Abstract. Post-Asymptotic Giant Branch (post-AGB) binary systems are binary interaction products. These stars have recently undergone a strong, but not well understood, binary interaction phase, leading to the formation of stable, compact circumbinary discs. These circumbinary discs are found to show many similar properties to protoplanetary discs around young stars. Here, we focus on one such system, namely IRAS 08544-4431 and resolve the inner regions of the complex circumstellar environment using multi-wavelength infrared interferometric techniques. The visibility data of PIONIER (H-band), GRAVITY (K-band), and MATISSE (L and N band) are analysed together using two families of geometric models, giving a good fit to all data.

Keywords. stars: AGB and post-AGB, binaries, circumstellar material, individual: (IRAS08544-4431), techniques: interferometric, high angular resolution

1. Introduction

Circumstellar discs are not only observed around young stellar objects (YSOs) but they are also present around various types of evolved stars. In this study, we focus on circumbinary discs around evolved low to intermediate mass ($0.8 - 8\ M_{\odot}$) binary systems. The presence of circumbinary discs around post-Asymptotic Giant Branch (post-AGB) binaries has been well established (De Ruyter *et al.* 2006; van Aarle *et al.* 2011; Van Winckel 2003, 2017). Photometric observations of such targets show an infrared excess in the spectral energy distributions (SEDs), the shape of which provides observational evidence of the presence of a stable circumbinary disc in the system. Fig. 1 shows an example of a SED of one of the post-AGB binary systems in the Galaxy.

Post-AGB binaries are believed to have undergone strong binary interaction during their evolution. These systems can, therefore, be considered as prime targets to study the outcome of binary interaction. While the presence of these discs is observationally well established, their formation process is not well understood yet (see e.g. the reviews by Van Winckel 2017 and De Marco & Izzard 2017). Moreover, the dynamical evolution of the central binary is poorly understood as the observed orbital properties of the total sample do not match the predicted ones. Many post-AGB binaries have periods and eccentricities unaccounted for in binary synthesis models (e.g. Oomen *et al.* 2018). Moreover, we do not yet identify these post-AGB systems as the progeny of AGB binary systems that have been found with e.g. the ALMA large program ATOMIUM (Decin *et al.* 2020). In this contribution we focus on the structure of the circumbinary discs as part of our longer-term goals to constrain the formation and evolution of these binary systems.

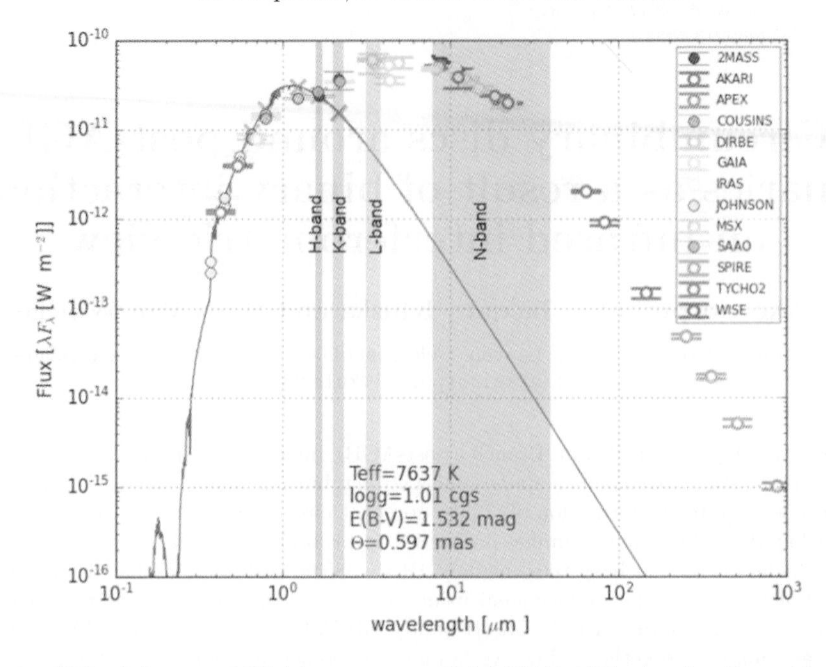

Figure 1. SED of the post-AGB binary system IRAS 08544-4431, showing the reddened photosphere of the star (full line) and the photometric data from several missions as well as the wavelength bands of interest for this study. Figure adapted from Oomen *et al.* (2018).

The circumbinary discs around these evolved binary systems show many similarities with protoplanetary discs around YSOs, despite their very different formation scenarios. Similarities include the stability and longevity (Bujarrabal *et al.* 2013a, 2015, 2017, 2018), disc mass (a few times $10^{-3}M_{\odot}$; e.g. Sahai *et al.* 2011; Hillen *et al.* 2014, 2015), and evidence for dust growth (e.g. Gielen *et al.* 2011). We use these discs as laboratories to study the disc structure and evolution at a very different stellar evolutionary phase than what YSOs provide. This includes the study of the formation of macrostructures, eventually leading to planet formation. Here we report the results of an observing campaign using all current VLTI instruments, to reveal the structure of the very inner disc regions of one post-AGB binary system, IRAS 08544-4431.

2. Multi-wavelength infrared interferometry

Infrared interferometric techniques are needed to spatially resolve the compact infrared emission of the very inner regions of the circumbinary disc. The current instruments on the VLTI, PIONIER, GRAVITY, and MATISSE, operating in the H, K, and L and N-bands, respectively, are perfectly suited for these observations. From Fig. 1 the corresponding wavelength ranges of these bands can be inferred. It shows that the SED of IRAS 08544-4431 is shifting from a star-dominated regime in the near-infrared (near-IR) to a disc-dominated regime in the mid-infrared (mid-IR). With interferometry in the near-IR, the structure of the optically thick disc inner rim is probed. Mid-IR interferometry, however, probes deeper into the disc, beyond the inner rim. We focused on modelling the visibility data, which is one of the interferometric observables. The visibility amplitude is the amplitude of the Fourier transform of the intensity or image on the sky.

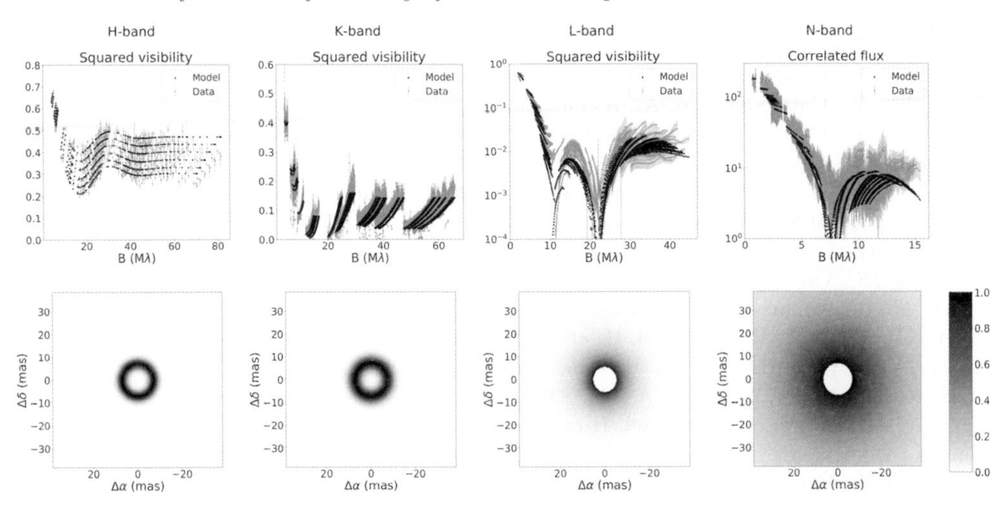

Figure 2. Top: Visibility data (grey dots) and the best-fitting geometric model (black diamonds) for the infrared interferometric bands. Bottom: images of the best-fitting geometric model of each band. Figure in gray-scale from Corporaal *et al.* (2021).

3. Results

We fitted geometric models to the visibility data of the different bands. Fig. 2 shows the visibility data overplotted with the best-fitting geometric model for the four bands and the corresponding model images. A family of geometric models was used to fit the data. Both a single temperature ring and a temperature gradient model were applied in our attempt to fit the data of each of the bands. From the χ^2 statistics, it was found that the near-IR bands were better fitted with a single temperature ring model while the mid-IR bands were better fitted with a temperature gradient model. The change of these best-fitting models from the near-IR to the mid-IR, in combination with information from the SED (Kluska *et al.* 2018), can be used to infer the structure of the inner rim of the disc. From the SED, it can be inferred that about one-third of the energy that is emitted by the post-AGB star, is captured by the circumbinary disc. To attain this amount of captured energy, a vertical extension somewhere in the disc is needed. From the difference in the best-fitting model between the near-IR and mid-IR, as well as from the amount of energy radiated by the disc in the H-band, we concluded that the vertical extension is at the inner rim. We propose that the inner rim is puffed-up and rounded, in a similar manner as protoplanetary discs around YSOs (Isella & Natta 2005).

4. Conclusions

Combining infrared interferometric instruments operating at different wavelengths, can give us an unprecedented, spatially-resolved view of the complex environment of the intriguing circumbinary discs around post-AGB binaries and of circumstellar discs in general. While in the near-IR we probe details of the inner rim of the disc, in the mid-IR we probe dust emission beyond the disc inner rim. By combining near-IR and mid-IR interferometric observations, we can conclude that the inner rim of the post-AGB binary system IRAS 08544-4431 is puffed-up and rounded. This work will be continued by modelling the disc with 3D radiative transfer models using MCMax3D (Min *et al.* 2009) with the goal of fitting both the photometric observations and the infrared interferometric observations of the four bands simultaneously.

References

Bujarrabal, V., Castro-Carrizo, A., Alcolea, J. *et al.*, 2013a, *A&A*, 557, L11
Bujarrabal, V., Castro-Carrizo, A., Alcolea, J. & Van Winckel, H, 2015, *A&A*, 575, L7
Bujarrabal, V., Castro-Carrizo, A., Alcolea, J. *et al.*, 2017, *A&A*, 597, L5
Bujarrabal, V., Castro-Carrizo, A., Van Winckel, H. *et al.*, 2018, *A&A*, 614, A58
Corporaal, A., Kluska, J., Van Winckel, H. *et al.*, 2021, *A&A*, 650, L13
Decin, L., Montargès, M., Richards, A., *et al.*, 2020, *Science*, 369, 6510
De Marco, O., & Izzard, 2017, R. G., *PASA*, 34, 1
De Ruyter, S., Van Winckel, H, Maas *et al.*, 2006, *A&A*, 448, 641
Gielen, C., Bouwman, J., Van Winckel, H, 2011, *A&A*, 533, A99
Hillen, M., Menu, J., Van Winckel, H. *et al.*, 2014, *A&A*, 568, A12
Hillen, M., de Vries, B.L̃., Menu, J. *et al.*, 2015, *A&A*, 578, A40
Isella, A. & Natta, A., 2005, *A&A*, 438, 899
Kluska, J., Hillen, M., Van Winckel, H *et al.*, 2018, *A&A*, 616, A153
Min, M., Dullemond, C.P., Dominik, C., de Koter, A. & Hovenier, J.W, 2009, *A&A*, 497, 155
Oomen, G-M., Van Winckel, H., Pols, O. *et al.*, 2018, *A&A*, 20, A85
Sahai, R., Claussen, M.J̃., Schnee, S. *et al.*, 2011, *ApJ*, 739, L3
van Aarle, E., Van Winckel, H., Lloyd Evans, T *et al.*, 2011, *A&A*, 530, A90
Van Winckel, H. 2003, *ARAA*, 41, 391
Van Winckel, H. 2017, Planetary Nebulae: Multi-Wavelength Probes of Stellar and Galactic Evolution, 323, 231

The Origin of Outflows in Evolved Stars
Proceedings IAU Symposium No. 366, 2022
L. Decin, A. Zijlstra & C. Gielen, eds
doi:10.1017/S1743921322000278

Pre-planetary nebulae: a context for principles, progress, and questions on how binaries and magnetic fields produce jets

Eric G. Blackman[1,2]

[1]Department of Physics and Astronomy, University of Rochester
Rochester, NY, 14621, USA
email: eric.blackman@rochester.edu

[2]Laboratory for Laser Energetics, University of Rochester
Rochester, NY, 14623, USA

Abstract. Astrophysical outflows treated initially as spherically symmetric often show evidence for asymmetry once seen at higher resolution. The preponderance of aspherical and multipolar planetary nebulae (PN) and pre-planetary nebulae (PPN) was evident after many observations from the Hubble Space Telescope. Binary interactions have long been thought to be essential for shaping asymmetric PN/PPN, but how? PPN are the more kinematically demanding of the two, and warrant particular focus. I address how progress from observation and theory suggests two broad classes of accretion driven PPN jets: one for wider binaries (PPN-W) where the companion is outside the outer radius of the giant and accretes via Roche lobe overflow, and the other which occurs in the later stages of CE for close binaries (PPN-C). The physics within these scenarios connects to progress and open questions about the role and origin of magnetic fields in the engines and in astrophysical jets more generally.

Keywords. (ISM:) planetary nebulae: general; ISM: jets and outflows; stars: magnetic fields; stars: winds, outflows; stars: AGB and post-AGB; (stars:) binaries (including multiple): close

1. Introduction and overview

Planetary nebulae (PN) are the penultimate evolutionary state of stars below $\sim 8M_{\odot}$ (e.g. Balick & Frank 2002) for stars massive enough to evolve off the main sequence during the age of the Universe. They are characterized by an ionization nebula sourced by photons from the hot exposed white dwarf (WD) at their core. Typically, the PN phase lasts of order 10^4 yr after which the remnant WD is left. Planetary nebulae have size scales $\sim 10^{17}$ cm and ages $\sim 10^4$ yr. Pre-Planetary nebulae (PPN) are reflection nebulae typically an order of magnitude smaller and an order of magnitude younger. They are more powerful in mechanical luminosity than PN, and are likely the earlier stage of a PN before sufficient mass clears to expose the ionizing core.

Early models of PN/PPN were based on the spherical interacting stellar wind (ISM) paradigm (Kwok et al. 1978) where a slow wind is followed by a fast wind and the interaction produces a shocked bubble. Evidence of asymmetry and bipolarity emerged later (Feibelman 1985; Gieseking et al. 1985; Miranda & Solf 1990; Lopez et al. 1993). This fostered generalizations of ISW models (GISW) with density asymmetries from equator to pole to explain the asphericity (Balick et al. 1987; Soker & Livio 1989). Basic ingredients of GISW models are present in more modern scenarios and simulations involving binaries, magnetic fields, and jets (Soker & Rappaport 2000; Blackman et al. 2001; Soker 2002; Balick et al. 2020; García-Segura et al. 2021; Ondratschek et al. 2021).

From the mid 1990s, high resolution HST observations revealed that asymmetry was the rule rather than the exception and the statistical categorization of morphologies became clearer. As a population, PN are overall 80% aspherical, up to 1/2 of the latter exhibiting jets (Balick & Frank 2002; De Marco & Soker 2011). The prevalence of asymmetric PPN among PPN is closer to 100%, with many showing multipolar structure (Borkowski et al. 1997; Sahai & Trauger 1998; Sahai et al. 2009).

The influence of binaries on shaping was proposed from the early days (Paczynski 1976; Soker & Livio 1989; Soker 1994, 1997; Soker & Livio 1994; Reyes-Ruiz & López 1999; Blackman et al. 2001) and is now considered essential to explain the high fraction of asymmetric PN/PPN, particularly when angular momentum (Soker 2006; Nordhaus et al. 2007) and outflow kinematics are considered. While PPN may represent a strongly ($\sim$ few100 yr) collimated jet phase due to close binary interaction, binaries likely also explain the asphericity evolution in the giant stellar wind phases that precede PPN (Decin et al. 2020). In general, binaries induce both equatorial and axial features. Examples of equatorial features include spiral arms and crystalline dust (Edgar et al. 2008; Mauron & Huggins 2006; Kim et al. 2017) and axial features include winds and jets perpendicular to the orbital plane (Hillwig et al. 2016).

Binaries also supply free energy for flows which can amplify magnetic fields (Blackman et al. 2001; Nordhaus et al. 2007; García-Segura et al. 2021; Ondratschek et al. 2021). This amplification may occur in an accretion disk onto a companion, a circumbinary disk, or a merger. The large-scale field that grows can in turn mediate launching and collimation of jets. Magnetic collimation acts to concentrate the pressure of a flow on axis, but the magnetic structures must themselves be collimated by ambient inertial envelopes or tori, as unbounded magnetic structures are unstable.

Observationally, > 200 PN have binaries (Boffin & Jones 2019; Jacoby et al. 2021) and an estimated $\sim 20\%$ of PN are preceded by close enough interaction to have incurred common envelope evolution (CEE) (Miszalski et al. 2009). The number of observed PN/PPN with binaries is less than the 50% fraction for all stars (Miszalski et al. 2009; Raghavan et al. 2010), but this is a lower limit on the prevalence of binary influence. In fact it is plausible that all aspherical PPN/PN are influenced by companions (Ciardullo et al. 1999; Bond 2000; Moe & De Marco 2006; Soker 2006; Corradi 2012; Jones 2020; Decin et al. 2020) including planets (Soker 1996; Reyes-Ruiz & López 1999; Nordhaus & Blackman 2006; De Marco & Soker 2011; Decin et al. 2020; Chamandy et al. 2021), particularly if we allow for the fact that the binary companion may be destroyed by tidal disruption during or after playing a dynamically significant role.

2. Kinematic demands on PPN/PN

PPN warrant special focus because they are kinematically more demanding than PN, and therefore provide stronger limits on source engines if indeed PPN are earlier stages of PN. The difference in kinematic demands is evident in comparing the approximate values for PPN and PN below. PPN fast wind durations are $\Delta t = 100 - 1000$ yr; outflow speeds $v \gtrsim 50$ km/s; mass in nebula $M = 0.5M_\odot$; mass-loss rate $\dot{M}_f = 5 \times 10^{-4}\ M_\odot$/yr; momentum injection rate $\Pi \sim 5 \times 10^{39}$ g $\cdot$ cm/s; mechanical luminosity $L_f \geqslant 8 \times 10^{35}$ erg/s. For the PPN slow winds, the corresponding values are $\Delta t = 6 \times 10^3$ yr, $v \sim 20$ km/s; $M = 0.5M_\odot$; $\dot{M}_f = 10^{-4} M_\odot$/yr; $\Pi \sim 2 \times 10^{29}$ g $\cdot$ cm/s; $L_s \sim 10^{34}$ erg/s.

For comparison, the PN fast wind properties are: $\Delta t = 10^4$ yr, $v \sim 2000$ km/s; $M = 10^{-4} M_\odot$; $\dot{M}_f = 10^{-8} M_\odot$/yr; $\Pi \sim 4 \times 10^{37}$ g $\cdot$ cm/s; $L_f \geqslant 1.3 \times 10^{34}$ erg/s. For the PN slow wind $\Delta t = 10^4$ yr, $v \sim 30$ km/s; $M = 0.1M_\odot$; $\dot{M}_f = 10^{-5} M_\odot$/yr; $\Pi \sim 6 \times 10^{38}$ g $\cdot$ cm/s; $L_s \sim 3 \times 10^{33}$ erg/s.

Most importantly, for PPN jets where sufficient data has been obtained, mostly all have momenta per unit time satisfying $d\left(Mv_j\right)_{PPN}/dt > 1000L/c > d\left(Mv_j\right)_{PN}/dt > L/c$ (Bujarrabal et al. 2001; Sahai et al. 2009, 2017). Thus, not only do PPN have higher momenta injection rates than the PN, these rates cannot be explained by optically thin radiative driving of jet outflows (Bujarrabal et al. 2001) and thus need another source, most likely involving accretion and close binary interaction (Blackman & Lucchini 2014).

2.1. *Momentum constraints and accretion modes*

The stringent momentum requirements of PPN outflows allow constraining the engine paradigm as follows (Blackman & Lucchini 2014). The mechanical luminosity of all non-relativistic astrophysical jets obey

$$L_m = \frac{1}{2}\dot{M}_{j,N}v_{j,N}^2 \leqslant \frac{1}{2}\frac{GM_a}{\dot{M}_a}r_{in} = \frac{1}{2}\dot{M}_a v_K^2 \tag{2.1}$$

where $\dot{M}_{j,N}$ is the "naked" jet mass ejection rate, $v_{J,N} \equiv Qv_K(r_{in})$, is the "naked" jet launch speed, $v_K = (GM_a/r_{in})^{1/2}$ is the Keplerian speed at the inner radius r_{in} of the assumed accretor, Q is a dimensionless number with a typical range $5 \gtrsim Q \gtrsim 1$ in MHD jet models (e.g. Blandford & Payne 1982; Pelletier & Pudritz 1992), and $\dot{M}_a$ is the accretion rate. Inequality (2.1) and the definition of Q imply

$$\dot{M}_a > Q^2\dot{M}_{j,N}. \tag{2.2}$$

As the jet runs into surrounding material, momentum conservation implies $\dot{M}_{j,N}Qv_k = \dot{M}_{j,obs}v_{j,obs}$, where $\dot{M}_{j,obs}$, and $v_{j,obs}$ are the observed jet mass ejection rate and speed that account for mass pileup. Solving for $\dot{M}_{j,N}$, and plugging into inequality (2.2) allows us to obtain a minimum for $\dot{M}_{j,N}$ and thus $\dot{M}_a$ given by

$$\dot{M}_a > Q^2\frac{M_{j,N}}{t_a} \simeq Q\frac{M_{j,obs}v_{j,obs}}{v_K t_a}, \tag{2.3}$$

where $M_{j,obs}$, $v_{j,obs}$ and t_a are the observed mass, speed and lifetime measured from observations. Numerically we have

$$\dot{M}_a \gtrsim 1.4\times10^{-4}\frac{M_\odot}{\text{yr}}\left(\frac{Q}{3}\right)\left(\frac{M_a}{M_\odot}\right)^{-1/2}\left(\frac{r_{in}}{R_\odot}\right)^{1/2}\left(\frac{M_{j,obs}}{0.1M_\odot}\right)\left(\frac{v_{j,obs}}{100\text{km/s}}\right)\left(\frac{t_{ac}}{500\text{yr}}\right)^{-1}. \tag{2.4}$$

So which modes of accretion satisfy this constraint? The values for a number of accretion modes together with the requirements for specific PPN are shown in Figure 1. Bondi-Hoyle-Lyttleton (BHL) onto secondary does not work because typical values would be $\dot{M}_{BH} = 1.15\times10^{-6}\dot{M}/\text{yr}$ (Huarte-Espinosa et al. 2013), for a primary wind of mass-loss rate $\dot{M}_W = 10^{-6}M_\odot/\text{yr}$, wind speed $v_W = 10$ km/s, primary mass $M_p = 1.5M_\odot$ and secondary mass $M_s = M_\odot$. Wind Roche lobe overflow (WRLOF) (Mohamed & Podsiadlowski 2012; Chen et al. 2017, 2018) also does not work for a typical separation $a \simeq 20$AU, dust acceleration radius $R_d \simeq 6R_p \simeq 10$AU, primary Roche lobe radius ~ 8.5AU, $\dot{M}_{WR} = 2\times10^{-5}M_\odot/\text{yr}$, and component masses $M_s = 0.6M_\odot$ and $M_p = M_\odot$.

The semi-empirically determined accretion rate for the Red Rectangle PPN is $\dot{M}_{RR} \gtrsim 5\times10^{-5}M_\odot/\text{yr}$, based on the luminosity of the far UV continuum of its HII region (Jura et al. 1997; Witt et al. 2009). This is also too small for most PPN.

There are two classes of accretion modes that do work. First, Roche lobe overflow (RLOF) (Meyer & Meyer-Hofmeister 1983; Ritter 1988) onto the companion from the primary envelope for a companion located outside the giant's envelope. For typical parameters, the RLOF rate is $\dot{M}_{RL} \simeq \rho_e R_e c_{s,e}^3/(GM_G) \gtrsim 5\times10^{-5}M_\odot/\text{yr}$ which is sufficient, where $\rho_e, R_e, c_{s,e}$ are the density, radius and sound speed of the outer giant envelope, and M_G is its mass.

A second viable accretion mode involves close binary interaction subsequent to when the secondary enters CE. This will only be successful after much of the envelope is unbound as I now explain. Accretion onto the core of the primary (Soker & Livio 1994) or the secondary within CEE can in principle be super-Eddington during the plunge phase of CE before the envelope is unbound, as estimates and simulations suggest $\dot{M}_{CE} \geqslant 10^{-3} M_\odot/\mathrm{yr}$, for secondary and primary masses $M_s = 0.6 M_\odot$ and $M_p = M_\odot$, respectively (Ricker & Taam 2012; Chamandy et al. 2018). But such modes require a "pressure valve" at the engine, otherwise accretion will be halted. Were the accretor a neutron star, neutrinos could supply this valve. For main sequence or WD accretors the jet itself must supply the valve. Estimates of the ram pressure and simulations show that a jet from a main sequence or WD companion during the plunge stage of CE is likely choked within the bound envelope (Lopez-Camara et al. 2021; Zou et al., in preparation). Accretion onto the primary core, or accretion from a shredded low-mass companion (Reyes-Ruiz & López 1999; Blackman et al. 2001; Nordhaus & Blackman 2006) or circumbinary accretion or merger are more effective (Ricker & Taam 2012; García-Segura et al. 2021; Ciolfi 2020; Ondratschek et al. 2021). These all take advantage of the deep gravitational potential well of an accretor and a substantial mass supply. The jet would be visible in the later stages when the CE envelope is unbound. The jet propagates along a reduced density axial channel (Zou et al. 2020; García-Segura et al. 2021; Ondratschek et al. 2021).

2.2. *Energy constraints and time sequence*

Energy constraints on PPN are also revealing. In cases measured, the energy in PPN outflows typically exceeds the envelope binding energy of the AGB host stars and exceeds the orbital energy from inspiral to observed radii from CE (Huggins 2012; Olofsson et al. 2015). This also points to the need for accretion or a merger to tap into the deeper gravitational potential wells.

While PN, unlike PPN, do not strongly constrain the energy or momentum, both PPN and PN do mutually provide a time sequence constraint on the jet and equatorial torus. In different sources, PPN/PN jets are observed to occur both before and after the equatorial dust tori form. In one sample, Huggins (2007) found that PPN jets follow tori by ~ 250 yr on average. That is consistent with equatorial ejecta helping to facilitate collimation, independent of whatever role magnetic fields might play. On the other hand, Tocknell et al. (2014) studied the kinematics of four post-common envelope PN. Three have jets that preceded CE ejection and one has 2 pairs of jets that follow the torus. Although CE is one natural way to get an equatorial torus, a torus may also form from an earlier RLOF phase (MacLeod et al. 2018a) as mass leaves through the L2 point and enters bound orbits.

2.3. *PN are plausibly later stages of accretion driven PPN*

If an accretion-like process onto the core of a companion of mass M_* powers PPN jets, then a connection between PPN and PN is kinematically consistent: the jet mechanical luminosity is

$$L_m \simeq \frac{GM_{*,\odot}\dot{M}_a\epsilon}{2R_i} = 4.5 \times 10^{36}\epsilon_{-1}\left(\frac{M_{*,\odot}\dot{M}_{a,-4}}{R_{i,10}}\right), \tag{2.5}$$

where ϵ_{-1} is a dimensionless efficiency from accretion to jet power in units of 0.1; $M_{*,\odot}$ is the accretor mass scaled in solar masses; $\dot{M}_{a,-4}$ is the mass accretion rate scaled in units of 10^{-4} $M_\odot/\mathrm{yr}$; and $R_{i,10}$ is the inner disk radius in units of 10^{10} cm. The naked jet speed is

$$v_{j,N} \sim Qv_K \simeq 1600Q \left(\frac{M_{*,\odot}}{R_{*,10}}\right)^{1/2} \text{km/s}, \tag{2.6}$$

so that the predicted observed PPN jet speed after mass pile-up from momentum conservation when the ejecta are optically thick, is given by

$$v_{j,obs} \simeq \frac{M_f v_f}{f_\Omega M_{env} + M_f} \sim 80\text{km/s}. \tag{2.7}$$

But once the outflow transitions to the PN stage the optical depth τ_d to dust scattering drops below unity, as estimated by

$$\tau_d = 2.5 \times 10^{-3} \left(\frac{n_d}{2.5 \times 10^{-13}\text{cm}^{-3}}\right) \left(\frac{\sigma_d}{10^{-8}\text{cm}^2}\right) \left(\frac{R}{10^{18}\text{cm}}\right), \tag{2.8}$$

where the dust number density n_d and scattering cross section σ_d are scaled to typical PN values. This is a simple explanation for the trend that PN have less power but faster winds. The naked jet, and thus the naked jet speed, is more exposed in PN as the optical depth decreases. More detailed transitions from PPN to PN speeds and powers can be predicted from engine models as a function of age and compared to individual sources or statistical observations.

3. Binary interactions: from weak to strong for low mass giants and low mass companions

Since the mechanisms of accretion that work to power PPN require binary interactions with orbital radii at least small enough for the primary to overflow its Roche lobe, the question of how sufficient numbers of binaries get close enough to produce the required number of PPN/PN arises. Observations suggest that at least 20% of PN have to have incurred CE (Miszalski et al. 2009). But since only 2.5% of PPN/PN should incur CE if tides alone are responsible for orbital decay to the RLOF phase (Madappatt et al. 2016), something else to tighten the orbits is needed.

For wide separations, analytic estimates of BHL accretion, which are too low to power PPN, match simulations and an accretion disk forms around the primary primarily from infall toward the retarded position of the secondary (Huarte-Espinosa et al. 2013; Blackman et al. 2013). But for this mode of accretion, such a high fraction of the mass lost from a typical AGB wind from which the BHL accretion draws, leaves without interacting much with the secondary. Therefore the orbit tends to increase (Chen et al. 2018; Decin et al. 2020). If however, for somewhat tighter orbits, WRLOF occurs (Mohamed & Podsiadlowski 2012), wind accelerated orbital decay (Chen et al. 2017, 2018) is possible and can greatly increase the number of systems that ultimately incur close enough interactions to produce PPN. More work is needed to make exact predictions.

Figure 2 shows the different consequences of initial binary separation, and where a PPN can be powered.

3.1. *Distinction between "PPN-W" and "PPN-C" and Example Cases*

Two classes of mechanisms work best to power PPN feature in the above discussion. In the symposium talk, I distinguished them as “preceding” and “succeeding” CE. In the discussion with R. Sahai, J. Kluska, and N. Soker afterward, it was suggested to me that a better distinction might be AGB and post-AGB since there are PPN objects which have 100 to 1000 day orbital periods (Bollen et al. 2021) and will never incur CE since their envelopes are mostly gone. However, what I wish to convey in a classification

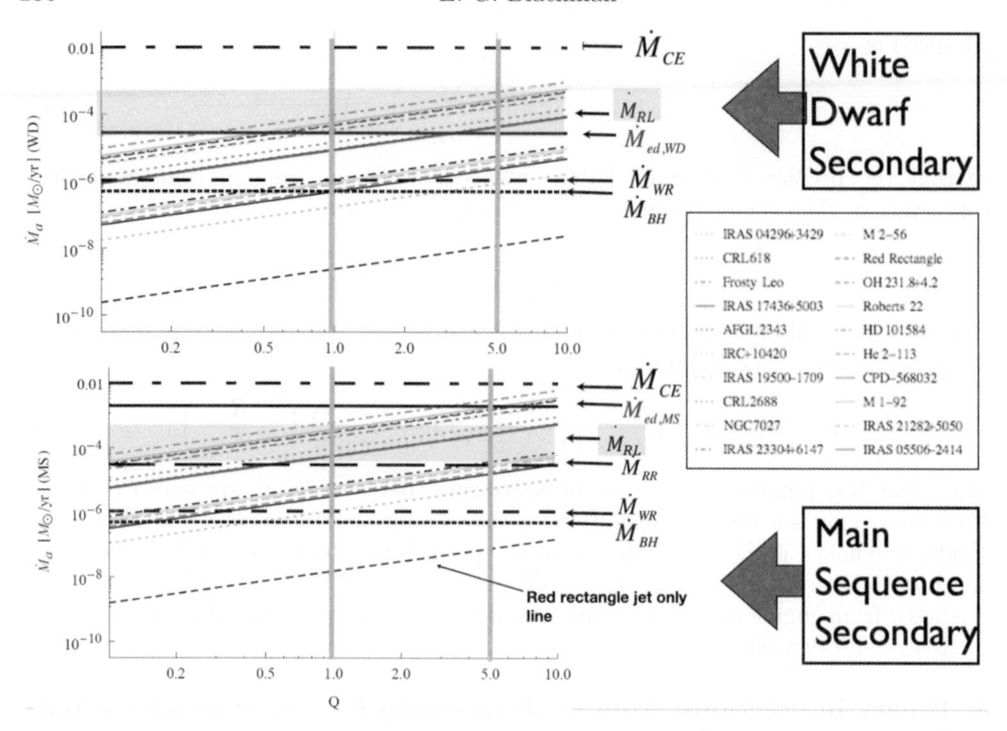

Figure 1. Adapted from Blackman & Lucchini (2014): Horizontal lines in each panel show examples of theoretically estimated accretion rates (BHL M_{BH}; Wind Roche lobe overflow M_{WR}; Red Rectangle based on HII region M_{RR}; Roche lobe overflow M_{RL}; Eddington M_{ed}; Accretion deep in common envelope M_{CE}) and the diagonal lines correspond to momentum requirements inferred from observations for each of the objects in the table as Q, the ratio of [illegible] jet speed to Keplerian speed at the disk inner radius, is varied. The vertical gold lines bound the typical range of Q from MHD jet models. The accretion mode of a given horizontal line is sufficient to power jets only where the points on a given horizontal line lie above those on a given diagonal line. The figure shows that RL and higher accretion rates are sufficient to power PPN but lower accretion rates are not generally sufficient.

scheme is distinguishing PPN mechanisms by their binary separation. This distinction can apply to systems with either AGB or RGB primaries.

Taking all of this into account, I label the two classes as PPN-W and PPN-C, where the W and C stand for "wide" and "close" and correspond, respectively, to orbital separations larger and smaller than the original giant envelope. Thus PPN-W would include RLOF accretion PPN, and PPN-C refers to close binary mechanisms. Inasmuch as the PPN in the aforementioned (Bollen et al. 2021) objects depend on the interaction of the observed binaries, these objects would be classified as PPN-W. I now discuss some specific example objects, and subsequently review some simulations in this context.

Figure 2 shows schematically the binary separation that distinguishes these two PPN classes, and Figure 3 shows two examples. Other examples are discussed below.

3.2. *Example PPN with the PPN-W and PPN-C distinction in mind*

HD 44179: Red Rectangle: The Red Rectangle PPN is best modeled as a main sequence secondary, accreting from the primary giant in an elliptical orbit, and moving in and out of the Roche Lobe of the primary (Witt et al. 2009) and is an example of a PPN-W. The outflow emanates within the central cavity of a cicumbinary torus of thickness 90 au and cavity diameter of 30 au, consistent with the formation of a torus from RLOF

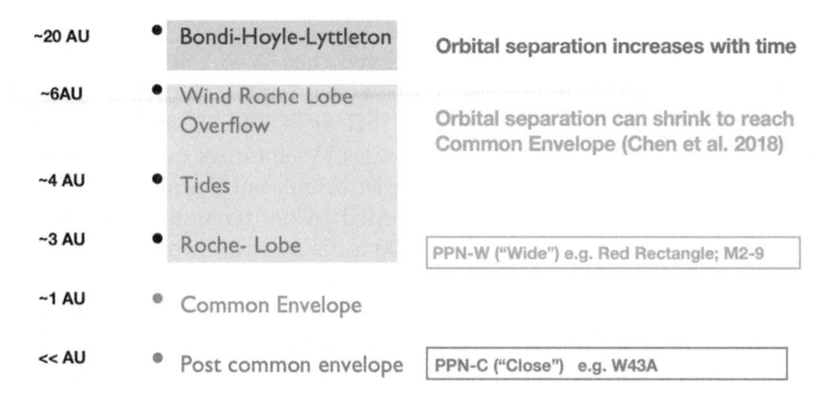

Figure 2. With decreasing orbital separation from top to bottom, the figure schematically indicates that separations that are too wide initially will widen further with BHL accretion. The red shading indicates no PPN will form from these systems. In the green region, indicated by the dominant mechanisms listed, the orbital separation can shrink (Chen et al. 2018). PPN can be produced at two key stages: in the RL regime indicated for this plot around 1-3 au (corresponding to PPN-W) and after much of the envelope is unbound in the post-CE stage (corresponding to PPN-C). Suggested examples of these two cases are shown in Figure 3.

(MacLeod et al. 2018a). The HST composite from Cohen et al. (2004) shows a bipolar axis length of $\sim 1.5 \times 10^4$ au. The jet produces a blue shift in Hα emission, modulating the primary's envelope emission. The accretion rate as constrained by the luminosity needed to source far-UV continuum for its HII region (Jura et al. 1997) assuming a distance of 710 pc, implies a maximum accretion disk temperature of 17,000K and a minimum accretion rate of $\dot{M}_a \geqslant 5 \times 10^{-5} M_\odot$/yr. This is is plotted in Figure 1 and is much larger than the constraint purely from the jet momentum calculation above, but can be accommodated by RLOF.

M2-9: M2-9 is a PPN-W with a jet from a companion orbiting in an 88-120 yr period, and injecting the flow into a hot bubble cavity. Early binary scenarios (Soker & Rappaport 2001) are now updated as Corradi et al. (2011) favored a fast jet induced illumination of the inner cavity (Doyle et al. 2000) rather than a photon source, based on delay time between knots. The jet produces mirror symmetry. There has been some ambiguity in interpreting this as potentially a symbiotic (RGB + WD companion instead of AGB+WD/MS) but the distinction is not important from a basic theoretical jet mechanism perspective since accretion from the RGB envelope onto the companion vs. accretion from the AGB envelope onto the companion both represent "W" nebulae in the classification above. Lykou et al. (2011) also identify a "disk/torus" from 15-900 au which may play a role in collimating the flow.

W43A: W43A is one of 15 "water fountain" sources exhibiting water maser emission (Diamond & Nyman 1988; Imai et al. 2002, 2005; Vlemmings et al. 2006; Amiri et al. 2010; Chong et al. 2015). Tafoya et al. (2020) interpreted W43A ALMA data in CO to reveal knots separated by a few years, a jet launched at 175 km/s, decelerating to 130 km/s, and collimated from 90 au to 1600 au. There is no binary detected, but Tafoya et al. (2020) suggest that the knots could indicate the influence of a binary period e.g. an eccentric orbit. That would make this a PPN-W. However, the tight collimation suggests that the outflow is produced from a tighter binary engine, suggesting that it may be a PPN-C outflow. The knots would then indicate a secular time scale (perhaps due to unsteady viscous accretion) compared to the much shorter orbital time scale at the base of the jet.

García-Segura et al. (2021) also argued that W43A is a post-CE object, also making it a PPN-C. In fact Khouri et al. (2021) observed that W43A and the other 14 known water fountain sources all likely incurred CEE, which implies that they are all PPN-C.

Magnetic fields measured in W43A are 85 mG at ~ 500 AU and are strong enough to collimate the measured outflow on those scales (Vlemmings et al. 2006; Amiri et al. 2010). These fields are quite far from the likely jet origin, but if scaled even linearly down to 10^{11} cm could provide the $>$ kG fields needed to be dynamically significant at the engine. Whether the source of the fields at 500 au is separate from, or an extension of, the fields generated in the jet engine is not yet clear.

IRAS16342-3814: This is another water maser fountain source with a collimated molecular jet and dust emission (Murakawa & Izumiura 2012; Sahai et al. 2017). As observed by Sahai et al. (2017), its high-speed jet exhibits 5 knots/blobs in each lobe, and gas of density $\sim 10^6/\mathrm{cm}^3$ is expanding in a 1300 au torus. There has been a rapid increase of mass-loss rate to $> 3.5 \times 10^{-4} M_\odot/\mathrm{yr}$ in the past ~ 455 years which suggests CEE.

Sahai et al. (2017) also constrain the circumstellar component ages for the AGB circumstellar envelope (~ 455 yr); extended high velocity outflow (EHVO) (130 to 305 yr with speed ~ 500 km/s); dust torus (160 yr); high velocity outflow (HVO) (~ 110 yr, with speed ~ 250 km/s). These indicate that the torus emerges several hundred years after the rapid AGB mass-loss increase, and the HVO appears very soon after torus formation. This is consistent with the time sequence in the Huggins (2012) sample mentioned earlier.

Although data from this object are not used in Figure 1, the inferred kinematics also require accretion rates as high as RLOF from the primary, or accretion operating within/after CEE. The high collimation, the absence of a detectable binary and the presence of a substantially increased AGB envelope all point to this source being classified as a PPN-C.

Calabash (Rotten Egg) Nebula OH232.84+4.22:

This object has a binary engine consisting of a Mira (AGB) variable (Cohen et al. 1981; Kastner et al. 1998), and an A0 main sequence companion in a likely $\gtrsim 50$ au orbit (Sánchez Contreras et al. 2004). The orbit is too wide to provide the needed accretion rate ($\sim 10^{-4} M_\odot/\mathrm{yr}$) to power the ~ 0.2 pc bipolar nebular lobes and explain the rate at which $\sim 1 M_\odot$ of circumstellar molecular gas from previous mass loss arose. Sánchez Contreras et al. (2004) speculate that an FU Orionis type outburst from accretion onto the companion might account for this. The source would then be a PPN-W.

However, another possibility is that there was a previous binary inspiral via CE which ejected envelope material and powered the PPN as a PPN-C. The outflow could then be sourced by circumbinary accretion or the release of free energy due to pre-merger core activity, with energy released as heating or mediated by self-generated magnetic fields along the lines of Ondratschek et al. (2021) discussed below. In this case, the presently observed wide binary would have little to do with what actually produces the collimated outflow. The Calibash nebula also has a rather collimated spine with little wobble. Sabin et al. (2015), using CARMA, found polarization that indicates a mostly toroidal ordered magnetic field perpendicular to the outflow. This is consistent with the orientation expected if some magnetic collimation were at work.

Dodson et al. (2018) found that H_2O maser emission aligned along bipolar lobes is perpendicular to an SiO maser disk and is ~ 40 yr old, confirming that mass loss is ongoing in the jet and that the history of mass loss is unsteady.

4. Common Envelope Evolution

CEE begins when the giant envelope engulfs the orbit of the secondary and the latter plunges in. The envelope could directly engulf the companion upon expansion to a giant phase for initially small enough orbital radii, but for most systems that incur CE, the

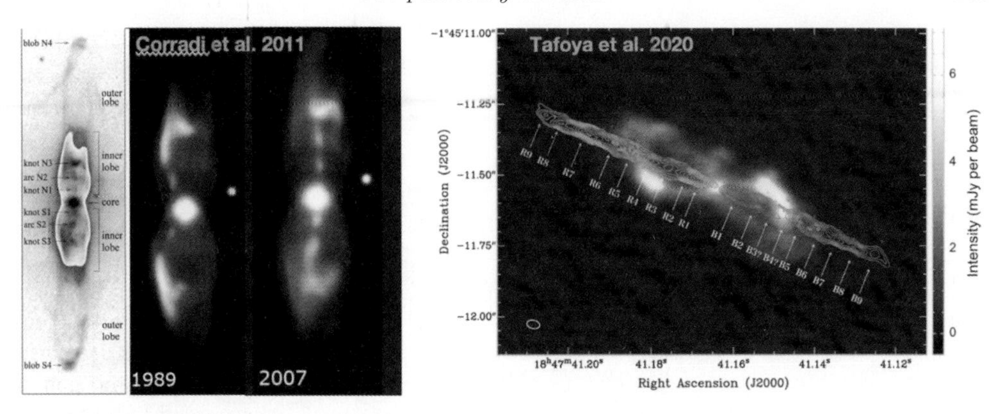

Figure 3. Left panel shows images of M2-9 from Corradi et al. (2011) and right panel shows W43A from Tafoya et al. (2020). M2-9 is likely a PPN-W, as the jet is produced from an accreting companion just outside the orbital radius of the giant, and consistent with RLOF accretion. Mirror symmetry is seen as the jet orbits and illuminates the surrounding bubble. In contrast, W43A is more likely a PPN-C, having a tightly collimated straight jet, plausibly produced deep within an CE, after a companion plunged close to the primary core.

process is likely preceded by a slow inward orbital migration that proceeds from wind induced orbital decay, tides, and RLOF. As discussed earlier, although the binaries that undergo BHL accretion expand because so little of the mass lost from the primary interacts with the secondary, closer binaries that incur stronger interaction via WRLOF can tighten. This tightening is further exacerbated by tides. CEE is ultimately important both for determining the properties and structure of mass loss as well as the orbital evolution of stellar systems that may include planets. The resulting effect on binary evolution and the efficacy with which angular momentum is removed from the system influences compact object merger rates and basic phenomenological properties of both high and low mass stellar systems (Paczynski 1976; Iben & Livio 1993; Taam & Sandquist 2000; Ivanova 2011; Ivanova et al. 2013a,b; Postnov & Yungelson 2014).

CEE is challenging to model accurately because of the wide range of temporal, spatial, and density scales from the 1 au envelope radius to the core dynamics $\leqslant 10^{10}$ cm, let alone the nebular outflow extending out to $\geqslant 0.1$ pc if one is to follow the full influence of CEE from giant star to PN. CE simulation efforts have, however, been progressing with substantial progress (Ricker & Taam 2012; Ivanova et al. 2013a,b, 2015; Ivanova & Nandez 2016; Ivanova 2018; Ohlmann et al. 2016a,b, 2017; MacLeod & Ramirez-Ruiz 2015a,b; MacLeod et al. 2017, 2018a,b; Staff et al. 2016a,b; Iaconi et al. 2017, 2018, 2019, 2020; Chamandy et al. 2018, 2019a,b, 2020, 2021; Ondratschek et al. 2021). Most simulations are run for ~ 100 days with the long end being ~ 4500 days (Ondratschek et al. 2021). Even that however, is short compared to the 100 to 1000 yr lifetimes of PPN, and the $\sim 10^4$ yr lifetimes of PN.

Important open questions include: how efficiently can a companion of given mass unbind the envelope upon inspiral? Does unbinding require recombination energy (Soker 2004; Ivanova & Nandez 2016; Glanz & Perets 2018)? What is the effect of convection? Convection might reduce the efficiency with which recombination might supply energy (Wilson & Nordhaus 2019) but might also redistribute energy more efficiently causing more mass to be unbound (Chamandy et al. 2018). For simulations without convection and without recombination however, CE may unbind the AGB envelope if the results from runs of 260 days for a $1.8M_{\odot}$ primary and $1M_{\odot}$ secondary were extrapolated to ~ 7 years (Chamandy et al. 2020), as shown in Figure 4.

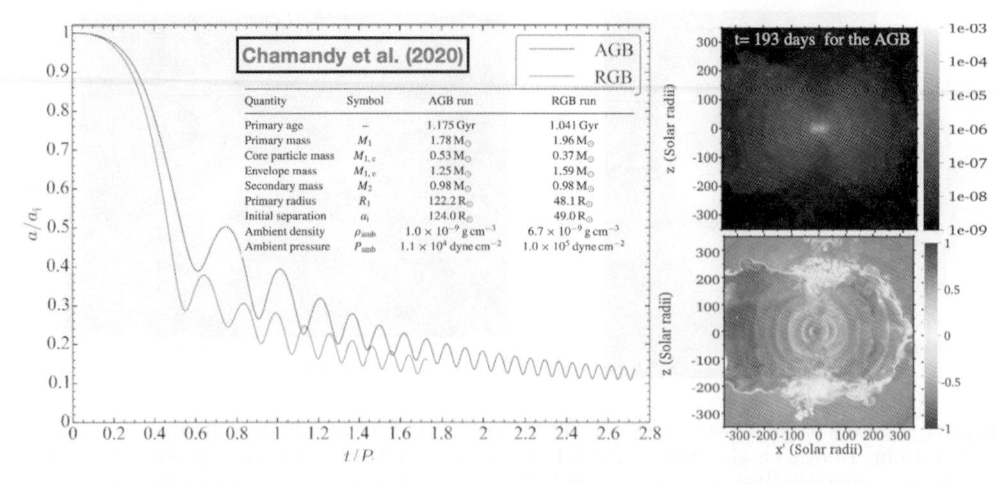

Quantity	Symbol	AGB run	RGB run
Primary age	–	1.175 Gyr	1.041 Gyr
Primary mass	M_1	1.78 $M_\odot$	1.96 $M_\odot$
Core particle mass	$M_{1,c}$	0.53 $M_\odot$	0.37 $M_\odot$
Envelope mass	$M_{1,e}$	1.25 $M_\odot$	1.59 $M_\odot$
Secondary mass	M_2	0.98 $M_\odot$	0.98 $M_\odot$
Primary radius	R_1	122.2 $R_\odot$	48.1 $R_\odot$
Initial separation	a_i	124.0 $R_\odot$	49.0 $R_\odot$
Ambient density	ρ_{amb}	1.0×10^{-9} g cm^{-3}	6.7×10^{-9} g cm^{-3}
Ambient pressure	P_{amb}	1.1×10^{4} dyne cm^{-2}	1.0×10^{5} dyne cm^{-2}

Figure 4. taken from Chamandy et al. (2020). Left panel: inter-particle separation for AGB and RGB runs with the rapid plunge phase, followed by slow inspiral. Time is normalized by orbital period: 96.5 days for the AGB curve and 23.2 days for the RGB. Separation is normalized by the initial orbital separation: 124 $R_\odot$ for AGB and 49 $R_\odot$ for the RGB. The AGB run is shown to ~ 260 days. The rate of unbinding at the end is steady at $0.17 M_\odot$/yr, and would unbind the full envelope within 7 yr if extrapolated. Right panel: vertical slices through the orbital plane of the AGB at $t = 193$ days for the AGB run. The left slice is the gas density in g/cm^3 and the right slice indicates bound (blue) and unbound (red) gas at this time quantified by a dimensionless measure ranging from 1 (strongly unbound) and -1 (strongly bound.

Armitage & Livio (2000) and Chevalier (2012) examined the ejection of CE by jets launched from a neutron star that inspirals inside the giant envelope. Others have considered different companions (Soker 2004; Papish et al. 2015; Soker 2015, 2016; Moreno Méndez et al. 2017; Shiber et al. 2017; Shiber & Soker 2018; Shiber et al. 2019; López-Cámara et al. 2019; Lopez-Camara et al. 2021). As discussed earlier, accretion onto a plunging main sequence or WD companion before the envelope is unbound, requires a pressure release valve in order to prevent thermal pressure from building up and abating the accretion (Chamandy et al. 2018). Jets from MS and WD companions are largely choked by the bound envelope after plunge-in for reasonable jet powers (Lopez-Camara et al. 2021). This can be estimated by comparing the ram pressure of the jet with the thermal pressure of the envelope. The total energy injected by the jet $L_m t_j$ where t_j is the jet lifetime, can also be compared with the binding energy to identify a minimum time scale over which this outflow could unbind the envelope, compared to other unbinding mechanisms. The efficiency with which the jet energy is used to unbind mass can also be studied in simulations. Although the jet may appear to be unimportant in a short simulation, over longer times its effect may be significant (Zou et al., in preparation).

Even if the jet starts during RLOF before the secondary plunges, it will be connected by an accretion stream to the envelope which will facilitate an already rapid drag-in to inspiral. Thus it seems unlikely that a jet would produce much unbound mass before plunge in. For a limited set of binary parameters which includes the case of a 1 au primary of mass 18 $M_\odot$ and a 5.4 $M_\odot$ secondary with no jet, MacLeod et al. (2018b) found that RLOF can last decades but that once the secondary enters the envelope the plunge is rapid. There have not yet been numerical studies that include accretion from the envelope and include a jet such that the envelope mass, accretion stream, and orbital inspiral are all self-consistently included, but the lessons learned so far do not suggest that the jet could prevent inspiral.

For CE, the list of topics with opportunity for more work includes: (i) the inclusion of convection; (ii) more complete treatments of the equation of state and better approximations to inclusion of recombination energy; (iii) more detailed radiative transfer; (iv) studying the sensitivity to initial conditions and initial binary separation; (v) increasing the duration of simulations; (vi) limiting the effects of softening of potential wells by smoothing and limited numerical resolution; (viii) ensuring conservation of energy and angular momentum in long term simulations.

5. Role of Magnetic Fields in Driving and Shaping

Because collimated jets from MS and WD companions are substantially choked during CEE plunge-in as described above, jets that produce PPN would be most likely visible either in the (i) RLOF phase before plunge producing a PPN-W or (ii) after substantial CE ejection has occurred producing a PPN-C.

So what about the collimation in each of these cases? Magnetic fields are likely important for jet launch and tight collimation as toroidal hoop stress can act to concentrate the pressure of the flow to an axial spine. But all magnetically collimated flows still require ambient pressure for stability. In fact there is no astrophysical context with a collimated jet for which an ambient wind or thermal pressure surrounding the jet is ruled out. In the present context, CE provides tori for inertial collimation both in the RLOF phase (MacLeod et al. 2018a,b) and in the post-CE ejection phase. We discuss the latter here.

García-Segura et al. (2018), ran 2-D simulations for 10, 000 yr, starting at 1 au with output from CEE simulation of Ricker & Taam (2012) at a time of 47 days as the initial conditions. Zou et al. (2020) ran 3-D simulations for 10,000 days, starting at 100 au with output from the CEE simulation of Reichardt et al. (2019). The results from both of these simulations exemplify the basic principle that a uncollimated hydrodynamic wind injected within the output of a CE simulation can be collimated by the ejecta. However the outflow of García-Segura et al. (2018) does not remain collimated and instead transitions to a wide barrel/elliptical structure in the absence of magnetic fields.

Importantly, magnetic fields are a "drive belt" not a "motor" and require a source of free energy (convection, orbital, accretion) supplied by binary interactions. Magnetic fields and binaries are therefore not competing mechanisms, but operate in symbiosis. The binary supplies the free energy and sets up the environment within which the magnetic field is amplified and functional. Analytic estimates for dynamically important disk engine field strengths in the PPN context give values $\geqslant 10$ kG (Blackman et al. 2001).

Most astrophysical MHD jet simulations have separated the detailed origin of the magnetic fields from the outflow dynamics. For example, the surface of the anchoring rotator has often been treated as boundary conditions with an imposed field. But progress is emerging across astrophysical disciplines in unifying field origin and jet formation self-consistently (Kathirgamaraju et al. 2019; Ruiz et al. 2021; Ondratschek et al. 2021). The range of approaches to the problem in the PPN/PN context has included analytic calculations (Pascoli 1997; Blackman et al. 2001; Tout & Regos 2003; Nordhaus & Blackman 2006; Nordhaus et al. 2007) focused on dynamo and/or power generation and three types of numerical approaches discussed in the subsections below.

5.1. *Shaping from imposed dynamically important magnetic fields*

Shaping of flows with an imposed magnetic field has been a long standing approach to model some aspects of outflow shaping (Chevalier & Luo 1994; Garcia-Segura 1997; García-Segura et al. 1999, 2005; Balick et al. 2020). Balick et al. (2020), for example, imposed a flow of 400 km/s with opening angle 40 deg injected normal to a sphere of radius 1000 au. A toroidal field was imposed with initial values varying between 0.003G $\leqslant$

$B \leqslant 0.3$G. The framework was tuned to produce resultant angular distributions of speeds and outflow geometry consistent with e.g. OH231.8+04.2 (Alcolea et al. 2001; Sánchez Contreras et al. 2018) and CRL618 (Balick et al. 2013; Riera et al. 2014) Hen2-104 (Corradi et al. 2001) and MyCn18 (Bryce et al. 1997; O'Connor et al. 2000).

Although such a method does not self-consistently generate jets because a strong magnetic field is imposed as an initial condition, one can still use the results to match observations and then infer what the field strength and geometry should be to inform further observations and theory. The method has not been used to study the possible differences between PPN-C and PPN-W which would would be valuable.

The next subsections focus on more steps toward self-consistent generation of collimated outflows, specifically in the PPN-C context.

5.2. *Multi-stage 2-D MHD simulations*

A second approach is to use output from 3-D CE simulations to set initial conditions for 2-D simulations, with only an imposed weak seed field. The weak field may grow dynamically and produce self-generated MHD outflows. Using this approach, García-Segura et al. (2020, 2021) started with the conditions of the Ricker & Taam (2012) CE after 56.7 days at which point 25% of the envelope is unbound and mass is being lost at the high rate of $2M_\odot$/yr. A weak seed magnetic field with toroidal and poloidal components was imposed on the scale of 1 au. The field grows, causes angular momentum loss, disk collapse, and a magneto-centrifugally launched wind. After 120 days, they took output for a second 2-D simulation, this time using an expanding grid envelope. They evolve the result for > 1000 years. They find that the resultant CE outflow forms from a circumbinary disk and a very tightly collimated outflow whose properties plausibly resemble W43A and the Calabash nebula. This computational model produces a PPN-C in the aforementioned classification scheme. The jet is produced after much of the envelope is ejected and the binary orbit is 7 times smaller than the RGB envelope at the time of the conditions used from Ricker & Taam (2012).

This approach is a step toward more self-consistency in that the imposed magnetic field is weak and the outflow is self-generated. There are certainly limitations to the fidelity of 2-D and the expanding grid, but the 2-D simulations can be run for > 1000yr which is orders of magnitude longer than what can be expected for 3-D simulations. There are always trade-offs, and precise realism is not always necessary to gain some insight.

5.3. *Simulations with "organic" 3-D magnetic field amplification and jet formation*

Complete modeling of a PPN-W formation requires simulating an RGB or AGB in RLOF with a companion accretor, and allowing the field to amplify within the accretion flow. A self-consistent magnetically mediated outflow should grow, along with any circumbinary molecular torus that might aid in collimation. For self-consistent PPN-C formation, a simulation would instead require the full CEE after the RLOF phase, starting with a stellar seed magnetic field in the giant and computing how the combination of CE, inspiral, field amplification, outflow, and collimation subsequently proceed.

There has not yet been a fully self-consistent simulation of a PPN-W by the above standard, and so here I focus on PPN-C. Indeed, Ondratschek et al. (2021) have broken new ground for the study of PPN-C with a full MHD CEE simulation that shows organic growth of a jet mediating magnetic field and collimated outflow production in 3-D. They use the AREPO code and include a prescription for recombination energy, which is important as it causes the simulated envelope to become unbound during the jet evolution. They start with a weak dynamically insignificant seed magnetic dipole field in the AGB star of mass 0.97 $M_\odot$ and include a companion of 0.243 $M_\odot$ as fiducial run. This mass ratio is then $q = 0.25$, but they also carried out runs for $q = 0.5$ and $q = 0.75$ that produced qualitatively similar results. They track the inspiral to ~ 4000 days.

The magnetic field is amplified some 15 orders of magnitude on the time scale of the simulations. Comparing runs with and without a magnetic field, they find that the field makes little difference to unbinding after 1000 days (also true of semi-analytical dynamos, Nordhaus et al.2007), which is expected: as discussed above, the magnetic field is not an extra source of energy but draws its energy from the sources already there, the orbital energy in this case. The field amplification arises from some combination of shear, and turbulence sourced by some combination of MRI, and Kelvin-Helmholtz instability and the emergent jet launches with cross sectional diameter of the circumbinary disk engine. The exact analytical modeling of the system has yet to be carried out, but this clearly demonstrates a type of PPN-C outflow, as the jet emanates from a circumbinary region which is a factor of ~ 8 times smaller than the initial AGB stellar radius. The most important lesson that this simulation highlights, is that the magnetic field facilitates formation of a magnetically driven and strongly collimated jet along a narrow axial channel. The collimated jet is absent without the field.

The simulation by Ondratschek et al. (2021) is a substantial step toward high fidelity global simulations of PPN-C. Under the hood, there are a number of issues that warrant further work and discussion. Convection in the AGB star has not been included. The treatment of recombination is approximate and non-local, and the extent to how this interfaces with convection is relevant in the broader CE context as discussed earlier. Convection will also add a disordered component to the initial seed dipole field and so the extent to which it influences the overall magnetic launch and collimation is also important. In the presence of convection, even the seed field may require a dynamo in the star since exponential generation is needed to compete with turbulent diffusion. There is also the question as to the relative role of a magnetic tower or a magneto-centrifugal launch, a distinction discussed further in the next section.

5.4. *Magnetic Tower, Magneto-centrifugal Launch, or Magnetic Bomb?*

Magnetically mediated launches that depend on the presence of large-scale magnetic fields are not all the same. There are essentially three types as described below. Although all share the fact that a gradient in toroidal field magnetic pressure helps to propel material and the hoop stress helps to collimate, they also differ in key aspects. They may not be mutually exclusive in a given source. For example, a magnetic tower may be embedded within a broader magneto-centrifugual launch, however, we do not yet know which mechanism dominates in any given source.

The magnetic tower (MT) (e.g. Lynden-Bell 1996, 2003; Uzdensky & MacFadyen 2006; Gan et al. 2017) can be initiated from magnetic field loops anchored between footpoints in relative differential rotation, for example loops that link a stellar core to a surrounding torus. The footpoint separation may be of comparable scale to the radius at which they are anchored, but both footpoints are contained in the engine itself. The differential rotation winds up the field, creating a toroidal component that establishes a magnetic pressure force which pushes material upward. The hoop stress collimates the pressure of this rising tower, but only when the ambient medium is surrounded by a balancing ambient pressure. The magnetic field is parallel to the axial flow on the jet axis, and becomes increasingly toroidal away from the axis toward the jet boundary. The outflows can remain marginally magnetically dominated inside the jet out to observable scales within the tower and maybe out to arbitrarily large-scales. Importantly, both signs of magnetic flux are contained within the structure that would appear as the jet tower because both of the original loops are within the jet. Their separation defines the diameter at the base of the tower. The jet contains zero vertical net magnetic flux.

The magneto-centrifugal launch (MCL) (Blandford & Payne 1982; Pelletier & Pudritz 1992) is a more widely invoked class of models for which the starting point is a large-scale open magnetic field with only one anchoring foot-point sign at the engine base and the other foot-point essentially at infinity. Plasma loaded onto quasi-rigid field lines at the base is centrifugally flung along these lines as angular momentum of the base is transferred via the field lines to the plasma. At the Alfvén radius, where the flow energy density becomes comparable to that of the field, the torodial field magnitude is comparable to that of the poloidal field and supplies some outward pressure and collimation. On larger scales farther from the engine, say $\gtrsim 100R_{in}$, the flow kinetic energy marginally dominates field energy within the jet, the opposite of the magnetic tower. Importantly, for the MCL, only one sign of poloidal magnetic flux resides within the jet, also in contrast to magnetic tower. For a source of magnetic field produced in a PPN accretion disk, MCL outflow scalings have been applied to PPN (Blackman et al. 2001). For the MCL like the MT, ambient collimation of the inner magnetic structures by some ambient flow or pressure is also needed and all simulations demonstrating collimation and steady jets have pressure equilibrium at the jet boundary.

One more magnetically mediated outflow model is a magnetic bomb (MB) (Matt et al. 2006). Here a wound magnetic field acts like a capacitor, suddenly releasing its energy in outflows. The starting point is an evolved star for which differential rotation has been established between the collapsed degenerate core, and the expanded envelope. The initial field anchored at the core be open, as per the MCL. However, this field can be weak initially, and there is no need to load mass from the core onto the field lines as the field is already mass loaded. Differential rotation between the envelope and core winds up the toroidal field. After the field reaches some threshold, it rapidly drives polar outflows from vertical magnetic pressure gradients, and equatorial outflows as the wound field in each hemisphere also squeezes flow outward from the equator. Like the MT, differential rotation is important from the start, but the primary differential rotation for the MB is vertical not radial. The MB for an initial dipole field would have just one sign of magnetic flux in the outflow, similar to the MCL and distinct from the MT.

5.5. *Measured Magnetic Fields*

Given the importance of magnetic fields, some further comments on what has been measured warrants mention. The measurement of magnetic fields in W43A (Vlemmings et al. 2006; Amiri et al. 2010) and the evidence for toroidal fields in the Calibash from polarization (Sabin et al. 2015) were mentioned above. Fields have also been measured in other AGB stars via masers (Herpin et al. 2006), masers in the OH shell of NML Cygni Etoka & Diamond (2004), and in Miras (Kemball & Diamond 1997; Kemball et al. 2009). Synchrotron emission in the post AGB star IRAS 15445-5449 has been measured at 7000 au, indicating a mG field which could collimate the jet (Perez-Sanchez et al. 2013). Leone et al. (2014) found no evidence for $>$ kG fields in the central stars of PN, but a field of $B \sim 655$G in NGC4361.

However, an important point is that none of the above measurements probe the field on the jet "launch" scales (< 0.01 au). Instead, they are measurements of the jet "propagation" (> 1 au) scales. That fields are dynamically important on propagation scales is important, but these do not directly constrain the jet formation at its engine.

Also, distinguishing the MT from MCL cannot be done with polarization measurements because they are ambiguous to 180 degrees in field orientation. Measurements that constrain whether the sign of the flux is uniform across the jet in a given (indicating MCL or MB) hemisphere or has two reversals across the jet (indicating MT) are needed.

6. Persistent challenges and distinguishing PPN-C from PPN-W

While PPN and CEE are likely associated, one challenge is to identify direct signatures of a CEE event (Khouri et al. 2021) and possibly to associate the two more directly. There remain some challenges that come with comparing theory, simulation and observation both in terms of the limits of numerical simulations and in predicting distinguishing features of PPN-C and PPN-W.

6.1. *Convection, magnetic fields, and the limits of numerical simulations*

In addition to potentially delocalising the deposition of recombination energy discussed earlier, convection also leads to turbulent diffusion of large-scale magnetic flux. In turbulent astrophysical systems, large-scale flux is almost never frozen. (Think of the Sun, where the large-scale field reverses every 11 years. This would be impossible if flux were frozen.) This raises the question of the initial field in the giant star used for simulations without convection (e.g. Ondratschek et al. 2021). In reality, the stellar field would not be exclusively ordered but may even be dominated by a random component. The total field would need to be sustained by large and small scale dynamos to overcome the exponential decay from turbulent diffusion. How various dynamos conspire in these engines to produce the dynamically significant large-scale fields, and how the turbulence affects the level of collimation of the jet remains to be studied.

A second open issue is that large scale transport rather than local isotropic turbulence may dominate in disks, whether or not the magneto-rotational instability (MRI) is the dominant instability (Blackman & Nauman 2015). The fraction of small scale, mesoscale, or large-scale transport is not well constrained. The scale of transport determines where energy is dissipated, which in turn determines the observed spectral signatures of accreting systems. Dissipation that occurs deep within an optically thick disk will produce thermal emission but dissipation that occurs in a corona or jet can be non-thermal. The fraction of thermal versus non-thermal emission can thus be used as a proxy for the scale of angular momentum transport (Blackman & Pessah 2009).

Another pervasive issue, is how sensitive the phenomenological output from simulations is as a function of initial and boundary conditions. Numerical simulations are useful to study small pieces of a physical system at high resolution or "kitchen sink" approaches at low resolution. But for kitchen sink simulations, the question of convergence is substantial. Dynamos and accretion disks are examples for which intermediate fidelity simulations can cause confusion. Suppose, for example, that analytic theory were to predict that in the asymptotic limit of large magnetic Reynolds number, a particular dynamo magnetic growth rate is independent of magnetic Reynolds number. Further suppose that the real astrophysical system of interest has a magnetic Reynolds number much larger than one could ever hope to simulate. If a simulation then exhibits a dependence on magnetic Reynolds number, it is not easy to determine whether the theory is wrong or whether the simulation is not in a sufficiently asymptotic regime to be realistic. In the case of the solar dynamo, intermediate fidelity simulations have indeed sometimes produced results that disagree with first generation theory, whilst higher fidelity simulations have shown more consistency with basic theory. In short, basic theory and computational simulations represent distinct approaches that are valuable both independently and in combination.

6.2. *Identifying distinct features of PPN-C and PPN-W*

The classification of PPN-W versus PPN-C, respectively, delineates whether the influential binary companion is inside or outside the outer radius of the initial giant star. Which observational consequences manifest from this distinction? In addition to subtle differences that may require simulations to identify, there are some conspicuous distinctions.

PPN-W would show the time variability of a larger binary orbit than PPN-C. Moreover, for PPN-W the jet has a diameter at launch that is much smaller than the orbital separation, so one might expect reflection symmetry to be common as the jet moves around the orbit. In contrast, for PPN-C, the jet cross section is closer to the size of the binary separation itself and less orbital motion of the jet is expected. There is also likely to be more surrounding torus mass for PPN-C than for PPN-W since the former happens inside the circumstellar envelope. This would suggest that more collimation is likely for PPN-C than PPN-W. A PPN-C jet may thus be narrower, possibly with point symmetry rather than reflection symmetry, if the overall jet precesses.

There may also be statistical population differences in duration between PPN-C and PPN-W as they are determined by different accretion processes. PPN-W durations would be determined by how long RLOF accretion occurs before the companion plunges into CEE. Predictions for this duration are not yet clear (MacLeod et al. 2018a). PPN-C would be determined by the time scale for circumbinary material to accrete after CEE but before enough envelope is lost to reduce the mass supply, or the time scale for accretion from a merger to run out of mass.

There may also be some influence of WD nuclear burning that is more common for PPN-C than PPN-W because all PPN-C would have at least 1 WD within its engine. Core X-rays and other signatures of dwarf novae could be more prevalent for PPN-C.

Another distinction is that although RLOF fueled PPN-W can produce a surrounding torus that precedes a jet, only for PPN-W can there also be a torus that follows the jet. This is because the PPN-W happens before CEE, if CEE is to happen. In contrast, the PPN-C would always occur only after all possible tori are produced in the system. This distinction could lead to statistical differences between the timing of jet and tori in the two populations. As emphasized earlier, even magnetized jets require an ambient wind or torus for stable collimation, so some minimum ambient material would be common to both PPN-C and PPN-W.

Finally there may be compositional differences. Since material supplying PPN-W jets comes from farther in the envelope, it may be less C-rich. The jet composition will be dependent on the jet material source, and this also depends on how effectively or ineffectively convective mixing homogenizes the composition.

7. Summary

PPN are more kinematically demanding than PN, and thus place tighter constraints on their mutual origin mechanisms if PN are a time evolved state of PPN. Kinematic constraints for PPN demand close binary interaction, at least as close as RLOF from the primary onto a secondary main-sequence star.

To produce PPN jets, binaries and magnetic fields likely act in symbiosis, with the free energy in orbital motion and accretion used to generate magnetic fields that drive and collimate outflows. The presence of circumstellar tori can in turn collimate and stabilize the magnetic structures, which is likely required to explain observed PPN jets. The classification of PPN jets as either PPN-W or PPN-C can be used to respectively distinguish those for which the binary separation in the engine is wider than, or less than the primary giant envelope radius. Jets produced by RLOF of the giant onto the secondary are examples of PPN-W, and jets produced after the companion plunges into CEE by a circumbinary disk or merger after the envelope largely unbinds would be PPN-C jets. Examples of both classes were discussed.

There has been significant progress over the past several decades in putting all of the pieces together. This is culminating in increasingly high fidelity simulations. Self-consistently generated magnetic fields, and the associated magnetically mediated jets are now seen to emerge organically in 3-D CE PPN-C simulations. There is open opportunity to achieve the equivalent for PPN-W.

Probing the physics of these simulations and their limitations remains an active effort. Convection is a fundamental feature of observed giants that is absent from most simulations, and presents an important frontier. The extent to which the observed phenomenology depends on initial binary parameters and boundary conditions needs to be explored. Understanding the different classes of PPN that can be produced and their observational signatures will benefit from further collaborations between observers and theorists to converge on specific predictions that can distinguish different mechanisms.

Acknowledgements

Thanks to B. Balick, L. Chamandy, O. De Marco, A. Frank, J. Kastner, J. Kluska, J. Nordhaus, R. Sahai, and N. Soker, and E. Wilson, and A. Zou for pertinent discussions. Thanks to L. Decin for meticulous editing and useful comments. Support from US Department of Energy grants DE-SC0020432 and DE-SC0020434, and US National Science Foundation grants AST-1813298, and PHY-2020249 are acknowledged.

References

Alcolea, J., Bujarrabal, V., Sánchez Contreras, C., Neri, R., & Zweigle, J. 2001, A & A, 373, 932
Amiri, N., Vlemmings, W., & van Langevelde, H. J. 2010, A & A, 509, A26
Armitage, P. J., & Livio, M. 2000, ApJ, 532, 540
Balick, B., & Frank, A. 2002, ARAA, 40, 439
Balick, B., Frank, A., & Liu, B. 2020, ApJ, 889, 13
Balick, B., Huarte-Espinosa, M., Frank, A., et al. 2013, ApJ, 772, 20
Balick, B., Preston, H. L., & Icke, V. 1987, AJ, 94, 1641
Blackman, E. G., Carroll-Nellenback, J. J., Frank, A., Huarte-Espinosa, M., & Nordhaus, J. 2013, MNRAS, 436, 904
Blackman, E. G., Frank, A., & Welch, C. 2001, ApJ, 546, 288
Blackman, E. G., & Lucchini, S. 2014, MNRAS, 440, L16
Blackman, E. G., & Nauman, F. 2015, Journal of Plasma Physics, 81, 395810505
Blackman, E. G., & Pessah, M. E. 2009, ApJl, 704, L113
Blandford, R. D., & Payne, D. G. 1982, MNRAS, 199, 883
Boffin, H. M. J., & Jones, D. 2019, The Importance of Binaries in the Formation and Evolution of Planetary Nebulae (Springer), doi:10.1007/978-3-030-25059-1
Bollen, D., Kamath, D., Van Winckel, H., De Marco, O., & Wardle, M. 2021, MNRAS, 502, 445
Bond, H. 2000, in Encyclopedia of Astronomy and Astrophysics, ed. P. Murdin (IOP), 2382
Borkowski, K. J., Blondin, J. M., & Harrington, J. P. 1997, ApJL, 482, L97
Bryce, M., López, J. A., Holloway, A. J., & Meaburn, J. 1997, ApJL, 487, L161
Bujarrabal, V., Castro-Carrizo, A., Alcolea, J., & Sánchez Contreras, C. 2001, A. & Ap., 377, 868
Chamandy, L., Blackman, E. G., Frank, A., Carroll-Nellenback, J., & Tu, Y. 2020, MNRAS, 495, 4028
Chamandy, L., Blackman, E. G., Frank, A., et al. 2019a, MNRAS, 490, 3727
Chamandy, L., Blackman, E. G., Nordhaus, J., & Wilson, E. 2021, MNRAS, 502, L110
Chamandy, L., Tu, Y., Blackman, E. G., et al. 2019b, MNRAS, 486, 1070

Chamandy, L., Frank, A., Blackman, E. G., et al. 2018, MNRAS, 480, 1898

Chen, Z., Blackman, E. G., Nordhaus, J., Frank, A., & Carroll-Nellenback, J. 2018, MNRAS, 473, 747

Chen, Z., Frank, A., Blackman, E. G., Nordhaus, J., & Carroll-Nellenback, J. 2017, MNRAS, 468, 4465

Chevalier, R. A. 2012, ApJL, 752, L2

Chevalier, R. A., & Luo, D. 1994, ApJ, 421, 225

Chong, S.-N., Imai, H., & Diamond, P. J. 2015, ApJ, 805, 53

Ciardullo, R., Bond, H. E., Sipior, M. S., et al. 1999, AJ, 118, 488

Ciolfi, R. 2020, General Relativity and Gravitation, 52, 59

Cohen, M., Van Winckel, H., Bond, H. E., & Gull, T. R. 2004, AJ, 127, 2362

Cohen, N. L., Hohfeld, R. G., Gorenstein, M. V., Pottash, R. I., & Willson, R. F. 1981, A & A, 95, 386

Corradi, R. L. M. 2012, Memorie della Societa Astronomica Italiana, 83, 811

Corradi, R. L. M., Livio, M., Balick, B., Munari, U., & Schwarz, H. E. 2001, ApJ, 553, 211

Corradi, R. L. M., Sabin, L., Miszalski, B., et al. 2011, MNRAS, 410, 1349

De Marco, O., & Soker, N. 2011, PASP, 123, 402

Decin, L., Montargès, M., Richards, A. M. S., et al. 2020, Science, 369, 1497

Diamond, P. J., & Nyman, L. Å. 1988, in The Impact of VLBI on Astrophysics and Geophysics, ed. M. J. Reid & J. M. Moran, Vol. 129, 249

Dodson, R., Rioja, M., Bujarrabal, V., et al. 2018, MNRAS, 476, 520

Doyle, S., Balick, B., Corradi, R. L. M., & Schwarz, H. E. 2000, AJ, 119, 1339

Edgar, R. G., Nordhaus, J., Blackman, E. G., & Frank, A. 2008, ApJL, 675, L101

Etoka, S., & Diamond, P. 2004, MNRAS, 348, 34

Feibelman, W. A. 1985, AJ, 90, 2550

Gan, Z., Li, H., Li, S., & Yuan, F. 2017, ApJ, 839, 14

Garcia-Segura, G. 1997, ApJL, 489, L189

García-Segura, G., Langer, N., Różyczka, M., & Franco, J. 1999, ApJ, 517, 767

García-Segura, G., López, J. A., & Franco, J. 2005, ApJ, 618, 919

García-Segura, G., Ricker, P. M., & Taam, R. E. 2018, ApJ, 860, 19

García-Segura, G., Taam, R. E., & Ricker, P. M. 2020, ApJ, 893, 150

—. 2021, ApJ, 914, 111

Gieseking, F., Becker, I., & Solf, J. 1985, ApJL, 295, L17

Glanz, H., & Perets, H. B. 2018, ArXiv e-prints, arXiv:1801.08130

Herpin, F., Baudry, A., Thum, C., Morris, D., & Wiesemeyer, H. 2006, A & A, 450, 667

Hillwig, T. C., Jones, D., De Marco, O., et al. 2016, ApJ, 832, 125

Huarte-Espinosa, M., Carroll-Nellenback, J., Nordhaus, J., Frank, A., & Blackman, E. G. 2013, MNRAS, 433, 295

Huggins, P. J. 2007, ApJ, 663, 342

Huggins, P. J. 2012, in American Astronomical Society Meeting Abstracts, Vol. 219, American Astronomical Society Meeting Abstracts #219, 239.01

Iaconi, R., De Marco, O., Passy, J.-C., & Staff, J. 2018, MNRAS, 477, 2349

Iaconi, R., Maeda, K., De Marco, O., Nozawa, T., & Reichardt, T. 2019, MNRAS, 489, 3334

Iaconi, R., Maeda, K., Nozawa, T., De Marco, O., & Reichardt, T. 2020, arXiv e-prints, arXiv:2003.06151

Iaconi, R., Reichardt, T., Staff, J., et al. 2017, MNRAS, 464, 4028

Iben, Jr., I., & Livio, M. 1993, PASP, 105, 1373

Imai, H., Nakashima, J.-i., Diamond, P. J., Miyazaki, A., & Deguchi, S. 2005, ApJL, 622, L125

Imai, H., Sasao, T., Obara, K., Omodaka, T., & Diamond, P. J. 2002, in Cosmic Masers: From Proto-Stars to Black Holes, ed. V. Migenes & M. J. Reid, Vol. 206, 80

Ivanova, N. 2011, ApJ, 730, 76

—. 2018, ApJl, 858, L24

Ivanova, N., Justham, S., Avendano Nandez, J. L., & Lombardi, J. C. 2013a, Science, 339, 433

Ivanova, N., Justham, S., & Podsiadlowski, P. 2015, MNRAS, 447, 2181

Ivanova, N., & Nandez, J. L. A. 2016, MNRAS, 462, 362
Ivanova, N., Justham, S., Chen, X., et al. 2013b, A. & Ap.r, 21, 59
Jacoby, G. H., Hillwig, T. C., Jones, D., et al. 2021, MNRAS, 506, 5223
Jones, D. 2020, Galaxies, 8, 28
Jura, M., Turner, J., Balm, & S. P. 1997, ApJ, 474, 741
Kastner, J. H., Weintraub, D. A., Merrill, K. M., & Gatley, I. 1998, AJ, 116, 1412
Kathirgamaraju, A., Tchekhovskoy, A., Giannios, D., & Barniol Duran, R. 2019, MNRAS, 484, L98
Kemball, A. J., & Diamond, P. J. 1997, ApJL, 481, L111
Kemball, A. J., Diamond, P. J., Gonidakis, I., et al. 2009, ApJ, 698, 1721
Khouri, T., Vlemmings, W. H. T., Tafoya, D., et al. 2021, Nature Astronomy, arXiv:2112.09689
Kim, H., Trejo, A., Liu, S.-Y., et al. 2017, Nature Astronomy, 1, 0060
Kwok, S., Purton, C. R., & Fitzgerald, P. M. 1978, ApJL, 219, L125
Leone, F., Corradi, R. L. M., Martínez González, M. J., Asensio Ramos, A., & Manso Sainz, R. 2014, A & A, 563, A43
Lopez, J. A., Meaburn, J., & Palmer, J. W. 1993, ApJL, 415, L135
López-Cámara, D., De Colle, F., & Moreno Méndez, E. 2019, MNRAS, 482, 3646
Lopez-Camara, D., De Colle, F., Moreno Mendez, E., Shiber, S., & Iaconi, R. 2021, arXiv e-prints, arXiv:2110.02227
Lykou, F., Chesneau, O., Zijlstra, A. A., et al. 2011, A & A, 527, A105
Lynden-Bell, D. 1996, MNRAS, 279, 389
—. 2003, MNRAS, 341, 1360
MacLeod, M., Antoni, A., Murguia-Berthier, A., Macias, P., & Ramirez-Ruiz, E. 2017, ApJ, 838, 56
MacLeod, M., Ostriker, E. C., & Stone, J. M. 2018a, ArXiv e-prints, arXiv:1808.05950
—. 2018b, ApJ, 863, 5
MacLeod, M., & Ramirez-Ruiz, E. 2015a, ApJ, 803, 41
—. 2015b, ApJl, 798, L19
Madappatt, N., De Marco, O., & Villaver, E. 2016, MNRAS, 463, 1040
Matt, S., Frank, A., & Blackman, E. G. 2006, ApJL, 647, L45
Mauron, N., & Huggins, P. J. 2006, A & A, 452, 257
Meyer, F., & Meyer-Hofmeister, E. 1983, A & A, 121, 29
Miranda, L. F., & Solf, J. 1990, Ap& SS, 171, 227
Miszalski, B., Acker, A., Moffat, A. F. J., Parker, Q. A., & Udalski, A. 2009, A & A, 496, 813
Moe, M., & De Marco, O. 2006, ApJ, 650, 916
Mohamed, S., & Podsiadlowski, P. 2012, Baltic Astronomy, 21, 88
Moreno Méndez, E., López-Cámara, D., & De Colle, F. 2017, MNRAS, 470, 2929
Murakawa, K., & Izumiura, H. 2012, A & A, 544, A58
Nordhaus, J., & Blackman, E. G. 2006, MNRAS, 370, 2004
Nordhaus, J., Blackman, E. G., & Frank, A. 2007, MNRAS, 376, 599
O'Connor, J. A., Redman, M. P., Holloway, A. J., et al. 2000, ApJ, 531, 336
Ohlmann, S. T., Röpke, F. K., Pakmor, R., & Springel, V. 2016a, ApJl, 816, L9
—. 2017, A. & Ap., 599, A5
Ohlmann, S. T., Röpke, F. K., Pakmor, R., Springel, V., & Müller, E. 2016b, MNRAS, 462, L121
Olofsson, H., Vlemmings, W., Maercker, M., et al. 2015, in Astronomical Society of the Pacific Conference Series, Vol. 499, Revolution in Astronomy with ALMA: The Third Year, ed. D. Iono, K. Tatematsu, A. Wootten, & L. Testi, 319
Ondratschek, P. A., Roepke, F. K., Schneider, F. R. N., et al. 2021, arXiv e-prints, arXiv:2110.13177
Paczynski, B. 1976, in IAU Symposium, Vol. 73, Structure and Evolution of Close Binary Systems, ed. P. Eggleton, S. Mitton, & J. Whelan, 75
Papish, O., Soker, N., & Bukay, I. 2015, MNRAS, 449, 288
Pascoli, G. 1997, ApJ, 489, 946

Pelletier, G., & Pudritz, R. E. 1992, ApJ, 394, 117

Perez-Sanchez, A. F., Vlemmings, W. H. T., Tafoya, D., & Chapman, J. M. 2013, MNRAS, 436, L79

Postnov, K. A., & Yungelson, L. R. 2014, Living Reviews in Relativity, 17, 3

Raghavan, D., McAlister, H. A., Henry, T. J., et al. 2010, ApJS, 190, 1

Reichardt, T. A., De Marco, O., Iaconi, R., Tout, C. A., & Price, D. J. 2019, MNRAS, 484, 631

Reyes-Ruiz, M., & López, J. A. 1999, ApJ, 524, 952

Ricker, P. M., & Taam, R. E. 2012, ApJ, 746, 74

Riera, A., Velázquez, P. F., Raga, A. C., Estalella, R., & Castrillón, A. 2014, A & A, 561, A145

Ritter, H. 1988, A & A, 202, 93

Ruiz, M., Tsokaros, A., & Shapiro, S. L. 2021, Phys. Rev. D., 104, 124049

Sabin, L., Hull, C. L. H., Plambeck, R. L., et al. 2015, MNRAS, 449, 2368

Sahai, R., Sugerman, B. E. K., & Hinkle, K. 2009, ApJ, 699, 1015

Sahai, R., & Trauger, J. T. 1998, AJ, 116, 1357

Sahai, R., Vlemmings, W. H. T., Gledhill, T., et al. 2017, ApJl, 835, L13

Sánchez Contreras, C., Alcolea, J., Bujarrabal, V., et al. 2018, A & A, 618, A164

Sánchez Contreras, C., Gil de Paz, A., & Sahai, R. 2004, ApJ, 616, 519

Shiber, S., Iaconi, R., De Marco, O., & Soker, N. 2019, MNRAS, 488, 5615

Shiber, S., Kashi, A., & Soker, N. 2017, MNRAS, 465, L54

Shiber, S., & Soker, N. 2018, MNRAS, arXiv:1706.00398

Soker, N. 1994, MNRAS, 270, 774

—. 1996, ApJ, 469, 734

—. 1997, ApJS, 112, 487

—. 2002, ApJ, 568, 726

—. 2004, New Astron., 9, 399

—. 2006, PASP, 118, 260

—. 2015, ApJ, 800, 114

—. 2016, New Astron. Revs, 75, 1

Soker, N., & Livio, M. 1989, ApJ, 339, 268

—. 1994, ApJ, [illegible], [illegible]

Soker, N., & Rappaport, S. 2000, ApJ, 538, 241

—. 2001, ApJ, 557, 256

Staff, J. E., De Marco, O., Macdonald, D., et al. 2016a, MNRAS, 455, 3511

Staff, J. E., De Marco, O., Wood, P., Galaviz, P., & Passy, J.-C. 2016b, MNRAS, 458, 832

Taam, R. E., & Sandquist, E. L. 2000, ARAA, 38, 113

Tafoya, D., Imai, H., Gómez, J. F., et al. 2020, ApJL, 890, L14

Tocknell, J., De Marco, O., & Wardle, M. 2014, MNRAS, 439, 2014

Tout, C. A., & Regos, E. 2003, in Astronomical Society of the Pacific Conference Series, Vol. 293, 3D Stellar Evolution, ed. S. Turcotte, S. C. Keller, & R. M. Cavallo, 100

Uzdensky, D. A., & MacFadyen, A. I. 2006, ApJ, 647, 1192

Vlemmings, W. H. T., Diamond, P. J., & Imai, H. 2006, Nature, 440, 58

Wilson, E. C., & Nordhaus, J. 2019, MNRAS, 485, 4492

Witt, A. N., Vijh, U. P., Hobbs, L. M., et al. 2009, ApJ, 693, 1946

Zou, Y., Frank, A., Chen, Z., et al. 2020, MNRAS, 497, 2855

The Origin of Outflows in Evolved Stars
Proceedings IAU Symposium No. 366, 2022
L. Decin, A. Zijlstra & C. Gielen, eds.
doi:10.1017/S1743921322000102

ALMA (finally!) discloses a rotating disk+bipolar wind system at the centre of the wind-prominent pPN OH 231.8+4.2

C. Sánchez Contreras[1], J. Alcolea[2], R. Rodríguez-Cardoso[1,2], V. Bujarrabal[3], A. Castro-Carrizo[4], L. Velilla-Prieto[5], G. Quintana-Lacaci[5], M. Santander-García[2], M. Agúndez[5] and J. Cernicharo[5]

[1]Centro de Astrobiología (CSIC-INTA), Camino Bajo del Castillo s/n, E-28691 Villanueva de la Cañada, Madrid, Spain
email: csanchez@cab.inta-csic.es

[2]Observatorio Astronómico Nacional (IGN), Alfonso XII No 3, 28014 Madrid, Spain

[3]Observatorio Astronómico Nacional (IGN), Ap 112, 28803 Alcalá de Henares, Madrid, Spain

[4]Institut de Radioastronomie Millimetrique, 300 rue de la Piscine, 38406 Saint Martin d'Heres, France

[5]Instituto de Fisica Fundamental (CSIC), C/ Serrano, 123, E-28006, Madrid, Spain

Abstract. We present interferometric continuum and molecular line emission maps obtained with the Atacama Large Millimeter/submillimeter Array (ALMA) of OH231.8+4.2, a well studied bipolar nebula around an asymptotic giant branch (AGB) star that is key to understand the remarkable changes in nebular morphology and kinematics during the short transition from the AGB to the Planetary Nebula (PN) phase. The excellent angular resolution of our maps ($\sim$20 mas $\sim$30 AU) allows us to scrutinize the central nebular regions of OH231.8+4.2, which hold the clues to understanding how this iconic object assembled its complex nebular architecture. We report, for the first time in this object and others of its kind (i.e. pre-PNe with massive bipolar outflows), the discovery of a rotating circumbinary disk of radius $\sim$30 AU traced by NaCl, KCl, and H_2O emission lines. The disk lies at the base of a young bipolar wind with signs of rotation as well. A compact spatially resolved dust disk is found perpendicular to the bipolar outflow. We also identify a point-like continuum source, which likely represents the central Mira star enshrouded by a $\sim 3\,R_\star$ shell or disk of hot ($\sim$1400 K) freshly formed dust. The point source is slightly off-centre from the disk centroid, enabling us for the first time to place constraints to the orbital separation of the central binary system.

Keywords. AGB and post-AGB stars, planetary nebulae, mass-loss, rotating disks, binaries, OH 231.8+4.2

---

1. Introduction

Currently, the PN/pre-PN (pPN) community is reaching full consensus that binaries are needed to help in the production of collimated fast winds (jets) that are in turn primary agents for the breaking of the spherical symmetry during the AGB to PN transition. Rotating structures are expected to form associated with the presence of stellar or substellar companions to mass-lossing stars, however, direct empirical confirmation and characterization of such structures is very difficult. To date, rotating circumbinary disks have been found in a population of binary post-AGB stars with near-infrared (NIR) excess, referred to as disk-prominent post-AGB stars, that curiously lack of massive fast outflows (see e.g. Van Winckel 2017; Bujarrabal et al. 2013 and references there in). In this contribution, we report on the first confirmed detection of a rotating disk in a pPN with massive bipolar outflows (referred to as wind-prominent pPN).

OH231.8+4.2 is a well known bipolar nebula around a mass-lossing AGB star, QX Pup. QX Pup is a Mira-type variable that as prematurely developed a massive ($\sim$1 $M_\odot$) pPN-like nebula with a spectacular bipolar morphology and very fast outflows, with velocities of up to a few hundred km s^{-1} that are reached at the tips of large-scale ($\approx$0.1-0.2 pc-sized) bipolar lobes. QX Pup is part of a binary system with (at least) one companion, an A0V star (Sánchez Contreras et al. 2004), whose presence is probably at the root of the seemingly premature evolution of this object to the next pPN stage, in which the spherical symmetry is broken and high-speed collimated outflows develop. OH231.8+4.2 is located at a distance of $d\sim$1500 pc.

The structure and kinematics of the molecular envelope of OH231.8+4.2 has been recently characterized with unprecedented detail based on $\sim$0.″2-0.″3-angular resolution continuum and molecular line maps obtained with ALMA (Sánchez Contreras et al. 2018). From these observations, we found that the molecular outflow has a structure much more complex than previously thought: we discovered an extravagant array of nested (but not always co-axial) small-to-large scale structures previously unknown that suggest a complex formation history. In the central regions of the nebula, we discovered two main structures: *i*) a compact parcel of gas and dust that surrounds the mass-lossing AGB star, referred to as clump *S*, that is selectively traced by certain species, including NaCl; and *ii*) a compact bipolar outflow that emanates from clump *S*, which is selectively traced by SiO. Unlike the large-scale CO outflow, the SiO-outflow is symmetric with respect to the equator and *slow* ($V_{\rm exp}\lesssim$20 km s^{-1}). The SiO-outflow is younger than the large-scale nebula, which is about $\sim$800 yr old.

2. Observations

With the goal of disecting the central regions of OH231.8+4.2, we observed this object again with the ALMA 12m interferometric array in its most extended configuration during cycle 5 (2017.1.00706.S). Two frequency settings within band 6 ($\sim$242-261 GHz and $\sim$217-234 GHz, respectively) were used to map molecular line and continuum emission. The data were obtained with 50-52 antennas with baselines ranging from 41.4 m to 16.2 km, resulting in a highest angular resolution of about $\sim$0.″02. The maximum recoverable scale (MRS) is $\sim$0.″3-0.″4.

In the following section, we present the maps of the continuum emission at 261 GHz and of NaCl and SiS/SiO transitions that selectively trace the central regions (down to $\sim$30 AU from the central star) of the molecular outflow. Full details on the observations, data reduction and image restoration procedures as well as our in-depth analysis of these and other molecules included in our ALMA dataset are presented in Sánchez Contreras et al. (in preparation).

3. Results: a compact rotating disk+outflow system in OH231.8+4.2

3.1. *Continuum maps*

As known from several previous works, the continuum emission in OH231.8+4.2 is due to dust thermal emission (e.g. SC+18 and references therein). The compact region around QX Pup identified in our pevious ALMA maps, clump *S*, is now spatially resolved in two main components: an extended disk-like component, elongated in the direction perpendicular to the bipolar nebula, and an unresolved component (Fig. 1). The dimensions of the extended component are consistent with a circular disk, of radius of $\sim$40 AU, inclined

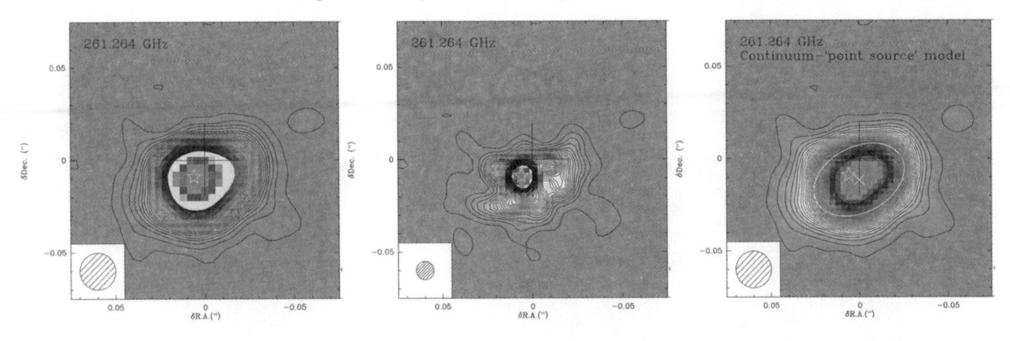

Figure 1. ALMA continuum emission maps at 261.264 GHz. In the left and middle pannels, the continuum maps with 20 and 10 mas resolution are shown; level steps are 0.15 mJy/beam $\sim$2.7σ. In the right panel, we show the continuum map with 20 mas resolution after fitting and substracting a point source model that is located at the position marked by the star-like symbol. The white ellipse represents the size and orientation of the extended disk model that best fits the residual continuum emission map after subtraction of the point source model. Note the offset between the position of the point-like continuum source (starlike symbol) and the center of the disk (white small cross). The shaded circular areas at the bottom-left corner of the maps represent the half-power-beam-width (HPBW). The large black cross marks the phase tracking center of the observations.

$\lesssim$40° with respect to the line of sight. This value is very similar to the inclination of the lobes with respect to the plane of the sky known from previous works (e.g. Kastner et al. 1992), indicating that the disk and the lobes are orthogonal.

The point-like continuum emission is consistent with being partially due to the stellar photosphere of QX Pup ($T_{\rm eff}\sim$2500 K, $R_\star\sim$2.1 AU) and to hot ($\sim$1400 K) dust in its vicinity (within a few stellar radii, $R_{\rm d}\sim$7.5 AU). We observe a small offset ($\sim$6.6 mas) between the centroid of the extended disk and the position of the point-like source. Assuming that the disk is circumbinary and adopting reasonable values for the mass of QX Pup ($m_1\sim0.7\,M_\odot$) and the A0V companion ($m_2\sim2\,M_\odot$) we derive from this offset an orbital separation of $a\sim$23 AU for the binary system.

3.2. *Molecular line maps*

We have observed a large number of molecular transitions from different species as part of this project. Here, we focus on line emission maps of NaCl and SiS/SiO, some of the molecules that selectively trace the central clump S and the compact SiO-outflow, respectively (SC+18).

3.2.1. NaCl

We have detected a total of 8 different transitions of NaCl in different v=0, 1, 2, and, tentatively 3, vibrational levels. The lines are weak but they all consistently show a very similar brightness distribution. Taking advantage of this, the lines have been combined to obtain a NaCl line-stacked emission cube with an increased signal-to-noise ratio. As shown in Fig. 2, the NaCl emission arises from a compact region surrounding the extended dust-continuum disk (indicated by the dashed ellipse) consistent with NaCl arising from the surface layers of the dust disk. These layers of the disk are in rotation as indicated by the clear velocity gradient along the equator noticeable in the first-moment map (right panel of Fig. 2): note that the emission from the east side is red-shifted while the west side is blue-shifted. The sense of the rotation of the circumbinary NaCl-disk, with its east (west) side receeding from (approaching to) us, is the same as that of the SiO-maser rotating torus found in the pulsating layers of QX Pup (Sánchez Contreras et al. 2002).

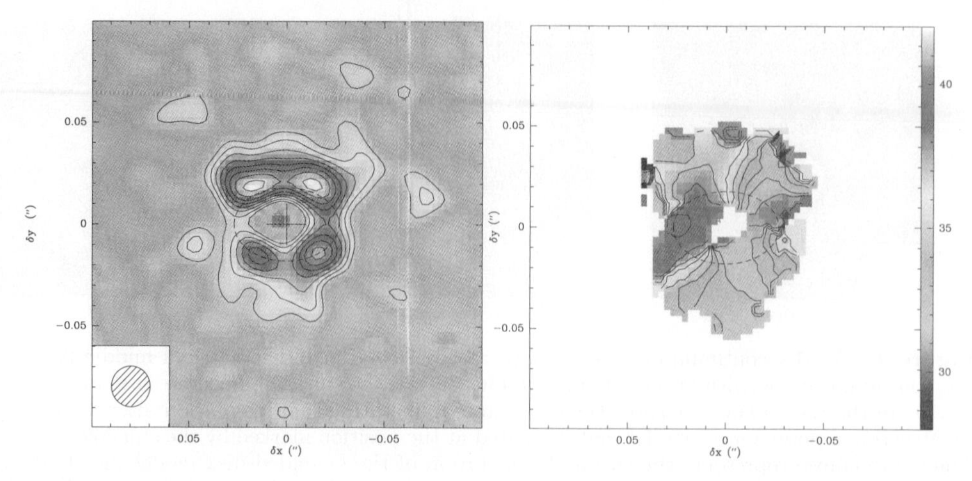

Figure 2. ALMA data of NaCl after stacking together the individual NaCl transitions detected in this work. Left) order-zero moment map over the $V_{\rm LSR}$=[24:46] km s^{-1} velocity range; contours are 2σ, 3σ,... by 1σ (σ=1.7mJy/beamkm s^{-1}). Right) first-moment map; contours go from $V_{\rm LSR}$=28 to 43 by 1 km s^{-1}. The wedge indicates the colour-$V_{\rm LSR}$ scale. Maps are rotated by 25° clockwise so the symmetry axis of the nebula is vertical. The center of the dust disk (dashed ellipse) inferred from the continuum emission maps is marked with a cross (R.A.=07^{h}42^{m}16^s.91500 and Dec.=−14°42′50″.0716, J20000) and has been adopted as the origin of positional offsets here and in Fig. 3.

The rotation velocity deduced from the NaCl maps is about $V_{\rm rot}\sim$4 km s^{-1} at a mean radial distance of ∼40 AU. In addition to rotation, there are also expansive motions in the circumbinary NaCl-disk with a strikingly low expansion speed of $V_{\rm exp}\sim$3 km s^{-1}.

To better constrain the geometry, kinematics, and physical conditions of the rotating disk layers probed by NaCl, we have compared the ALMA NaCl maps with the predictions of a LTE radiative transfer model for this species. We have approximated the geometry of the NaCl-emitting volume as two co-axial tori of radius ∼30-35 AU displaced along the nebula axis from the continuum disk midplane by ∼20 AU, emulating the two surface layers (above and below) the disk where the NaCl is detected. The overall kinematics is reasonably well described with a composite velocity fied that includes rotation and expansion in the equatorial plane. For a uniform temperature of $T_{\rm rot}\sim$400-500 K, deduced from an independent analysis of the different NaCl transitions detected, and adopting a fractional NaCl-to-H_2 abundace of X(NaCl)$\sim 5\times10^{-9}$ (SC+18), we derive average H_2 densities of $\approx$10^9 cm^{-3}, resulting in a total mass of the disk's surface layers of about 2×10^{-3} $M_\odot$. The angular resolution of our ALMA data is still insufficient to study the radial variations of the model parameters across the disk layers, in particular, we cannot infer whether the rotation follows a Keplerian or a sub-Keplerian velocity law. The presence of expansion suggests sub-Keplerian rotation (e.g. Kervella et al. 2016), which would then imply a lower limit to the central dynamic mass of ∼0.7 $M_\odot$.

In addition to NaCl, we have found two other molecular species that selectively trace the rotating equatorial structure at the core of OH231.8+4.2, namely, potassium chloride (KCl, i.e. another salt) and water (H_2O); these species are potentially unique tools for identifying disks and measuring their kinematics in pPNe.

3.2.2. SiS v=0 and SiO v=1

The compact (∼ 1″ × 4″) bipolar outflow discovered in SC+18 is traced by several rotational transitions in the v=0 and v=1 vibrational states of SiO and SiS (including

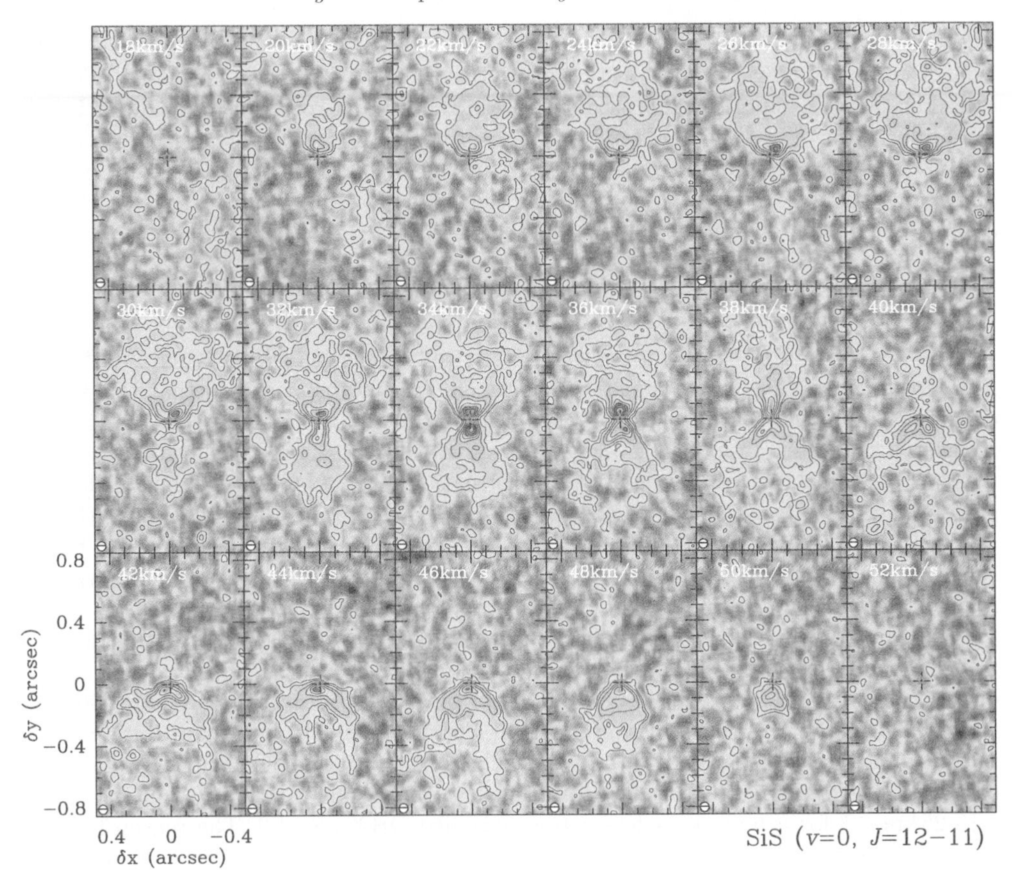

Figure 3. SiS v=0 (J=12-11) velocity-channel maps (contours: 2σ, 4σ,... by 4σ; σ=0.57 mJy/beam). Natural weigthing and tapering (with a tapering distance of 5700 m) have been used to restore the emission maps with a half-power clean beam width of HPBW=$0\farcs06\times0\farcs06$.

some isotopologues). We refer to this component as the SiO/SiS-outflow (hereafter, SS-outflow). In this contribution, we discuss the SiS v=0, J=12-11 line, which best delineates the dense walls of the bipolar SS-outflow, and the vibrationally excited SiO v=1, J=6-5 line, which traces the innermost layers of the SS-outflow (i.e. closer to the center) where the gas is presumably denser and hotter.

The SiS v=0, J=12-11 maps (Fig. 3) show that the lobes have a conical geometry at their base, consistent with a wide opening angle ($\theta\sim90^{\circ}$) wind, and a more rounded morphology at their tips. The SS-outflow emerges from the surface layers of the rotating equatorial disk, where the SiS v=0, J=12-11 and NaCl emission partially overlap. We find low-to-moderate expansion velocities throughout the SS-outflow. From a simple a spatio-kinematic model of the SS-outflow to fit the SiS v=0, J=12-11 data we deduce a gradual outward acceleration of the gas along the lobes up to a terminal expansion speed of about $\sim$17 km s^{-1}, which is reached at a radial distance of $\sim$250 AU. The radial expansion continues at constant velocity beyond this point. We have constrained the inclination of the SS-outflow to values around $\sim$30°-35° and its kinematical age to about one hundred years at the base (to be compared with the 800 yr age of the large-scale CO outflow).

Our maps from the vibrationally excited SiO v=1, J=6-5 line show emission arising from a compact elongated region of dimensions $\sim 0''.05 \times 0''.12$ oriented along PA$\sim$25°, consistent with a bipolar wind running inside the hollow rotating disk traced by the salts and water. The kinematics at the base of the SS-outflow as traced by SiO v=1, J=6-5 emission is predominantly expansive but the signature of rotation is also present, particularly in regions close to the equator. This was somehow expected given the SiO and NaCl emission overlap at the base of the SS-outflow.

4. Discussion

NaCl beyond the dust formation zone. It is well known that in normal AGB stars, NaCl forms by equilibrium chemistry near the stellar photosphere and that this species disappears from the gas phase rapidly as it gets incorporated into dust grains given its refractory character. In OH231.8+4.2, we observe NaCl on the surface of the dust disk, which is beyond the region of massive dust formation where NaCl should be significantly gas depleted as a consequence of condensation onto the grains. At the base of the SS-outflow, NaCl is co-spatial with SiO (and SiS), which is a well known shock tracer, suggesting that shocks are probably efficiently extracting NaCl from grains and returning it to the gas phase.

Low expansion velocity in the disk. Another surprising result from this work is the extremely low expansion velocity ($V_{\rm exp}$$\sim$3 km s^{-1}) measured at the surface layers of the rotating disk, at relatively large radial distances from the center of $\sim$40 AU$\sim$ 6×10^{14}cm$\sim$20 $R_\star$. At these distant regions, clearly beyond the massive dust condensation (wind acceleration) zone, the wind velocity of a normal O-rich AGB star (like, e.g., IK Tau, Decin et al. 2010) should be close to the terminal velocity, that is, close to $\sim$15-25 km s^{-1} for a high-mass loss rate object like OH231.8+4.2. The reason for such innefficient wind acceleration in OH231.8+4.2 is unknown (it may be caused by the presence of very optically thick dust or very large grains along the equator), but, as it is shown in the next paragraph, a low wind velocity turns out to be essential to promote the formation of an equatorially dense circumbinary structure as a result from wind Roche Lobe Overflow (wRLOF) mass transfer (Mohamed & Podsiadlowski 2007) in the central binary system.

Formation of an equatorial density enhanced (EDE) region. In addition to an extremely low wind velocity, OH231.8+4.2 brings together other favorable conditions for the formation of a dense equatorial structure as a result of wRLOF mass transfer, namely: a very massive companion (q=$m_1/m_2$$\sim$0.4<1) and comparable sizes of the dust condensation radius ($R_{\rm d}$$\sim$7.5 AU) and the Roche lobe radius (r$_{\rm L}$$\sim$7 AU, for $a$$\sim$23 AU; Eggleton 1983). As shown e.g. in the recent work by El Mellah et al. (2020), these are key parameters for effective mass transfer from QX Pup to the companion and the eventual compression of the AGB wind on the orbital plane to form an EDE.

The SS-outflow. Given its properties (low-velocity and wide opening angle), the SS-outflow could simply result from the confinement of the on-going AGB wind by the dense equatorial torus (as in the so-called Generalized Interacting Stellar Winds scenario of PN-shaping, see e.g. Balick and Frank 2002). The absence of fast jet-like ejections (together with the lack of classical accretion indicators, such as Hα emission from the nucleus) rules out high-rate accretion by a compact object *at present*. The situation was clearly different $\sim$800 yr ago, when the large-scale bipolar nebula was shaped and accelerated up to velocities of $\approx$400 km s^{-1}, a process that necessarily required accretion on (and jet-launching from) a main-sequence companion. Perhaps the accretion disk around the companion has been exhausted in recent times. Alternatively, as we proposed in Sánchez Contreras et al. (2004), OH231.8+4.2 could be at present in a low-rate accretion (or 'quiescent') state in which the disk around the companion is steadily building up its mass but there is no effective disk-to-companion accretion (and no jet launching), a situation that occurs in other astrophyiscal systems known to go through alternating accretion ourtburst and post-outburst (quiescent) states (like FU Ori objects and symbiotics stars, e.g. Hartmann & Kenyon 1996).

Acknowledgements

This work is part of the I+D+i projects PID2019-105203GB-C22 and PID2019-105203GB-C21, funded by the Spanish MCIN/ AEI/10.13039/501100011033.

References

Alcolea, J., Bujarrabal, V., Sánchez Contreras, C., Neri, R., and Zweigle, J. 2001, *A&A*, 373, 932

Balick, B., and Frank, A. 2002, *ARA&A*, 40, 439

Bujarrabal, V., Alcolea, J., Van Winckel, H., et al. 2013, *A&A*, 557, A104

Decin, L., Justtanont, K., De Beck, E., et al. 2010, *A&A*, 521, L4

Eggleton, P. P. 1983, *ApJ*, 268, 368

El Mellah, I., Bolte, J., Decin, L., et al. 2020, *A&A*, 637, A91.

Hartmann, L. & Kenyon, S. J. 1996, *ARA&A*, 34, 207

Kastner, J. H., Weintraub, D. A., Zuckerman, B., et al. 1992, *ApJ*, 398, 552

Kervella, P., Homan, W., Richards, A. M. S., et al. 2016, *A&A*, 596, A92

Mohamed, S. & Podsiadlowski, P. 2007, 15th European Workshop on White Dwarfs, 372, 397

Sánchez Contreras, C., Desmurs, J. F., Bujarrabal, V., et al. 2002, *A&A*, 385, L1

Sánchez Contreras, C., Gil de Paz, A., and Sahai, R. 2004, *ApJ*, 616, 519

Sánchez Contreras, C., Alcolea, J., Bujarrabal, V., et al. 2018, *A&A*, 618, A164

van Winckel, H. 2017, Planetary Nebulae: Multi-Wavelength Probes of Stellar and Galactic Evolution, Proceedings of the International Astronomical Union, IAU Symposium, 323, 231

The Origin of Outflows in Evolved Stars
Proceedings IAU Symposium No. 366, 2022
L. Decin, A. Zijlstra & C. Gielen, eds.
doi:10.1017/S1743921322000205

Lessons from the ionised and molecular mass of post-CE PNe

Miguel Santander-García[1], David Jones[2,3], Javier Alcolea[1], Valentín Bujarrabal[4] and Roger Wesson[5]

[1]Observatorio Astronómico Nacional (OAN-IGN), Alfonso XII, 3, 28014, Madrid, Spain
email: m.santander@oan.es

[2]Instituto de Astrofísica de Canarias, 38205, La Laguna, Spain

[3]Departamento de Astrofísica, Universidad de La Laguna, 38206, La Laguna, Spain

[4]Observatorio Astronómico Nacional (OAN-IGN), Apdo. 112, 28803, Alcalá de Henares, Spain

[5]Department of Physics and Astronomy, University College London, Gower St, London, UK

Abstract. Close binary evolution is widely invoked to explain the formation of axisymmetric planetary nebulae, after a brief common envelope phase. The evolution of the primary would be interrupted abruptly, its still quite massive envelope being fully ejected to form the PN, which should be more massive than a planetary nebula coming from the same star, were it single. We test this hypothesis by investigating the ionised and molecular masses of a sample consisting of 21 post-common-envelope planetary nebulae, roughly one fifth of their known total population, and comparing them to a large sample of regular planetary nebulae (not known to host close-binaries). We find that post-common-envelope planetary nebulae arising from single-degenerate systems are, on average, neither more nor less massive than regular planetary nebulae, whereas post-common-envelope planetary nebulae arising from double-degenerate systems are considerably more massive, and show substantially larger linear momenta and kinetic energy than the rest. Reconstruction of the common envelope of four objects further suggests that the mass of single-degenerate nebulae actually amounts to a very small fraction of the envelope of their progenitor stars. This leads to the uncomfortable question of where the rest of the envelope is, raising serious doubts on our understanding of these intriguing objects.

Keywords. planetary nebulae: general, planetary nebulae: individual: NGC 6778, circumstellar matter, binaries: close, Stars: mass-loss, Stars: winds, outflows

---

1. Introduction

Planetary nebulae (PNe) display beautiful, complex morphologies often showing high degrees of symmetry, mostly elliptical or bipolar. Among the mechanisms usually invoked to explain their intriguing shaping (Balick & Frank, 2002), angular momentum transfer from a companion is currently considered a key ingredient (Jones & Boffin 2017; Decin *et al.* 2020). In the case of a sufficiently close companion, the system undergoes a common envelope (CE) event when the primary star evolves along the giant branch(es), expanding before eventually overflowing its Roche lobe and engulfing its companion (Paczynski 1976). In this very brief stage (~1 year), the orbit of the secondary star quickly shrinks due to drag forces, providing angular momentum for the system to eject the CE, which we will observe as a PN.

From an observational point of view, the number of confirmed post-CE PNe has been quickly growing over the last decades, amounting to around 100 objects so far†. On theoretical grounds, nevertheless, the mechanism of CE ejection remains elusive, with most hydrodynamic models unable to gravitationally unbind the whole envelope without recurring to additional energy reservoirs such as recombination energy from the ionised regions (e.g. Ricker & Taam 2012; Jones 2020; Ohlmann *et al.* 2016; Ivanova 2018; Chamandy *et al.* 2020). To sum up, simulations collectively show that the CE has a major role in shaping PNe, but we are far from fully understanding the physics behind the death of a significant fraction of stars in the Universe.

Carefully estimating the actual mass of post-CE PNe could help towards a better understanding of the physics of CE ejection. In this respect, it can be helpful to establish comparisons between these objects and the general population of PNe, encompassing nebulae arising not only from close binaries but also from single stars and longer period binary stars that did not experience a CE. The only existing previous work covering this topic studied the ionised mass of a sample of post-CE PNe, suggesting these objects are slightly less massive, on average, than the general population of PNe (Frew & Parker 2007).

In this work we systematically analyse the mass of post-CE PNe, extending it to cover molecular masses as well as ionised ones, on a sample comprising 21 post-CE PNe (roughly 1/5th of the total known population of these objects), and putting it into context by comparing it to a larger sample of 97 'regular' PNe (nebulae not known to arise from close-binary systems). This proceeding summarises the main results of our study. Please consult the main publication for additional details, information on the methods followed and general discussion on the mass of post-CE PNe (Santander-García *et al.* 2022).

2. Sample and Observations

We used the IRAM 30m radiotelescope to carry out mm-wavelength observations in order to probe the molecular content of an initial sample of nine post-CE PNe of the northern sky which were previously unobserved in ^{12}CO and ^{13}CO J=1−0 and J=2−1. Only one of them, NGC 6778, was detected, and its molecular emission profiles were indicative of the presence of a thin ring-like structure along the equator of the nebula, located outwards from the equatorial ionised, broken ring visible in optical images (Guerrero & Miranda 2012). See Santander-García *et al.* (2022) for a spatiokinematical model of this structure including radiative-transfer of CO species.

The sample was later extended to 21 post-CE PNe, including objects with molecular observations published in the literature (Huggins & Healy 1989; Huggins *et al.* 1996, 2005; Guzman-Ramirez 2018). Note that only 3 of these 21 objects, NGC 6778, NGC 2346, and NGC 7293, show molecular emission at all, hence the molecular masses computed below are upper limits in most cases.

In addition to the molecular data, we also gathered dereddenned Hα fluxes and optical sizes from Frew *et al.* (2016), as well as literature-based values of the densities and electronic temperatures of every object in the sample.

3. The mass of post-CE PNe

We computed the ionised and molecular masses of the whole sample of post-CE PNe in a systematic way. We here describe the analyses performed, present our results on the masses, and compare them to the masses of a large sample of 'regular' PNe estimated in the same way.

† See updated list with references to discovery papers in `http://www.drdjones.net/bcspn/`

3.1. *Ionised masses*

The total ionised masses of the PNe were calculated as:

$$M_{\rm ion} = \frac{4\ \pi\ D^2\ F({\rm H}\beta)\ m_p}{h\nu_{{\rm H}\beta}\ n_e\ \alpha^{\rm eff}_{{\rm H}\beta}}, \tag{3.1}$$

where D is the distance, $F(\mathrm{H}\beta)$ is the dereddened, spatially integrated Hβ flux, m_p is the mass of the proton, $h\nu_{\mathrm{H}\beta}$ is the energy of an Hβ photon, n_e is the electron density, and $\alpha^{\rm eff}_{{\rm H}\beta}$ is the effective recombination coefficient of Hβ (Corradi *et al.* 2015).

For the results to be as standardised as possible, we only utilised Hβ fluxes derived from the dereddened Hα surface brightness tabulated by Frew *et al.* (2016), integrated over the ellipse defined by the minor and major axes tabulated in the same work. With respect to electron temperatures, we used T[O III] determinations when possible, and assumed T_e=10 000 K in those objects where no determination was available. Similarly, we used [S II] line doublet determinations of the electron densities (except for NGC 246, where only an estimate based on [O II] was available). Note that, if part of the nebulae consist of dense condensations above the density traced by [S II], the resulting ionised masses would in principle be lower (adding to the problem outlined below). As for distances, we prioritised GAIA eDR3 determinations by Gaia *et al.* (2020), and their associated errors were $< 33\%$. In the absence of these, we used distances by Frew *et al.* (2016).

3.2. *Molecular masses*

We computed the molecular mass for the three objects in our sample which show molecular emission, as well as conservative (3-σ) upper limits to the molecular mass of the rest of the objects in the sample. We assume that: *(i)* the CO level populations are in local thermodynamic equilibrium (LTE), and are characterised by an excitation temperature $T_{\rm ex}$; *(ii)* the CO abundance X relative to molecular hydrogen is constant throughout the nebula; and *(iii)* the selected CO transition is optically thin. Under these conditions, the molecular mass $M_{\rm mol}$ of the nebula is:

$$M_{\rm mol} = \frac{4\ \pi\ m_{\rm H_2}\ D^2}{A_{\rm ul}\ X\ h\ \nu\ g_u}\ e^{\frac{h\nu}{kT_{\rm ex}}}\ Z(T_{\rm ex})\ f_{\rm He}\ S_\nu, \tag{3.2}$$

where $m_{\rm H_2}$ is the mass of the hydrogen molecule, h and k are the Planck and Boltzmann constants, ν is the frequency of the transition, $A_{\rm ul}$ its Einstein coefficient, g_u the degeneracy of its upper state, Z the partition function, D the distance to the nebula, $\rm f_{He}$ the correction factor to account for helium abundance (assumed to be He/H=0.1 and thus resulting in $\rm f_{He}$=1.2, because we also assume the majority of particles to be of molecular hydrogen), and S_ν the flux density:

$$S_\nu = \frac{2\ k\ \nu^2\ F}{c^2}, \tag{3.3}$$

where c is the speed of light in vacuum, and F the total flux of the nebula in the given transition, integrated both spatially and spectrally. We assumed an excitation temperature of $T_{\rm ex}$=50 K, and a CO abundance X=2$\times$10^{-4} for every object in our calculations, and used primarily the ^{12}CO J=2$-$1 transition, more ubiquous and better detected in the literature. In order to overcome the clear limitation that this transition (as well as the ^{12}CO J=3$-$2 one) is usually not optically thin, we introduce a correction factor for the underestimated masses resulting from J=2$-$1 (and J=3$-$2) transitions in order to match masses found via the J=1$-$0 transition in a sample of PNe in which both transitions are available in the literature (Huggins & Healy 1989; Huggins *et al.* 1996, 2005; Guzman-Ramirez 2018). These resulted in a factor 3.65 to calculations using ^{12}CO J=2$-$1 and a factor 5.0 for those using ^{12}CO J=3$-$2. We recall that the best correction

Table 1. Computed ionised and molecular masses of the post-CE sample.

PN G	Common name	D (kpc)	$M_{\rm ion}$ ($M_\odot$)	$M_{\rm mol}$ ($M_\odot$)
		Single-Degenerate post-CE PNe		
G034.5-06.7	NGC 6778	2.79±0.79	$0.19^{+0.14}_{-0.10}$	0.02±0.02
G036.1-57.1	NGC 7293	0.200±0.002	$0.09^{+0.13}_{-0.05}$	0.3±0.2
G053.8-03.0	Abell 63	2.703±0.219	$0.012^{+0.04}_{-0.009}$	<0.006
G054.2-03.4	Necklace	4.6±1.1	$0.009^{+0.017}_{-0.006}$	<0.007
G068.1+11.1	ETHOS 1	4.2±0.0	$0.008^{+0.3}_{-0.008}$	<0.007
G086.9-03.4	Ou 5	5.0±1.0	$0.18^{+0.4}_{-0.12}$	<0.012
G118.8-74.7	NGC 246	0.556±0.025	$0.07^{+0.12}_{-0.05}$	<0.02
G208.5+33.2	Abell 30	2.222±0.148	$0.015^{+0.02}_{-0.009}$	<0.20
G215.6+03.6	NGC 2346	1.389±0.039	$0.09^{+0.09}_{-0.04}$	0.7±0.5
G221.8-04.2	PM 1-23	5.2±2.0	$0.015^{+1.2}_{-0.014}$	<0.17
G307.5-04.9	MyCn 18	4.000±1.280	$0.07^{+0.10}_{-0.04}$	<0.06
G338.1-08.3	NGC 6326	5.000±1.500	$0.6^{+0.5}_{-0.3}$	<0.06
G338.8+05.6	Hen 2-155	4.348±1.323	$0.3^{+0.2}_{-0.14}$	<0.10
G342.5-14.3	Sp 3	$2.22^{+0.61}_{-0.48}$	$0.09^{+0.08}_{-0.04}$	<0.06
G349.3-04.2	Lo 16	1.818±0.132	$0.4^{+0.7}_{-0.3}$	<0.013
		Double-Degenerate post-CE PNe		
G009.6+10.5	Abell 41	4.89±1.4	$0.16^{+0.15}_{-0.09}$	<0.011
G049.4+02.4	Hen 2-428	4.545±1.446	$0.7^{+0.8}_{-0.4}$	<0.010
G058.6-03.6	V458 Vul	12.5±2.0	$0.11^{+0.19}_{-0.07}$	<0.08
G197.8+17.3	NGC 2392	1.818±0.165	$0.4^{+0.4}_{-0.19}$	<0.09
G290.5+07.9	Fg 1	2.564±0.197	$0.4^{+0.2}_{-0.15}$	<0.09
G307.2-03.4	NGC 5189	1.471±0.043	$0.11^{+0.03}_{-0.03}$	<0.09

factor will vary from nebula to nebula, and that this method is meant to be statistically meaningful in order to allow for comparisons with the ionised mass of these objects, and among subclasses of post-CE PNe.

3.3. *Results*

The resulting masses of the studied post-CE PNe are displayed in Table 1. Interestingly, a trend arises when we divide the sample in the categories of single-degenerate (SD) and double-degenerate (DD) systems, according to one or both components of the binary pair being a post-AGB star: PNe hosting DD systems are considerably more massive, on average, than their SD counterparts. Thus, the geometric mean of the ionised+molecular mass for the SD sample is 0.15 $M_\odot$, with a geometric standard deviation (GSD) factor of 3.4, whereas for the DD sample the geometric mean is substantially larger, 0.31 $M_\odot$, with a narrower GSD of 1.7.

Considering the linear momenta and kinetic energies of these objects can provide additional insight. We gathered the expansion velocities from systematic works such as Weinberger (1989). These seem to follow a similar trend to ionised+molecular mass, with values somewhat larger in DD systems than in SD ones. The resulting linear momenta have substantially different geometric means of 6.3×10^{38} g cm s^{-1} (GSD factor 3.5) and 2.2×10^{39} g cm s^{-1} (GSD factor 2.3), for SD and DD systems respectively. The kinetic energies differ in a more pronounced way, with SD systems having a geometric mean of 8.1×10^{44} erg (GSD factor 3.7), whereas DD systems show a much larger 3.9×10^{45} erg (GSD factor 4.2) for DD ones.

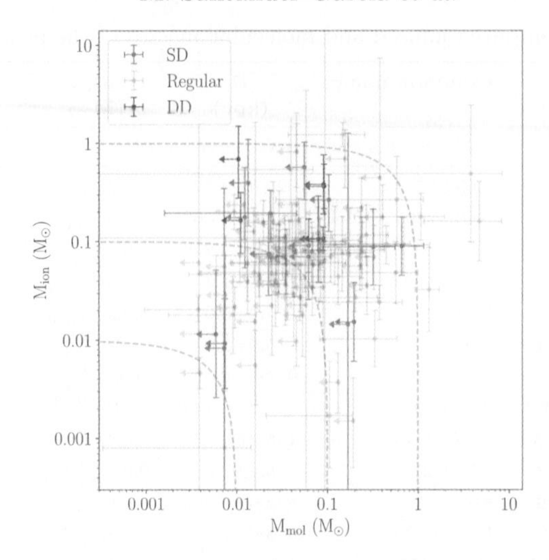

Figure 1. Ionised vs. molecular mass of our post-CE PNe sample and the comparison 'regular' PNe sample. The further to the top and to the right a nebula is, the more massive it is. Dashed lines represent 'isomasses', indicating equal ionised+molecular mass; if neutral atomic mass is neglected, individual nebulae run along these lines as their gas content is progressively ionised.

3.4. *Comparison with regular PNe*

We can put these results in the context of the general population of PNe. Are post-CE PNe more or less massive on average than the general population of PNe? The answer to this question may have strong implications for theories of CE interaction and ejection.

We therefore reviewed the literature in order to derive the ionised and molecular masses of an additional sample of 97 'regular' PNe (not known to arise from binary systems). All of them have dereddened Hα flux and diameters obtained in a systematic way by Frew *et al.* (2016), as well as available ^{12}CO observations and precise distance determinations (either accurate GAIA eDR3 determinations, or being listed as a 'distance calibrator' by Frew *et al.* (2016)). Note that this sample is not limited by volume and is not exempt from selection biases, although the effect of those biases, once filtered by the inclusion criteria, is unclear and difficult to predict. In any case, we stress the intrinsic limitation of the comparison provided in this section, which should be taken with a pinch of salt for the the time being.

The resulting ionised+molecular masses (or their upper limits) of the 'regular' PNe are displayed in Figure 1 along with the results for the SD and DD post-CE PNe analysed in section 3.3. The geometric mean of the mass is the same as that for SD systems, 0.15 $M_\odot$, with a geometric standard deviation (GSD) factor of 3.1. An analysis including the expansion velocities suggests that the characteristic linear momenta of the SD and regular sample are also very similar, although the slightly higher expansion velocities shown by post-CE systems make the kinetic energies of SD post-CE PNe somewhat higher than those of regular PNe. DD systems, on the other hand, show substantially more massive and faster expanding nebulae (larger linear momenta and kinetic energy) than either SD systems and 'regular' PNe.

4. Discussion and conclusions

Our results suggest post-CE PNe hosting SD systems are, on average, neither more nor less massive than regular PNe, whereas post-CE PNe from DD systems are considerably more massive than both groups. This discrepancy is larger when considering the linear momentum and kinetic energy of these nebulae. In other words, DD systems seem more capable of unbinding and accelerating a larger amount of material than SD systems.

Furthermore, our work suggests a profound mismatch between observations and modelling. On the one hand, models of CE ejection tend to fail to unbind the whole AGB envelope. On the other hand, the observed ionised+molecular mass of these systems seem substantially lower than the expected mass of the AGB envelope at the time of CE occurence, especially in the case of SD systems. In order to quantitatively account for this discrepancy, at least on a first order approximation, we attempted the reconstruction of the CE of two SD systems, Abell 63 and Hen 2-155, and two DD systems, Fg 1 and Hen 2-428. Following the methodology given in Iaconi & De Marco (2019); De Marco (2011), we derived the percentage of the AGB envelope mass that the observed nebula accounts for, as well as the percentage of the orbital shrinking energy budget actually spent on unbinding and accelerating the observed nebula. We find that the mass of the two SD nebulae are $\sim$1% and $\sim$22% of their respective AGB envelopes, while for the two DD systems the corresponding figures are $\sim$64% and $\sim$61%. Given the large uncertainties, this suggests that DD systems could be unbinding the whole AGB envelope, whereas in the case of SD systems it is hard to reconcile the idea that the observed nebulae is actually the AGB envelope at the time of CE. Examining the energy budget also procures a surprising conclusion: the four analysed systems seem to have spent only between $\sim$1% and $\sim$11% of their available orbital shrinking energy budget in unbinding and accelerating their nebulae.

This leads to some uncomfortable questions: If the primary star is of similar mass to normal post-AGB stars, and thus the mass of the nebula amounts to just a tiny fraction of the star's envelope, then where is the rest of the envelope? Why are we unable to detect it somewhere in the vicinity of the star?

References

Balick. B., & Frank, A. 2002, *ARAA*, 40, 439
Chamandy, L., et al. 2020 *MNRAS*, 495, 4028
Corradi, R. L. M., et al. 2015, *ApJ*, 803, 99
De Marco, O. 2011, *MNRAS*, 411, 2277
Decin, L., et al. 2020, *Science*, 369, 1497
Frew, D. J., & Parker, Q. A. 2007, *Proceedings of the APN IV conference*, 475
Frew, D. J., Parker, Q. A., & Bojičić, I. S. 2016, *MNRAS*, 455, 1459
Gaia Collaboration, et al. 2020, *A&A*, 649, 1
Guerrero, M. A., & Miranda, L. F. 2012, *A&A*, 539, 47
Guzman-Ramirez, L. et al. 2018, *A&A*, 618, 91
Huggins, P. J., & Healy, A. P. 1989, *ApJ*, 346, 201
Huggins, P. J., Bachiller, R., Cox, P., & Forveille, T. 1996 *A&A*, 315, 284
Huggins, P. J., Bachiller, R., Planesas, P., Forveille, T., & Cox, P. 2005, *APJS*, 160, 272
Iaconi, R., & De Marco, O. 2019, *MNRAS*, 450, 2550
Ivanova, N. 2018, *ApJL*, 858, 24
Jones, D., & Boffin, H. M. J. 2017, *Nature Astronomy*, 1, 117
Jones, D. 2020, *Reviews in Frontiers of Modern Astrophysics; From Space Debris to Cosmology* (Springer), p. 123
Ohlmann, S. T., Röpke, F. K., Pakmor, R., Springel, V. 2016 *ApJL*, 816, 90
Paczynski, B. 1976, *Proceedings of the IAU Symposium 73*, 75.
Ricker, P. M., & Taam, R. E. 2012, *ApJ*, 746, 74
Santander-García, M., Jones, D., Alcolea, J., Bujarrabal, V., & Wesson, R. 2021, *A&A*, 658, 17
Weinberger, R. 1989, *A&AS*, 78, 301

The Origin of Outflows in Evolved Stars
Proceedings IAU Symposium No. 366, 2022
L. Decin, A. Zijlstra & C. Gielen, eds.
doi:10.1017/S1743921322000643

Outflows and disks in post-RGB objects – products of common envelope ejection?

Geetanjali Sarkar[1] **and Raghvendra Sahai**[2]

[1]Department of Physics, Indian Institute of Technology, Kanpur U.P., India
email: gsarkar@iitk.ac.in

[2]Jet Propulsion Laboratory Pasadena, CA, USA

Abstract. Interacting binaries within a common envelope, wherein the primary is a red giant are believed to result in a recently identified evolutionary class – the dusty post-RGB stars. Our SED modeling of eight post-RGBs in the LMC indicates the presence of geometrically thick disks with substantial opening angle in addition to the outer shells. We estimated the total dust mass (and gas mass assuming gas-to-dust ratio) in the disks and shells and set constraints on the dust grain compositions and sizes. The only known Galactic object of this class is the Boomerang nebula. Additionally, we present a DUSTY model of the Boomerang that can serve as a template for 2D modeling of the object using RADMC-3D. 2D modeling is essential to dissect the morphology of the spatially-unresolved post-RGBs in the LMC. These models may then be tested with future HST and ALMA imaging, together with JWST spectroscopy of these objects.

Keywords. (stars:) circumstellar matter, (stars:) binaries (including multiple): close, stars: evolution, stars: AGB and post-AGB

1. Introduction

Most proto-planetary nebulae (PPNs) are characterised by collimated bipolar lobes separated by a dusty disk or torus (Sahai et al. 1998; Kwok, Su & Hrivnak 1998; Bujarrabal et al. 1998; García-Lario, Riera & Manchado 1999; Kwok Hrivnak & Su 2000; Hrivnak et al. 2008). Following heavy mass loss on the AGB, the luminous ($L \sim 5000 - 10,000\,L_\odot$), cool ($T_{eff} < 3000\,K$) AGB stars are transformed into post-AGBs evolving towards higher temperatures at constant luminosity. The spherical (AGB) envelopes are transformed into aspherical bipolar/multipolar morphologies via the action of collimated high-velocity outflows that sculpt these envelopes from the inside out, as the central stars evolve to the post-AGB/PPN phase (Sahai & Trauger 1998; Sahai, Morris & Villar 2011). Phillips & Ramos-Larios 2010 and references therein have tried to explain the formation of the jet engines that produce such outflows as also the bipolar/multipolar structures in PPNe. Common envelope evolution (CEE) in close binary systems is by far the most popular amongst these.

Binarity may also be responsible for the rapid and unexpected evolution of Red Giant Branch (RGB) stars to the PPN stage (post-RGB stars). Bipolar morphology similar to post-AGB PPNe was discovered in the Boomerang Nebula (Wegner & Glass 1979; Sahai & Nyman 1997). However, the luminosity of the central star in the Boomerang is much lower ($L \sim 300\,L_\odot$) than is possible for a post-AGB star. Sahai et al. (2017) show that the Boomerang is most likely a post-RGB star and that merger with a binary companion may have triggered its extreme mass loss ($\sim 10^{-3}\ M_\odot yr^{-1}$) at a very high ejection velocity ($165\ km\,s^{-1}$), over a relatively short period (3500 yr) via CEE.

Using optical spectroscopy and *Spitzer* photometry, Kamath et al. (2014, 2015, 2016) identified post-RGB objects similar to the Boomerang in the Large and Small Magellanic Clouds (LMC and SMC). These objects were found to have large mid-IR excesses. Distance modulus for the LMC/18.54 and SMC/18.93 (Keller & Wood 2006) enabled reliable luminosity estimates for these stars; the low luminosities ($< 2500\ L_\odot$) indicate that these stars have not yet reached the AGB phase. Post-RGB stars are thus separated from post-AGB ones by their low luminosities.

2. Post-RGB stars in the LMC

We selected a sample of eight post-RGB stars in the LMC with mid and far-IR excess from Kamath et al. (2015: hereafter KWVW15). Our sample consists of an equal number of "shell" (dust peak beyond 10 μm) and "disk" (peak around 10 μm or bluer) sources from KWVW15. On a colour-colour plot, the shell sources have $[3.6]-[4.5] < 0.5$ and $[8]-[24] > 4.0$. Disk sources, on the other hand, have $[3.6]-[4.5] > 0.5$, or in some cases, $[3.6]-[4.5] < 0.5$ in combination with $[8]$-$[24] < 3.0$. Additionally, to ensure that the objects are indeed post-RGB (and not post-AGB) we selected objects with $L < 1000\ L_\odot$. The observed SEDs extend from the U-band to 24μm.

The SEDs were modeled using the one-dimensional radiative transfer code, DUSTY (Ivezić et al. 2012) (Fig. 1, Table 1). The dust density was assumed to be proportional to r^{-2}, where r is the radial distance from the star. Whenever a single shell model (one-component model) was found to be insufficient to obtain a fit to the observed SED, we used a two-component fit, in which we added an inner component, representative of a hot, compact disk. We approximated the inner disk to be an axially symmetric wedge-shaped fraction of a sphere; this fraction is hereafter referred to as the "disk fraction"†. The free parameters for the modeling include the dust temperature (T_d at the inner boundary of the disks/shells, the dust grain sizes, grain composition, optical depth (τ) at 0.55μm and relative shell (disk) thickness (Y = ratio of the outer to the inner shell/disk radius).

The DUSTY code outputs the SED, normalized to the bolometric flux, F_{bol}, which was determined by scaling the model SED to match the de-reddened SED of each source. We estimated the luminosity for each model as $L = 4\pi d^2 F_{bol}$, assuming a distance, d=50 kpc, for the LMC. The dust mass M_d in the circumstellar component is estimated using:

$$M_d = 4\pi R_{in}^2 Y(\tau_{100}/\kappa_{100})$$

that applies to objects obeying a r^{-2} density distribution; the total mass (gas+dust), $M_{gd} = M_d\delta$ (Sarkar & Sahai 2006). R_{in} is the inner radius of the dust shell, Y is the shell relative thickness specified in the DUSTY input (R_{out}/R_{in}), τ_{100} is the shell optical depth at 100 μm, κ_{100} is the dust mass absorption coefficient at 100 μm and δ is the gas-to-dust ratio. We assume $\delta = 200$, and $\kappa_{100} = 34\,cm^2g^{-1}$, as in (Sarkar & Sahai 2006). The derived parameters from our best-fit models are summarised in Table 1. For comparison, we show our model dust mass values, M_d(outer shell), overplotted (horizontal lines, Fig. 2) on a plot of theoretically estimated dust masses in the ejecta of common envelope systems versus initial stellar mass, taken from Fig. 2 of Lü et al. (2013).

3. Dissecting post-RGB morphology using RADMC-3D

Not much is known about the as yet unresolved dusty post-RGBs in the LMC and SMC. We may dissect their morphology by using the code RADMC-3D (Dullemond et al. 2012). Using the latter, observed SEDs of the post-RGBs may be modeled to estimate the

† A disk with an opening angle of $\theta°$, intercepts a fraction $\sin(\theta)$ of the radiation emitted by the star within a 4π solid angle.

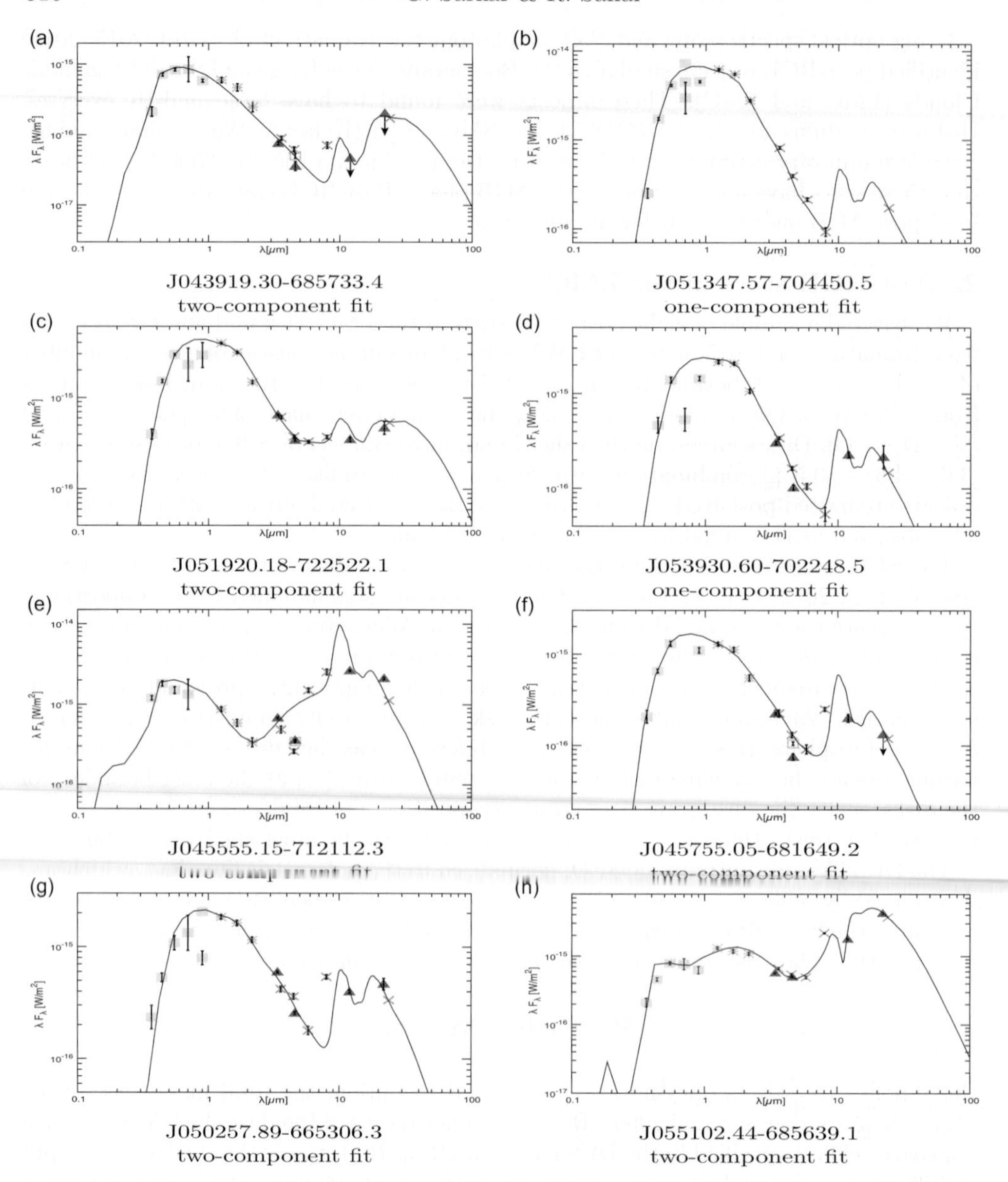

Figure 1. The best-fit models (black curves) to the observed SEDs of the post-RGB sources. The observed fluxes are de-reddened for Galactic and LMC reddening. U,B,V,R,I (yellow), 2MASS J,H,K (cyan) data are plotted along with WISE photometry (purple) and data from the SAGE-LMC Survey (green) which covers the IRAC and MIPS bands. The error bars and upper limits (arrows) are indicated in black.

disk mass, the flaring angle of the disk, the grain sizes in the disk, the disk inclinations, the disk scale height (when the the inclinations are closer to face-on) and the extent of the disk. We can then produce 2D-images of the objects at a particular wavelength for comparison with future HST and ALMA images. This approach may be tested using the well-studied Boomerang nebula in our Galaxy as a template.

ALMA images of the Boomerang (Sahai et al. 2017) revealed three distinct components in its morphology – compact disk, circumstellar envelope and bipolar lobes.

Table 1. Important parameters derived from the best-fit pRGB models.

Object	Inner disk					Outer shell					
	T^a_d(in) (K)	τ^b	a^c_{min} (μm)	a^d_{max} (μm)	M^e_{gd} ($M_\odot$)	T^a_d(in) (K)	τ^b	a^c_{min} (μm)	a^d_{max} (μm)	M^e_{gd} ($M_\odot$)	L^f ($L_\odot$)
shell sources											
J043919.30-685733.4	1000	0.5	0.005	0.25	2.19×10^{-8}	130	0.65	0.005	0.25	5.2×10^{-3}	116
J051347.57-704450.5	...	...	...	...	...	250	0.35	0.1	0.25	4×10^{-5}	776
J051920.18-722522.1	500	0.4	0.3	20	2.59×10^{-5}	110	0.65	0.005	0.25	3.44×10^{-2}	582
J053930.60-702248.5	...	...	...	...	...	300	0.70	0.005	0.25	5.81×10^{-5}	295
disk sources											
J045555.15-712112.3	800	0.7	0.005	0.25	2.67×10^{-6}	500	1.8	0.005	0.25	8.73×10^{-5}	621
J045755.05-681649.2	1300	0.5	0.005	2.0	9.64×10^{-9}	400	0.6	0.1	1.0	5.73×10^{-5}	217
J050257.89-665306.3	1200	0.5	0.3	5.0	5.77×10^{-8}	250	0.75	0.005	1.0	2.68×10^{-4}	303
J055102.44-685639.1	2000	1.0	0.005	0.05	1.99×10^{-8}	350	12.0	0.005	0.07	3.05×10^{-3}	621

a: The (input) dust temperature at shell (disk) inner radius; **b:** The dust shell's (disk's) optical depth at 0.55μ; **c:** The minimum dust grain size; **d:** The maximum dust grain size; **e:** The (inferred) circumstellar (gas+dust) mass ($M_{gd} = M_d\delta$); **f:** The (inferred) luminosity

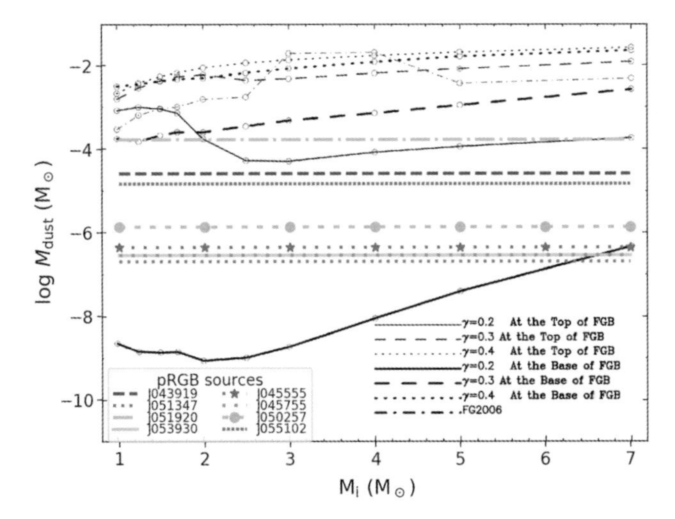

Figure 2. The circumstellar dust mass (M_d) of our post-RGB objects are shown (horizontal lines) on a plot of theoretically estimated dust masses in the ejecta of common envelope systems versus initial stellar mass, taken from Fig. 2 of Lü et al. (2013). FGB refers to the first red giant branch. FG2006 refers to results by Ferrarotti & Gail (2006), showing the dust masses produced in the dust-driven outflows of AGB stars.

We reconstructed its SED from the U-band to 140μm using the Vizier database (`https://vizier.u-strasbg.fr/viz-bin/VizieR`). A DUSTY model fit to the SED (Fig. 3) required an inner disk, a warm inner shell and an ultra-cold outer shell in agreement with Sahai et al. (2017). The values of the physical parameters charararacterizing the model components are given in Table 2; these parameters will provide initial estimates for the 2D modeling of the object using RADMC-3D.

4. Discussion

When the primary in a binary system overflows its Roche lobe and the resulting mass transfer proceeds too rapidly to be accreted by the compact companion, a CE system results. CEE (Paczynski 1976; Ivanova et al. 2013) may give rise to close binary systems. Two scenarios may exist — a rapid plunge in of the companion or a slow spiral-in

Table 2. Model of the Boomerang Nebula.

Component	Parameter	Value
Cool Shell	R_{in}	6.52×10^{16}cm
	Y	20
	T_d(in)	50 K
	τ	4.0
	grain size	$0.1 - 0.25\ \mu$m
	M_d	$3.26 \times 10^{-2} M_\odot$
Warm Shell	R_{in}	3.23×10^{16}cm
	Y	1.5
	T_d(in)	145 K
	τ	1.0
	grain size	MRN
	M_d	$2.85 \times 10^{-4} M_\odot$
Inner Disk	R_{in}	6.08×10^{14}cm
	Y	1.5
	T_d(in)	1100 K
	τ	40.0
	grain size	MRN
	M_d	$4.03 \times 10^{-6} M_\odot$

R_{in}=inner radius of dust shell(disk); Y=thickness of dust shell (disk); T_d=dust temperature at inner shell (disk) boundary; τ=optical depth at 0.55μm; MRN: Mathis, Rumpl, Nordsieck grain size distribution (Mathis et al. 1977); M_d=circumstellar dust mass

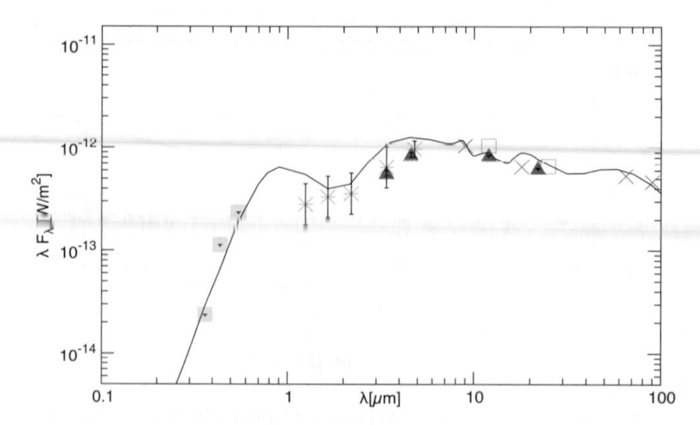

Figure 3. DUSTY model fit (black) to the observed SED of the Boomerang nebula (disk fraction = 0.5). The observed fluxes are de-reddened using $A_v = 0.28$. U,B,V (yellow), J,H,K,L,M (blue) data are plotted along with IRAS fluxes (orange), WISE photometry (purple) and AKARI data (green). The error bars are indicated in black. The Boomerang is assumed to be at a distance of 1.5 kpc (Sahai & Nyman 1997).

phase. In the former, if enough orbital energy is deposited to the CE via dynamical friction, the whole envelope is ejected on a dynamical timescale (Paczynski 1976). The second case provides an alternate route for the envelope ejection (Ivanova et al. 2002; Podsiadlowski et al. 2010) over "several dynamical time scales" (Ivanova et al. 2013; Clayton et al. 2017). Recently, a third scenario has been proposed by Glanz & Perets (2018) wherein dust driven winds similar to those observed in AGB stars may lead to the ejection of the CE.

Dust formation may occur in the expanding gas (Lü et al. 2013, Iaconi et al. 2020) which may explain the presence of circumstellar dust shells in these systems. Some fraction of the ejected mass may also fall back and interact with the binary leading to the formation of circumbinary disks (Kashi & Soker 2011). Our derived dust masses may be taken as a lower limit since a substantial mass may lie longwards of 24μm. Comparison of our model dust shell masses with theoretical models of Lü et al. (2013) (Fig. 2) suggests that CEE occurred near or at the tip of the RGB. Further, we find that the disk fractions are surprisingly large (typically 0.3–0.4), implying that these are geometrically thick, flared structures with a substantial opening angle. Our modeling shows that the KWVW15 "shell"/"disk" classification is not robust. All "disk" sources in our sample require the presence of "shells". In addition, we find the presence of a disk in some "shell" sources (J043919.30-685733.4 and J051920.18-722522.1). We also find evidence that for some post-RGB sources the ejected matter may be carbon-rich, even though it is expected to be oxygen-rich. This provides independent support for the hypothesis of binary interaction leading to the formation of dusty post-RGBs. Modeling the unresolved post-RGBs using RADMC-3D will pave the way for future observations with HST, JWST, and ALMA that will lead to a better understanding of the evolutionary processes that produce post-RGB objects.

Acknowledgements

G.S. would like to acknowledge financial support from the Department of Science and Technology (DST), Government of India, through a grant numbered SR/WOS-A/PM-93/2017. R.S.'s contribution to the research described here was carried out at the Jet Propulsion Laboratory, California Institute of Technology, under a contract with NASA, and funded in part by NASA via ADAP awards, and multiple HST GO awards from the Space Telescope Science Institute.

References

Bujarrabal, V., Alcolea, J., Sahai, R., Zamorano, J. & Zijlstra, A. A. 1998, *A&A*, 331, 361
Clayton, M., Podsiadlowski, P., Ivanova, N., Justham, S. 2017, *MNRAS*, 470, 1788
Dullemond, C. P., Juhasz, A., Pohl, A., et al. 2012, *RADMC-3D: A multi-purpose radiative transfer tool, Astrophysics Source Code Library*. http://ascl.net/1202.015
García-Lario, P., Riera, A. & Manchado, A. 1999, *ApJ*, 526, 854
Glanz, H. & Perets, H.B. 2018, *MNRAS*, 478, L12
Hrivnak, B. J., Smith, N., Su, K. Y. L. & Sahai, R. 2008, *ApJ*, 688, 327
Iaconi, R., Maeda, K., Nozawa, T., De Marco, O., Reichardt, T., 2020, *MNRAS*, 497, 3
Ivanova, N., Podsiadlowski, P., Spruit, H. 2002, *MNRAS*, 334, 819
Ivanova, N., Justham, S., Chen, X., et al. 2013, *The Astronomy and Astrophysics Review*, 21, 59
Ivezić, Z., Nenkova, M., Heymann, F. & Elitzur, M. 2012, *U*ser Manual for DUSTY (V4)
Kamath, D., Wood, P. R. & Van Winckel, H. 2014, *MNRAS*, 439, 2211
Kamath, D., Wood, P. R. & Van Winckel, H. 2015, *MNRAS*, 454, 1468
Kamath, D., Wood, P. R., Van Winckel, H., & Nie, J.D. 2016, *A&A* (Letters), 586, L5
Kashi, Amit & Soker, Noam 2011, *MNRAS*, 417, 1466
Keller S. C. & Wood P. R. 2006, *ApJ*, 642, 834
Kwok, S., Su, K. Y. L. & Hrivnak, B. J. 1998, *ApJ* (Letters), 501, L117
Kwok, S., Hrivnak, B. J., & Su, K. Y. L. 2000, *ApJ* (Letters), 544, L149
Lü, G., Zhu, C. & Podsiadlowski, P. 2013, *ApJ*, 768, 193
Mathis, J. S., Rumpl, W. & Nordsieck, K. H. 1977, *ApJ*, 217, 425
Paczynski, B. 1976, in *I*AU Symposium 73, *Structure and Evolution of Close Binary Systems*, eds. Eggleton, P., Mitton, S., Whelan, J., p. 75
Phillips, J. P., Ramos-Larios, G. 2010, *MNRAS*, 405, 2179
Podsiadlowski, P., Ivanova, N., Justham, S., Rappaport, S. 2010, *MNRAS*, 406, 840
Sahai, R. & Nyman, Lars-Ake 1997 *ApJ*, 487, L155

Sahai, R., et al. 1998, *ApJ*, 493, 301
Sahai, R. & Trauger, J. T. 1998, *AJ*, 116, 1357
Sahai, R., Morris, M. R. & Villar, G. G. 2011, *AJ*, 141, 134
Sahai, R., Vlemmings, W.H.T. & Nyman, L.A. 2017, *ApJ*, 841, 110
Sarkar, G. & Sahai, R. 2006, *ApJ*, 644, 1171
Wegner, G. & Glass, I. S. 1979, *MNRAS*, 188, 327

The Origin of Outflows in Evolved Stars
Proceedings IAU Symposium No. 366, 2022
L. Decin, A. Zijlstra & C. Gielen, eds.
doi:10.1017/S1743921322000825

RGB mass loss: inferences from CMD-fitting and asteroseismology

Marco Tailo

Dipartimento di Fisica e Astronomia Augusto Righi, Università degli Studi di Bologna,
Via Gobetti 93/2, I-40129 Bologna, Italy
email: mrctailo@gmail.com, marco.tailo@unibo.it

Abstract. The amount of mass lost by stars during the red-giant branch (RGB) phase is one of the main parameters needed to fully understand later stages of stellar evolution. In spite of its importance, a fully-comprehensive physical understanding of this phenomenon is still missing, and we, mostly, rely on empirical formulations. The Galactic Globular Clusters are ideal targets to derive such formulations, but, until recently, the presence of multiple populations has been a major challenge.

We will discuss the insights on RGB mass loss that can be obtained from the study of the horizontal branch stars in such stellar associations. The estimates obtained via the study of the photometric data will be compared with recent and newly obtained estimates derived for few high metallicity open clusters and a large sample of field stars with asteroseismic techniques.

Keywords. stars: Hertzsprung-Russell diagram; stars: horizontal-branch; stars: mass loss; stars: fundamental parameters

---

1. Introduction

Determining the mass loss along the Red Giant Branch (RGB) is a crucial step to fully understand later stages of stellar evolution and by some extent understand the fate of planetary systems. The comparison between the stellar mass of the helium-burning- and the RGB-stars provides an efficient approach to infer the RGB mass loss in simple stellar populations.

In the old globular clusters (GCs) the helium burning stars are hosted on the Horizontal Branch(HB). These stars reach their position along the branch, at higher effective temperatures than their progenitors, after the degenerate helium in their core has ignited during the Helium flash at the Tip of the RGB. The total mass deficit, if any, of the resulting stars with respect to the RGB ones represents the sought-after mass loss. Recent works (Tailo et al. 2020, 2021; T20,T21) have introduced an innovative approach, based on the comparison of the photometric data with state of the art stellar population models, to infer the mass loss of the distinct stellar populations in GCs and applied this method to a large sample of more than 50 old, low metallicity GCs; also including those GCs without evidence of multiple stellar populations.

In younger and in metal rich ([Fe/H]>-0.3) clusters the helium burning stars are clustered in a smaller portion of the colour magnitude diagram; this new locus is dubbed Red Clump (RC) to distinguish it from the HB (see Girardi 2016 for a general review). The relatively large interval of mass values of the RC stars, compared to the small colour extension of this locus, makes difficult to get accurate mass and mass loss estimates with photometric tools alone. Therefore, an approach involving asteroseismic techniques is more appropriate and can lead to better mass estimates.

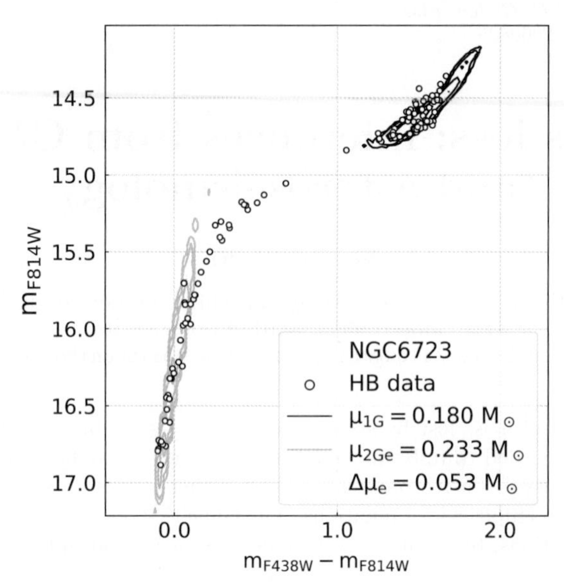

Figure 1. An example of the simulation realized in our study to obtain an estimate of the mass loss in the stellar population in the Galactic GCs. The photometric data of NGC 6723 are represented as white points. The black and the grey contour plots represent the best fit simulations for the 1G and the 2Ge stars in this cluster. An additional mass loss ($\Delta\mu$) is required to correctly describe the latter population.

The comparison of the results coming from the old, metal poor GCs and the asteroseismic study of the mass loss in the young metal rich clusters can give new insight on the complexity of the mass loss in RGB stars.

2. Mass loss in old GCs

The large majority of the old Galactic GCs host two kinds of stellar populations: a first generation (1G) of stars, characterized by abundance patterns similar to the field stars of comparable metallicity, and a second generation (2G), showing different patterns and enhanced helium abundance (see Milone et al. 2017, 2018 and Marino et al. 2019 for a review of the observational framework). The mechanism and the evolutionary path leading to the formation of these multiple populations are still largely debated (see Renzini et al. 2015 and Bastian & Lardo 2018 for a review of the theoretical framework) and no conclusive theoretical explanation and unified evolutionary scenario has been developed.

On the HB these two kinds of population can easily be distinguished. On the basis of the spectroscopic results (from e.g. Marino et al. 2008) and seminal theoretical works (D'Antona et al. 2002), in most GCs it is possible to identify the 1G stars with the reddest group of stars along the HB, and the 2G with the others. Among the various groups forming the 2G, the extreme part (2Ge), corresponding to the most enhanced sub-stellar populations occupies the bluest end of the HB.

The parameter degeneracy, traditionally associated to the effects of mass loss and helium, can be broken for both groups of stars adopting either the primordial helium abundance, for the 1G, or the helium values provided by e.g. Milone et al. (2018), in the case of the 2Ge. Therefore the last parameter to be evaluated is the integrated mass loss of the two stellar populations.

By comparing the photometric data with stellar populations models such mass loss estimates can be achieved. An example is shown in Fig. 1 for the cluster NGC 6723. The

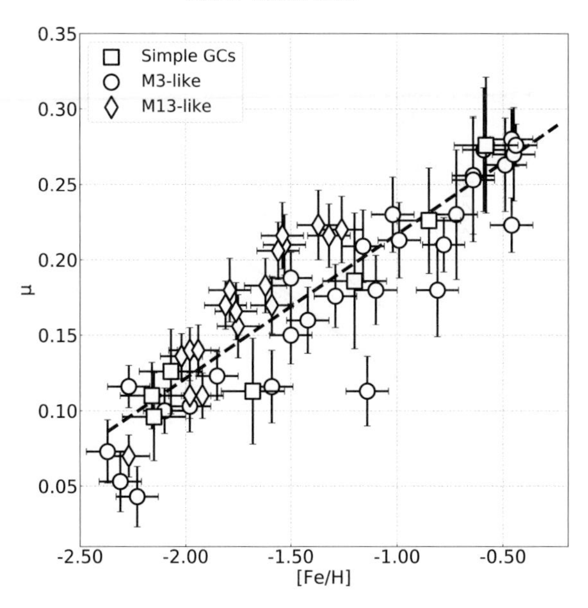

Figure 2. *Left panel:* Integrated mass loss of the first generation stars in the GC sample analysed in our works (T20,T21). The three series of points represent simple population clusters, M3-like and M13-like GCs, respectively as squares, circles and diamonds.

best fit simulations for the 1G and the 2Ge are selected with the T20 and T21 method, to whom we refer for further details, and are represented with black and grey contour plots in the figure. For the 1G stars we found a mean mass of $M^{1G}_{RGB} = 0.827 \pm 0.014 M_\odot$ and $M^{1G}_{HB} = 0.647 \pm 0.023 M_\odot$, respectively for the RGB and HB stars, whereas for the 2Ge stars we found $M^{2Ge}_{RGB} = 0.793 \pm 0.016 M_\odot$ and $M^{2Ge}_{HB} = 0.560 \pm 0.024 M_\odot$. These values give integrated RGB mass loss values of $\mu_{1G} = 0.180 \pm 0.023 M_\odot$ and $\mu_{2Ge} = 0.233 \pm 0.024 M_\odot$ respectively, for the 1G and 2Ge stars.

If we look at the results for all the clusters analysed in our works we see that a relation between the integrated mass loss and the parameters of the host clusters emerge. While we refer to T20 and T21 for the full discussion, we examine here what we deem to be the most important one, as it is connected to the evolution of the early cluster. Indeed, if we look at the mass loss of the 1G stars as a function of cluster metallicity, left panel of Figure 2, wee see that a strong relation is present. In the figure we represent the least square best fit line that can be derive from the data as the dashed black line. The relation can also be described via the following equation: $\mu = (0.095 \pm 0.006) \times [Fe/H] + (0.312 \pm 0.011)$. If we convert these value in mass loss rate via the most commonly used Reimers (1975) formula (η_R), we get that the free parameter in the model has to vary from a minimum of $\eta_R \sim 0.05$ at [Fe/H]$\sim$-2.4 to a maximum of $\eta_R \sim 0.60$ at [Fe/H]$\sim$-0.5 (T20).

In Figure 3 we compare the integrated mass loss obtained with our relation with the others present in the literature. For the old Galactic GCs, we reported the results from Gratton et al. (2010), black and Origlia et al. 2014). We report the results from Savino et al. (2019) for the Tucana dSph galaxy and from Salaris et al. (2013) for the Sculptor dSph galaxy. Finally more example of similar trends, albeit for a smaller sample of GCs, can be found in VandenBerg et al. (2016); Denissenkov et al. (2017); VandenBerg & Denissenkov (2018). Furthermore the mass loss rate we obtain here can also be compared with the ones in McDonald & Zijlstra (2015, and references therein), where the authors, using the results from Gratton et al. (2010), found an almost constant $\eta_R = 0.477 \pm 0.070$ for the entire range of metallicities.

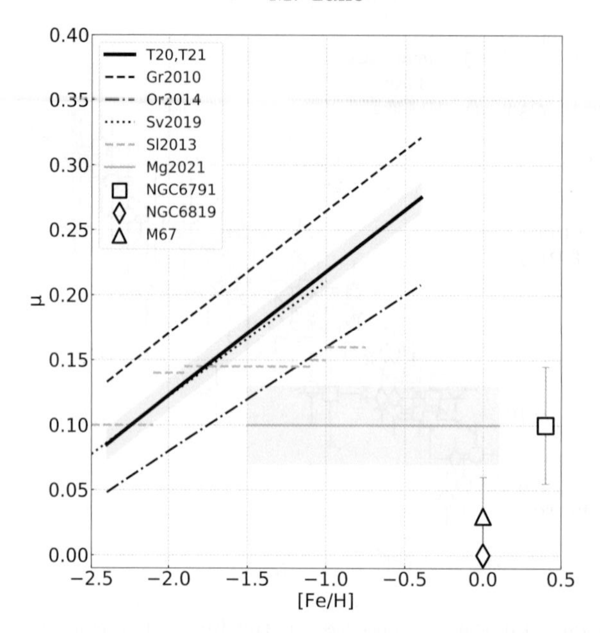

Figure 3. Comparison of the mass loss law obtained from our results (solid black) and few other relations and estimate found in the literature. T20,T21: Miglio et al. 2021; Tailo et al. 2021; Gr2010: Gratton et al. 2010; Or2014: Origlia et al. 2014; Sv2019: Savino et al. 2019; Sl2013: Salaris et al. 2013; Mg2021: Miglio et al. 2021; NGC6791 from Miglio et al. 2012 and NGC6819 from Miglio et al. 2012 and Handberg et al. 2017. M67 from Stello et al. (2016) See text for details.

The fact that most of these relations, obtained for the old stellar populations and with different methods, share the same slope, suggests that RGB evolution could be universal in old stellar associations like GCs and dSph.

3. Extending to higher metallicity range

The analysis of the helium burning stars, and, consequently, the estimate of the mass loss, in higher metallicity or younger GCs is trickier. When the metallicity is high enough, or the age low enough, the HB collapse in a small region of the CMD and is dubbed Red Clump (RC), to distinguish it from the HB.

The tracks hosted in this region are clumped together in a small area but cover a relatively large interval of mass values (see e.g. Girardi 2016) compared to their HB counterparts. This makes obtaining accurate mass estimate difficult with photometric tools only, and a combination of tools from different fields is necessary to reach good precision.

Among the tools available, asteroseismology is able to provide precise estimates of the mass of these stars while at the same time it allows to avoid most of the difficulties encountered with photometric techniques. We refer to Chaplin & Miglio (2013) and Aerts (2021) for a general overview. In a nutshell, continuous photometric observations of cool stars (giants, in particular) can be used to detect solar-like oscillations and spectra. Constraints from these spectra can be combined with surface properties to infer precise stellar masses. If we apply then these techniques to both helium burning and RGB stars we also get a more precise estimate of mass loss. This has been indeed performed on few young, high metallicity Galactic open clusters.

In an early work, Miglio et al. (2012) performed the asteroseismological analysis of the two young, high metallicity clusters NGC6791 and NGC6819. For NGC6791 Miglio and collaborators estimated the integrated mass loss to be $\sim$0.10$\pm$0.02 $M_\odot$. For NGC6819 by Miglio and collaborators a mean mass of 1.61$\pm$0.02 $M_\odot$ and 1.62$\pm$0.03 $M_\odot$, respectively for the RGB and RC stars. Therefore they estimate an almost absent mass loss. Recently the mass and mass loss estimates for NGC6819 have been updated by Handberg et al. (2017) confirming the results by Miglio and collaborators.

Another example of the evaluation of mass loss done via asteroseismic techniques can be found for the open cluster M67 (Stello et al. 2016). Stello and collaborators analysed a large number of RGB stars and a conspicuous sample of RC stars finding no evidence of strong mass loss.

An extensive analysis of the mass and the mass loss of a large number of stars can be found in Miglio et al. (2021). In this paper the authors analyse a large number of field stars in the Kepler field. When combining the mass and age information with the [Fe/H] and [α/Fe] values coming from the APOGEE DR14 (Abolfathi et al. 2018), their main results an average 0.1 $M_\odot$ of mass loss for the old (older than 4Gyr), high [α/Fe] stars ([α/Fe]>0.1) in the sample.

We included the results from Miglio et al. 2021 for the stars in the Kepler field in the right panel of Figure 3 as the grey solid line; the shaded area approximately represent the 1σ interval. In similar way the results for NGC 6791 from Miglio et al. 2012 and for NGC6819 from Miglio et al. 2012 and Handberg et al. 2017 are also reported. Finally we included the results from Stello et al. (2016).

4. A missing piece of the puzzle?

The comparison between the results coming from the high metallicity stars with the ones obtained for the old, low metallicity stellar populations in GCs or dSph galaxies shows a discrepancy.

Indeed, it seems to be tension between the mass loss predicted by the empirical relations described in Figure 3 at solar and higher metallicity values and the direct observations carried out with asteroseismic techniques. This may be the result of our incomplete understanding of the phenomenon of mass loss or of still unknown interaction with the formation environment. Finally, as younger, more massive stars have lower RGB Tip luminosities (for fixed metallicity and helium values, see e.g. Serenelli et al. 2017) it is also possible that the low, integrated mass loss is consequence of the higher surface gravity of these stars.

This would make the relations presented in Figure 3 valid only in those range of metallicity or age values or even valid only for these specific class of stellar associations. However investigating the origin of such discrepancy is beyond the scope of this work as it would need accurate simulation and observation of early clusters.

References

Abolfathi B., Aguado D. S., Aguilar G., Allende Prieto C., Almeida A., Ananna T. T., Anders F., et al., 2018, ApJS, 235, 42. doi:10.3847/1538-4365/aa9e8a

Aerts C., 2021, RvMP, 93, 015001. doi:10.1103/RevModPhys.93.015001

Bastian N., Lardo C., 2018, ARA&A, 56, 83. doi:10.1146/annurev-astro-081817-051839

Chaplin W. J., Miglio A., 2013, ARA&A, 51, 353. doi:10.1146/annurev-astro-082812-140938

D'Antona F., Caloi V., Montalbán J., Ventura P., Gratton R., 2002, A&A, 395, 69. doi:10.1051/0004-6361:20021220

Denissenkov P. A., VandenBerg D. A., Kopacki G., Ferguson J. W., 2017, ApJ, 849, 159. doi:10.3847/1538-4357/aa92c9

Girardi L., 2016, ARA&A, 54, 95. doi:10.1146/annurev-astro-081915-023354

Gratton R. G., Carretta E., Bragaglia A., Lucatello S., D'Orazi V., 2010, A&A, 517, A81. doi:10.1051/0004-6361/200912572

Handberg R., Brogaard K., Miglio A., Bossini D., Elsworth Y., Slumstrup D., Davies G. R., et al., 2017, MNRAS, 472, 979. doi:10.1093/mnras/stx1929
Marino A. F., Villanova S., Piotto G., Milone A. P., Momany Y., Bedin L. R., Medling A. M., 2008, A&A, 490, 625. doi:10.1051/0004-6361:200810389
Marino A. F., Villanova S., Milone A. P., Piotto G., Lind K., Geisler D., Stetson P. B., 2011, ApJL, 730, L16. doi:10.1088/2041-8205/730/2/L16
Marino A. F., Milone A. P., Renzini A., D'Antona F., Anderson J., Bedin L. R., Bellini A., et al., 2019, MNRAS, 487, 3815. doi:10.1093/mnras/stz1415
McDonald I., Zijlstra A. A., 2015, MNRAS, 448, 502. doi:10.1093/mnras/stv007
Miglio A., Brogaard K., Stello D., Chaplin W. J., D'Antona F., Montalbán J., Basu S., et al., 2012, MNRAS, 419, 2077. doi:10.1111/j.1365-2966.2011.19859.x
Miglio A., Chiappini C., Mackereth J. T., Davies G. R., Brogaard K., Casagrande L., Chaplin W. J., et al., 2021, A&A, 645, A85. doi:10.1051/0004-6361/202038307
Milone A. P., Piotto G., Renzini A., Marino A. F., Bedin L. R., Vesperini E., D'Antona F., et al., 2017, MNRAS, 464, 3636. doi:10.1093/mnras/stw2531
Milone A. P., Marino A. F., Renzini A., D'Antona F., Anderson J., Barbuy B., Bedin L. R., et al., 2018, MNRAS, 481, 5098. doi:10.1093/mnras/sty2573
Origlia L., Ferraro F. R., Fabbri S., Fusi Pecci F., Dalessandro E., Rich R. M., Valenti E., 2014, A&A, 564, A136. doi:10.1051/0004-6361/201423617
Reimers D., 1975, MSRSL, 8, 369
Renzini A., D'Antona F., Cassisi S., King I. R., Milone A. P., Ventura P., Anderson J., et al., 2015, MNRAS, 454, 4197. doi:10.1093/mnras/stv2268
Salaris M., de Boer T., Tolstoy E., Fiorentino G., Cassisi S., 2013, A&A, 559, A57. doi:10.1051/0004-6361/201322501
Savino A., Tolstoy E., Salaris M., Monelli M., de Boer T. J. L., 2019, A&A, 630, A116. doi:10.1051/0004-6361/201936077
Serenelli A., Weiss A., Cassisi S., Salaris M., Pietrinferni A., 2017, A&A, 606, A33. doi:10.1051/0004-6361/201731004
Stello D., Vanderburg A., Casagrande L., Gilliland R., Silva Aguirre V., Sandquist E., Leiner D., et al., 2016, ApJ, 832, 133. doi:10.3847/0004-637X/832/2/133
Tailo M., Milone A. P., Lagioia E. P., D'Antona F., Jang S., Vesperini E., Marino A. F., et al., 2021, MNRAS, 503, 694. doi:10.1093/mnras/stab568
Tailo M., Milone A. P., Lagioia E. P., D'Antona F., Marino A. F., Vesperini E., Caloi V., et al., 2020, MNRAS, 498, 5745. doi:10.1093/mnras/staa2639
VandenBerg D. A., Denissenkov P. A., 2018, ApJ, 862, 72. doi:10.3847/1538-4357/aaca9b
VandenBerg D. A., Denissenkov P. A., Catelan M., 2016, ApJ, 827, 2. doi:10.3847/0004-637X/827/1/2

The Origin of Outflows in Evolved Stars
Proceedings IAU Symposium No. 366, 2022
L. Decin, A. Zijlstra & C. Gielen, eds.
doi:10.1017/S1743921322000382

Towards the identification of carriers of the unidentified infrared (UIR) bands in novae

Izumi Endo[1], Itsuki Sakon[1], Takashi Onaka[2,1], Yuki Kimura[3], Seiji Kimura[4], Setsuko Wada[4], L. Andrew Helton[5], Ryan M. Lau[6], Yoko Kebukawa[7], Yasuji Muramatsu[8], Nanako O. Ogawa[9], Naohiko Ohkouchi[9], Masato Nakamura[10] and Sun Kwok[11]

[1]University of Tokyo, 7-3-1 Hongo Bunkyo-ku, Tokyo 113-0033, Japan

[2]Meisei University, 2-1-1 Hodokubo, Hino, Tokyo 191-8506, Japan

[3]Institute of Low Temperature Science, Hokkaido University, Kita-19, Nishi-8, Kita-ku, Sapporo 060-0819, Japan

[4]The University of Electro-Communications, 1-5-1, Chofugaoka, Chofu, Tokyo 182-8585, Japan

[5]SOFIA Science Center/NASA Ames Research Center, MS 211-1, P.O. Box 1, Moffett Field, CA 94035-0001, USA

[6]Institute of Space and Astronautical Science, Japan Aerospace Exploration Agency, 3-1-1 Yoshinodai, Sagamihara, 229-8510, Japan

[7]Yokohama National University, 79-5 Tokiwadai, Hodogaya-ku, Yokohama 240-8501, Japan

[8]University of Hyogo, 2167 Shosha, Himeji-shi, Hyogo, 671-2280, Japan

[9]Japan Agency for Marine-Earth Science and Technology, 2-15 Natsushima-Cho, Yokosuka, 237-0061, Japan

[10]Nihon University, Narashinodai, Funabashi 274-8501, Japan

[11]The University of British Columbia, 2329 West Mall Vancouver, V6T 1Z4, Canada

Abstract. The unidentified infrared (UIR) bands, whose carriers are thought to be organics, have been widely observed in various astrophysical environments. However, our knowledge of the detailed chemical composition and formation process of the carriers is still limited. We have synthesized laboratory organics named Quenched Nitrogen-included Carbonaceous Composite (QNCC) by quenching plasma produced from nitrogen gas and hydrocarbon solids. Infrared and X-ray analyses of QNCC showed that infrared properties of QNCC well reproduce the UIR bands observed in novae and amine structures contained in QNCC play an important role in the origin of the broad 8 μm feature, which characterizes the UIR bands in novae. QNCC is at present the best laboratory analog of organic dust formed around dusty classical novae, which carries the UIR bands in novae via thermal emission process (Endo et al. 2021).

Keywords. (stars:) novae, cataclysmic variables, ISM: lines and bands, infrared: ISM

---

1. Introduction

The breakthrough brought by the space-based infrared observations, unaffected by atmospheric absorption, made it clear that organics ubiquitously exist throughout the universe from the solar system to distant galaxies. The unidentified infrared (UIR) bands, which consist of a series of emission features arising from aromatic and/or aliphatic C-C and C-H bonds (Allmandola et al. 1989; Tielens 2008), have been widely observed in various astrophysical environments (e.g., Tokunaga 1997). The organics which carry the UIR bands must, therefore, be a major constituent of the circumstellar and interstellar medium of the Galaxy. However, our knowledge of their exact chemical nature is still quite limited.

Polycyclic aromatic hydrocarbons (PAHs; Allmandola et al. 1989) and PAH-like molecular species have been commonly used to interpret the properties of the UIR bands observed in the interstellar medium. However, no individual PAH have successfully reproduced the observed properties of UIR bands. Other than PAHs, bulk carbonaceous grains have also been proposed as possible candidates of the carriers of the UIR bands observed in circumstellar environment. Quenched Carbonaceous Composite (Sakata et al. 1984, 1987) and Hydrogenated Amorphous Carbons (HAC; Jones et al. 1990) are laboratory analogues, while coal (Guillois et al. 1996) and kerogen (Papoular 2001) are examples of organic solids in the terrestrial environments. The mixed aromatic-aliphatic organics nanoparticles (MAON; Kwok & Zhang 2011), which contain heteroatoms including nitrogen in addition to hydrocarbon models have recently been suggested as a more realistic interpretation and the challenges to understand the nature of the carries of the UIR bands are still ongoing.

Classical novae, the final evolutional stage of binary systems harboring a white dwarf, eject heavy elements through outburst events and a part of them exhibit signs of dust formation. Dusty classical novae offer a valuable opportunity to investigate the formation process of dust and organics in space thanks to their higher occurrence. Past studies have shown that the UIR bands observed around novae are characterized by the presence of a broad 8 μm feature (Helton et al. 2011; Sakon et al. 2016).

2. Quenched Nitrogen-included Carbonaceous Composite (QNCC); a laboratory analogue of organics around novae

We have synthesized a laboratory organic, Quenched Nitrogen included Carbonaceous Composite (QNCC), which can well reproduce the characteristics of the UIR bands observed in dusty classical novae. Figure 1 shows comparison of the infrared absorption spectrum of QNCC with the UIR bands observed in the nova V2361 Cyg on Day 116 after the outburst. The overall spectral properties of the spectrum of QNCC, especially the profiles of the broad 8 μm feature, are in good agreement with the UIR bands observed in V2361 Cyg. QNCC is synthesized by quenching plasma generated from nitrogen gas and hydrocarbon solids, including filmy QCC (Sakata et al. 1990) and PAHs, by a 2.45 GHz microwave discharge. Any QNCCs produced from different hydrocarbon solids (e.g., coronene, anthracene) consistently exhibit a characteristic broad feature at around 8 μm, as well as other major features at 3.3, 6.3, and 11.4 μm. The synthesis method of QNCC qualitatively mimic a possible formation process of organics around novae, where a nitrogen-rich nova wind (e.g., $N/N_{\odot} \sim 201$ for V84s Cen; Gehrz et al. 1998, $N/N_{\odot} \sim 219$ for V2361 Cyg; Munari et al. 2008) interacts with pre-existing carbonaceous dust in the circumstellar medium.

We performed X-ray Absorption Near-edge Structure (XANES) analyses of QNCC with the measurement station for X-ray absorption spectroscopy installed in the beamline BL10 at the NewSUBARU synchrotron radiation facility at the University of Hyogo (Kuki et al. 2015). The result of the XANES analyses show the presence of amine structures, which we conclude to be responsible for the broad 8 μm UIR band observed in novae. The N/C ratio of QNCC is 3-5% based on the measurement using the modified elemental analyzer/isotope ratio mass spectrometer (EA/IRMS; Ogawa et al. 2010). We conclude that QNCC is at present the best laboratory analog of organic dust formed in circumstellar environment of dusty classical novae (Endo et al. 2021).

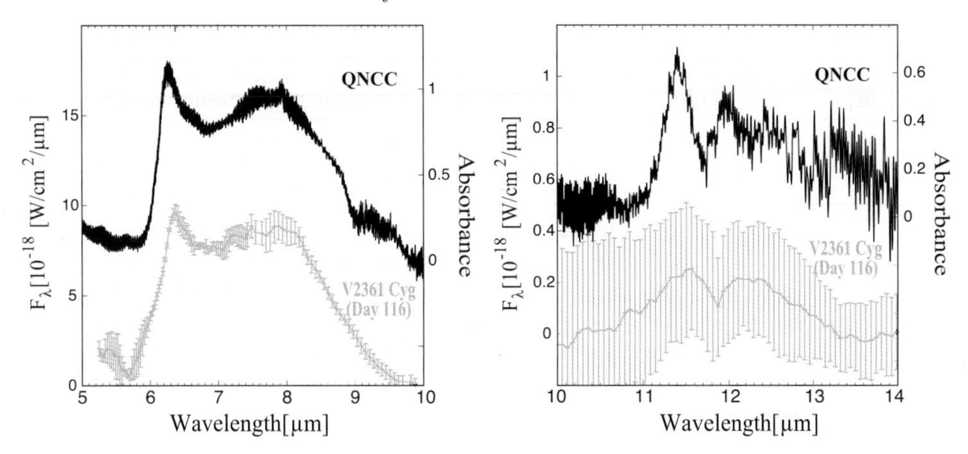

Figure 1. Comparison of the absorption spectrum of QNCC with the UIR bands observed in the classical nova V2361 Cyg on Day 116 after the outburst.

3. Can bulk organic dust carry the UIR bands in novae?

Two major emission mechanisms have been considered to explain the UIR bands and their carriers. Gas-phase PAH molecules emit the UIR bands through the process of UV-pumped IR fluorescence (Leger & Puget 1984; Allmandola et al. 1985). On the other hand, bulk organic dust in the vicinity of a heating source can reach the energy equilibrium and carry the UIR bands via a thermal emission process (Dwek et al. 1980; Duley & Williams 2011).

In the circumstellar environment of novae, bulk organic dust is likely to carry the UIR bands via a thermal emission process because gas-phase PAH molecules are not supposed to survive such harsh conditions around the white dwarf in a nova remnant (Evans & Rawlings 1994). Additionally, Kwok & Zhang (2011) indicated that the carriers of the UIR bands observed in novae V2361 Cyg and V2362 Cyg are expected to be complex organic dust that contains a mixture of miscellaneous aliphatic branches attached to the newly formed ring clusters rather than gas-phase pure PAH molecules.

In order to further clarify whether bulk organic dust can carry the UIR bands in novae, we examined the results of multi-epoch observations of V2361 Cyg with the Infrared Spectrograph (IRS) on Spitzer Space Telescope. The blackbody temperatures (T_d) of the continuum emission of V2361 Cyg are estimated as 660 K on $\tau = 102$ days, 610 K on $\tau = 116$, and 380 K on $\tau = 251$ after the outburst according to Helton et al. (2011). Based on the very simple optically-thin dust emission model, the temporal evolution of the dust temperatures $T_d(\tau)$ is determined by the energy balance between the input energy that the dust particle receives and the output energy that the dust particle radiates:

$$\pi a^2 \bar{Q}(a, T_*) L / 4\pi r^2 = 4\pi a^2 \bar{Q}(a, T_d) \sigma T_d^4, \tag{3.4}$$

where a is a grain size of spherical dust, T_* is the effective temperature of the heating source, L is the effective luminosity of the heating sources, σ is the Plank constant, $\bar{Q}(a, T_*)$ and $\bar{Q}(a, T_d)$ are the blackbody mean of absorption coefficient at $T = T_*$ and $T = T_d$, respectively, and r is the distance between the heating source and dust. Assuming the constant expansion velocity of dust as v, r can be given by $r = v\tau$.

In the case of a blackbody, the temporal evolution of the dust temperature as a function of the epoch from the nova outburst, τ, is given by

$$T_d \propto \tau^{-0.5} \tag{3.5}$$

Table 1. The temporal evolution of the height of 8 μm feature in V2361 Cyg.

τ [days]	T_d [K]	Height of the 8 μm feature [Jy]	Evolution of the 8 μm feature relative to Day 102: Observed	Evolution of the 8 μm feature relative to Day 102: Thermal equilibrium model
102	660	1.8	100 %	100 %
116	610	1.6	88.9 %	78.9 %
251	380	0.19	10.6 %	12.6 %

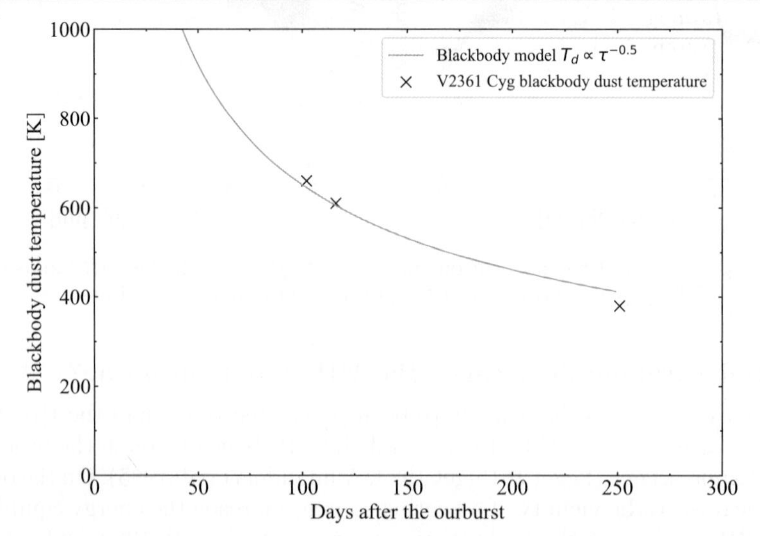

Figure 2. The evolution of the blackbody temperature of the continuum emission in V2361 Cyg as a function of the epoch τ after the outburst.

As shown in Figure 2, the behavior of the blackbody temperatures of continuum emission observed at $\tau = 102$, 116 and 251 days in V2361 Cyg (Helton et al. 2011) roughly agree with this very simple model.

The height of the 8μm feature, which characterize the UIR bands in novae, measured at $\tau = 102$, 116 and 251 days are quoted from Helton et al. (2011) and are summarized in Table 1. If the thermal equilibrium model can be applied to the UIR bands in V2361 Cyg, the evolution of the height of the 8 μm feature should be proportional to $[\exp\{hc/\lambda k T_d\} - 1]^{-1}$, where $\lambda = 8$ μm, and h, c, and k are the Plank constant, light speed, and the Boltzman constant, respectively. The observed values of the 8 μm feature height and those of the thermal equilibrium model relative to that at $\tau = 102$ are summarized in Table 1. The observed height of the 8 μm feature at $\tau = 251$ drops to 10.6% of that at $\tau = 102$. The thermal equilibrium model indicates that the value of 8 μm feature height at $\tau = 251$ drops to 12.6% of that at $\tau = 102$. This is roughly consistent with the observed value (i.e., 10.6%).

4. Conclusion

We have synthesized QNCC by the quenched condensation of plasma generated from nitrogen gas and hydrocarbon solids. We conclude that QNCC is at present the best laboratory analog of organic dust formed around dusty classical novae, which carries the UIR bands via a thermal emission process. In the present discussion, we neglect the changing in the luminosity of the heating source (L) as well as the possible destruction of the carriers of the emission. Although long-term monitoring observations are needed to clearly conclude which emission process is dominant, at least, our interpretation that the 8 μm feature is emitted via the thermal emission process consistently explain the observed temporal variations of the 8 μm feature strength well.

References

Allmandola, L. J., Tielens, A. G. G. M., & Barker, J. R. 1985, *ApJL*, 290, 25

Allmandola, L. J., Tielens, A. G. G. M., & Barker, J. R. 1989, *ApJS*, 71, 733

Duley, W. W., & Williams, D. A. 2011, *ApJL*, 737, L44

Dwek, E., Sellgren, K., Soifer, B. T., & Werner, M. W. 1980, *ApJ*, 238, 140

Endo, I., Sakon, I., Onaka, T., et al. 2021, *ApJ*, 917, 103

Evans, A., & Rawlings, J. M. C. 1994, *MNRAS*, 269, 427

Gehrz, R. D., Truran, J. W., Williams, R. E., & Starrfield, S. 1998, *PASP*, 110, 3

Guillois, O., Nenner, I., Papoular, R., & Reynaud, C. 1996, *ApJ*, 464, 810

Helton, L. A., Evans, A., Woodward, C. E., & Gehrz, R. D. 2011, *EAS Publications Series*, 46, 407

Jones, A. P., Duley, W. W., & Williams, D. A. 1990, *QJRAS*, 31, 567

Kuki, M., Uemura, T., Yamaguchi, M., et al. 2015, *Journal of Photopolymer Science and Technology*, 28, 531

Kwok, S., & Zhang, Y. 2011, *Nature*, 479, 80

Kwok, S., & Zhang, Y. 2013, *ApJ*, 771, 5

Leger, A., & Puget, J. L. 1984, *A&A*, 137, L5

Munari, U., Siviero, A., Henden, A., et al. 2008, *A&A*, 492, 145

Ogawa, N. O., Nagata, T., Kitazato, H., & Ohkouchi, N. 2010, *Earth, Life and Isotopes*, 339

Papoular, R. 2001, *A&A*, 378, 597

Sakata, A., Wada, S., Tanabe, T., & Onaka, T. 1984, *ApJ*, 287, L51

Sakata, A., Wada, S., Onaka, T., & Tokunaga, A. T. 1987, *ApJL*, 320, L63

Sakata, A., Wada, S., Onaka, T., & Tokunaga, A. T. 1990, *ApJ*, 353, 543

Sakon, I., Sako, S., Onaka, T., et al. 2016, *ApJ*, 817, 145

Tielens, A. G. G. M. 2008, *ARAA*, 46, 289

Tokunaga A. T. 1997, *Diffuse Infrared Radiation and the IRTS, ASP Conference Series*, 124, 149

Author index